A TEXTBOOK OF VECTOR ANALYSIS

A TEXTBOOK OF

VECTOR ANALYSIS

(With applications of Geometry and Statics)

VECTOR → ALGEBRA
VECTOR → CALCULUS

[For the Students of B.A. and B.Sc. (Pass & Hons. Courses)]

SHANTI NARAYAN
Former Dean of Colleges,
and Principal, Hans Raj College,
University of Delhi,
DELHI

Revised by

P.K. MITTAL
M.Sc., Ph.D.
Head of Mathematics Deptt.
Govt. Post Graduate College
RISHIKESH
(UTTARANCHAL)

S. CHAND PUBLISHING

S Chand And Company Limited
(ISO 9001 Certified Company)

S Chand And Company Limited

(ISO 9001 Certified Company)

Head Office: Block B-1, House No. D-1, Ground Floor, Mohan Co-operative Industrial Estate, New Delhi – 110 044 | Phone: 011-66672000

Registered Office: A-27, 2nd Floor, Mohan Co-operative Industrial Estate, New Delhi – 110 044 | Phone: 011-49731800

www.**schandpublishing.com**; e-mail: **info@schandpublishing.com**

Branches

Chennai	:	Ph: 23632120; chennai@schandpublishing.com
Guwahati	:	Ph: 2738811, 2735640; guwahati@schandpublishing.com
Hyderabad	:	Ph: 40186018; hyderabad@schandpublishing.com
Jalandhar	:	Ph: 4645630; jalandhar@schandpublishing.com
Kolkata	:	Ph: 23357458, 23353914; kolkata@schandpublishing.com
Lucknow	:	Ph: 4003633; lucknow@schandpublishing.com
Mumbai	:	Ph: 25000297; mumbai@schandpublishing.com
Patna	:	Ph: 2260011; patna@schandpublishing.com

First Edition 1955
Subsequent Editions and Reprints 1958, 62, 66, 69, 72, 75, 76, 82, 83, 85, 86, 87, 91, 94, 95, 96, 98, 2003, 2005, 2007, 2009, 2010, 2011, 2012, 2013, 2015, 2016, 2017, 2018, 2020 (Twice), 2022 (Twice)

Reprint 2022

ISBN: 978-81-219-2243-2 **Product Code:** H6VAN61MATH10ENZX0XR

PRINTED IN INDIA

By Vikas Publishing House Private Limited, Plot 20/4, Site-IV, Industrial Area Sahibabad, Ghaziabad – 201 010 and Published by S Chand And Company Limited, A-27, 2nd Floor, Mohan Co-operative Industrial Estate, New Delhi – 110 044.

BOOKS BY THE SAME AUTHOR

- Differential Calculus
- Integral Calculus
- Elements of Real Analysis
- Theory of Functions of A Complex Variable
- A Textbook of Calculus Part–I
- A Textbook of Calculus Part–II
- A Textbook of Calculus Part–III

BOOKS REVISED BY P.K. MITTAL

- A Course of Mathematical Analysis
- Analytical Solid Geometry
- A Textbook of Matrices
- A Textbook of Vector Algebra
- A Textbook of Vector Calculus

PREFACE TO THE REVISED EDITION

The book originally about 50 years ago, has during the intervening period been revised and re-printed several times. Due to the changing trends of University studies and the demand of the students a thorough revision of the book was overdue. I very humbly took the challenge of revising a perfect work of Dr. Shanti Narayan and tried to meet the demands, of those who always loved and liked the books of Dr. Shanti Narayan.

To meet out the changing trends in the course structure of different Indian Universities, a part of Vector Calculus has been incorporated with Vector Algebra to bring out the book of Vector Analysis in the present form. I am sure that in the present form the book will be of immense use not only to the undergraduate students of different Indian Universities but also to those preparing for competitive examinations.

The book, in the present form, is an humble effort to make it more useful to the students and teachers. I owe my special gratitude to Sri Ravindra Kumar Gupta, Managing Director, S. Chand & Company Ltd., for giving me opportunity to revise the books of late Dr. Shanti Narayan, an eminent Indian Mathematician.

I also acknowledge my sincere thanks to Sri Navin Joshi, General Manager (Sales & Marketing) and Shri Dharmendra Jha, editor, M/s. S. Chand & Company Ltd., for providing necessary assistance in this revision work.

As the need for improvement is never ending, I will look forward to receive valuable and useful suggestions from our users for further improvements in the book.

Rishikesh **P.K. Mittal**

PREFACE

The present book is a companion volume to *Vector Algebra* by one of us and deals with the theory and applications of Vector Calculus of Differential Geometry and various branches of Applied Mathematics.

Chapter I of the book deals with vector functions of one and more scalar variables and introduces in a systematic and rigorous manner the notions of limit, continuity and derivability in relation to the same. A section on Indefinite integration is also given. Chapters II and III deal with Differential geometry and consider curves as well as surfaces. Cartesian Formulation of the various results arrived at is also given.

Chapter IV and V on Mechanics special attention has been paid to *Moving axes* and the *Kinematics* of a rigid body involving the existence of Angular velocity of a rigid body moving about a fixed point and reduction to *Screw Motion* in the general case have been given a systematic treatment. Euler's equations of motion have been obtained. The various problems in relation to Dynamics of a particle have also been dealt with.

Chapter VI deals with gradient, divergence and curl of point functions in terms of cartesian rectangular as well as curvilinear co-ordinate axes.

Chapter VII deals with different types of line, surface and volume integrals and the two integral transformation theorems of Gauss and Stokes.

In Chapter VIII on Hydrodynamics the notion of pressure in relation to perfect fluids as a scalar point function has been arrived at and the equation of continuity and Euler's equation of motion have been obtained. The equations for the steady irrotational motion of incompressible fluids have been deduced.

The final chapter on Electromagnetic Theory covers most of the topics in Electrostatics, Magnetism and Electromagnetism included in the post-graduate courses. The two final sections deal with Maxwell's equations and electromagnetic waves.

The text has been throughout profusely illustrated by means of a large number of solved examples. A large number of exercises has also been given.

In all applications it has been our constant endeavour to see that the treatment is systematic and as much self-contained as possible.

It is earnestly hoped that as a result of the study of the book a serious student will be enabled to acquire facility and confidence in the use of vector calculus so that he could employ the same in an effective manner to various applications of a suitable nature.

The exercises needing answers are such that they require only direct application of formulas and accordingly, taking into account the possible standard of the readers of the book, the answers have not been given.

We are grateful to our esteemed friend Dr. P.L. Bhatnagar, D. Sc., who kindly went through a part of the manuscript and made a number of useful suggestions.

The authors would gratefully welcome notices of mistakes and suggestions for improvements.

Delhi **AUTHOR**

A TEXT BOOK OF VECTOR CALCULUS (with applications)

First publication : 1958 *— Latest edition:*

The book has been written jointly with Prof. J.N. Kapur, Department of Mathematics, Indian Institute of Technology, Kanpur.

It is a companion volume to 'Vector Algebra' and deals with the theory and applications of vector calculus and differential geometry and various branches of Applied Mathematics. They text has been throughout profusely illustrated by means of a large number of solved examples. A large number of exercises has also been given.

CONTENTS

1

Multiplication of Vectors by Scalars and Addition of Vectors

Introduction. Vectors constitute one of the several *Mathematical systems* which can be usefully employed to provide mathematical handling for certain types of problems in *Geometry, Mechanics* and other branches of Applied Mathematics. Other such systems are those of *Matrices, Tensors, Quaternions* etc. The system of vectors alone, however, will be the subject of study in this book.

Application of mathematics to a problem or a body of problems consists in constructing a system of entities and equipping the same with some structure [order or/and Algebraic or/and Topologic] such that the system of entities and the corresponding structure have a close correspondence with the objects of study in the problem.

Vectors facilitate mathematical study of such physical objects as possess *Direction* in addition to *Magnitude.* Velocity of a particle, for example, is one such object.

It is true that the set of real numbers also provides mathematical tool for the study of various types of physical problems for which vectors are found useful but the use of vectors is more direct and natural. As a result of limitation to the set of real numbers, we associate not one number but a *set* of numbers to a physical entity involving direction, for we have to split up the entity into components and associate a number with each. The use of vectors, however, avoids this splitting up and leads to a direct study of the objects in question.

In this book, the Algebra of Vectors will be dealt with and applied to the study of *Geometry* and *Statics* alongwith an elementary idea of vector calculus. Vectors Calculus involving topological notions of *Limit, Continuity,* and topological algebraic notions of *Derivability* and *Integrability* has been treated in a *second book.

Note. The study of vectors naturally leads to the notion of directed line segments which we shall now introduce.

---

* Vector Calculus by Shanti Narayan and J.N. Kapur.

1.1. DIRECTED LINE SEGMENTS

Any given portion of a given straight line where for the two end points are distinguished as Initial and Terminal is called a Directed Line Segment.

The directed line segment with *initial* point A and *terminal* point B is denoted by the symbol

$$\overrightarrow{AB}.$$

The two end points of a directed line segment are not interchangeable and the directed line segments

$$\overrightarrow{AB} \text{ and } \overrightarrow{BA}$$

must be thought of as different.

1.1.1. Length, Support and Sense of a Directed Line Segment

Associated with every directed line segment

$$\overrightarrow{AB},$$

we have its

Length, Support and Sense

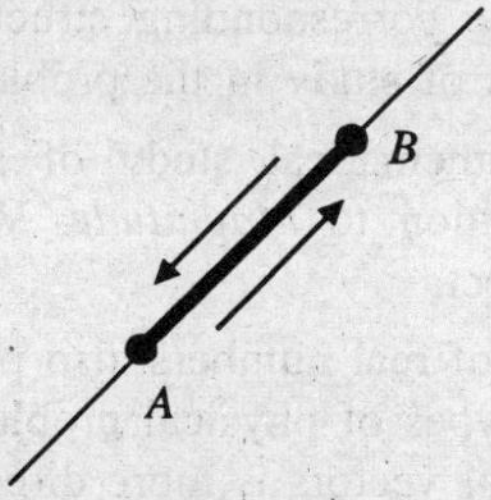

Fig. 1.1

(*i*) **Length.** The length of $\overrightarrow{AB}$ will be denoted by the symbol

$$|\overrightarrow{AB}|$$

Clearly, we have

$$|\overrightarrow{AB}| = |\overrightarrow{BA}|.$$

(*ii*) **Support.** The line of unlimited length of which a directed line segment is a part is called its line of support or simply the Support.

(*iii*) **Sense.** The sense of $\overrightarrow{AB}$ is from A to B and that of $|\overrightarrow{BA}|$ from B to A so that the sense of a directed line segment is from its initial to the terminal point.

The directed line segments

$$\overrightarrow{AB} \text{ and } \overrightarrow{BA}$$

have the same lengths and supports but different senses (Fig. 1.1).

The question of comparison of the senses of two directed line segments arises only when they have the same or parallel support (Fig. 1.2).

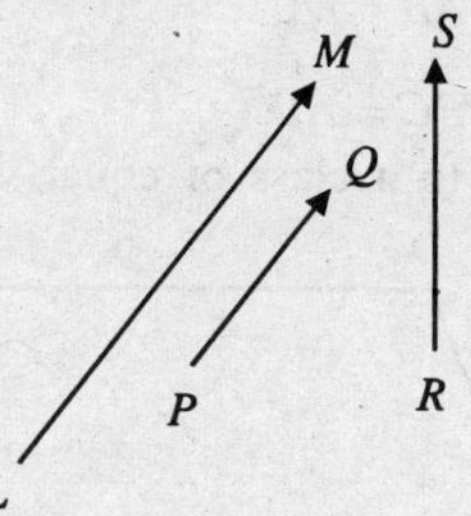

Fig. 1.2

Two directed line segments having the same or parallel supports may have the same or opposite senses.

Ex. How many Directed Line Segments are determined by 2, 3 and 4 given points.

1.2. VECTORS AND SCALARS

1.2.1. Vector

Def. *A directed line segment is called vector.*

1.2.2. Equality of Two Vectors

Def. *Two vectors are said to be equal if they have*

(*i*) *the same length,*

(*ii*) *the same* **or** *parallel supports, and*

(*iii*) *the same sense.*

It may thus be seen that two different Directed Line Segments may correspond to the same vector.

Thus, the vectors

$$\overrightarrow{AB}, \overrightarrow{CD}, \overrightarrow{EF}$$

are equal (Fig. 1.3).

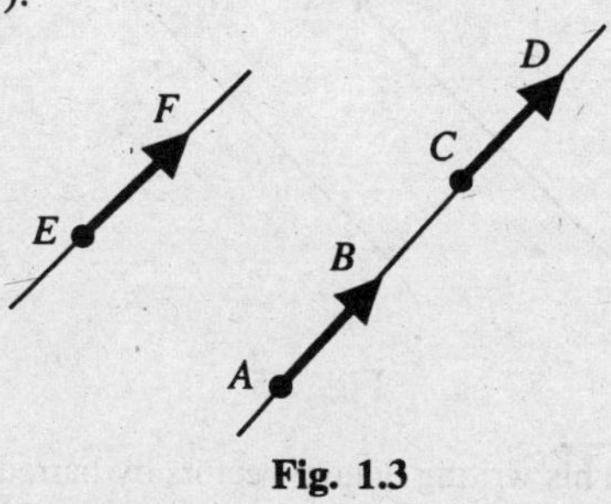

Fig. 1.3

Two vectors will *not* be equal if they have different lengths or inclined supports or again, they will not be equal even if they have the same lengths and parallel supports but different senses (Fig. 1.2).

If $ABCD$ is a parallelogram, we have

$$\vec{AB} = \vec{DC}$$

and

$$\vec{BC} = \vec{AD}$$

Every vector belongs to a class of equal vectors.

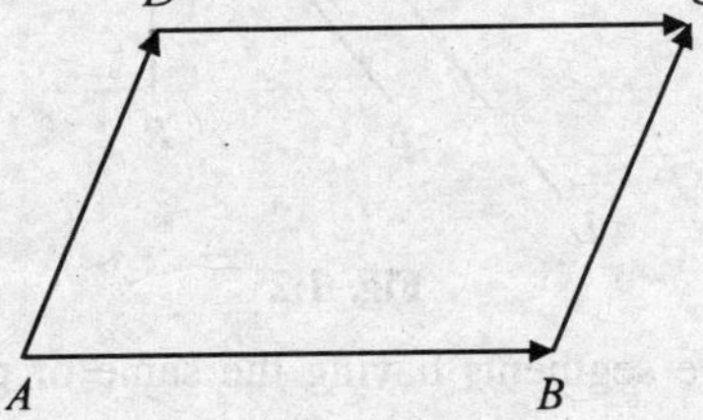

Fig. 1.4

1.2.3. Notation for a Vector

A vector is also denoted by a single letter such that

a, b, c, etc.

by using **bold face*** type so that we may write

$$\mathbf{a} = \vec{AB}.$$

Then the symbol

$$|\mathbf{a}|$$

denotes the length of the vector, **a**, also called the **Magnitude** of the vector.

1.2.4. Co-initial Vectors

It is possible to replace a given vector by another equal vector having any given point as its initial point.

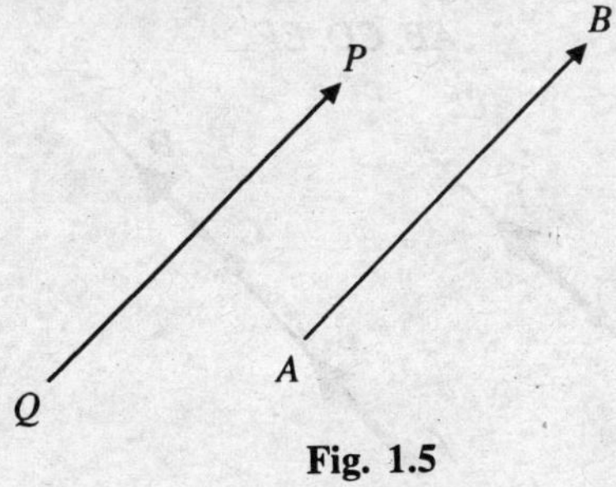

Fig. 1.5

* The reader may, in his writing denote vectors by barred letters such as $\vec{a}, \vec{b}$.

Thus, if $\overrightarrow{AB}$ be any given vector and O, any given point, then by drawing through O, a line OP parallel to AB in the same sense as AB of length equal to that of AB, we obtain a vector $\overrightarrow{OP}$ equal to $\overrightarrow{AB}$ and with the given point, O as the initial point (Fig. 1.5).

Vectors with the same initial point may be called Co-initial Vectors.

1.2.5. Zero Vector

A vector whose initial and terminal points are coincident is called the **Zero Vector.**

The length of the zero vector is zero but it can be thought of as having any line as its line of support.

The zero vector is denoted by the bold face type $\mathbf{0}$. It will be seen that the zero vector has many properties similar to those of the zero number.

1.2.6. Scalars

In the following, real numbers will be called scalars. The absolute value of a scalar, m, will as usual, be denoted by the symbol $|m|$ so that

$$|m| = m \text{ if } m \text{ ³ } 0 \quad \text{and} \quad -m \quad \text{if } m < 0.$$

1.3. ALGEBRA OF VECTORS

It is possible to develop an *Algebra of Vectors* which proves useful in the study of Geometry, Mechanics and other branches of Applied Mathematics.

By the '*Algebra of Vectors*' will be meant a body of work which prescribes various useful manners of combining vectors and scalars satisfying some laws which may be called *Laws of Composition.*

The following manners of composition will be introduced in appropriate places in the book :

I. Multiplication of Vectors by Scalars.

II. Addition of Vectors.

III. Scalar Multiplication of Vectors.

IV. Vector Multiplication of Vectors.

The first two will be introduced in this chapter and the last two in Chapter III and Chapter IV respectively. The first two compositions which will be dealt with in this chapter are also called *Linear Compositions.*

1.3.1. Multiplication of Vectors by Scalars

Let, $\mathbf{a}$, be any given vector and, m, be any given scalar. Then the symbol

$$m\,\mathbf{a},$$

called the product of the vector, **a**, by the scalar, m, is a vector such that

(*i*) the *length* of $m\,\mathbf{a}$ is given by

$$|\,m\,\mathbf{a}\,| = |\,m\,|\;\;|\,\mathbf{a}\,|$$

i.e., the length of the vector, $m\,\mathbf{a}$ is m times or, $-m$, times that of, **a**, according as m is positive (including zero) or negative.

Thus, $\quad \mathbf{b} = m\,\mathbf{a} \quad \Rightarrow \quad |\,m\,| = |\,\mathbf{b}\,| \,/\, |\,\mathbf{a}\,|.$

(*ii*) the *support* of $m\,\mathbf{a}$, is the same or parallel, to that of **a**.

(*iii*) the *sense* of $m\,\mathbf{a}$, is the same or opposite, to that of **a**, according as m is positive or negative.

The symbol, $m\,\mathbf{a}$, is also sometimes written as, $\mathbf{a}\,m$, so that the scalar, m, appears on the right of the vector instead of on the left.

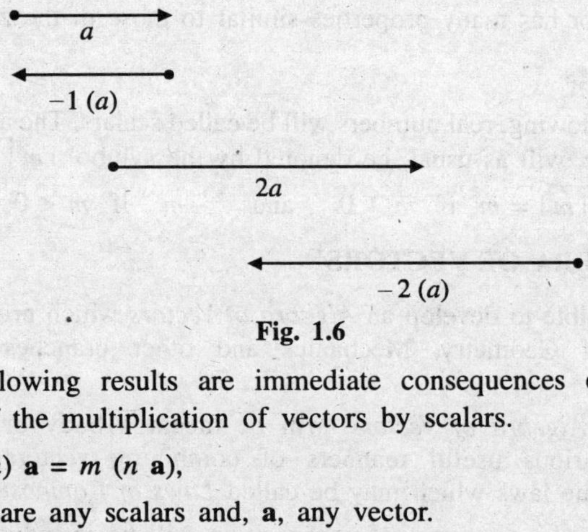

Fig. 1.6

The following results are immediate consequences of the above definition of the multiplication of vectors by scalars.

I. $(m\,n)\,\mathbf{a} = m\,(n\,\mathbf{a})$,

where m, n are any scalars and, **a**, any vector.

II. $0\,\mathbf{a} = \mathbf{0}$,

so that the product of a vector, **a**, by the zero scalar is the zero vector.

Here, 0, on the left stands for the zero scalar and **0**, on the right for the zero vector.

III. If two vectors have the same or parallel supports, then each can be thought of as a product of the other by a suitable scalar, the absolute value of the scalar being the ratio of the lengths of the vectors taken in an appropriate order.

Thus, for example, if C be the midpoint of a line AB, we have

$$\overrightarrow{AB} = 2\,\overrightarrow{AC},\;\; \overrightarrow{BC} = -\frac{1}{2}\,\overrightarrow{AB}$$

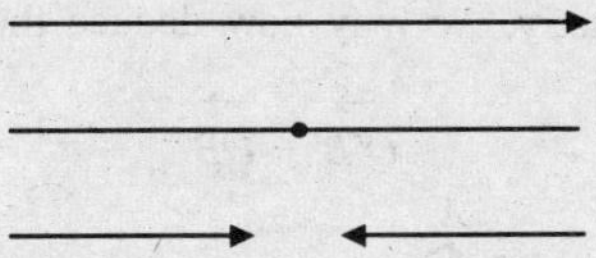

Fig. 1.7

Two parallel vectors may also be described as collinear.

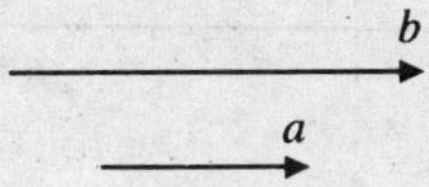

Fig. 1.8

1.3.2. Addition Composition

Let **a, b** be two given vectors. Take a point O.

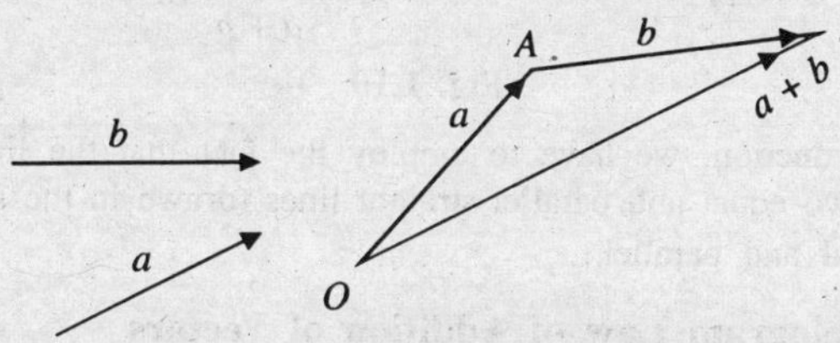

Fig. 1.9

Let

$$\overrightarrow{OA} = \mathbf{a},\ \overrightarrow{AB} = \mathbf{b},$$

so that the terminal point of the vector **a** is the initial point of the vector **b**.

The vector

$$\overrightarrow{OB}$$

is said to be the sum of the vectors, **a** and **b**, and we write

$$\overrightarrow{OB} = \overrightarrow{OA} + \overrightarrow{AB} = \mathrm{a} + \mathrm{b}.$$

We have $\mathbf{a} + \mathbf{0} = \mathbf{a}$, for every **a**; **0**, being the zero vector.

Note. As a matter of logical necessity, we must show that the sum of two vectors is independent of the choice of the point O.

Let O, O¢ be any two points and let

$$\overrightarrow{OA} = \mathbf{a} = \overrightarrow{O'A'},$$

$$\overrightarrow{AB} = \mathbf{b} = \overrightarrow{A'B'}.$$

By elementary geometry, we may now deduce that

$$\overrightarrow{O'B'} = \overrightarrow{OB}.$$

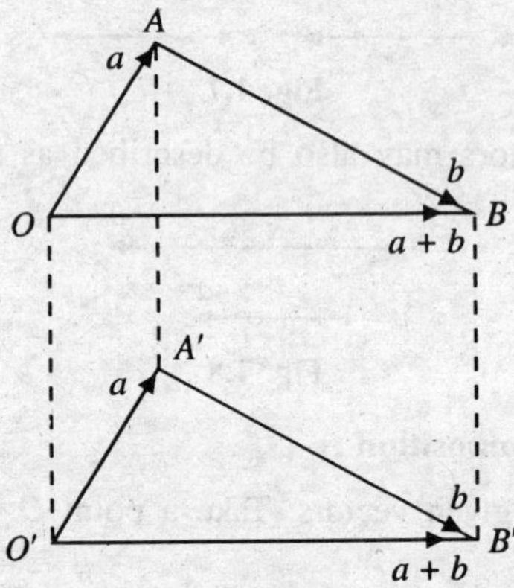

Fig. 1.10

For this deduction, we have to employ the fact that the lines joining the extremities of two equal and parallel straight lines (drawn in the same sense) are themselves equal and parallel.

1.3.3. Parallelogram Law of Addition of Vectors

Consider a parallelogram $OABC$.

The the sum of the vectors

$\overrightarrow{OA}$ and $\overrightarrow{OC}$ is the vector $\overrightarrow{OB}$.

for $\overrightarrow{OB} = \overrightarrow{OA} + \overrightarrow{AB} = \overrightarrow{OA} + \overrightarrow{OC}$.

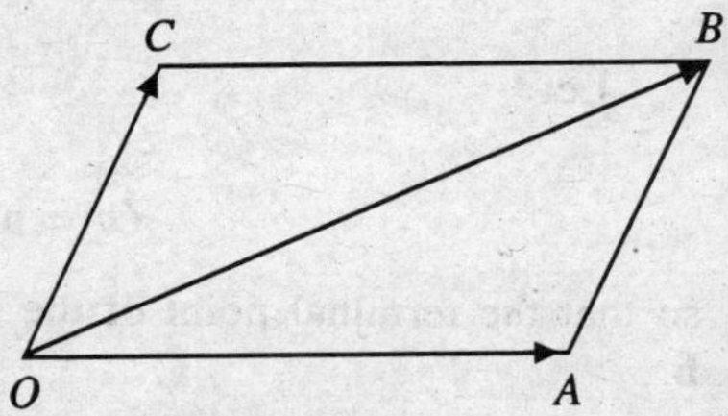

Fig. 1.11

1.4. LAWS OF ADDITION COMPOSITION

1.4.1. *Addition of vectors is Commutative, i.e.,*

$$\mathbf{a} + \mathbf{b} = \mathbf{b} + \mathbf{a}$$

for any pair of vectors, **a**, **b**.

Let $\overrightarrow{OA} = \mathbf{a},\ \overrightarrow{AB} = \mathbf{b}.$

We have

$$\mathbf{a} + \mathbf{b} = \overrightarrow{OB}.$$

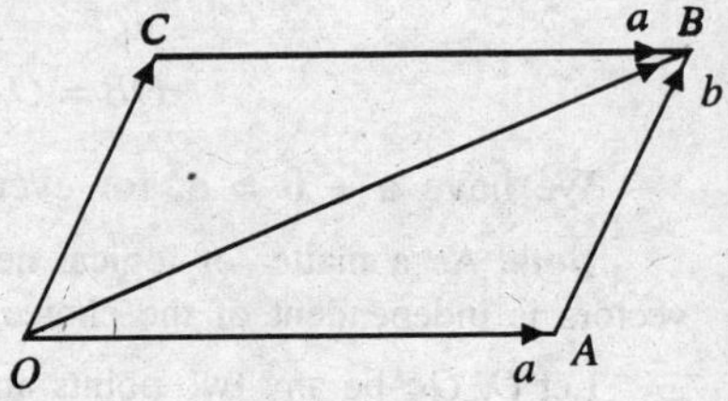

Fig. 1.12

Complete the parallelogram $OABC$ having OA and OC as adjacent sides. Then

$$\vec{OC} = \vec{AB} = \mathbf{b},\ \vec{CB} = \vec{OA} = \mathbf{a}$$

so that we have

$$\vec{OB} = \vec{OC} + \vec{CB} = \mathbf{b} + \mathbf{a}$$

Hence, $\mathbf{a} + \mathbf{b} = \mathbf{b} + \mathbf{a}$.

1.4.2. *Addition of vectors is Associative, i.e.,*

$$\mathbf{a} + (\mathbf{b} + \mathbf{c}) = (\mathbf{a} + \mathbf{b}) + \mathbf{c}$$

where, **a, b, c** are any three vectors.

Take any point O. Let

$$\vec{OA} = \mathbf{a},\ \vec{AB} = \mathbf{b},\ \vec{BC} = \mathbf{c}.$$

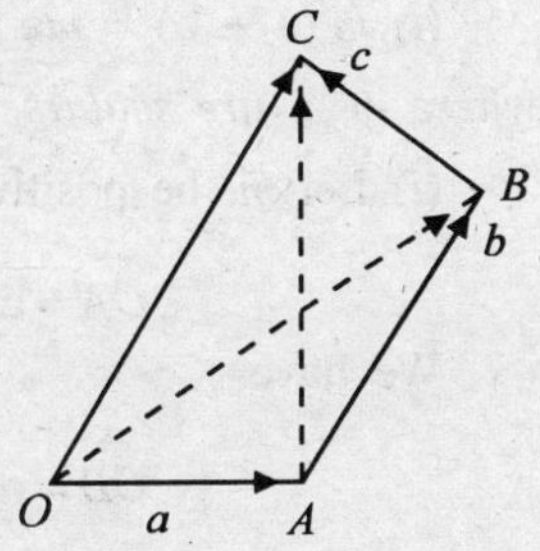

Fig. 1.13

We have

$$\mathbf{b}+\mathbf{c} = \vec{AB} + \vec{BC} = \vec{AC}$$

$$\Rightarrow \qquad \mathbf{a}+(\mathbf{b}+\mathbf{c}) = \vec{OA} + \vec{AC} = \vec{OC}.$$

Again

$$\mathbf{a}+\mathbf{b} = \vec{OA} + \vec{AB} = \vec{OB}$$

$$\Rightarrow \qquad (\mathbf{a}+\mathbf{b})+\mathbf{c} = \vec{OB} + \vec{BC} = \vec{OC}.$$

Thus

$$(\mathbf{a}+\mathbf{b})+\mathbf{c} = \vec{OC} = \mathbf{a}+(\mathbf{b}+\mathbf{c}).$$

In view of the equality of the vectors.

$$(\mathbf{a} + \mathbf{b}) + \mathbf{c},\ \ \mathbf{a} + (\mathbf{b} + \mathbf{c})$$

We denote each of these equal vectors by

$$\mathbf{a} + \mathbf{b} + \mathbf{c}.$$

1.4.3. *Negative of a vector*

For every vector **a**,

$$\mathbf{a} + (-1)\,\mathbf{a} = \mathbf{0}.$$

If $\vec{OA} = \mathbf{a}$, we have, according to the definition of the multiplication of vectors by scalars,

$$\vec{AO} = (-1)\,\mathbf{a}.$$

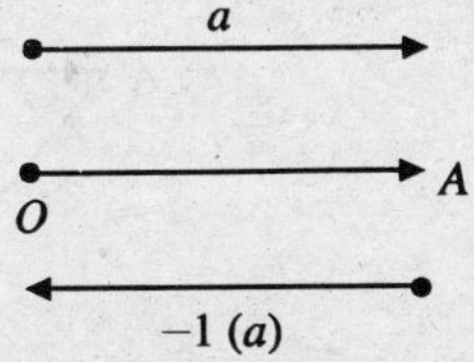

Fig. 1.14

Thus,

$$\mathbf{a}+(-1)\,\mathbf{a} = \vec{OA}+\vec{AO} = \vec{OO} = \mathbf{0}.$$

On account of this property, the vector (– 1) **a** is called the negative of the vector **a**, and we write

$$-\mathbf{a} = (-1)\,\mathbf{a}$$

so that the relation $\mathbf{a} + (-1)\,\mathbf{a} = \mathbf{0}$, may also be re-written as

$$\mathbf{a} + (-\mathbf{a}) = \mathbf{0}.$$

1.5. RELATIONS BETWEEN THE TWO COMPOSITIONS

We shall now consider the relations between the two linear compositions considered in § 3.1. and § 3.2.

(*i*) $m(\mathbf{a} + \mathbf{b}) = m\mathbf{a} + m\mathbf{b}$. (*ii*) $(m + n)\,\mathbf{a} = m\mathbf{a} + n\mathbf{a}$,

where m, n are scalars and **a, b** *are vectors.*

(*i*) Let *m*, be positive and let

$$\vec{OA} = \mathbf{a},\ \vec{AB} = \mathbf{b},$$

We have

$$\vec{OB} = \mathbf{a} + \mathbf{b}.$$

Let $$\vec{OA'} = m\mathbf{a}.$$

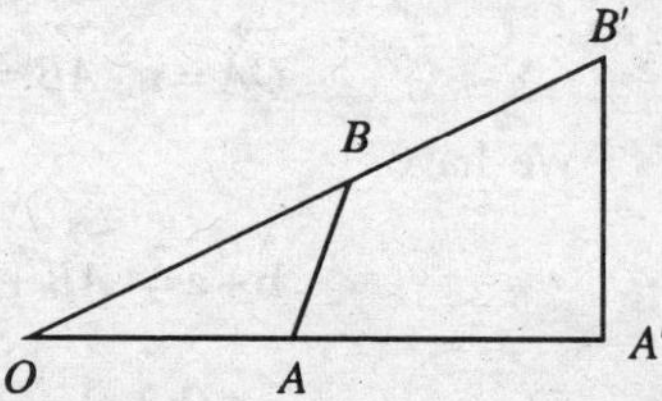

Fig. 1.15

Through *A′* draw a line parallel to *AB* and in the same sense as $\vec{AB}$ to meet OB in *B′*. With the help of Elementary Geometry, we have

$$A'B' = m\,AB \quad \text{and} \quad OB' = m\,OB.$$

Thus, we have

$$\vec{A'B'} = m\,\vec{AB} = m\mathbf{a} + m\mathbf{b}$$

and $$\vec{OB'} = m\,\vec{OB} = m(\mathbf{a} + \mathbf{b}).$$

It follows that

$$m(\mathbf{a} + \mathbf{b}) = \vec{OB'} = \vec{OA'} + \vec{A'B'} = m\mathbf{a} + m\mathbf{b}.$$

The result may similarly be proved with the help of the accompanying figure 16, when *m* is negative.

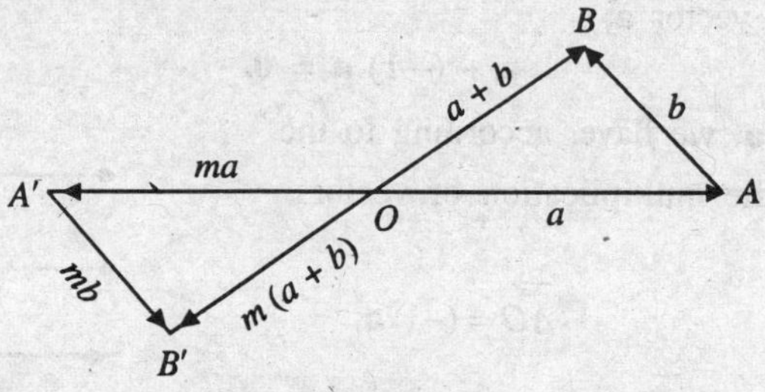

Fig. 1.16

(*ii*) The proof is a simple development of the definitions.

1.6. SUBTRACTION

Difference of Two Vectors. Def. *If* **a, b,** *be any two given vectors, then we write*

$$\mathbf{a} + (-\mathbf{b}) = \mathbf{a} - \mathbf{b}.$$

and call the composition Subtraction.

Thus, we have, in particular

$$\mathbf{a} - \mathbf{a} = \mathbf{a} + (-\mathbf{a}) = \mathbf{a} + (-1)\,\mathbf{a} = \mathbf{0}.$$

It is important to see that

$$\vec{OA} = \mathbf{a},\ \vec{OB} = \mathbf{b} \ \Rightarrow\ \vec{AB} = \mathbf{b} - \mathbf{a}.$$

In fact, we have

$$\mathbf{b} - \mathbf{a} = \mathbf{b} + (-\mathbf{a})$$

$$= \vec{OB} + \vec{AO} = \vec{AO} + \vec{OB} = \vec{AB}.$$

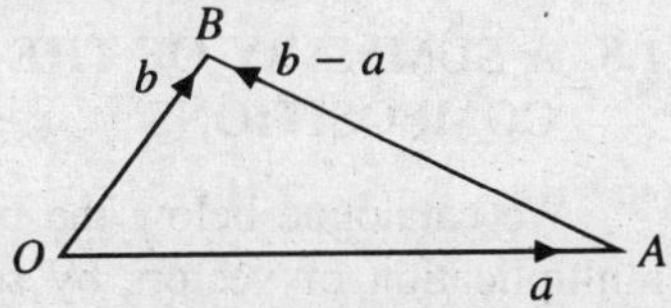

Fig. 1.17

1.6.1. $-(\mathbf{a} + \mathbf{b}) = -\mathbf{a} - \mathbf{b}$. We have

$$[(-\mathbf{a}) + (-\mathbf{b})] + [\mathbf{b} + \mathbf{a}]$$
$$= (-\mathbf{a}) + [(-\mathbf{b}) + \mathbf{b}] + \mathbf{a}$$
$$= -\mathbf{a} + \mathbf{0} + \mathbf{a}$$
$$= (-\mathbf{a} + \mathbf{a}) + \mathbf{0} = \mathbf{0}.$$

Thus, $(-\mathbf{a}) + (-\mathbf{b}) = -(\mathbf{a} + \mathbf{b})$

$\Rightarrow \qquad (-\mathbf{a} + \mathbf{b}) = -\mathbf{a} - \mathbf{b}.$

The proof of the result may be also directly obtained with the help of figure 18 given below.

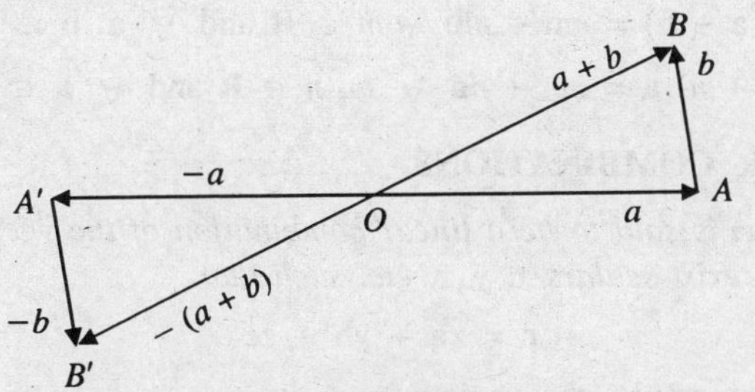

Fig. 1.18

1.7. A VECTOR EQUATION

If **a, b** *be two given vectors, then the vector equation*

$$\mathbf{a} + x = \mathbf{b}.$$

is satisfied by one and only one vector, viz.,

$$x = \mathbf{b} - \mathbf{a}.$$

We have

$$\mathbf{a} + x = \mathbf{b}$$

$$\Leftrightarrow \quad (-\mathbf{a}) + (\mathbf{a} + x) = (-\mathbf{a}) + \mathbf{b}$$

$$\Leftrightarrow \quad [(-\mathbf{a}) + \mathbf{a}] + x = \mathbf{b} + (-\mathbf{a})$$

$$\Leftrightarrow \quad \mathbf{0} + x = \mathbf{b} - \mathbf{a}$$

$$\Leftrightarrow \quad x = \mathbf{b} - \mathbf{a}.$$

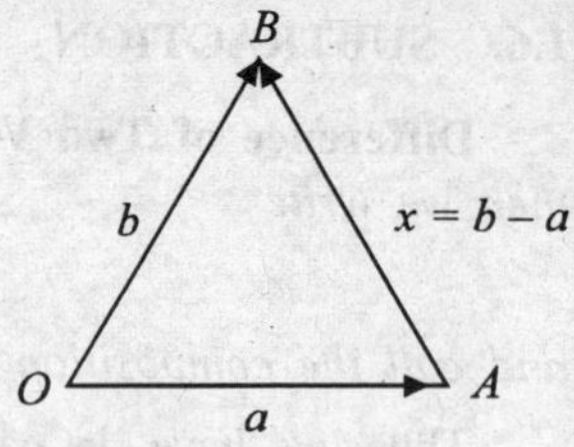

Fig. 1.19

1.8. A SUMMARY OF THE BASIC PROPERTIES OF THE LINEAR COMPOSITIONS

We catalogue below the basic laws of the Addition of vectors and of multiplication of vectors by scalars, *i.e.*, of the linear compositions. The set of all the vectors will be denoted by **V** and the set of all real numbers by **R**.

I. $\mathbf{a} + \mathbf{b} = \mathbf{b} + \mathbf{a} \ \forall \ \mathbf{a}, \mathbf{b} \in \mathbf{V}$;

II. $\mathbf{a} + (\mathbf{b} + \mathbf{c}) = (\mathbf{a} + \mathbf{b}) + \mathbf{c} \ \forall \ \mathbf{a}, \mathbf{b}, \mathbf{c} \in \mathbf{V}$;

III. $\exists$ **a** vector, *viz.*, **0** such that

$$\mathbf{a} + \mathbf{0} = \mathbf{a} = \mathbf{0} + \mathbf{a} \ \forall \ \mathbf{a} \in \mathbf{V};$$

IV. To each $\mathbf{a} \in \mathbf{V}$ there corresponds $\mathbf{b} \in \mathbf{V}$ such that

$$\mathbf{a} + \mathbf{b} = \mathbf{0} = \mathbf{b} + \mathbf{a};$$

The vector **b** is denoted by $-\mathbf{a}$.

V. $(mn)(\mathbf{a}) = m(n\mathbf{a}) \ \forall \ \mathbf{a} \in \mathbf{V}$ and $\forall \ m, n \in \mathbf{R}$;

VI. $1(\mathbf{a}) = \mathbf{a} \ \forall \ \mathbf{a} \in \mathbf{V}$;

VII. $m(\mathbf{a} + \mathbf{b}) = m\mathbf{a} + m\mathbf{b} \ \forall \ m \in \mathbf{R}$ and $\forall \ \mathbf{a}, \mathbf{b} \in \mathbf{V}$;

VIII. $(m + n)\,\mathbf{a} = m\mathbf{a} + n\mathbf{a} \ \forall \ m, n \in \mathbf{R}$ and $\forall \ \mathbf{a} \in \mathbf{V}$.

1.9. LINEAR COMBINATIONS

A vector, **r**, *is said to be a linear combination of the vectors,* **a, b, c,** *... etc. if there exist scalars x, y, z, etc. such that*

$$\mathbf{r} = x\mathbf{a} + y\mathbf{b} + z\mathbf{c}$$

Thus, for example, the vectors

$$2\mathbf{a} + \mathbf{b} - 4\mathbf{c}, \quad \mathbf{a} + 2\mathbf{b} - 3\mathbf{c}$$

are linear combinations of the vectors **a, b, c.**

A linear combination of vectors involves the two linear compositions of the addition of vectors and the multiplication of vectors by scalars.

In the following linear combinations

$$x\mathbf{a},\ x\mathbf{a} + y\mathbf{b},\quad x\mathbf{a} + y\mathbf{b} + z\mathbf{c}$$

will be of special interest to us.

1.9.1. Parallel Vectors. Collinear Vectors

Of the two vectors having the same or parallel supports, each is a linear combination of the other. Thus, if **a, b** be two parallel vectors, then there exists a scalar x such that $\mathbf{b} = x\mathbf{a}$ or $\mathbf{a} = (1/x)\,\mathbf{b}$.

Parallel vectors are also called collinear vectors, for the support of parallel coinitial vectors is the same line.

1.9.2. Def. Coplanar Vectors.

A set of vectors is said to be coplanar, if their supports are parallel to the same plane, *i.e.*, if there exists a plane parallel to the supports of each of the vectors.

The support of coplanar co-initial vectors are coplanar.

The vector $x\mathbf{a} + y\mathbf{b}$ which is a linear combination of the vectors **a, b** is coplanar with **a** and **b.**

EXAMPLE

ABCDEF is a regular hexagon. Let $\vec{AB} = \mathbf{a}$ *and* $\vec{BC} = \mathbf{b}$. *Find the vectors determined by the other four sides taken in order. Also express the vectors* $\vec{AC}, \vec{AD}, \vec{AF}, \vec{AE}, \vec{CE}$ *in terms of* **a** *and* **b.**

Solution. $\vec{AC} = \vec{AB} + \vec{BC} = \mathbf{a} + \mathbf{b}$

$\because$ AD is parallel and double of BC,

$\therefore$ $\vec{AD} = 2\mathbf{b}$.

In $\Delta\, ACD$,

$$\vec{AC} + \vec{CD} = \vec{AD}$$

$$\Rightarrow \quad \vec{CD} = \vec{AD} - \vec{AC} = 2\mathbf{b} - (\mathbf{a} + \mathbf{b}) = \mathbf{b} - \mathbf{a}.$$

$$\vec{AF} = \vec{CD} = \mathbf{b} - \mathbf{a}.$$

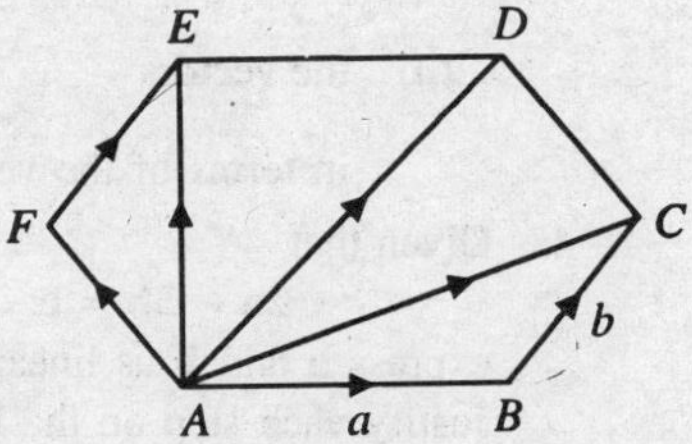

Fig. 1.20

Now, $\vec{DE} = \vec{BA} = -\mathbf{a}$

$\vec{EF} = \vec{CB} = -\mathbf{b}$

$\vec{FA} = \vec{DC} = -(\mathbf{b} - \mathbf{a}) = \mathbf{a} - \mathbf{b}$

Again, $\vec{AE} = \vec{AD} + \vec{DE} = 2\mathbf{b} + (-\mathbf{a}) = 2\mathbf{b} - \mathbf{a}$

and $\vec{CE} = \vec{CD} + \vec{DE} = \mathbf{b} - \mathbf{a} + (-\mathbf{a})$

$$= \mathbf{b} - 2\mathbf{a}$$

EXERCISES

1. ABC is any triangle and D, E, F are the middle points of its sides BC, CA, AB respectively; express

   (*i*) the vectors $\vec{BC}, \vec{AD}, \vec{BE}$ and $\vec{CF}$

   as linear combinations of the vectors $\vec{AB}$ and $\vec{AC}$.

   (*ii*) the vectors $\vec{AB}, \vec{BC}, \vec{CA}, \vec{AD}$

   as linear combinations of the vectors $\vec{BE}$ and $\vec{CF}$.

   (*iii*) the vectors $\vec{AC}, \vec{BC}, \vec{AD}$ and $\vec{CF}$

   as linear combinations of the vectors $\vec{AB}$ and $\vec{BE}$.

   Show that $\vec{AD} + \vec{BE} + \vec{CF} = 0$.

2. $ABCD$ is a parallelogram and AC, BD are its diagonals. Express

   (*i*) $\vec{AC}$ and $\vec{BD}$ in terms of $\vec{AB}$ and $\vec{AD}$,

   (*ii*) $\vec{AB}$ and $\vec{AD}$ in terms of $\vec{AC}$ and $\vec{BD}$,

   (*iii*) $\vec{AB}$ and $\vec{AC}$ in terms of $\vec{AD}$ and $\vec{BD}$.

   Show that

   $$\vec{AC} + \vec{DB} = 2\vec{DC}, \quad \vec{AC} - \vec{BD} = 2\vec{AB}.$$

3. $OABC$ is a tetrahedron express

   (*i*) the vectors $\vec{BC}, \vec{CA},$ and $\vec{AB}$

   in terms of the vectors $\vec{OA}, \vec{OB}$ and $\vec{OC}$;

   (*ii*) the vectors $\vec{OA}, \vec{OB}$ and $\vec{CA}$

   in terms of the vectors $\vec{OC}, \vec{AB}$ and $\vec{BC}$

4. Given that

   $$2\mathbf{a} - 3\mathbf{b} = \mathbf{c}; \quad -3\mathbf{a} + 5\mathbf{b} = \mathbf{d},$$

   express **a** and **b** as linear combinations of **c** and **d**.
   Justify each step on the basis of different laws of compositions.

5. Same question as 5 above for the following pairs of equations :

   (*i*) $-\mathbf{a} + 2\mathbf{b} = 2\mathbf{c}; \quad 5\mathbf{a} - 2\mathbf{b} = 3\mathbf{d}.$

   (*ii*) $-\mathbf{a} + \mathbf{b} = -\mathbf{c}; \quad 2\mathbf{a} - \mathbf{b} = \mathbf{c} + \mathbf{d}.$

6. OB and OC are two lines and D is a point on BC such that

   $$\frac{BD}{DC} = \frac{m}{n},$$

show that

$$\vec{OD} = \frac{n\vec{OB} + m\vec{OC}}{m+n}$$

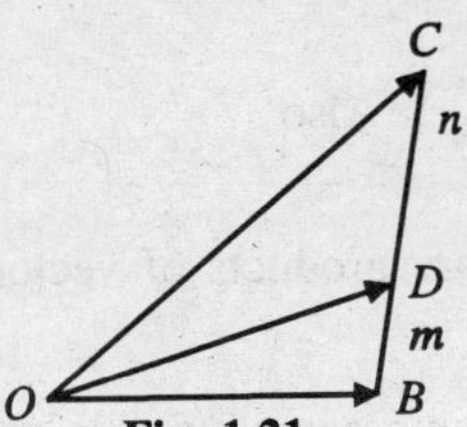

Fig. 1.21

7. Which of the following statements are correct :
If in triangle OAC, B is the mid-point of AC and

$\vec{OA} = \mathbf{a}$ and $\vec{OB} = \mathbf{b}$, then

(*a*) $\vec{OC} = \frac{1}{2}(\mathbf{a}+\mathbf{b})$, (*b*) $\vec{OC} = 2\mathbf{b} - 2\mathbf{a}$,

(*c*) $\vec{OC} = 2\mathbf{b} - \mathbf{a}$, (*d*) $\vec{OC} = 3\mathbf{a} - 2\mathbf{b}$,

(*e*) $\vec{OC} = 3\mathbf{b} - 2\mathbf{a}$.

8. Which of the following statements are correct :
If M is the mid-point of AB and O is any point, then

(*a*) $\vec{OM} = \vec{OA} + \vec{MA}$, (*b*) $OM = \vec{OA} - \vec{MA}$

(*c*) $\vec{OM} = \frac{1}{2}(\vec{OA} - \vec{OB})$, (*d*) $OM = \frac{1}{2}(\vec{OB} + \vec{OA})$,

(*e*) $\vec{OM} = \frac{1}{2}(OB - \vec{OA})$.

1.10. EXPRESSIONS AS LINEAR COMBINATIONS

1.10.1. Coplanar Vectors

If **a**, **b** *be two given non-collinear vectors, then every vector* **r**, *coplanar with* **a** *and* **b** *can be represented as a linear combination*

$$x\mathbf{a} + y\mathbf{b}$$

x, y being some scalars. Also this representation is unique.

Existence. Take any point O.

Let $\vec{OA} = \mathbf{a}$, $\vec{OB} = \mathbf{b}$.

Let $\vec{OP} = \mathbf{r}$ be any vector coplanar with $\vec{OA}$ and $\vec{OB}$.

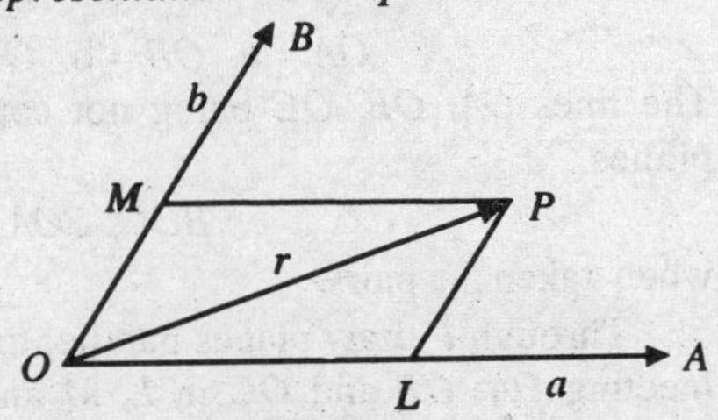

Fig. 1.22

The lines OA, OB and OP are coplanar.

Through P, draw lines parallel to OB and OA to meet OA and OB respectively in L and M. We have

$$\mathbf{r} = \vec{OP} = \vec{OL} + \vec{LP} = \vec{OL} + \vec{OM}.$$

Also

$$\vec{OL} \text{ and } \vec{OM}$$

are products of vectors

$$\vec{OA} \text{ and } \vec{OB}$$

with some suitable scalars. Let

$$\vec{OL} = x\vec{OA} = x\mathbf{a}, \quad \vec{OM} = y\vec{OB} = y\mathbf{b}.$$

Thus, we have

$$\mathbf{r} = x\mathbf{a} + y\mathbf{b}.$$

Uniqueness. Let, if possible,

$$\mathbf{r} = x'\mathbf{a} + y'\mathbf{b}.$$

We have

$$x\mathbf{a} + y\mathbf{b} = x'\mathbf{a} + y'\mathbf{b}$$

$$\Rightarrow \quad (x - x')\,\mathbf{a} + (y - y')\,\mathbf{b} = \mathbf{0}.$$

If $x - x' \neq 0$, we have

$$\mathbf{a} = \frac{y - y'}{x - x'}\mathbf{b},$$

so that **a** is collinear with **b**. We thus arrive at a contradiction.

Thus, $x = x'$. Similarly $y = y'$.

Hence, the uniqueness of the representation.

1.10.2. Arbitrary System of Vectors

If **a, b, c** *be three given non-coplanar vectors, then any vector,* **r**, *can be represented as a linear combination*

$$x\mathbf{a} + y\mathbf{b} + z\mathbf{c}$$

x, y, z being some scalars. Also this representation is unique.

Take any point O. Let

$$\vec{OA} = \mathbf{a}, \ \vec{OB} = \mathbf{b}, \ \vec{OC} = \mathbf{c}, \ \vec{OP} = \mathbf{r}.$$

The lines OA, OB, OC being not coplanar, they determine three different planes

$$BOC, \ COA, \ AOB$$

when taken in pairs.

Through P draw planes parallel to the three planes BOC, COA and AOB meeting OA, OB and OC in L, M and N respectively so that we obtain a parallelopiped having OP as a diagonal. We have

$$\begin{aligned} \mathbf{r} &= \vec{OP} \\ &= \vec{OL} + \vec{LP} \\ &= \vec{OL} + \vec{LN'} + \vec{N'P} = \vec{OL} + \vec{OM} + \vec{ON}. \end{aligned}$$

There exists scalars x, y and z such that

$$\vec{OL} = x\vec{OA} = x\mathbf{a},\ \vec{OM} = y\vec{OB} = y\mathbf{b},\ \vec{ON} = z\vec{OC} = z\mathbf{c}.$$

Thus, we have

$$\mathbf{r} = x\mathbf{a} + y\mathbf{b} + z\mathbf{c}.$$

Uniqueness. Let, if possible

$$\mathbf{r} = x'\mathbf{a} + y'\mathbf{b} + z'\mathbf{c}.$$

We have

$$x\mathbf{a} + y\mathbf{b} + z\mathbf{c} = x'\mathbf{a} + y'\mathbf{b} + z'\mathbf{c}$$

$$\Rightarrow \quad (x - x')\,\mathbf{a} + (y - y')\,\mathbf{b} + (z - z')\,\mathbf{c} = \mathbf{0}.$$

If $x - x' \neq 0$, we have

$$\mathbf{a} = \frac{y - y'}{x' - x}\mathbf{b} + \frac{z - z'}{x - x'}\mathbf{c}$$

$\Rightarrow$ **a** is coplanar with 1, **b** and **c** and we arrive at a contradiction of the given hypothesis.

Thus, $x - x' = 0 \Leftrightarrow x = x'$. Similarly $y = y'$ and $z = z'$.

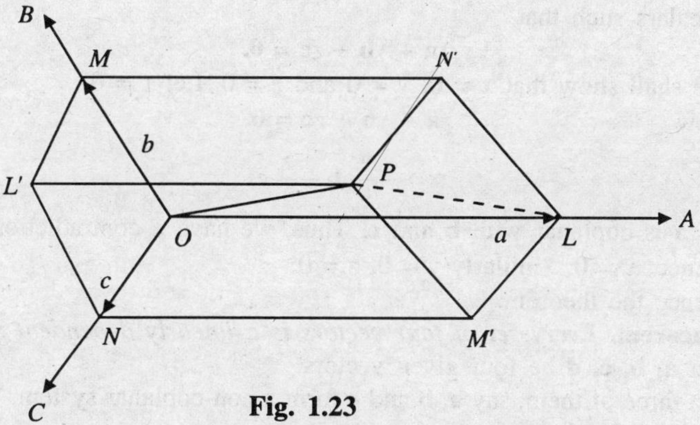

Fig. 1.23

The uniqueness of the representation is thus proved.

Ex. Show that the vector $x\mathbf{a} + y\mathbf{b}$ is coplanar with the vectors **a** and **b**.

1.11. LINEARLY INDEPENDENT AND DEPENDENT SYSTEMS OF VECTORS

Def. *A system of vectors*

$$\mathbf{a}, \mathbf{b}, \mathbf{c}, \ldots$$

is said to be linearly dependent, if there exists a system of scalars

$$x, y, z, \ldots$$

not all zero, such that

$$0 = x\mathbf{a} + y\mathbf{b} + z\mathbf{c} + \ldots, \qquad \ldots(i)$$

A system which is not linearly dependent is said to be linearly independent. For such a system, every relation of the type (i) implies that

$$x = 0,\ y = 0,\ z = 0, \ldots$$

Note. A pair of collinear vectors is a linearly dependent system. Also a triad of coplanar vectors is a linearly dependent system.

Theorem. *A pair of non-zero, non-collinear vectors is a linearly dependent system.*

Let **a, b** be a pair of non-zero, non-collinear vectors and let x, y be two scalars such that

$$x\mathbf{a} + y\mathbf{b} = \mathbf{0}.$$

We shall show that $x = 0$ and $y = 0$. Let $x \neq 0$. Now

$$x\mathbf{a} + y\mathbf{b} = \mathbf{0} \Rightarrow \mathbf{a} = -\frac{y}{x}\mathbf{b}$$

so that **a** appears as a product of the vector **b** with a scalar. Thus **a** and **b** are collinear and as such we arrive at a contradiction.

Hence, $x = 0$. Similarly $y = 0$.

Hence the theorem.

Theorem. *A triad of non-zero, non-coplanar vectors is a linearly independent system.*

Let **a, b, c** be three non-zero, non-coplanar vectors and let x, y, z be three scalars such that

$$x\mathbf{a} + y\mathbf{b} + z\mathbf{c} = \mathbf{0}.$$

We shall show that $x = 0$, $y = 0$ and $z = 0$. Let $x \neq 0$.

Now $$x\mathbf{a} + y\mathbf{b} + z\mathbf{c} = \mathbf{0}.$$

$$\Rightarrow \quad \mathbf{a} = -\frac{y}{x}\mathbf{b} - \frac{z}{x}\mathbf{c},$$

so that **a** is coplanar with **b** and **c**. Thus, we have a contradiction.

Hence, $x = 0$. Similarly $y = 0$, $z = 0$.

Hence the theorem.

Theorem. *Every set of four vectors is a linearly dependent system.*

Let **a, b, c, d** be four given vectors.

Let three of them, say **a, b** and **c** form a non-coplanar system. We then have a relation of the form

$$\mathbf{d} = x\mathbf{a} + y\mathbf{b} + z\mathbf{c} \quad \Rightarrow \quad x\mathbf{a} + y\mathbf{b} + z\mathbf{c} + (-\mathbf{d}) = \mathbf{0}$$

so that **a, b, c, d** form a linearly dependent system.

Now suppose that no three of the given vectors forms a non-coplanar system. Consider the triad of vectors **a, b, c** which are coplanar. Supposing that no pair of these three vectors is collinear, we have a relation of the form

$$\mathbf{c} = x\mathbf{a} + y\mathbf{b} \quad \Rightarrow \quad x\mathbf{a} + y\mathbf{b} + (-\mathbf{c}) + (0)\,\mathbf{d} = 0$$

so that **a, b, c, d** form a linearly dependent system.

We may similarly consider the case when a, b, c are all collinear.

1.12. BASES. COMPONENTS AND CO-ORDINATES OF A VECTOR

Let *a, b, c* be three given non-coplanar vectors so that if *r* be any given vector, we have the relation

$$\mathbf{r} = x\mathbf{a} + y\mathbf{b} + z\mathbf{c}$$

The vectors

$$x\mathbf{a},\quad y\mathbf{b},\quad z\mathbf{c}$$

are said to be the *components* and the scalars x, y, z the *co-ordinates* of the vector $\mathbf{r}$, relatively to the triad of non-coplanar vectors **a, b, c.**

The set of non-coplanar vectors **a, b, c** relatively to which we may decompose any given vector **r** is called a **Base.** Every triad of non-coplanar vectors can thus be thought of as a Base.

Of course, the components and the co-ordinates of a given vector will be different for different Bases.

1.12.1. Co-ordinates of the Sum of Two Vectors and the Product of a Vector with a Scalar

Let **r, r′** be two given vectors with co-ordinates (x, y, z) and (x', y', z') respectively relatively to a given base $\{\mathbf{a}, \mathbf{b}, \mathbf{c}\}$ so that we have

$$\mathbf{r} = x\mathbf{a} + y\mathbf{b} + z\mathbf{c}; \qquad \mathbf{r}' = x'\mathbf{a} + y'\mathbf{b} + z'\mathbf{c}.$$

Thus,

$$\mathbf{r} + \mathbf{r}' = (x + x')\,\mathbf{a} + (y + y')\,\mathbf{b} + (z + z')\,\mathbf{c}$$

so that the co-ordinates of $\mathbf{r} + \mathbf{r}'$ are

$$(x + x', y + y', z + z')$$

obtained by co-ordinate-wise addition of the co-ordinates of **r** and **r′**.

Again, if m be any scalar, we have

$$m\mathbf{r} = mx\mathbf{a} + my\mathbf{b} + mz\mathbf{c}$$

so that the co-ordinates of the vector $m\mathbf{r}$ are

$$(mx, my, mz)$$

obtained on co-ordinate-wise multiplication of the co-ordinates of **r** with m.

Note 1. It is easy to see that if (x, y, z) be the co-ordinates of a vector r relatively to a base (**a, b, c**) so that we have a relation

$$\mathbf{r} = x\mathbf{a} + y\mathbf{b} + z\mathbf{c},$$

then

$$\mathbf{r} \text{ is coplanar with } \mathbf{a}, \mathbf{b} \iff z = 0,$$

$$\mathbf{r} \text{ is coplanar with } \mathbf{a} \iff y = 0, z = 0.$$

Note 2. If **a, b** be two non-collinear vectors in a plane, then every vector in that plane is expressible in the form $x\mathbf{a} + y\mathbf{b}$ so that we can think of x and y as the co-ordinates of this vector. We may thus see that a system of coplanar vectors has a base consisting of only two members. Similarly, for a system of collinear vectors, we have a base consisting of only **one** member.

EXAMPLES

Example 1. *Show that if a triad* **a, b, c** *is a base, then the triad*

$$\mathbf{a},\ \mathbf{a} + \mathbf{b},\ \mathbf{a} + \mathbf{b} + \mathbf{c}$$

is also a base.

Given that x, y, z are the co-ordinates of a vector **u** *relatively to the base* **a, b, c,** *what are its co-ordinates relatively to the base* **a, a + b, a + b + c,**

Solution. We write

$$\mathbf{p} = \mathbf{a},\ \mathbf{q} = \mathbf{a} + \mathbf{b},\ \mathbf{r} = \mathbf{a} + \mathbf{b} + \mathbf{c}.$$

and show that **p, q, r** is a linearly independent set.

Now

$$l\mathbf{p} + m\mathbf{q} + n\mathbf{r} = \mathbf{0}$$

$$\Rightarrow \quad l\mathbf{a} + \mathbf{m}\,(\mathbf{a} + \mathbf{b}) + n\,(\mathbf{a} + \mathbf{b} + \mathbf{c}) = \mathbf{0}$$

$$\Rightarrow \quad (l + m + n)\,\mathbf{a} + (m + n)\,\mathbf{b} + n\mathbf{c} = \mathbf{0}$$

$$\Rightarrow \quad l + m + n = 0, \quad m + n = 0, \quad n = 0$$

[**a, b, c** being a base and therefore a linearly independent set]

$$\Rightarrow \quad n = 0,\ m = 0 = l = 0$$

$\Rightarrow$ **p, q, r** is a linearly independent set

$\Rightarrow$ **p, q, r** is a base.

Again we are given that

$$\mathbf{u} = x\mathbf{a} + y\mathbf{b} + z\mathbf{c}. \qquad ...(i)$$

Let

$$\mathbf{u} = x'\mathbf{a} + y'\,(\mathbf{a} + \mathbf{b}) + z'\,(\mathbf{a} + \mathbf{b} + \mathbf{c})$$

$$= (x' + y' + z')\,\mathbf{a} + (y' + z')\,\mathbf{b} + z'\mathbf{c}. \qquad ...(ii)$$

The expression of **u** as a linear combination of **a, b, c** being unique, we obtain

$$\left.\begin{aligned} x' + y' + z' &= x \\ y' + z' &= y \\ z' &= z \end{aligned}\right\} \Leftrightarrow \left\{\begin{aligned} z' &= z, \\ y' &= y - z, \\ x' &= x - y. \end{aligned}\right.$$

Thus, $x - y$, $y - z$, z are the required co-ordinates relatively to the new base. Thus, we have

$$\mathbf{u} = x\mathbf{a} + y\mathbf{b} + z\mathbf{c}.$$

and $\quad \mathbf{u} = (x - y)\,\mathbf{a} + (y - z)\,(\mathbf{a} + \mathbf{b}) + z\,(\mathbf{a} + \mathbf{b} + \mathbf{c}).$

Example 2. *Examine whether the vectors* 5**a** + 6**b** + 7**c**, 7**a** – 8**b** + 9**c** *and* 3**a** + 20**b** + 5**c**, (**a, b, c** *being non-coplanar vectors*) *are linearly independent or dependent.*

Solution. If possible, let txe linearly dependent. Then there exist scalars x_1, x_2, x_3, not all zero, such that

$$x_1\,(5\mathbf{a} + 6\mathbf{b} + 7\mathbf{c}) + x_2\,(7\mathbf{a} - 8\mathbf{b} + 9\mathbf{c}) + x_3\,(3\mathbf{a} + 20\mathbf{b} + 5\mathbf{c}) = \mathbf{0} \qquad ...(i)$$

$$\Rightarrow \quad (5x_1 + 7x_2 + 3x_3)\,\mathbf{a} + (6x_1 - 8x_2 + 20x_3)\,\mathbf{b} + (7x_1 + 9x_2 + 5x_3)\,\mathbf{c} = 0$$

As **a, b, c** are non-coplanar vectors,

$$\Rightarrow \quad \begin{aligned} 5x_1 + 7x_2 + 3x_3 &= 0 \\ 6x_1 - 8x_2 + 20x_3 &= 0 \\ 7x_1 + 9x_2 + 5x_3 &= 0 \end{aligned}$$

From first two equations, we get

$$\frac{x_1}{2} = \frac{x_2}{-1} = \frac{x_3}{-1} = k \quad \text{(say)}$$

$$\therefore \quad x_1 = 2k, \quad x_2 = -k, \quad x_3 = -k.$$

These values also satisfy the third equation.

Hence, there exist scalars x_1, x_2, x_3 such that (1) holds. Hence, given vectors are linearly dependent.

Example 3. *If* **a, b, c** *be three non-zero, non-coplanar vectors, find a relation between the vectors* **a** + 3**b** + 4**c**, **a** – 2**b** + 3**c**, **a** + 5**b** – 2**c**, 6**a** + 14**b** + 4**c**.

Solution. Let

$$\mathbf{a} + 3\mathbf{b} + 4\mathbf{c} = x_1 (\mathbf{a} - 2\mathbf{b} + 3\mathbf{c}) + x_2 (\mathbf{a} + 5\mathbf{b} - 2\mathbf{c}) + x_3 (6\mathbf{a} + 14\mathbf{b} + 4\mathbf{c})$$

$$\Rightarrow \quad \mathbf{a} + 3\mathbf{b} + 4\mathbf{c} = (x_1 + x_2 + 6x_3)\,\mathbf{a} + (-2x_1 + 5x_2 + 14x_3)\,\mathbf{b} + (3x_1 - 2x_2 + 4x_3)\,\mathbf{c}$$

Equating the coefficients of **a, b, c** on both sides, we get

$$x_1 + x_2 + 6x_3 = 1, \quad -2x_1 + 5x_2 + 14x_3 = 3, \quad 3x_1 - 2x_2 + 4x_3 = 4$$

$$\Rightarrow \qquad x_1 = -2, \; x_2 = -3, \; x_3 = 1.$$

$$\therefore \quad \mathbf{a} + 3\mathbf{b} + 4\mathbf{c} = -2 (\mathbf{a} - 2\mathbf{b} + 3\mathbf{c}) - 3 (\mathbf{a} + 5\mathbf{b} - 2\mathbf{c}) + (6\mathbf{a} + 14\mathbf{b} + 4\mathbf{c}).$$

EXERCISES

1. Given that the vectors **a, b, c** form a base, find the co-ordinates of the vectors.
   (*i*) 3**u** – **v** + **w** if **u** = **a** + **c**, **v** = **b** + **c**, **w** = **a** – **b**;
   (*ii*) 2**u** – 3**v** + 4**w** if **u** = 2**b** + 3**c**, **v** = – 2**a** + **c**, **w** = **a** – **b** + **c**;
   (*iii*) 2**u** – 3**v** + 4**w** if **u** = 2**a** + 3**b** + 4**c**, **v** = 3**a** – 2**b** + 4**c**, **w** = 2**a** – 4**b** + 3**c**.
2. Show that the following set of vector is linearly dependent :
   a – 2**b** + 3**c**, – 2**a** + 3**b** – 4**c**, – **b** + 2**c**.
   Find also the linear relations.
3. **a, b, c** is a system of linearly independent vectors; show that the following sets of vectors are also linearly independent.
   (*i*) **a** – **b** + **c**, **b** + **c** – **a**, 2**a** – 3**b** + 4**c**.
   (*ii*) 2**a** – **b** + 3**c**, **a** + **b** – 2**c**, **a** + **b** – 3**c**.
4. Given that *a, b, c* is a linearly independent set of vectors show that :
   (*i*) **a** + 2**b** + **c**, 2**a** – **b** + **c**, 3**a** + **b** + 2**c**
   is a linearly dependent set.
   (*ii*) 2**a** + **b** + **c**, **a** – **b** + 2**c**, – **a** + **b** – **c**,
   is a linearly independent set.
   (*iii*) *m***a** + **b** + **c**, **a** + *m***b** + **c**, **a** + **b** + *m***c**
   is a linearly independent set if and only if $m \neq -2$.
5. **u, v, w** being a linearly independent set of vectors; examine the set **p, q, r** where

$$\mathbf{p} = \mathbf{u}\cos a + \mathbf{v}\cos b + \mathbf{w}\cos c,$$
$$\mathbf{q} = \mathbf{u}\sin a + \mathbf{v}\sin b + \mathbf{w}\sin c,$$
$$\mathbf{r} = \mathbf{u}\sin (x + a) + \mathbf{v}\sin (x + b) + \mathbf{w}\sin (x + c),$$

   for linear independence.

6. Consider a base **a, b, c** and a vector

$$-2\mathbf{a} + 3\mathbf{b} - \mathbf{c}.$$

Compute the co-ordinates of this vector relatively to the base **p, q, r** where

$$\mathbf{p} = 2\mathbf{a} - 3\mathbf{b},\quad \mathbf{q} = \mathbf{a} - 2\mathbf{b} + \mathbf{c},\quad \mathbf{r} = -3\mathbf{a} + \mathbf{b} + 2\mathbf{c}.$$

7. With reference to the base, **u, v, w,** show that the vectors

$$\mathbf{u} + \mathbf{w},\quad -\mathbf{u} + \mathbf{w},\quad \mathbf{u} + \mathbf{v} + \mathbf{w}$$

also constitute a base. Give the formulae for the transformation of the co-ordinates of a vector.

8. (Fig. 23). Given that x, y, z are the co-ordinates of a vector in terms of tbe base $\overrightarrow{OL}, \overrightarrow{OM}, \overrightarrow{ON}$, find its co-ordinates in terms of the base $\overrightarrow{OL'}, \overrightarrow{OM'}, \overrightarrow{ON'}$.

9. Find the linear relations between the following system of vectors; **a, b, c**; being any three non-coplanar vectors;

(*i*) $\mathbf{a} - \mathbf{b} + \mathbf{c},\ \mathbf{b} + \mathbf{c} - \mathbf{a},\ \mathbf{c} + \mathbf{a} + \mathbf{b},\ 2\mathbf{a} - 3\mathbf{b} + 4\mathbf{c}.$

(*ii*) $\mathbf{a} - 2\mathbf{b} + \mathbf{c},\ -\mathbf{b} + 2\mathbf{c} + \mathbf{a},\ \mathbf{c} - \mathbf{a} + 3\mathbf{b},\ \mathbf{a} + \mathbf{b} + \mathbf{c}.$

10. Given that **a, b** is a pair of non-collinear vectors such that

(*i*) $(1 + 2h - k)\,\mathbf{a} + (2 - h + 2k)\,\mathbf{b} = \mathbf{0},$

(*ii*) $(3h + k)\,\mathbf{a} + (1 - 2h)\,\mathbf{b} = h\mathbf{a} + 2k\mathbf{b}.$

find h and k in each case.

11. Given that **a, b, c** is a triad of three non-coplanar vectors such that

(*i*) $(1 - h - k + l)\,\mathbf{a} + (2h - 3k - l)\,\mathbf{b} + (3 + h + k)\,\mathbf{c} = \mathbf{0},$

(*ii*) $(2h + k)\,\mathbf{a} + (3 - 4h + l)\,\mathbf{b} + (1 + h + k)\,\mathbf{c} = h\mathbf{a} + k\mathbf{b} + l\mathbf{c}.$

find h, k and l in each case.

12. Show that the three vectors with co-ordinates

$$(x_1, y_1, z_1),\quad (x_2, y_2, z_2),\quad (x_3, y_3, z_3),$$

relatively to any given base will be linearly dependent if and only if

$$\begin{vmatrix} x_1 & y_1 & z_1 \\ x_2 & y_2 & z_2 \\ x_3 & y_3 & z_3 \end{vmatrix} = 0.$$

SUMMARY

1. Properties of multiplication of vectors by scalars and of addition of vectors.

I. $\mathbf{a} + \mathbf{b} = \mathbf{b} + \mathbf{a}\ \forall\ \mathbf{a}, \mathbf{b} \in \mathbf{V}.$

II. $\mathbf{a} + (\mathbf{b} + \mathbf{c}) = (\mathbf{a} + \mathbf{b}) + \mathbf{c}\ \forall\ \mathbf{a}, \mathbf{b}, \mathbf{c} \in \mathbf{V}.$

III. $\exists$ **a** vector, *viz.*, **0** such that

$$\mathbf{a} + \mathbf{0} = \mathbf{a} = \mathbf{0} + \mathbf{a}\ \forall\ \mathbf{a} \in \mathbf{V}.$$

IV. To each $\mathbf{a} \in \mathbf{V}$ there corresponds $\mathbf{b} \in \mathbf{V}$ such that

$$\mathbf{a} + \mathbf{b} = \mathbf{0} = \mathbf{b} + \mathbf{a}.$$

The vector **b** is denoted by – **a.**

V. $(mn)\,\mathbf{a} = m\,(n\mathbf{a})\ \forall\ \mathbf{a} \in \mathbf{V}$ and $\forall\ m, n \in \mathbf{R}$.

VI. $1\,(\mathbf{a}) = \mathbf{a}\ \forall\ \mathbf{a} \in \mathbf{V}$;

VII. $m\,(\mathbf{a} + \mathbf{b}) = m\mathbf{a} + m\mathbf{b}\ \forall\ m \in \mathbf{R}$ and $\forall\ \mathbf{a}, \mathbf{b} \in \mathbf{V}$.

VIII. $(m + n)\,\mathbf{a} = m\mathbf{a} + n\mathbf{a}\ \forall\ m, n\ \mathbf{a} \in \mathbf{R}$ and $\forall\ \mathbf{a} \in \mathbf{V}$.

2. A vector **r** is a linear combination of

(*i*) two vectors **a, b** if ∃ scalars *x, y* such that

$$\mathbf{r} = x\mathbf{a} + y\mathbf{b}.$$

(*ii*) three vectors **a, b, c** if ∃ scalars *x, y, z* such that

$$\mathbf{r} = x\mathbf{a} + y\mathbf{b} + z\mathbf{c}.$$

3. (*i*) If two vectors **a, b** are linearly independent, then

$$x\mathbf{a} + y\mathbf{b} = 0, \quad \Rightarrow \quad x = 0,\ y = 0.$$

Two collinear vectors are linearly dependent and two non-collinear vectors are linearly independent.

(*ii*) If three vectors **a, b, c** are linearly independent, then

$$x\mathbf{a} + y\mathbf{b} + z\mathbf{c} = 0, \quad \Rightarrow \quad x = 0,\ y = 0,\ z = 0.$$

Three coplanar vectors are linearly dependent and three non-coplanar vectors are linearly independent. Every system of four vectors is linearly dependent.

4. If a vector **r** is coplanar with two vectors **a, b,** ∃ scalars *x, y* such that

$$\mathbf{r} = x\mathbf{a} + y\mathbf{b}$$

If **a, b, c** be three given non-coplanar vectors and **r** is any vector, ∃ scalars *x, y, z* such that

$$\mathbf{r} = x\mathbf{a} + y\mathbf{b} + z\mathbf{c}.$$

5. The set of linearly independent vectors such that any given vector is a linear combination of the members of the set is called a Base.

OBJECTIVE QUESTIONS

For each of the following questions, four alternatives are given for the answer. Only one of them is correct. Choose the correct alternative.

1. If **a** and **b** represent the sides $\overrightarrow{AB}$ and $\overrightarrow{BC}$ of a regular hexagon *ABCDEF*, than $\overrightarrow{FA} =$

(*a*) $\mathbf{b} - \mathbf{a}$ (*b*) $\mathbf{a} - \mathbf{b}$

(*c*) $\mathbf{a} + \mathbf{b}$ (*d*) None of these

2. If *c* is the middle point of *AB* and *P* is any point outside *AB*; then

(*a*) $\overrightarrow{PA} + \overrightarrow{PB} = \overrightarrow{PC}$ (*b*) $\overrightarrow{PA} + \overrightarrow{PB} = 2\overrightarrow{PC}$

(*c*) $\overrightarrow{PA} + \overrightarrow{PB} + \overrightarrow{PC} = \mathbf{0}$ (*d*) $\overrightarrow{PA} + \overrightarrow{PB} + 2\overrightarrow{PC} = \mathbf{0}$

3. One of the following is not a vector :
(*a*) displacement (*b*) work
(*c*) centrifugal (*d*) gravitational field

4. Which one of the following is not a scalar :
(*a*) temperature (*b*) density
(*c*) mass (*d*) weight

5. Let $\alpha = (x + 4y)\,\mathbf{a} + (2x + y + 1)\,\mathbf{b}$ and

$$\beta = (y - 2x + 2)\,\mathbf{a} + (2x - 3y - 1)\,\mathbf{b},$$

where **a** and **b** are non-zero, non-collinear. If $3\alpha + 2\beta$, then
(*a*) $x = 1$, $y = 2$ (*b*) $x = 2$, $y = 1$
(*c*) $x = -1$, $y = 2$ (*d*) $x = 2$, $y = -1$

6. Direction of zero vector
(*a*) does not exist (*b*) towards origin
(*c*) indeterminate (*d*) None of these

7. If **a** be an unit vector, then
(*a*) direction of **a** is constant
(*b*) magnitude of **a** is constant
(*c*) direction and magnitude of **a** is constant
(*d*) any one of direction or magnitude is constant.

8. If *a, b, c* are three non-coplanar, non-zero vectors and *r* is any vector, then *r* can be represented as $x\mathbf{a} + y\mathbf{b} + z\mathbf{c}$
(*a*) always (*b*) never
(*c*) is one and only one way (*d*) None of these

9. If vectors $(x - 2)\,\mathbf{a} + \mathbf{b}$ and $(2x + 1)\,\mathbf{a} - \mathbf{b}$ are parallel, then $x =$
(*a*) 1/3 (*b*) 3
(*c*) – 3 (*d*) – 1/3

10. If **a** is a vector and *x* is a non-zero scalar, then
(*a*) $x\mathbf{a}$ is a vector in the direction of **a**
(*b*) $x\mathbf{a}$ is a vector collinear to **a**
(*c*) $x\mathbf{a}$ and **a** have independent directions
(*d*) None of these.

ANSWERS

1. (*b*) **2.** (*b*) **3.** (*b*) **4.** (*d*) **5.** (*d*)
6. (*c*) **7.** (*c*) **8.** (*c*) **9.** (*a*) **10.** (*b*)

2

Geometry with Vectors. Affine Geometry

Introduction. The application to Geometry of the Vector Algebra so far developed will be considered in this chapter. It should be of some importance to notice that geometrical problems of *Affine nature* alone can be treated with the help of the compositions of the addition of vectors and the multiplication of vectors by scalars. These two compositions are inadequate for an unrestricted treatment of Metric relations involving comparisons of lengths, angles, areas and volumes and the compositions of scalars and vector products are needed for the purpose which will be considered in the following chapters.

It is not proposed to consider the detailed meanings and significance of the terms, **Affine** and **Metric** here. It may only suffice to say that the relations which involve comparisons of distances lying along the *same* or *parallel* lines only or deal with the comparison of directions only so far as parallelism is concerned are called *Affine.*

Thus, the scope of Affine geometry in so far as it deals with lengths and angles is very limited.

Some well-known theorems of plane Geometry such as Pappu's theorem, Desargue's theorem on triangles etc., will also be proved with the help of vector methods and included in on appendix to the chapter.

2.1. ORIGIN OF REFERENCE, POSITION VECTOR

The application of vector methods of Geometry depends upon the concept of *Position Vector or* the *Co-ordinate Vector* of a point which will now be introduced.

2.1.1. We take any arbitrary point, O to be called the **Origin of reference.**

The position vector of any point P, with respect to the origin O is the vector $\overrightarrow{OP}$.

Thus, with the choice of any point, O as the origin of reference, we can associate a vector to every point P and, conversely, to every given vector, **r**, there corresponds a point, *viz.*, the point P such that

$$\vec{OP} = \mathbf{r}.$$

In fact P is the terminal point of the vector **r**, whose initial point is the origin O.

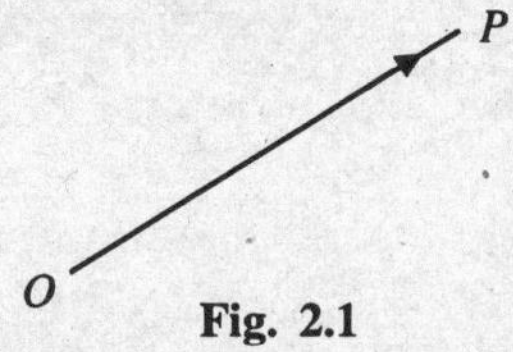

Fig. 2.1

2.1.2. Choice of the Origin of Reference

Origin of reference may be chosen arbitrarily but a suitable choice of the same often facilitates the solution of geometrical problems a good deal.

The origin of reference having been once chosen, a point whose position vector is **r** will usually be referred to as the point **r**.

2.1.3. *A vector expressed in terms of the position vectors of its end points.* The equality

$$\vec{AB} = \vec{OB} - \vec{OA}$$

which expresses any vector

$$\vec{AB}$$

in terms of the position vectors

$$\vec{OA} \text{ and } \vec{OB}$$

of its end points will prove very useful in what follows.

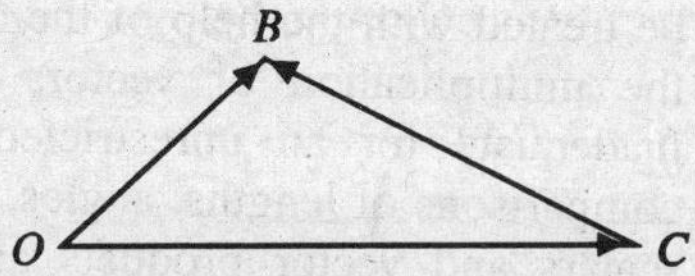

Fig. 2.2

2.2. SECTION FORMULA

To find the position vector of the point which divides the line joining two given points in a given ratio.

Let O, be the origin of reference and let **a, b** be the position vectors of the given points, A, B so that we have

$$\vec{OA} = \mathbf{a}, \ \vec{OB} = \mathbf{b}.$$

Let P divide AB so that

$$\frac{AP}{PB} = \frac{m}{n}. \quad ...(i)$$

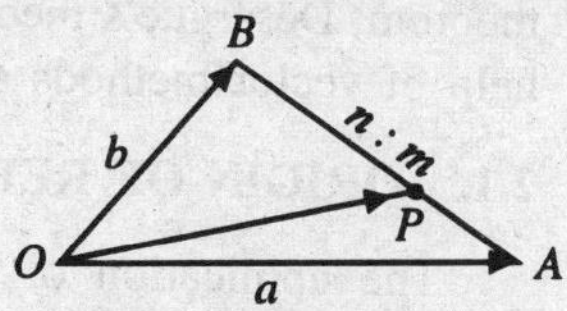

Fig. 2.3

Here m/n is positive or negative according as, P, divides AB internally or externally. We have to express the position vector $\vec{OP}$ of the point P in terms of the position vectors $\vec{OA}$ and $\vec{OB}$ of the points A and B.

We re-write (i) as

$$nAP = mPB$$

and obtain the vector equality

$$n\vec{AP} = m\vec{PB}.$$

Expressing the vector $\vec{AP}$ and $\vec{PB}$ in terms of the position vectors of the end points, we obtain

$$n(\vec{OP} - \vec{OA}) = m(\vec{OB} - \vec{OP})$$

$$\Rightarrow \quad (n+m)\vec{OP} = m\vec{OB} + n\vec{OA}$$

$$\Rightarrow \quad \vec{OP} = \frac{m\vec{OB} + n\vec{OA}}{m+n} = \frac{m\mathbf{b} + n\mathbf{a}}{m+n},$$

This result is analogous to the corresponding one in Cartesian Geometry.

Cor. Middle point. The position vector of the mid-point of the join of two points with position vectors, **a**, and **b**, is

$$\frac{1}{2}(\mathbf{a}+\mathbf{b})$$

obtained on taking $m = n$.

2.3. APPLICATIONS TO GEOMETRY

We shall now consider the applications to geometry of the subject so far developed.

EXAMPLES

Example 1. *Show that the medians of a triangle are concurrent.*

Solution. Let the position vectors of the vertices A, B, C of a triangle ABC with respect to any origin be **a, b, c.**

The position vectors of the mid-points D, E, F of the sides are respectively,

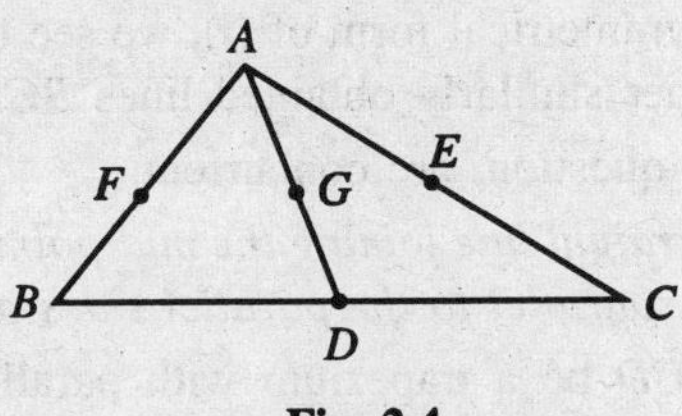

Fig. 2.4

$$\frac{1}{2}(\mathbf{b}+\mathbf{c}),\ \frac{1}{2}(\mathbf{c}+\mathbf{a}),\ \frac{1}{2}(\mathbf{a}+\mathbf{b})$$

Position vector of the point G dividing AD in the ratio 2 : 1 is

$$\frac{2\cdot\frac{1}{2}(\mathbf{b}+\mathbf{c})+1.\mathbf{a}}{2+1}=\frac{1}{3}(\mathbf{a}+\mathbf{b}+\mathbf{c}) \qquad ...(i)$$

By symmetry, we see that this point also lies on the other two medians.

Thus, the medians of a triangle are concurrent. Also the position vector of the point of concurrence, is $\frac{1}{3}(\mathbf{a}+\mathbf{b}+\mathbf{c})$; **a, b, c;** being the position vectors of the vertices of the triangle.

The point of concurrence of the medians of a triangle is called its *Centroid.*

Example 2. *Show that the lines joining the vertices of a tetrahedron to the centroids of the opposite faces meet in a point.*

Solution. Let the position vectors of the vertices of a tetrahedron *ABCD*, with respect to any origin of reference *O* be

a, b, c, d.

The position vectors of the centroids

G_1, G_2, G_3, G_4

of the triangular faces

BCD, CDA, DAB, ABC are

$$\frac{1}{3}(\mathbf{b}+\mathbf{c}+\mathbf{d}),\quad \frac{1}{3}(\mathbf{c}+\mathbf{d}+\mathbf{a}),$$

$$\frac{1}{3}(\mathbf{d}+\mathbf{a}+\mathbf{b}),\quad \frac{1}{3}(\mathbf{a}+\mathbf{b}+\mathbf{c}),$$

respectively.

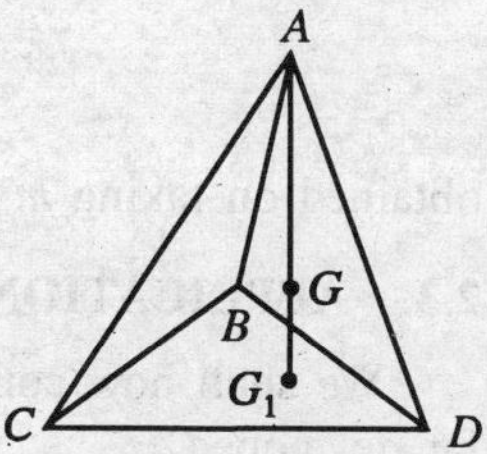

Fig. 2.5

The position vector of the point dividing AG_1 in the ratio 3 : 1 is

$$\frac{3\cdot\frac{1}{3}(\mathbf{b}+\mathbf{c}+\mathbf{d})+1\cdot\mathbf{a}}{3+1}=\frac{\mathbf{a}+\mathbf{b}+\mathbf{c}+\mathbf{d}}{4}. \qquad ...(i)$$

By virtue of the symmetrical form of (*i*), we see that this point as well lies on each of the other similarly obtained lines BG_2, CG_3, DG_4.

Thus the lines, in question, are concurrent.

Example 3. *The straight line joining the mid-points of two non-parallel sides of a trapezium is parallel to the parallel sides and half of their sum.*

Solution. Let *ABCD* be a trapezium with parallel sides *AB, CD.*

A as the origin of reference. Let

$$\vec{AB}=\mathbf{b},\quad \vec{AD}=\mathbf{d},$$

so that **b, d** are the position vectors of the points *B* and *D* respectively with the point *A* as the origin of reference.

As DC is parallel to AB, the vector $\overrightarrow{DC}$ must be a product of the vector $\overrightarrow{AB}$ by some scalar t. Let

$$\overrightarrow{DC} = t\,\overrightarrow{AB} = t\mathbf{b} \qquad ...(i)$$

Now the position vector of C is

$$\overrightarrow{AC} = \overrightarrow{AD} + \overrightarrow{DC} = \mathbf{d} + t\mathbf{b}.$$

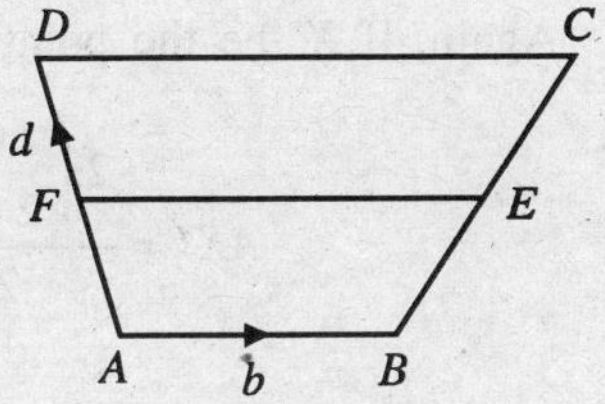

Fig. 2.6

The position vectors of the mid-points E and F of BC and AD are $\frac{1}{2}(\mathbf{b}+\mathbf{d}+t\mathbf{b})$ and $\frac{1}{2}\mathbf{d}$ respectively.

We have

$$\overrightarrow{FE} = \overrightarrow{AE} + \overrightarrow{AF}$$

$$= \frac{1}{2}(\mathbf{b}+\mathbf{d}+t\mathbf{b}) - \frac{1}{2}\mathbf{d}$$

$$= \frac{1}{2}(1+t)\,\mathbf{b} = \frac{1}{2}(1+t)\,\overrightarrow{AB}$$

$\Rightarrow$ the vector $\overrightarrow{FE}$ is the product of the vector $\overrightarrow{AB}$ by a scalar $\frac{1}{2}(1+t)$

$$\Rightarrow \qquad FE \parallel AB \text{ and } FE = \frac{1}{2}(1+t)\,AB.$$

Also from (*i*), we have

$$DC = t\,AB.$$

It follows that

$$AB + DC = (1 + t)\,AB = 2\,EF.$$

Example 4. *If M, N are the mid-points of the sides AB, CD of a parallelogram ABCD, prove that DM and BN cut the diagonal AC at its points of trisection which are also the points of trisection of DM and BN respectively.*

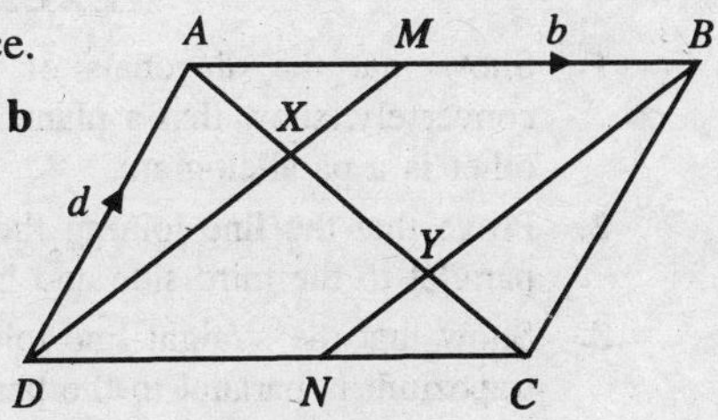

Fig. 2.7

Solution. A as origin of reference. Take position vectors of B and D as $\mathbf{b}$ and $\mathbf{d}$ respectively.

$ABCD$ is a parallelogram.

$$\Rightarrow \quad \overrightarrow{BC} = \overrightarrow{AD} = \mathbf{d}.$$

$$\vec{AC} = \vec{AB} + \vec{BC} = \mathbf{b} + \mathbf{d}$$

If X is a point which trisects AC, then

$$\vec{AX} = \frac{1}{3}(\mathbf{b} + \mathbf{d}) \quad ...(i)$$

Again, if X' be the point which divides DM in the ratio 2 : 1, then

$$\vec{AX'} = \frac{2.\left(\frac{1}{2}\mathbf{b}\right) + 1.\mathbf{d}}{2+1} = \frac{1}{3}(\mathbf{b} + \mathbf{d})$$

$$= \vec{AX}$$

Implying that X and X' must coincide. Thus, DM cuts the diagonal AC at its point of trisection X which is also the point of trisection of DM.

Similarly other result can be established.

Example 5. *The middle points of the adjacent sides of any quadrilateral are joined; prove that the figure so formed is a parallelogram.*

Solution. $ABCD$ is a quadrilateral and E, F, G, H be the mid-points of AB, BC, CD and DA respectively. Let A be the origin of reference and $\mathbf{b}$, $\mathbf{c}$, $\mathbf{d}$ be the position vectors of B, C and D.

Position vectors of E, F, G, H are

$\frac{1}{2}\mathbf{b}$, $\frac{1}{2}(\mathbf{b} + \mathbf{c})$, $\frac{1}{2}(\mathbf{c} + \mathbf{d})$ and $\frac{1}{2}\mathbf{d}$

respectively.

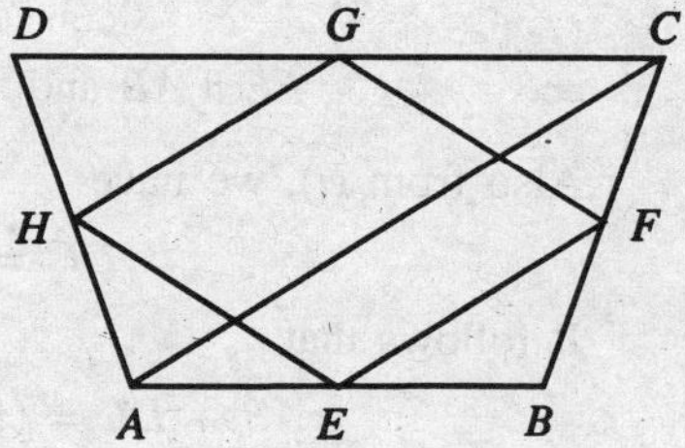

Fig. 2.8

$$\therefore \quad \vec{EF} = \frac{1}{2}(\mathbf{b} + \mathbf{c}) - \frac{1}{2}\mathbf{b} = \frac{1}{2}\mathbf{c}$$

and $$\vec{HG} = \frac{1}{2}(\mathbf{c} + \mathbf{d}) - \frac{1}{2}\mathbf{d} = \frac{1}{2}\mathbf{c}$$

$\Rightarrow$ EF is equal and parallel to HG. Hence, $EFGH$ is a parallelogram.

EXERCISES

1. Show that the diagonals of a parallelogram bisect each other. Also conversely, show that a plane quadrilateral whose diagonals bisect each other is a parallelogram.
2. Prove that the line joining the mid-points of two sides of a triangle is parallel to the third side and half of it.
3. Show that the straight line joining the mid-points of the diagonals of a trapezium is parallel to the parallel sides and is half of their difference.

4. Prove that the straight lines joining the mid-points of pairs of opposite edges of a tetrahedron are concurrent.

5. G is the centroid of a tetrahedron $ABCD$; $A'B'C'D'$ is another tetrahedron such that AA', BB', CC', and DD', are all bisected at G; show that G is also the centroid of the tetrahedron $A'B'C'D'$.

6. OP is a diagonal of the parallelopiped with coterminous edges OL, OM, ON, and OQ, OR, OS are the diagonals of the parallelopiped constructed with

$$OM,\ ON,\ OP;\ ON,\ OL,\ OP;\ OL,\ OM,\ OP$$

as coterminous edges. Finally OM is a diagonal of the parallelopiped with OQ, OR, OS as coterminous edges. Show that OM lies along OL and is five times OL.

7. Show that the four diagonals of any parallelopiped are concurrent and are bisected at the point of concurrence.

8. The diagonals of the three faces of the parallelopiped drawn from the same vertex are prolonged half their lengths; show that the three points thus obtained are coplanar with the opposite vertex.

2.4. PARAMETRIC VECTORIAL EQUATIONS

It is possible to express the position vectors of points on given lines and planes in terms of some fixed vectors and variable scalars, called *parameters*, such that

(*i*) for arbitrary values of the parameters, the resulting position vectors represent points on the locus in question, and

(*ii*) conversely, the position vector of each point on the locus arises for some suitable values of the parameters.

Parametric Vectorial Equation of a Line. To find the parametric vectorial equation of a line which passes through a given point and is parallel to a given line.

Take any point, O, as the origin of reference.

Let, **a**, be the position vector of the given point, A, and let **b**, be a vector parallel to the given line.

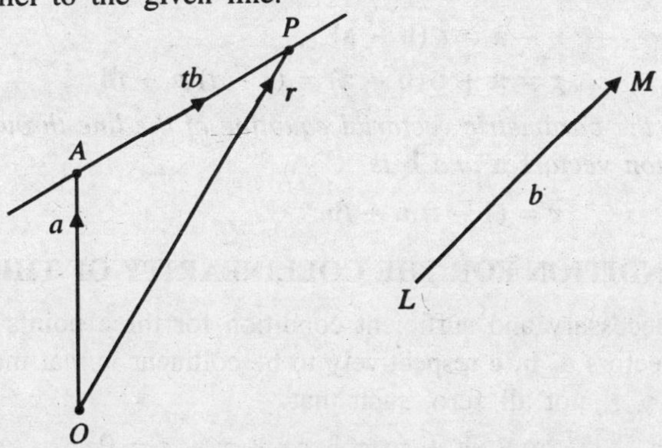

Fig. 2.9

Let, **r**, be the position vector of any point P, on the given line.

The vector $\overrightarrow{AP} = \mathbf{r} - \mathbf{a}$ and **b** being collinear, we have

$$\mathbf{r} - \mathbf{a} = t\mathbf{b} \quad \Rightarrow \quad \mathbf{r} = \mathbf{a} + t\mathbf{b}. \qquad \ldots(i)$$

Each point, P, of the line arises for some value of the scalar t. Also, conversely, for each value of the scalar, t,

$$\mathbf{a} + t\mathbf{b}$$

is the positive vector of a point of the line.

Hence, (*i*) is the required parametric vectorial equation of the given line so that the *parametric vectorial equation of the line through a point with a position vector* **a** *and is parallel to a vector*, **b**, *is*

$$\mathbf{r} = \mathbf{a} + t\mathbf{b},$$

where, t, is a scalar parameter. (Kanpur, 94)

Cor. Line through two given points. Let **a, b** be the position vectors of the two given points A, B with reference to any origin O and let **r** be the position vector of any point P on the line.

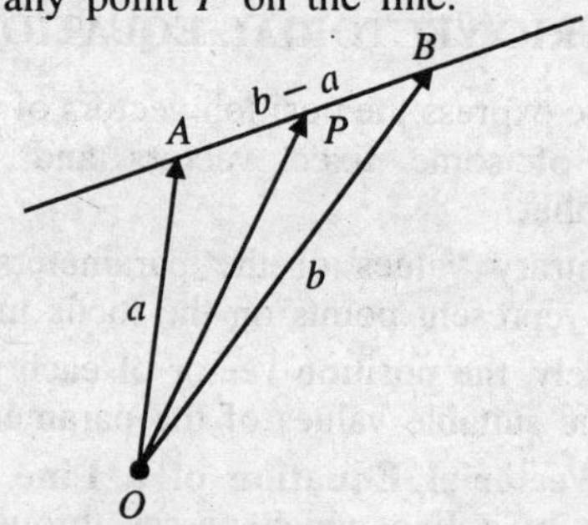

Fig. 2.10

Now the vectors $\overrightarrow{AP} = \mathbf{r} - \mathbf{a}$ and $\overrightarrow{AB} = \mathbf{b} - \mathbf{a}$ being collinear the required equation of the line AB is

$$\mathbf{r} - \mathbf{a} = t\,(\mathbf{b} - \mathbf{a})$$

$$\mathbf{r} = \mathbf{a} + t\,(\mathbf{b} - \mathbf{a}) = (1 - t)\,\mathbf{a} + t\mathbf{b}$$

Thus, *the parametric vectorial equation of the line through two points with position vectors* **a** *and* **b** *is*

$$\mathbf{r} = (1 - t)\,\mathbf{a} + t\mathbf{b}.$$

2.5. CONDITION FOR THE COLLINEARITY OF THREE POINTS

The necessary and sufficient condition for three points A, B, C with position vectors **a, b, c** respectively to be collinear is that there exist three scalars x, y, z, not all zero, such that

$$x\mathbf{a} + y\mathbf{b} + z\mathbf{c} = \mathbf{0}, \quad x + y + z = 0.$$

We have $\vec{AB} = \mathbf{b} - \mathbf{a}$, $\vec{AC} = \mathbf{c} - \mathbf{a}$.

The points A, B and C will be collinear if and only if the vectors $\mathbf{b} - \mathbf{a}$ and $\mathbf{c} - \mathbf{a}$ are collinear.

The condition is necessary. Let the points A, B, C be collinear so that the vectors $\vec{AB}$ and $\vec{AC}$ are collinear. Thus, there exists a scalar k such that

$$\mathbf{b} - \mathbf{a} = k(\mathbf{c} - \mathbf{a}) \Rightarrow (k - 1)\mathbf{a} + \mathbf{b} - k\mathbf{c} = \mathbf{0}.$$

Taking $x = k - 1$, $y = 1$, $z = -k$, we see that if the points A, B, C are collinear, then there exists scalars x, y, z not all zero such that

$$x\mathbf{a} + y\mathbf{b} + z\mathbf{c} = \mathbf{0} \quad \text{and} \quad x + y + z = 0.$$

The condition is sufficient. Let there exist three scalars x, y, z are not zero such that

$$x\mathbf{a} + y\mathbf{b} + z\mathbf{c} = \mathbf{0},\ x + y + z = 0.$$

From these we obtain

$$-(y + z)\mathbf{a} + y\mathbf{b} + z\mathbf{c} = \mathbf{0}$$

$$\Rightarrow y(\mathbf{b} - \mathbf{a}) + z(\mathbf{c} - \mathbf{a}) = \mathbf{0}$$

$$\Rightarrow \mathbf{c} - \mathbf{a} = -yz^{-1}(\mathbf{b} - \mathbf{a}) \text{ if } z \neq 0.$$

Thus, the vectors $\mathbf{c} - \mathbf{a}$ and $\mathbf{b} - \mathbf{a}$ are collinear and as such the points A, B, C are collinear.

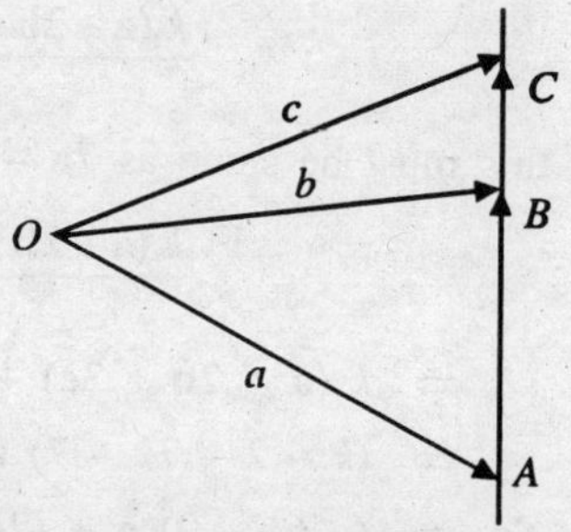

Fig. 2.11

Note. Rewriting the relations $x\mathbf{a} + y\mathbf{b} + z\mathbf{c} = \mathbf{0}$, $x + y + z = 0$ in the form $\mathbf{c} = \dfrac{x\mathbf{a} + y\mathbf{b}}{x + y}$, we may see that the point C divides AB in the ratio $y : x$.

EXAMPLES

Example 1. *Show that the three points A, B, C with position vectors*

$$-2\mathbf{a} + 3\mathbf{b} + 5\mathbf{c},\ \mathbf{a} + 2\mathbf{b} + 3\mathbf{c},\ 7\mathbf{a} - \mathbf{c}$$

are collinear. Find the ratio in which C divides AB. Obtain also the equation of the line.

Solution. Let O be the origin of reference.

We have

$$\vec{AB} = \vec{OB} - \vec{OA} = (\mathbf{a} + 2\mathbf{b} + 3\mathbf{c}) - (-2\mathbf{a} + 3\mathbf{b} + 5\mathbf{c}) = 3\mathbf{a} - \mathbf{b} - 2\mathbf{c}.$$

$$\vec{AC} = \vec{OC} - \vec{OA},$$

$$= (7\mathbf{a} - \mathbf{c}) - (-2\mathbf{a} + 3\mathbf{b} + 5\mathbf{c})$$

$$= (9\mathbf{a} - 3\mathbf{b} - 6\mathbf{c}) = 3\vec{AB}$$

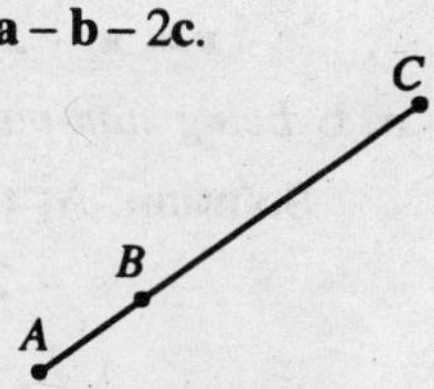

Fig. 2.12

Thus, the vectors $\vec{AB}$, $\vec{AC}$ are collinear. These vectors being also coterminous, the points *A*, *B*, *C*, are collinear.

Also, we have

$$2(-2\mathbf{a}+3\mathbf{b}+5\mathbf{c})-3(\mathbf{a}+2\mathbf{b}+3\mathbf{c})+(7\mathbf{a}-\mathbf{c})=0$$

where $$2-3+1=0.$$

so that the truth of the result is also seen by following the result of § 2.5.

By § 2.5 note, we may see that *C* divides *AB* in the ratio 3 : 2.

Another Method. Let the point *C* divide *AB* in the ratio *k* : 1.

The point dividing *AB* in the ratio *k* : 1 being

$$\frac{k(\mathbf{a}+2\mathbf{b}+3\mathbf{c})+1(-2\mathbf{a}+3\mathbf{b}+5\mathbf{c})}{k+1}$$

this must be same as $7\mathbf{a}-\mathbf{c}$ so that

$$\frac{k(\mathbf{a}+2\mathbf{b}+3\mathbf{c})+1(-2\mathbf{a}+3\mathbf{b}+5\mathbf{c})}{k+1}=7\mathbf{a}-\mathbf{c}$$

$$\Rightarrow\ k(\mathbf{a}+2\mathbf{b}+3\mathbf{c})+1.(-2\mathbf{a}+3\mathbf{b}+5\mathbf{c})=(k+1)(7\mathbf{a}-\mathbf{c})$$

$$\Rightarrow\ (k-2-7k-7)\,\mathbf{a}+(2k+3)\,\mathbf{b}+(3k+5+k+1)\,\mathbf{c}=\mathbf{0}$$

$$\Rightarrow\ (-6k-9)\,\mathbf{a}+(2k+3)\,\mathbf{b}+(4k+6)\,\mathbf{c}=\mathbf{0}$$

$$\Rightarrow\ -6k-9=0,\quad 2k+3=0,\quad 4k+6=\mathbf{0}.$$

These equations are constant in *k* and

$$k=-\frac{3}{2}.$$

Thus, the points *A*, *B*, *C* are collinear and *C* divides *AB* in the ratio

$$-3:2.$$

The equation of the line through the two points *A*, *B* is

$$\mathbf{r}=-2\mathbf{a}+3\mathbf{b}+5\mathbf{c}+t\,[(\mathbf{a}+2\mathbf{b}+3\mathbf{c})-(-2\mathbf{a}+3\mathbf{b}+5\mathbf{c})]$$
$$=-2\mathbf{a}+3\mathbf{b}+5\mathbf{c}+t\,(3\mathbf{a}-\mathbf{b}-2\mathbf{c}).$$

It may be seen that the point *C* arises for $t=3$.

Example 2. *Find the point of intersection of the lines*

$$\mathbf{r}=\mathbf{a}-2\mathbf{b}+\lambda(\mathbf{b}+2\mathbf{a}),\ \mathbf{r}=2\mathbf{a}-\mathbf{b}+\mu(\mathbf{a}+2\mathbf{b}),$$

a, b *being non-parallel vectors.*

Solution. At the point of intersection of the lines, we have

$$\mathbf{a}-2\mathbf{b}+\lambda(\mathbf{b}+2\mathbf{a})=2\mathbf{a}-\mathbf{b}+\mu(\mathbf{a}+2\mathbf{b})$$

$$\Rightarrow\quad (2\lambda-\mu-1)\,\mathbf{a}+(\lambda-2\mu-1)\,\mathbf{b}=\mathbf{0}.$$

As, **a**, **b** are non-parallel vectors, we have

$$2\lambda - \mu - 1 = 0, \quad \lambda - 2\mu - 1 = 0$$

giving $\lambda = \frac{1}{3}$ and $\mu = -\frac{1}{3}$.

Putting $\lambda = \frac{1}{3}$, in the equation of the line with λ parameter, we see that the position vector of the point of intersection, is $\frac{5}{2}(\mathbf{a} - \mathbf{b})$. It may be verified that for $\mu = -\frac{1}{3}$, we get the same point.

Example 3. *Through the middle point M of the side AD of a parallelogram ABCD, the straight line BM is drawn cutting AC at R and CD produced at Q; prove that*

$$QR = 2RB.$$

Solution. Take A as the origin of reference.

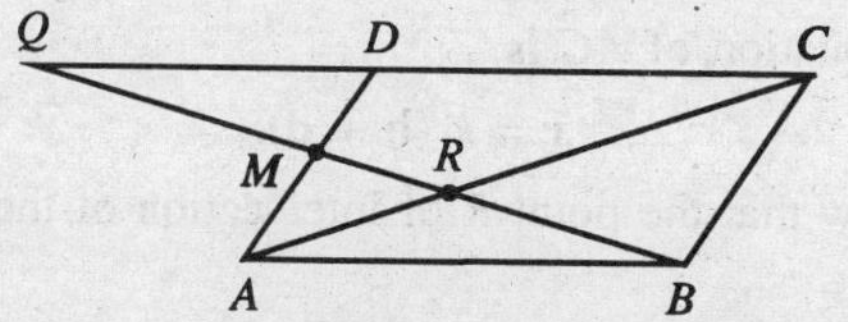

Fig. 2.13

Let **b**, **d** be the position vectors of the point *B* and *D* so that

$$\vec{AB} = \mathbf{b}, \ \vec{AD} = \mathbf{d}.$$

The position vector of *M* is $\frac{1}{2}\mathbf{d}$ and that of *C* is

$$= \vec{AB} + \vec{BC} = \vec{AB} + \vec{AD} = \mathbf{b} + \mathbf{d}.$$

The equation of the line *BM* joining the points, B, M with position vectors **b**, $\frac{1}{2}\mathbf{d}$ is

$$\mathbf{r} = \mathbf{b} + t\left(\frac{1}{2}\mathbf{d} - \mathbf{b}\right) \qquad \text{...}(i)$$

and the equation of the line *DC* through the point *D* with position vector **d** and parallel to $\vec{AB} = \mathbf{b}$ is

$$\mathbf{r} = \mathbf{d} + p\mathbf{b}. \qquad \text{...}(ii)$$

At the point of intersection of the lines (*i*) and (*ii*), we have

$$\mathbf{b}+t\left(\frac{1}{2}\mathbf{d}-\mathbf{b}\right)=\mathbf{d}+p\mathbf{b}$$

$$\Rightarrow \quad (1-t-p)\,\mathbf{b}+\left(\frac{1}{2}t-1\right)\mathbf{d}=0$$

$$\Rightarrow \quad 1-t-p=0,\ \frac{1}{2}t-1=0$$

$$\Rightarrow \quad t=2,\ \ p=-1;$$

the vectors **b, d** being non-collinear.

Substituting the value of t in (*i*) or of p in (*ii*), we see that the position vector of the point Q of intersection of the lines BM and DC is

$$\mathbf{d}-\mathbf{b}=\overrightarrow{AQ}.$$

We shall now find the position vector of the point R of intersection of the lines AC and BM.

Again the equation of AC is

$$\mathbf{r}=k\,(\mathbf{b}+\mathbf{d}).$$

We may show that the point R of intersection of the lines AC and BM is

$$\frac{1}{3}(\mathbf{b}+\mathbf{d})=\overrightarrow{AR}$$

It follows that

$$\begin{cases}\overrightarrow{QR}=\overrightarrow{AR}-\overrightarrow{AQ}=\frac{4}{3}\mathbf{b}-\frac{2}{3}\mathbf{d}=2\left(\frac{2}{3}\mathbf{b}-\frac{1}{3}\mathbf{d}\right)\\ \overrightarrow{RB}=\overrightarrow{AB}-\overrightarrow{AR}=\mathbf{b}-\frac{1}{3}(\mathbf{b}+\mathbf{d})=\frac{2}{3}\mathbf{b}-\frac{1}{3}\mathbf{d},\end{cases}$$

Thus $$\overrightarrow{QR}=2\overrightarrow{RB}\Rightarrow QR=2RB.$$

Example 4. *In a triangle ABC, D divides BC in the ratio 3 : 2 and E divides CA in the ratio 1 : 3. The lines AD and BE meet at H and CH meets AB in F. Find the ratio in which F divides AB.*

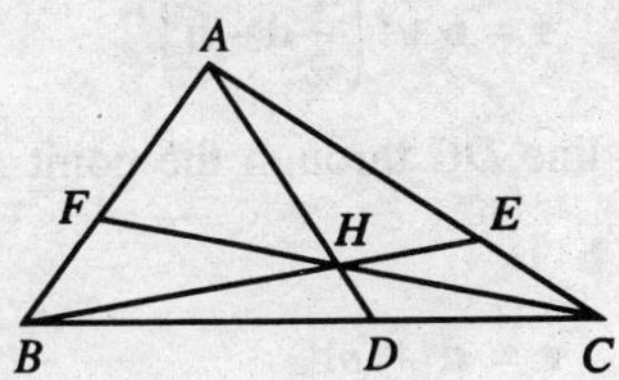

Fig. 2.14

Solution. Take C as the origin of reference. Let **a, b** be the position vectors of the points A and B respectively. Now the point D is $\frac{2}{5}\mathbf{b}$ and the point E is $\frac{1}{4}\mathbf{a}$.

The equation of AD and BE are

$$\mathbf{r} = \mathbf{a} + k\left(\frac{2}{5}\mathbf{b} - \mathbf{a}\right), \text{ and } \mathbf{r} = \mathbf{b} + t\left(\frac{1}{4}\mathbf{a} - \mathbf{b}\right) \text{ respectively.}$$

For the point of intersection H, we have

$$\mathbf{a} + k\left(\frac{2}{5}\mathbf{b} - \mathbf{a}\right) = \mathbf{b} + t\left(\frac{1}{4}\mathbf{a} - \mathbf{b}\right) \qquad ...(i)$$

$$\Rightarrow \quad 1 - k = \frac{t}{4},\ \frac{2}{5}k = 1 - t \Rightarrow k = \frac{5}{6};\ t = \frac{2}{3}.$$

Putting these values of k and t in the equations of the lines AD and BE, we see that the point H is

$$\frac{1}{6}\mathbf{a} + \frac{1}{3}\mathbf{b}.$$

The equations of CH and AB being

$$\mathbf{r} = t\left(\frac{\mathbf{a}}{6} + \frac{\mathbf{b}}{3}\right), \text{ and } \mathbf{r} = \mathbf{a} + k\,(\mathbf{b} - \mathbf{a}),$$

we have for the point of intersection F,

$$t\left(\frac{\mathbf{a}}{6} + \frac{\mathbf{b}}{3}\right) = \mathbf{a} + k\,(\mathbf{b} - \mathbf{a})$$

$$\Rightarrow \quad \left(\frac{t}{6} - 1 + k\right) = 0,\ \left(\frac{t}{3} - k\right) = 0.$$

These give $t = 2;\ k = \frac{2}{3}$.

Thus, the point F is

$$\frac{\mathbf{a} + 2\mathbf{b}}{3},$$

so that F divides AB in the ratio 2 : 1.

Example 5. *Find the equations of the bisectors of the angles between the lines*

$$\mathbf{r} = \mathbf{a} + t\mathbf{b}, \quad \mathbf{r} = \mathbf{a} + p\mathbf{c}.$$

Solution. Let AL, AM be the given lines so that $\mathbf{a}$ is the position vector of the point A. Let P, Q, R be three points at unit distances from A as in the figure. Let E, F be the mid-points of PQ, QR.

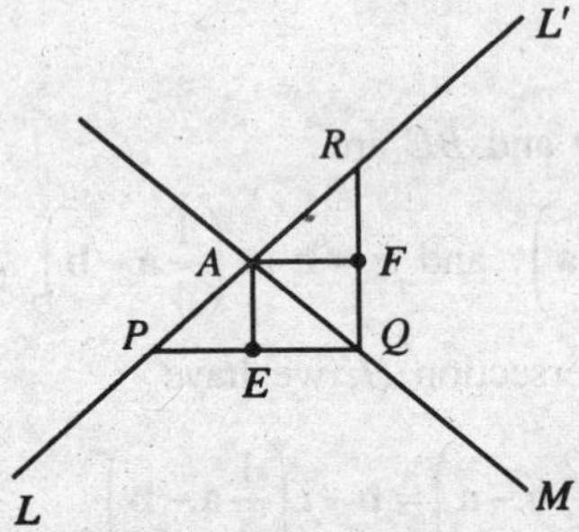

Fig. 2.15

Then AE, AF are the bisectors of the angles between the given lines.

The position vectors of P, Q, R are

$$\mathbf{a}+\frac{\mathbf{b}}{|\mathbf{b}|},\ \mathbf{a}+\frac{\mathbf{c}}{|\mathbf{c}|},\ \mathbf{a}-\frac{\mathbf{b}}{|\mathbf{b}|}$$

respectively, implying that the position vectors of E, F are

$$\mathbf{a}+\frac{1}{2}\left(\frac{\mathbf{b}}{|\mathbf{b}|}+\frac{\mathbf{c}}{|\mathbf{c}|}\right),\ \mathbf{a}+\frac{1}{2}\left(\frac{\mathbf{c}}{|\mathbf{c}|}-\frac{\mathbf{b}}{|\mathbf{b}|}\right).$$

Thus, the required equations of the bisectors AE, AF are

$$\mathbf{r}=\mathbf{a}+t\left(\frac{\mathbf{b}}{|\mathbf{b}|}+\frac{\mathbf{c}}{|\mathbf{c}|}\right),\ \mathbf{r}=\mathbf{a}+p\left(\frac{\mathbf{c}}{|\mathbf{c}|}-\frac{\mathbf{b}}{|\mathbf{b}|}\right).$$

respectively, where t and p are scalar parameters.

Example 6. *Show that the internal (external) bisector of any angle of a triangle divides the base internally (externally) in the ratio of the sides containing the angle.*

Solution. Let the lengths of the sides BC, CA, AB of a triangle ABC be denoted by α, β, γ respectively.

Take A as the origin of reference. Let $\mathbf{b}$, $\mathbf{c}$ be the position vectors of the points B and C so that

$$\vec{AB}=\mathbf{b},\ \vec{AC}=\mathbf{c}.$$

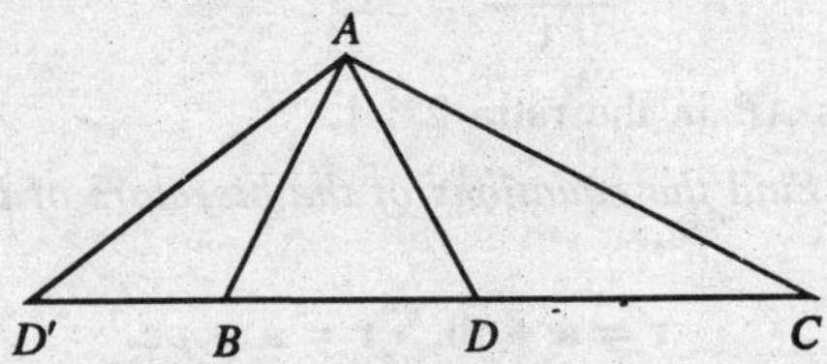

Fig. 2.16

We have

$$|\vec{AB}| = \gamma, \ |\vec{AC}| = \beta.$$

Now $\dfrac{\mathbf{b}}{\gamma}, \dfrac{\mathbf{c}}{\beta}$ are unit vectors along *AB* and *AC* so that the equation of the internal bisector *AD* is

$$\mathbf{r} = t\left(\frac{\mathbf{b}}{\gamma} + \frac{\mathbf{c}}{\beta}\right). \qquad ...(i)$$

Also the equation of the line BC joining the points B and C with position vectors **b** and **c** is

$$\mathbf{r} = \mathbf{b} + p\,(\mathbf{c} - \mathbf{b}). \qquad ...(ii)$$

At the point of intersection *D* of the lines *AD* and *BC*, we have

$$t\left(\frac{\mathbf{b}}{\gamma} + \frac{\mathbf{c}}{\beta}\right) = \mathbf{b} + p\,(\mathbf{c} - \mathbf{b})$$

$$\Rightarrow \qquad \left(\frac{t}{\gamma} - 1 + p\right) = \mathbf{b} + \left(\frac{t}{\beta} - p\right)\mathbf{c} = 0.$$

As the vector **b, c** are non-parallel, we have

$$\frac{t}{\gamma} - 1 + p = 0, \ \frac{t}{\beta} - p = 0$$

$$\Rightarrow \qquad t = \frac{\beta\gamma}{\beta + \gamma}, \ p = \frac{\gamma}{\beta + \gamma}.$$

Substituting the value of *t* in (*i*) or of *p* in (*ii*), we see that the position vector of the point of intersection *D* of *AD* and *BC* is

$$\frac{\beta\mathbf{b} + \gamma\mathbf{c}}{\beta + \gamma}$$

which divides *BC* in the ratio $\gamma : \beta$, *i.e., AB : AC.*

Considering the external bisector *AD′*

$$\mathbf{r} = t\left(\frac{\mathbf{c}}{\beta} - \frac{\mathbf{b}}{\gamma}\right),$$

we may similarly prove the part corresponding to the external bisector.

Example 7. *ABC is a triangle; AD, AD′ are the bisectors of the angle A meeting BC in D, D′ respectively; A′ is the mid-point of DD′; B′, C′, are the points on CA and AB similarly obtained. Show that the points A′, B′, C′, are collinear.*

Let **a, b, c** be the position vectors of the vertices *A, B, C* with respect to any origin of reference *O*. Let the lengths of the sides *BC, CA, AB* of the triangle be α, β, γ.

The position vectors of *D, D′* are

$$\frac{\beta\mathbf{b}+\gamma\mathbf{c}}{\beta+\gamma},\ \frac{\beta\mathbf{b}-\gamma\mathbf{c}}{\beta-\gamma}$$

respectively, so that the position vector of the mid point *A′* of *D D′* is

$$\frac{1}{2}\left(\frac{\beta\mathbf{b}+\gamma\mathbf{c}}{\beta-\gamma}+\frac{\beta\mathbf{b}-\gamma\mathbf{c}}{\beta-\gamma}\right)=\frac{\beta^2\mathbf{b}-\gamma^2\mathbf{c}}{\beta^2-\gamma^2}=a',\ \text{say.}$$

By symmetry, the position vectors of *B′, C′* are

$$\frac{\gamma^2\mathbf{c}-\alpha^2\mathbf{a}}{\gamma^2-\alpha^2}=\mathbf{b}',\ \text{say, and}\ \frac{\alpha^2\mathbf{a}-\beta^2\mathbf{b}}{\alpha^2-\beta^2}=\mathbf{c}',\ \text{say.}$$

We now have

$$(\beta^2-\gamma^2)\,\mathbf{a}'+(\gamma^2-\alpha^2)\,\mathbf{b}'+(\alpha^2+\beta^2)\,\mathbf{c}'=0$$

where

$$(\beta^2-\gamma^2)+(\gamma^2-\alpha^2)+(\alpha^2+\beta^2)=0$$

It follows that the points *A′, B′, C′*, with position vectors **a′**, **b′**, **c′** are collinear (§ 2.5 Page 32).

Example 8. *Show that the internal bisectors of the angles of a triangle are concurrent.*

Solution. Take any point *O* as the origin of reference. Let **a, b, c** be the position vectors of the vertices *A, B, C* and α, β, γ the lengths of the sides of the triangle. The position vector of the point *D* where the internal bisector of the angle *A* meets *BC* is

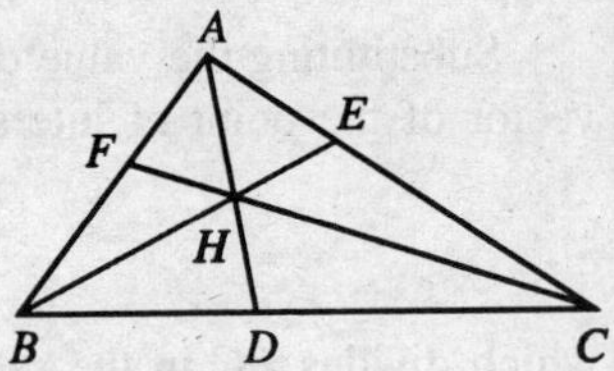

Fig. 2.17

$$\frac{\beta\mathbf{b}+\gamma\mathbf{c}}{\beta+\gamma}.$$

The position vector of the point dividing *AD* in the ratio (β + γ) / α is

$$\frac{\alpha\mathbf{a}+\beta\mathbf{b}+\gamma\mathbf{c}}{\alpha+\beta+\gamma}. \quad ...(i)$$

By symmetry, we see that this point lies on the other two bisectors also.

Thus, the bisectors are concurrent and the point of concurrence is given by (*i*).

EXERCISES

1. Find the points of intersection of the following pairs of lines, assuming that the vectors **a** and **b** are not parallel.

   (*i*) $\mathbf{r} = \gamma\,(\mathbf{b} - \mathbf{a}),\ \ \mathbf{r} = 2\mathbf{b} + \mu\mathbf{a}.$

   (*ii*) $\mathbf{r} = \gamma\,(\mathbf{b} + \mathbf{a}),\ \ \mathbf{r} = \mu\,(\mathbf{b} - \mathbf{a}).$

   (*iii*) $\mathbf{r} = (6\mathbf{b} - \mathbf{a}) + \gamma\,(2\mathbf{a} - \mathbf{b}),\ \ \mathbf{r} = (\mathbf{b} - \mathbf{a}) + \mu\,(\mathbf{b} + 3\mathbf{a})$

   (*iv*) $\mathbf{r} = 2\mathbf{a} + 2\mathbf{b} + \gamma\,(\mathbf{a} - \mathbf{b}),\ \ \mathbf{r} = 3\mathbf{a} + \mu\,(\mathbf{b} - \mathbf{a}).$

   (*v*) $\mathbf{r} = \mathbf{a} + \mu\mathbf{b},\ \ \mathbf{r} = \mathbf{b} + \gamma\mathbf{a}.$

2. Test for collinearity the sets of points with the following position vectors : (Kanpur 93)

   (*i*) $\mathbf{a} - 2\mathbf{b} + 3\mathbf{c}$, $\quad 2\mathbf{a} + 3\mathbf{b} - 4\mathbf{c}$, $\quad -7\mathbf{b} + 10\mathbf{c}$.

   (*ii*) $3\mathbf{a} - 4\mathbf{b} + 3\mathbf{c}$, $\quad -4\mathbf{a} + 5\mathbf{b} - 6\mathbf{c}$, $\quad 4\mathbf{a} - 7\mathbf{b} + 6\mathbf{c}$.

   (*iii*) $2\mathbf{a} + 5\mathbf{b} - 4\mathbf{c}$, $\quad \mathbf{a} + 4\mathbf{b} - 3\mathbf{c}$, $\quad 4\mathbf{a} + 7\mathbf{b} - 6\mathbf{c}$.

   (*iv*) $5\mathbf{a} + 4\mathbf{b} + 2\mathbf{c}$, $\quad 6\mathbf{a} + 2\mathbf{b} - \mathbf{c}$, $\quad 7\mathbf{a} + \mathbf{b} - \mathbf{c}$.

3. Show that the lines

$$\mathbf{r} = \mathbf{a} + t\,(\mathbf{b} + \mathbf{c}), \quad \mathbf{r} = \mathbf{b} + t\,(\mathbf{c} + \mathbf{a})$$

   intersect and find also their point of intersection.

4. In a $\Delta\,OAB$, E is the mid-point of AB and F is the point in OA such that $OF = 2\,FA$. Calculate the ratios $OC : CE$ and $BC : CF$ where C is the point of intersection of OE and BF.

5. Find the position vector of the point P of intersection of the lines l_1 and l_2 with vector equations

$$r = 2\mathbf{a} + \lambda\,(\mathbf{b} - 3\mathbf{a}) \ \text{ and } \ r = 3\,(\mathbf{a} - \mathbf{b}) - \mu\,(\mathbf{a} + \mathbf{b}),$$

   where **a, b** are non-parallel vectors. Q is the point in l_1 when $\lambda = 1$ and R is the point in l_2 when $\mu = 1$. If $PQSR$ is a parallelogram, give the vector equations of the lines QS and RS as also the position vector of the point S.

6. In a $\Delta\,OAB$, C is the mid-point of AB and D is the mid-point of OB. The line OC meets AD at G. Calculate the ratio in which G divides OC and AD. If E is the point in which OA meets BG, calculate the ratio in which G divides BE and show that E is the mid-point of OA.

7. The points O, A, B, X and Y are such that $\overrightarrow{OA} = \mathbf{a}$, $\overrightarrow{OB} = \mathbf{b}$, $\overrightarrow{OX} = 3\mathbf{a}$ and $\overrightarrow{OY} = 3\mathbf{b}$. Express $\overrightarrow{BX}$ and $\overrightarrow{AY}$ in terms of **a** and **b**.

   The point P divides AY in the ratio $1 : 3$. Express $\overrightarrow{BP}$ in terms of **a** and **b** and show that P lies on BX.

8. In a triangle ABC, D divides BC in the ratio $2 : 1$ and E divides CA in the ratio $3 : 1$. The line DE meets the line AB in F. Find the ratio in which the point F divides AB.

9. The median AD of a triangle ABC is bisected at E and BE is produced to meet the side AC in F; show that

$$AF = \frac{1}{3}AC \text{ and } EF = \frac{1}{4}BF.$$

10. $ABCD$ is a parallelogram and L, M are the mid-points of AB and CD respectively; show that DL and BM trisect AC and are trisected by AC.

11. A line EF drawn parallel to the base BC of a triangle ABC meets AB and AC in F and E respectively; BE and CF meet in L. Show that AL bisects BC.

12. The sides CA, AB of a triangle ABC are divided internally in the same ratio at E and F; show that EF divides BC externally in the square of this ratio.

13. Points Y and Z are taken on the sides CA, AB of a triangle ABC such that $CY = YA$ and $AZ = 2\,ZB$; BY and CZ meet at P, show that $CP = 3\,PZ$. Also find the ratio in which AP divides BC.

14. A, B, C and A', B', C' are two sets of points on two skew lines such that

$AB : BC = A'B' : B'C'$.

Show that the middle points of AA', BB', CC are collinear.

2.6. PARAMETRIC VECTORIAL EQUATION OF A PLANE

To find the parametric vectorial equation of the plane which passes through a given point and is parallel to two given lines.

Let AB, AC be two lines through the point A parallel to the vector $\mathbf{b}$ and $\mathbf{c}$.

Take any point O, as the origin of reference. Let, $\mathbf{a}$, be the position vector of the given point A. Let, $\mathbf{r}$, be the position vector of any point $\mathbf{P}$ on the given plane.

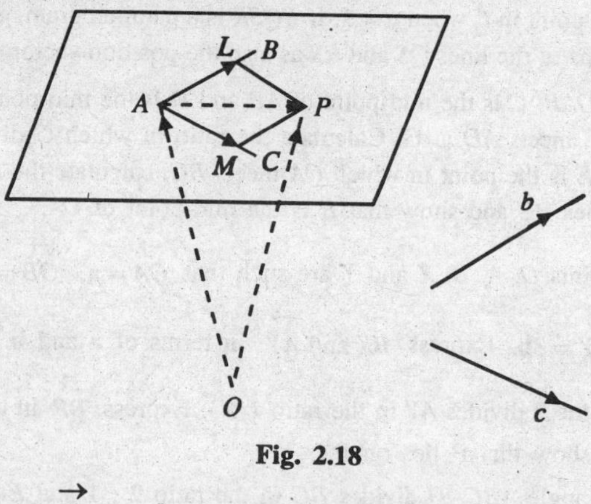

Fig. 2.18

The vectors $\overrightarrow{AP} = \mathbf{r} - \mathbf{a}, \mathbf{b}, \mathbf{c}$ being coplanar, we have a relation of the form

$$\mathbf{r} - \mathbf{a} = t\mathbf{b} + p\mathbf{c}$$

$$\Rightarrow \qquad \mathbf{r} = \mathbf{a} + t\mathbf{b} + p\mathbf{c}$$

where t and p are scalars.

Here each point of the plane arises for some values of the scalars t, p. Also conversely, it is clear that for arbitrary scalar values of t and p, $\mathbf{a} + t\mathbf{b} + p\mathbf{c}$ is the position of a point on the plane.

Thus, the parametric vector equation of the plane which passes through the point with position vector **a**, *and which is parallel to the vectors* **b** *and* **c** *is*

$$\mathbf{r} = \mathbf{a} + t\mathbf{b} + p\mathbf{c},$$

where t, p are scalars parameters.

Cor. 1. Plane through two given points and parallel to a given line. If **a, b** are the position vectors of the given points A, B and **c** is a vector parallel to the given plane, then the vectors $\overrightarrow{AP} = r - a$, $\overrightarrow{AB} = b - a$, and **c** being coplanar, we see that

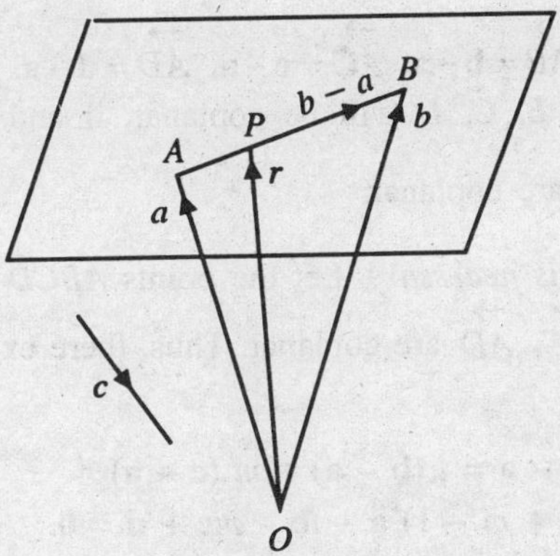

Fig. 2.19

$$\mathbf{r} = \mathbf{a} + t(\mathbf{b} - \mathbf{a}) + p\mathbf{c} = (1 - t)\mathbf{a} + t\mathbf{b} + p\mathbf{c}$$

is the required equation. (Kanpur 95)

Cor. 2. Plane through three given points. If **a, b, c** be the position vectors of three given points A, B, C, then the vectors

$$\overrightarrow{AP} = \mathbf{r} - \mathbf{a},\ \overrightarrow{AB} = \mathbf{b} - \mathbf{a},\ \overrightarrow{AC} = \mathbf{c} - \mathbf{a},$$

being coplanar, the equation of the plane is

$$\mathbf{r} = \mathbf{a} + t(\mathbf{b} - \mathbf{a}) + p(\mathbf{c} - \mathbf{a}) = (1 - t - p)\mathbf{a} + t\mathbf{b} + p\mathbf{c}$$

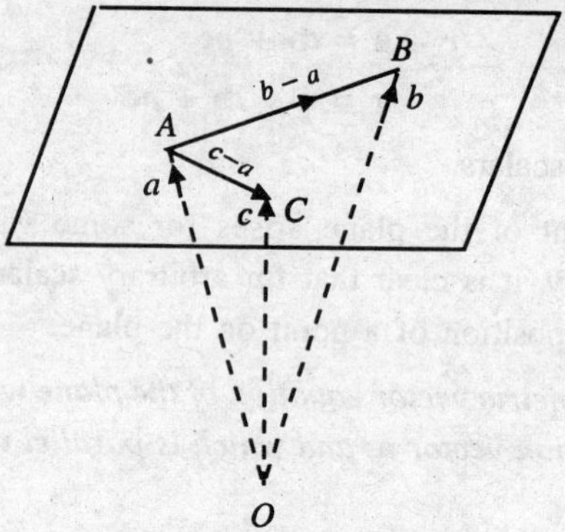

Fig. 2.20

2.7. CONDITION FOR THE COPLANARITY OF FOUR POINTS

The necessary and sufficient condition for four points A, B, C, D with position vectors **a, b, c, d** *to be coplanar is that there exists four scalars x, y, z, t not all zero, such that*

$$x\mathbf{a} + y\mathbf{b} + z\mathbf{c} + t\mathbf{d} = \mathbf{0}, \quad x + y + z + t = 0.$$

We have

$$\vec{AB} = \mathbf{b} - \mathbf{a}, \ \vec{AC} = \mathbf{c} - \mathbf{a}, \ \vec{AD} = \mathbf{d} - \mathbf{a}.$$

The points A, B, C, D will be coplanar, if and only if the vectors $\vec{AB}$, $\vec{AC}$ and $\vec{AD}$ are coplanar.

The condition is necessary. Let the points $ABCD$ be coplanar so that the vectors $\vec{AB}, \vec{AC}, \vec{AD}$ are coplanar. Thus, there exist scalars l, m such that

$$\mathbf{d} - \mathbf{a} = l\,(\mathbf{b} - \mathbf{a}) + m\,(\mathbf{c} - \mathbf{a})$$

$$\Rightarrow \qquad (l + m - 1)\,\mathbf{a} - l\mathbf{b} - m\mathbf{c} + \mathbf{d} = \mathbf{0}.$$

Taking $x = l + m - 1$, $y = -l$, $z = -m$, $t = 1$, we see that if the points A, B, C, D are coplanar, there exist scalars x, y, z, t not all zero such that $x\mathbf{a} + y\mathbf{b} + z\mathbf{c} + t\mathbf{d} = \mathbf{0}$ and $x + y + z + t = 0$.

The condition is sufficient. Let there exist scalars x, y, z, t not all zero such that

$$x\mathbf{a} + y\mathbf{b} + z\mathbf{c} + t\mathbf{d} = \mathbf{0}, \quad x + y + z + t = 0.$$

From these we obtain

$$-(y + z + t)\,\mathbf{a} + y\mathbf{b} + z\mathbf{c} + t\mathbf{d} = \mathbf{0}$$

$$\Rightarrow \qquad \mathbf{d} - \mathbf{a} = (-y\,t^{-1})\,(\mathbf{b} - \mathbf{a}) + (-z\,t^{-1})\,(\mathbf{c} - \mathbf{a}) \ \text{ if } y \neq 0$$

so that **d − a, b − a, c − a** are coplanar vectors and as such A, B, C, D are coplanar points.

EXAMPLES

Example 1. *Show that the points with position vectors*

$-6\mathbf{a}+3\mathbf{b}+2\mathbf{c},\ \ 3\mathbf{a}-2\mathbf{b}+4\mathbf{c},\ \ 5\mathbf{a}+7\mathbf{b}+3\mathbf{c},\ \ -13\mathbf{a}+17\mathbf{b}-\mathbf{c}$

are coplanar.

Solution. Denoting the given points by A, B, C, D respectively, we obtain

$$\overrightarrow{AB}=\overrightarrow{OB}-\overrightarrow{OA}=(3\mathbf{a}-2\mathbf{b}+4\mathbf{c})-(-6\mathbf{a}+3\mathbf{b}+2\mathbf{c})$$
$$=9\mathbf{a}-5\mathbf{b}+2\mathbf{c}.$$
$$\overrightarrow{AC}=11\mathbf{a}+4\mathbf{b}+\mathbf{c}.$$
$$\overrightarrow{AD}=-7\mathbf{a}+14\mathbf{b}-3\mathbf{c}.$$
$$\overrightarrow{AB}-2\overrightarrow{AC}=-13\mathbf{a}-13\mathbf{b}.$$
$$\overrightarrow{AD}+3\overrightarrow{AC}=26\mathbf{a}+26\mathbf{b}=-2(\overrightarrow{AB}-2\overrightarrow{AC})$$
$$\Rightarrow\quad \overrightarrow{AD}+3\overrightarrow{AC}=-2\overrightarrow{AB}+4\overrightarrow{AC}$$
$$\Rightarrow\quad -2\overrightarrow{AB}+\overrightarrow{AC}=\overrightarrow{AD}.$$

Thus, the vectors $\overrightarrow{AB}, \overrightarrow{AC}, \overrightarrow{AD}$ are coplanar. These three vectors, being also coterminous, we see that the four points A, B, C, D are coplanar.

This result may also be proved with the help of the result of § 2.7. The student may find the equation of the plane through the points A, B, C and show that the point D lies on it.

Example 2. *Show that the lines*

$$\mathbf{r}=8\mathbf{a}-9\mathbf{b}+10\mathbf{c}+t\,(3\mathbf{a}-16\mathbf{b}+7\mathbf{c}),$$
$$\mathbf{r}=15\mathbf{a}+29\mathbf{b}+5\mathbf{c}+p\,(3\mathbf{a}+8\mathbf{b}-5\mathbf{c}),$$

are non-coplanar; **a, b, c** *being non-coplanar vectors.*

Solution. We shall show that the two lines have no point in common.

Assuming that the lines have a point in common, we have

$$8\mathbf{a}-9\mathbf{b}+10\mathbf{c}+t\,(3\mathbf{a}-16\mathbf{b}+7\mathbf{c}),$$
$$=15\mathbf{a}+29\mathbf{b}+5\mathbf{c}+p\,(3\mathbf{a}+8\mathbf{b}-5\mathbf{c})$$
$$\Rightarrow\quad (8+3t-15-3p)\,\mathbf{a}+(-9-16t-29-8p)\,\mathbf{b}$$
$$+(10+7t-5+5p)\,\mathbf{c}=\mathbf{0}$$
$$\Rightarrow\quad (-7+3t-3p)\,\mathbf{a}+(-38-16t-8p)\,\mathbf{b}+(5+7t+5p)\,\mathbf{c}=\mathbf{0}.$$

As **a, b, c** are non-coplanar vectors, it follows that

$$-7+3t-3p=0,\quad -38-16t-8p=0,\quad 5+7t+5p=0.$$

It may be shown that these three equations in the two unknowns t and p are not consistent.

Thus, the two given lines have no point in common and hence are non-coplanar. These lines are also not parallel.

Example 3. *Assuming that the vectors* **a** *and* **c** *are not collinear, examine the following lines for intersection :*

$$\mathbf{r} = 6\mathbf{a} - \mathbf{c} + \lambda\,(2\mathbf{c} - \mathbf{a});\quad \mathbf{r} = \mathbf{a} - \mathbf{c} + \mu\,(\mathbf{a} + 3\mathbf{c}).$$

Solution. Assuming that the lines intersect, there must exist values of λ and μ such that

$$6\mathbf{a} - \mathbf{c} + \lambda\,(2\mathbf{c} - \mathbf{a}) = \mathbf{a} - \mathbf{c} + \mu\,(\mathbf{a} + 3\mathbf{c})$$

$$\Rightarrow \quad (6 - \lambda - \mu - 1)\,\mathbf{a} + (2\lambda + 1 + 1 - 3\mu)\,\mathbf{c} = \mathbf{0}$$

$$\Rightarrow \quad (5 - \lambda - \mu)\,\mathbf{a} + (2\lambda - 3\mu)\,\mathbf{c} = \mathbf{0}$$

$$\Rightarrow \quad \left.\begin{array}{l} 5 - \lambda - \mu = 0 \\ 2\lambda - 3\mu = 0 \end{array}\right\}$$

$$\Rightarrow \quad \left.\begin{array}{l} -\lambda - \mu = -5 \\ 2\lambda - 3\mu = 0 \end{array}\right\}$$

These give $\lambda = 3$; $\mu = 2$.

Putting the values of λ and μ in the given equations, we see that the lines intersect and the point of intersection is

$$3\mathbf{a} + 5\mathbf{c}.$$

Also therefore the lines are coplanar.

Example 4. *Find the point of intersection of the line*

$$\mathbf{r} = 2\mathbf{a} + \mathbf{b} + t\,(\mathbf{b} - \mathbf{c})$$

and the plane

$$\mathbf{r} = \mathbf{a} + \lambda\,(\mathbf{b} + \mathbf{c}) + \mu\,(\mathbf{a} + 2\mathbf{b} - \mathbf{c})$$

Solution. At the point of intersection of the line and the plane, we have

$$2\mathbf{a} + \mathbf{b} + t\,(\mathbf{b} - \mathbf{c}) = \mathbf{a} + \lambda\,(\mathbf{b} + \mathbf{c}) + \mu\,(a + 2\mathbf{b} - \mathbf{c})$$

$$\Rightarrow \quad (2 - 1 - \mu)\,\mathbf{a} + (1 + t - \lambda - 2\mu)\,\mathbf{b} + (-t + \lambda + \mu)\,\mathbf{c} = \mathbf{0}$$

$$\Rightarrow \quad 1 - \mu = 0,\; 1 + t - \lambda - 2\mu = 0,\; -t - \lambda + \mu = 0$$

These give $\mu = 1$; $\lambda = 0$; $t = 1$.

Putting the value of t in the given equation of the line, we see that the position vector of the point of intersection is

$$2\mathbf{a} + 2\mathbf{b} - \mathbf{c}.$$

The same result may be obtained by putting the values of λ and μ in the equation of the plane.

Example 5. *Obtain the vector equation of the line of intersection of the planes*

$$\mathbf{r} = \mathbf{b} + \lambda_1\,(\mathbf{b} - \mathbf{a}) + \mu_1\,(\mathbf{a} + \mathbf{c});\quad \mathbf{r} = \mathbf{c} + \lambda_2\,(\mathbf{b} - \mathbf{c}) + \mu_2\,(\mathbf{a} + \mathbf{b});$$

a, b, c *being non-coplanar vectors.*

Solution. At the point of intersection of the two planes, we have

$$\mathbf{b} + \lambda_1\,(\mathbf{b} - \mathbf{a}) + \mu_1\,(\mathbf{a} + \mathbf{c}) = \mathbf{c} + \lambda_2\,(\mathbf{b} - \mathbf{c}) + \mu_2\,(\mathbf{a} + \mathbf{b})$$

$$\Rightarrow \quad (-\lambda_1 + \mu_1 - \mu_2)\,\mathbf{a} + (1 - \lambda_1 + \lambda_2 + \mu_2)\,\mathbf{b}$$

$$+ (\mu_1 - 1 + \lambda_2)\,\mathbf{c} = \mathbf{0}.$$

As **a, b, c** form a linearly independent system, we have

$$-\lambda_1 + \mu_1 - \mu_2 = 0,\quad 1 + \lambda_1 - \lambda_2 + \mu_2 = 0,\quad \mu_1 - 1 + \lambda_2 = 0.$$

We have 3 equations in four unknowns $\lambda_1, \mu_1, \lambda_2, \mu_2$.

We now eliminate λ_2, μ_2 and find a relation between λ_1 and μ_1. We may obtain

$$\lambda_1 = 0.$$

Writing $\lambda_1 = 0$, we may see that the line in equation is $\mathbf{r} = \mathbf{b} + \mu_1 (\mathbf{a} + \mathbf{c})$ where μ_1 is the parameter.

We also have $\lambda_2 = 1 - \mu_1$.

Writing $\lambda_2 = 1 - \mu_2$ and $\mu_2 = \mu_1$ we see that the vectorial equation of the line may be described as

$$\mathbf{r} = \mathbf{c} + (1 - \mu_1)(\mathbf{b} - \mathbf{c}) + \mu_1 (\mathbf{a} + \mathbf{b})$$

$$\Rightarrow \quad \mathbf{r} = \mathbf{b} + \mu_2 (\mathbf{a} + \mathbf{c}).$$

which is the same result as obtained by putting $\lambda_1 = 0$ in the first equation of the plane.

Example 6. *Find the vector equation of the line of intersection of the planes*

$$\mathbf{r} = \lambda_1 (\mathbf{a} + \mathbf{b}) + \mu_1 (\mathbf{a} - \mathbf{c})$$

and $$\mathbf{r} = -2\mathbf{b} + \lambda_2 (\mathbf{a} + 2\mathbf{b} - \mathbf{c}) + \mu_2 \mathbf{a}.$$

Solution. For points on the line of intersection, we have

$$\lambda_1 (\mathbf{a} + \mathbf{b}) + \mu_1 (\mathbf{a} - \mathbf{c}) = -2\mathbf{b} + \lambda_2 (\mathbf{a} + 2\mathbf{b} - \mathbf{c}) + \mu_2 \mathbf{a}$$

$$\Rightarrow \quad (\lambda_1 + \mu_1 - \lambda_2 - \mu_2)\,\mathbf{a} + (\lambda_1 + 2 - 2\lambda_2)\,\mathbf{b} + (-\mu_1 + \lambda_2)\,\mathbf{c} = \mathbf{0}.$$

$$\Rightarrow \quad \left.\begin{aligned} \lambda_1 + \mu_1 - \lambda_2 - \mu_2 &= 0 \\ \lambda_1 - 2\lambda_2 + 2 &= 0 \\ -\mu_1 + \lambda_2 &= 0 \end{aligned}\right\}$$

Here we have 3 linear equations in four unknowns $\lambda_1, \lambda_2, \mu_1$ and μ_2. Solving these we get

$$\lambda_2 = k\,;\ \lambda_1 = 2k - 2\,;\ \mu_2 = 2k - 2. \quad \text{Here } \mu_1 = k.$$

Putting the values of l_1 and m_1 in the first equation of the plane, we have

$$\mathbf{r} = (2k - 2)(\mathbf{a} + \mathbf{b}) + k(\mathbf{a} - \mathbf{c})$$

$$= (3k - 2)\,\mathbf{a} + (2k - 2)\,\mathbf{b} - k\mathbf{c} \qquad ...(i)$$

Also putting the values of λ_2 and μ_2 in the second equation of the plane, we obtain

$$\mathbf{r} = -2b + k(\mathbf{a} + 2\mathbf{b} - \mathbf{c}) + (2k - 2)\,\mathbf{a}$$

$$= (3k - 2)\,\mathbf{a} + (2k - 2)\,\mathbf{b} - k\mathbf{c}.$$

Thus, the required vector equation of the line of the intersection of the given planes is

$$\mathbf{r} = -2\mathbf{a} - 2\mathbf{b} + k\,(3\mathbf{a} + 2\mathbf{b} - \mathbf{c}).$$

Example 7. *Show that four points* $\mathbf{r}_i = l_i\mathbf{a} + m_i\mathbf{b} + n_i\mathbf{c},\ i = 1, 2, 3, 4$ *are coplanar if and only if*

$$\begin{vmatrix} l_1 & l_2 & l_3 & l_4 \\ m_1 & m_2 & m_3 & m_4 \\ n_1 & n_2 & n_3 & n_4 \\ 1 & 1 & 1 & 1 \end{vmatrix} = 0.$$

Solution. If the given points are coplanar, then there must exist a relation of the form

$$x\,(l_1\mathbf{a} + m_1\mathbf{b} + n_1\mathbf{c}) + y\,(l_2\mathbf{a} + m_2\mathbf{b} + n_2\mathbf{c}) + z\,(l_3\mathbf{a} + m_3\mathbf{b} + n_3\mathbf{c}) + t\,(l_4\mathbf{a} + m_4\mathbf{b} + n_4\mathbf{c}) = 0 \quad ..(1)$$

where $$x + y + z + t = 0 \quad ...(2)$$

$$(1) \Rightarrow \quad (xl_1 + yl_2 + zl_3 + tl_4)\,\mathbf{a} + (xm_1 + ym_2 + zm_3 + tm_4)\,\mathbf{b} + (xn_1 + yn_2 + zn_3 + tn_4)\,\mathbf{c} = 0$$

Since **a, b, c** are non-coplanar vectors, hence above equation holds only when

$$xl_1 + yl_2 + zl_3 + tl_4 = 0 \quad ...(3)$$
$$xm_1 + ym_2 + zm_3 + tm_4 = 0 \quad ...(4)$$
$$xn_1 + yn_2 + zn_3 + tn_4 = 0 \quad ...(5)$$

Eliminating x, y, z and t between (2), (3), (4) and (5), we get the required result in the determinant form.

EXERCISES

(In the following **a, b, c** denote a set of non-coplanar vectors.)

1. $OABC$ is a tetrahedron such that

$$\vec{OA} = \mathbf{a},\ \vec{OB} = \mathbf{b},\ \vec{OC} = \mathbf{c}.$$

Find the equations of its faces.

2. $O\,A\,B\,C\,D\,E\,F\,G$ is a parallelopiped such that

$$\vec{OA} = \mathbf{a},\ \vec{OB} = \mathbf{b},\ \vec{OC} = \mathbf{c}.$$

Obtain the equations of its six faces.

3. Find the equations of the planes through the following triads of points :

(*i*) $2\mathbf{a} + 2\mathbf{b} - 2\mathbf{c}$, $\quad 3\mathbf{a} + 4\mathbf{b} + 2\mathbf{c}$, $\quad 7\mathbf{a} + 6\mathbf{c}$,

(*ii*) $\mathbf{a} + \mathbf{b} + \mathbf{c}$, $\quad \mathbf{a} - \mathbf{b} + \mathbf{c}$, $\quad -7\mathbf{a} - 3\mathbf{b} - 5\mathbf{c}$,

4. Examine for coplanarity the following sets of points :

(*i*) $-6\mathbf{a} + 3\mathbf{b} + 2\mathbf{c}$, $3\mathbf{a} - 2\mathbf{b} + 4\mathbf{c}$, $5\mathbf{a} + 7\mathbf{b} + 3\mathbf{c}$, $-13\mathbf{a} + 17\mathbf{b} -$

c.

(ii) $6\mathbf{a} - 4\mathbf{b} + 4\mathbf{c}, \; -\mathbf{a} - 2\mathbf{b} - 3\mathbf{c}, \; \mathbf{a} + 2\mathbf{b} - 5\mathbf{c}, \; -4\mathbf{c}.$

(iii) $3\mathbf{a} + 2\mathbf{b} - 5\mathbf{c}, \; 3\mathbf{a} + 8\mathbf{b} + 5\mathbf{c}, \; -3\mathbf{a} + 2\mathbf{b} + \mathbf{c}, \; \mathbf{a} + 4\mathbf{b} - 3\mathbf{c}.$

5. Show that the four points with position vectors

$$-\mathbf{a} + 4\mathbf{b} - 3\mathbf{c}, \; 3\mathbf{a} + 2\mathbf{b} - 5\mathbf{c}, \; -3\mathbf{a} + 8\mathbf{b} - 5\mathbf{c}, \; -3\mathbf{a} + 2\mathbf{b} + \mathbf{c}$$

are coplanar.

6. Show that the line joining the points $6\mathbf{a} + 4\mathbf{b} + 4\mathbf{c}$, $-4\mathbf{c}$, intersects the join of the points $\mathbf{a} + 2\mathbf{b} - 5\mathbf{c}$ and $-\mathbf{a} - 2\mathbf{b} - 3\mathbf{c}$.

7. Find the position vectors of the points of intersection of the line

$$\mathbf{r} = \mathbf{a} + 2\mathbf{b} + t\,(\mathbf{a} - \mathbf{c})$$

with each of the following planes :

(i) $\mathbf{r} = \lambda\,(\mathbf{a} + 2\mathbf{b}) + \mu\,(2\mathbf{b} + \mathbf{c}).$

(ii) $\mathbf{r} = \mathbf{b} + \lambda\,(\mathbf{a} + \mathbf{c}) + \mu\,(2\mathbf{a} + \mathbf{b} - \mathbf{c}).$

(iii) $\mathbf{r} = 2\mathbf{a} + \lambda\,(\mathbf{a} + \mathbf{c}) + \mu\,(\mathbf{a} + \mathbf{b}).$

8. Find the point of intersection of the line

$$\mathbf{r} = 2\mathbf{a} + 3\mathbf{b} + t\mathbf{c}$$

with the plane

$$\mathbf{r} = \mathbf{a} - \mathbf{b} + p\,(\mathbf{a} + \mathbf{b} - \mathbf{c}) + k\,(\mathbf{a} + \mathbf{c} - \mathbf{b}).$$

9. Find the points of intersection, whenever they exist, of the following pairs of lines :

(i) $\mathbf{r} = \mathbf{a} + \mathbf{b} + \lambda_1\,(\mathbf{b} - \mathbf{c}), \quad \mathbf{r} = -(\mathbf{a} + \mathbf{c}) + \lambda_2\,(\mathbf{a} + \mathbf{b}).$

(ii) $\mathbf{r} = 3\,(\mathbf{a} + \mathbf{c}) + \lambda\,(\mathbf{b} - \mathbf{c}), \quad \mathbf{r} = 2\,(\mathbf{b} - \mathbf{c}) + \mu\,(\mathbf{a} + \mathbf{b}).$

(iii) $\mathbf{r} = \mathbf{b} - 2\mathbf{c} + \lambda\,(\mathbf{a} + \mathbf{b}), \quad \mathbf{r} = 2\mathbf{b} - \mathbf{c} + \mu\,(\mathbf{b} + \mathbf{c}).$

10. Obtain the vector equations of the lines of intersection of the following pairs of planes :

(i) $\mathbf{r} = \lambda\,(\mathbf{a} + \mathbf{b}) + \mu\,(\mathbf{b} - \mathbf{c}),$
$\mathbf{r} = -2\mathbf{a} + \lambda_1\,(2\mathbf{a} + \mathbf{b} - \mathbf{c}) + \mu_1\mathbf{b}.$

(ii) $\mathbf{r} = \mathbf{a} + \lambda_1\,(\mathbf{b} + \mathbf{c}) + \mu_1\,(\mathbf{a} - \mathbf{b}),$
$\mathbf{r} = \lambda_2\,(\mathbf{a} + \mathbf{c}) + \mu_2\,(2\mathbf{a} - \mathbf{b} + \mathbf{c}).$

(iii) $\mathbf{r} = \mathbf{b} + \lambda_1\,(\mathbf{a} + \mathbf{b}) + \mu_1\,(\mathbf{a} - \mathbf{c}),$
$\mathbf{r} = \lambda_2\,(\mathbf{a} + \mathbf{b}) + \mu_2\,(\mathbf{b} + \mathbf{c}).$

(iv) $\mathbf{r} = -2\mathbf{a} + \lambda_1\,(\mathbf{b} + \mathbf{c}) + \mu_1\,(\mathbf{a} - \mathbf{c}),$
$\mathbf{r} = \mathbf{a} + 3\mathbf{c} + \lambda_2\,(\mathbf{a} - \mathbf{b}) + \mu_2\,(\mathbf{a} + \mathbf{b} + 2\mathbf{c}).$

11. Show that the following lines intersect :

$$\mathbf{r} = (\mathbf{a} - \mathbf{b} - 10\mathbf{c}) + t\,(2\mathbf{a} - 3\mathbf{b} + 8\mathbf{c}),$$
$$\mathbf{r} = (4\mathbf{a} - 3\mathbf{b} - \mathbf{c}) + k\,(\mathbf{a} - 4\mathbf{b} + 7\mathbf{c}).$$

Find also the point of intersection.

12. Show that the lines

$$\mathbf{r} = -3\mathbf{a} + 6\mathbf{b} + t\,(-4\mathbf{a} + 3\mathbf{b} + 2\mathbf{c}),$$
$$\mathbf{r} = -2\mathbf{a} + 7\mathbf{c} + t\,(-4\mathbf{a} + \mathbf{b} + \mathbf{c}),$$

do not intersect.

13. Examine the intersection of the following pair of lines :

$$\mathbf{r} = -\mathbf{a} - 3\mathbf{b} - 5\mathbf{c} + t\,(3\mathbf{a} + 5\mathbf{b} + 7\mathbf{c}),$$

$$\mathbf{r} = 2\mathbf{a} + 4\mathbf{b} + 6\mathbf{c} + t\,(\mathbf{a} + 3\mathbf{b} + 5\mathbf{c}),$$

14. Show that a linear relation

$$x_1\mathbf{a}_1 + \ldots + x_i\mathbf{a}_i + \ldots + x_n\mathbf{a}_n = \mathbf{0}$$

connecting the position vectors

$$\mathbf{a}_1, \ldots, \mathbf{a}_i, \ldots, \mathbf{a}_n$$

of any n points

$$\mathrm{A}_1, \ldots, \mathrm{A}_i, \ldots, \mathrm{A}_n$$

will be independent of the choice of origin if and only if the sum

$$x_1 + \ldots + x_i + \ldots + x_n$$

of the scalar coefficients of the vectors is zero.

[Compare with the conditions for coplanarity of four given points (§ 2.7, Page 45) and the collinearity of three given points, (see § 2.5, Page 32)]

15. Verify the condition obtained in Ex. 14 in each of the following cases :

(*i*) The point C divides the line AB in the ratio $m : n$.

(*ii*) A, B, C, D are four points such that the line AB is parallel to the line CD.

(*iii*) A, B, C, D are the vertices of a parallelogram.

(*iv*) G is the centroid of the triangle with vertices A, B, C.

(*v*) G is the centroid of the tetrahedron with vertices A, B, C, D.

SUMMARY

1. If **a, b** be the position vectors of two points A, B, then the position vector $\overrightarrow{OP}$ of the point P which divides AB in the ratio $m : n$ is

$$\frac{m\mathbf{b} + n\mathbf{a}}{m + n};$$

O being the origin of reference.

2. The parametric vectorial equation of the line

(*i*) which passes through the point with position vector **a** and which is parallel to the vector **b** is

$$\mathbf{r} = \mathbf{a} + t\mathbf{b}.$$

(*ii*) which passes through two points A, B with position vectors **a, b** is

$$\mathbf{r} = \mathbf{a} + t\,(\mathbf{b} - \mathbf{a}),$$

t is the parameter in each case.

3. The parametric vectorial equation of the plane

(*i*) which passes through a point with position vector **a** and which is parallel to the vectors **b, c** is

$$\mathbf{r} = \mathbf{a} + t\,(\mathbf{b} - \mathbf{a}) + p\,(\mathbf{c} - \mathbf{a}).$$

(*ii*) which passes through the points A, B with position vectors **a, b** and which is parallel to the vector **c** is

$$\mathbf{r} = \mathbf{a} + t\,(\mathbf{b} - \mathbf{a}) + p\mathbf{c}.$$

(*iii*) which passes through the three points A, B, C with position vectors **a, b, c** is

$$\mathbf{r} = \mathbf{a} + t(\mathbf{b} - \mathbf{a}) + p(\mathbf{c} - \mathbf{a}).$$

Here t, p are the parameters in each case.

4. (*i*) Three points A, B, C with position vectors **a, b, c** are collinear if there exist three scalars x, y, z such that

$$x\mathbf{a} + y\mathbf{b} + z\mathbf{c} = \mathbf{0} \quad \text{and} \quad x + y + z = 0.$$

Also in this case the point C divides AB in the ratio $y : x$.

(*ii*) Four points A, B, C, D with position vectors **a, b, c, d** are coplanar if there exist scalars x, y, z, t such that

$$x\mathbf{a} + y\mathbf{b} + z\mathbf{c} + t\mathbf{d} = \mathbf{0}, \quad x + y + z + t = 0.$$

Also in this case the position vector of the point of intersection of the lines AB and CD is

$$\frac{x\mathbf{a} + y\mathbf{b}}{x + y} \text{ which is the same as } \frac{z\mathbf{c} + t\mathbf{d}}{z + t}.$$

OBJECTIVE QUESTIONS

For each of the following questions, four alternatives are given for the answer. Only one of them is correct. Choose the correct alternative.

1. If **a, b, c** are position vectors of the vertices of a ΔABC, then $\overrightarrow{AB} + \overrightarrow{BC} + \overrightarrow{CA} =$

(*a*) $\mathbf{0}$ (*b*) $2\mathbf{a}$ (*c*) $2\mathbf{b}$ (*d*) $3\mathbf{c}$

2. If **a, b, c** are the position vectors of the points A, B, C respectively, then $\overrightarrow{AB} + \overrightarrow{BC} + \overrightarrow{AC}$ is equal to

(*a*) $\mathbf{0}$ (*b*) $2(\mathbf{b} - \mathbf{a})$

(*c*) $2(\mathbf{c} - \mathbf{a})$ (*d*) $\mathbf{a} + \mathbf{b} + \mathbf{c}$

3. The position vectors of A and B are **a** and **b** respectively. Then the position vector of a point D dividing AB in the ratio $2 : 3$ externally is

(*a*) $\frac{1}{5}(2\mathbf{a} + 3\mathbf{b})$ (*b*) $\frac{1}{5}(2\mathbf{b} + 3\mathbf{a})$

(*c*) $\frac{1}{6}(\mathbf{a} + \mathbf{b})$ (*d*) $(3\mathbf{a} - 2\mathbf{b})$

4. Point A is $\mathbf{a} + 2\mathbf{b}$, and **a** divides AB in the ratio $2 : 3$. The position vector of B is

(*a*) $2\mathbf{a} - \mathbf{b}$ (*b*) $\mathbf{b} - 2\mathbf{a}$

(*c*) $\mathbf{a} - 3\mathbf{b}$ (*d*) $\mathbf{b}$

5. P, Q, R, S have position vectors $\mathbf{p}, \mathbf{q}, \mathbf{r}, \mathbf{s}$, such that $\mathbf{p} - \mathbf{q} = 2(\mathbf{s} - \mathbf{r})$ then

(*a*) PQ and RS bisect each other

(*b*) PQ and PR bisect each other

(*c*) PQ and RS trisect each other

(*d*) QS and PR trisect each other

6. $\mathbf{a}, \mathbf{b}, \mathbf{c}$ are three non-zero vectors, no two of which are collinear and the vector $\mathbf{a} + \mathbf{b}$ is collinear with $\mathbf{c}$, $\mathbf{b} + \mathbf{c}$ is collinear with $\mathbf{a}$, then $\mathbf{a} + \mathbf{b} + \mathbf{c}$ is equal to

(*a*) $\mathbf{a}$ (*b*) $\mathbf{b}$ (*c*) $\mathbf{c}$ (*d*) None of these

7. Given that the vectors $\mathbf{a}$ and $\mathbf{b}$ are non-collinear, the value of x and y for which the vector quantity $2\mathbf{u} - \mathbf{v} = \mathbf{w}$ holds true if $\mathbf{u} = x\mathbf{a} + 2y\mathbf{b}$, $\mathbf{v} = -2y\mathbf{a} + 3x\mathbf{b}$, $\mathbf{w} = 4\mathbf{a} - 2\mathbf{b}$ are

(*a*) $x = 4/7, \; y = 6/7$ (*b*) $x = 10/7, \; y = 4/7$

(*c*) $x = 8/7, \; y = 2/7$ (*d*) $x = 2, \; y = 3$

8. If a, b, c are position vectors of three-collinear points such that $x\mathbf{a} + y\mathbf{b} + z\mathbf{c} = 0$ and atleast one scalar $x, y, z \neq 0$, then

(*a*) $x + y + z = 0$ (*b*) $x + y + z \neq 0$

(*c*) there exists no relation between x, y and z

(*d*) None of these

9. If $x\mathbf{u} + y\mathbf{v} + z\mathbf{w} = x'(\mathbf{u} + \mathbf{w}) + y'(-\mathbf{u} + \mathbf{w}) + z'(\mathbf{u} + \mathbf{v} + \mathbf{w})$, $\mathbf{u}, \mathbf{v}, \mathbf{w}$ being non-zero, non-coplanar vectors, then x' is equal to

(*a*) $-\frac{1}{2}x + y - \frac{1}{2}z$ (*b*) $-\frac{1}{2}x + \frac{1}{2}z$

(*c*) $-\frac{1}{2}x - y + \frac{1}{2}z$ (*d*) $x - \frac{1}{2}y + \frac{1}{2}z$

10. The points O, A, B, X and Y are such that $\overrightarrow{OA} = \mathbf{a}$, $\overrightarrow{OB} = \mathbf{b}$, $\overrightarrow{OX} = 3\mathbf{a}$, $\overrightarrow{OY} = 3\mathbf{b}$; and the point P divides AY in the ratio 1 : 3 internally, then the vector $\overrightarrow{PB}$ is equal to

(*a*) $\frac{1}{4}(\mathbf{a} - 3\mathbf{b})$ (*b*) $\frac{1}{4}(3\mathbf{a} - 2\mathbf{b})$

(*c*) $\frac{1}{4}(3\mathbf{a} - \mathbf{b})$ (*d*) $\frac{1}{4}(2\mathbf{a} - 3\mathbf{b})$

ANSWERS

1. (*a*)	**2.** (*c*)	**3.** (*d*)	**4.** (*c*)	**5.** (*d*)
6. (*d*)	**7.** (*b*)	**8.** (*a*)	**9.** (*c*)	**10.** (*c*)

APPENDIX I

1.1. VECTORIAL PROOFS OF SOME WELL-KNOWN CLASSICAL THEOREMS

1. Ceva's Theorem and its Converse. *If D, E, F are three points on the sides BC, CA, AB respectively of a triangle ABC such that the lines AD, BE and CF are concurrent, then*

$$\frac{BD}{CD} \cdot \frac{CE}{AE} \cdot \frac{AF}{BF} = -1$$

and conversely.

Let AD, BE and CF concur at H. Take any point, O, as the origin of reference. Let $\mathbf{a}$, $\mathbf{b}$, $\mathbf{c}$, $\mathbf{h}$ be the position vectors of the points A, B, C, H respectively.

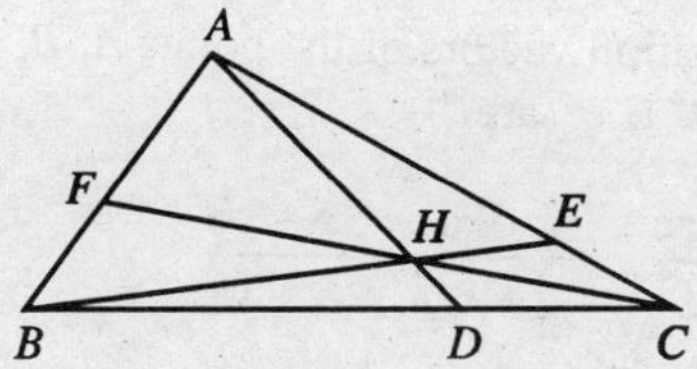

Fig. 1

These four points being coplanar, there exist four scalars x, y, z, t such that

$$x\mathbf{a} + y\mathbf{b} + z\mathbf{c} + t\mathbf{h} = \mathbf{0}, \quad x + y + z + t = 0.$$

These give

$$\frac{x\mathbf{a} + y\mathbf{b}}{x+y} = \frac{z\mathbf{c} + t}{z+t}.$$

Now $$\frac{x\mathbf{a} + y\mathbf{b}}{x+y} \text{ and } \frac{z\mathbf{c} + t\mathbf{h}}{z+t}$$

are points of the lines AB and CH respectively.

It follows that

$$(x\mathbf{a} + y\mathbf{b}) \,/\, x + y$$

is the position vector of the point F of intersection of the lines AB and CH so that F divides AB in the ratio

$$AF \,/\, FB = y/x \quad \Leftrightarrow \quad AF \,/\, BF = -\, y/x.$$

Similarly

$$\frac{BD}{CD} = -\frac{z}{y}, \quad \frac{CE}{AE} = -\frac{x}{z}$$

$$\therefore \quad \frac{BD}{CD} \cdot \frac{CE}{AE} \cdot \frac{AF}{BF} = \left(-\frac{z}{y}\right) \cdot \left(-\frac{x}{z}\right) \cdot \left(-\frac{y}{x}\right) = -1.$$

Conversely, let D, E, F be three points on the sides BC, CA and AB respectively such that

$$\frac{BD}{CD} \cdot \frac{CE}{AE} \cdot \frac{AF}{BF} = -1. \quad \text{...}(i)$$

Suppose that

$$\frac{BD}{CD} = -\frac{z}{y}, \quad \frac{CE}{AE} = -\frac{x}{z} \text{ so that by } (i)$$

$$\frac{AF}{BF} = -\frac{y}{x}.$$

$$\therefore \quad \frac{BD}{DC} = \frac{z}{y}, \quad \frac{CE}{EA} = \frac{x}{z}, \quad \frac{AF}{FB} = \frac{y}{x}.$$

If **a, b, c** be the position vectors of the points A, B, C, then the position vectors of the points D, E, F are

$$\frac{y\mathbf{b} + z\mathbf{c}}{y+z}, \quad \frac{z\mathbf{c} + x\mathbf{a}}{z+x}, \quad \frac{x\mathbf{a} + y\mathbf{b}}{x+y}$$

respectively.

Therefore the position vectors of the points dividing AD, BE, CF in the ratios

$$y + z : x, \quad z + x : y, \quad x + y : z$$

are all equal to

$$\frac{x\mathbf{a} + y\mathbf{b} + z\mathbf{c}}{x+y+z}.$$

Thus, the lines AD, BE and CF are concurrent.

2. Menelau's Theorem and its Converse. *If D, E, F are three points on the sides BC, CA, AB respectively of a triangle ABC such that the points D, E, F are collinear, then*

$$\frac{BD}{CD} \cdot \frac{CE}{AE} \cdot \frac{AF}{BF} = 1.$$

and conversely.

Let BE, CF meet at H.

Let **a**, **b**, **c**, **h** be the position vectors of the four points A, B, C, H relative to any origin O of reference.

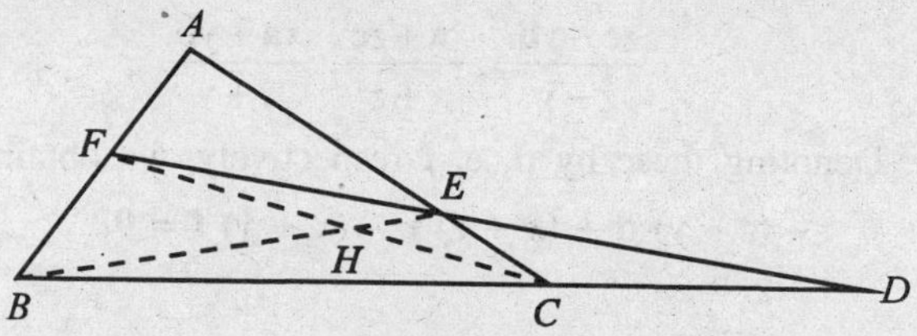

Fig. 2

These four points being coplanar, there exist four scalars x, y, z, t such that

$$x\mathbf{a} + y\mathbf{b} + z\mathbf{c} + t\mathbf{h} = \mathbf{0}, \quad x + y + z + t = 0.$$

The position vectors of the points E and F, therefore, are

$$\frac{x\mathbf{a}+z\mathbf{c}}{x+z}, \quad \frac{x\mathbf{a}+y\mathbf{b}}{x+y}$$

respectively.

We now require the position vector of the point D. Writing

$$\mathbf{e} = \frac{x\mathbf{a}+z\mathbf{c}}{x+z}, \quad \mathbf{f} = \frac{x\mathbf{a}+y\mathbf{b}}{x+y}$$

and eliminating, **a**, we have

$$(x + z)\,\mathbf{e} - (x + y)\,\mathbf{f} = z\mathbf{c} - y\mathbf{b},$$

$$\Rightarrow \quad \frac{(x+z)\,\mathbf{e}-(x+y)\,\mathbf{f}}{(x+z)-(x+y)} = \frac{z\mathbf{c}-y\mathbf{b}}{z-y}.$$

This equality shows that $(z\mathbf{c} - y\mathbf{b}) / (z - y)$ is the position vector of the point D. Thus

$$\frac{BD}{CD} = -\frac{z}{y}, \quad \frac{CE}{EA} = \frac{x}{z}, \quad \frac{AF}{FB} = \frac{y}{x}.$$

$$\Rightarrow \quad \frac{BD}{CD}\cdot\frac{CE}{AE}\cdot\frac{AF}{BF} = 1.$$

Conversely, let D, E, F be three points on the sides BC, CA, AB such that

$$\frac{BD}{CD}\cdot\frac{CE}{AE}\cdot\frac{AF}{BF} = 1.$$

Suppose that

$$\frac{BD}{CD} = \frac{z}{y}, \quad \frac{CE}{AE} = -\frac{x}{z}, \text{ so that } \frac{AF}{BF} = -\frac{y}{z}.$$

$$\Rightarrow \quad \frac{BD}{DC} = -\frac{z}{y}, \quad \frac{CE}{EA} = \frac{x}{z}, \quad \frac{AF}{EB} = \frac{y}{x}.$$

Thus, if **a, b, c** be the position vectors of the vertices A, B, C, then the position vectors of the points D, E, F, are

$$\frac{z\mathbf{c}-y\mathbf{b}}{z-y},\quad \frac{x\mathbf{a}+z\mathbf{c}}{x+z},\quad \frac{x\mathbf{a}+y\mathbf{b}}{x+y}$$

respectively. Denoting these by **d, e, f** respectively, we obtain

$$-(z-y)\,\mathbf{d}+(x+z)\,\mathbf{e}-(x+y)\,\mathbf{f}=\mathbf{0},$$

where

$$-(z-y)+(x+z)-(x+y)=0.$$

Thus, the points D, E, F are collinear.

3. Pappu's Theorem. *If A_1, A_2, A_3 and B_1, B_2, B_3 are two sets of collinear points, then the points of intersection of the pairs of lines*

$$A_1B_2,\ A_2B_1;\ A_2B_3,\ A_3B_2;\ A_3B_1,\ A_1B_2.$$

are collinear.

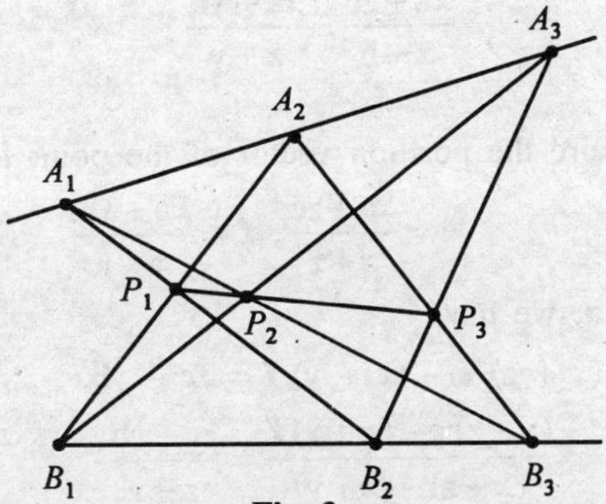

Fig. 3

The two lines must of course be coplanar.

Take O, the point of intersection of the two lines as the origin of reference. Let **a, b** be the two vectors along the two lines.

Let

$$\lambda_1\mathbf{a},\ \lambda_2\mathbf{a},\ \lambda_3\mathbf{a};\ \mu_1\mathbf{b},\ \mu_2\mathbf{b},\ \mu_3\mathbf{b},$$

be the position vectors of the two sets of points, A_1, A_2, A_3 and B_1, B_2, B_3 respectively.

The equations of the lines A_1B_2 and A_2B_1 are

$$\mathbf{r}=\lambda_1\mathbf{a}+t(\lambda_1\mathbf{a}-\mu_2\mathbf{b}),\quad \mathbf{r}=\lambda_2\mathbf{a}+p(\lambda_2\mathbf{a}-\mu_1\mathbf{b}).$$

At their point of intersection, we have

$$\lambda_1\mathbf{a}+t(\lambda_1\mathbf{a}-\mu_2\mathbf{b})=\lambda_2\mathbf{a}+p(\lambda_2\mathbf{a}-\mu_1\mathbf{b})$$

$$\Rightarrow\quad (\lambda_1+t\lambda_1-\lambda_2-p\lambda_2)\,\mathbf{a}+(-t\mu_2+p\mu_1)\,\mathbf{b}=\mathbf{0}$$

$$\Rightarrow\quad \lambda_1+t\lambda_1-\lambda_2-p\lambda_2=0,\quad -t\mu_2+p\mu_1=0,$$

the vectors **a, b** being non-parallel.

These give

$$p = \frac{\mu_2(\lambda_2 - \lambda_1)}{(\lambda_1\mu_1 - \lambda_2\mu_2)}.$$

Making substitution, we see that the position vector of the point of intersection of the lines A_1B_2, A_2B_1 is

$$\frac{\lambda_1\lambda_2(\mu_1 - \mu_2)}{\lambda_1\mu_1 - \lambda_2\mu_2}\mathbf{a} + \frac{\mu_1\mu_2(\lambda_1 - \lambda_2)}{\lambda_1\mu_1 - \lambda_2\mu_2}\mathbf{b} = \mathbf{p}_3, \text{ say.}$$

Changing the suffixes, 1, 2 to 2, 3 respectively and 1, 2 to 3, 1 respectively, we see that the position vectors of the other two points of intersection are

$$\frac{\lambda_2\lambda_3(\mu_2 - \mu_3)}{\lambda_2\mu_2 - \lambda_3\mu_3}\mathbf{a} + \frac{\mu_2\mu_3(\lambda_2 - \lambda_3)}{\lambda_2\mu_2 - \lambda_1\mu_1}\mathbf{b} = \mathbf{p}_1, \text{ say}$$

$$\frac{\lambda_3\lambda_1(\mu_3 - \mu_1)}{\lambda_3\mu_3 - \lambda_1\mu_1}\mathbf{a} + \frac{\mu_3\mu_1(\lambda_3 - \lambda_1)}{\lambda_3\mu_3 - \lambda_2\mu_2}\mathbf{b} = \mathbf{p}_2, \text{ say}$$

Now we may see that

$$\Sigma\lambda_1\mu_1(\lambda_2\mu_2 - \lambda_3\mu_3)\,\mathbf{p}_1 = \mathbf{0},$$

where

$$\Sigma\lambda_1\mu_1(\lambda_2\mu_2 - \lambda_3\mu_3) = 0.$$

Thus, the points are collinear.

Or directly, we may show that the vector $P_2 - P_1$ is the product of the vector $P_3 - P_1$ by a scalar.

4. Desargue's Theorem. *If ABC, $A_1B_1C_1$ are two triangles such that the three lines AA_1, BB_1 and CC_1 are concurrent, then the points of intersection of the three pairs of sides*

$$BC,\ B_1C_1;\quad CA,\ C_1A_1;\quad AB,\ A_1B_1$$

are collinear and conversely.

Let AA_1, BB_1 and CC_1 are concur at O. Take O as the origin of reference. Let

$$\vec{OA} = \mathbf{a},\ \vec{OB} = \mathbf{b},\ \vec{OC} = \mathbf{c}.$$

Then

$$\vec{OA_1} = \lambda\mathbf{a},\ \vec{OB_1} = \mu\mathbf{b},\ \vec{OC_1} = \nu\mathbf{c}$$

where λ, μ, ν are some scalars. The equations of BC and B_1C_1 respectively are

$$\mathbf{r} = \mathbf{b} + t\,(\mathbf{b} - \mathbf{c}),\quad \mathbf{r} = \mu\mathbf{b} + p\,(\mu\mathbf{b} - \nu\mathbf{c}),$$

so that at the point of intersection P of BC and B_1C_1, we have

$$\mathbf{b} + t\,(\mathbf{b} - \mathbf{c}) = \mu\mathbf{b} + p\,(\mu\mathbf{b} - \nu\mathbf{c}),$$

$$\Rightarrow \qquad (1 + t - \mu - p\mu)\,\mathbf{b} + (-t + p\nu)\,\mathbf{c} = \mathbf{0}.$$

$$\Rightarrow \qquad 1 + t - \mu - p\mu = 0, \quad -t + p\nu = 0.$$

the vectors **b** and **c** being non-parallel.

These give

$$p = (1 - \mu) / (\mu - \nu).$$

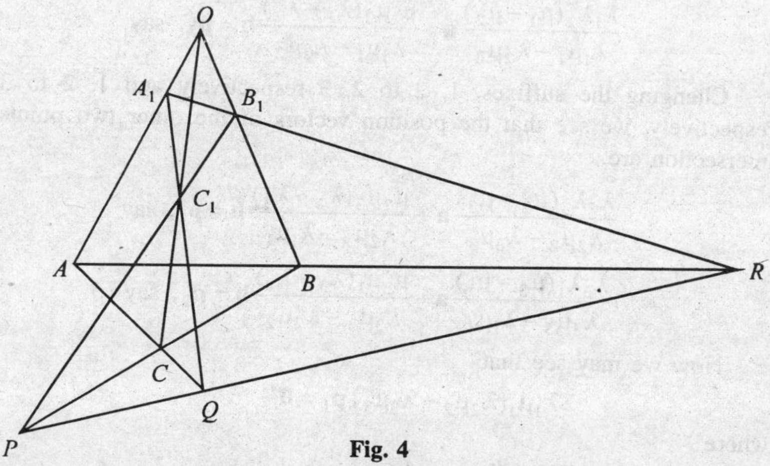

Fig. 4

Making substitution, we see that the position vector of the point P of intersection of BC, B_1C_1 is

$$\mu\mathbf{b} + \frac{1-\mu}{\mu-\nu}(\mu\mathbf{b} - \nu\mathbf{c}) = \frac{1-\nu}{\mu-\nu}\,\mu\mathbf{b} + \frac{1-\mu}{\nu-\mu}\,\nu\mathbf{c} = \mathbf{p}, \text{ say.}$$

By considerations of symmetry, the two other points of intersection Q, R are

$$\frac{1-\lambda}{\nu-\lambda}\,\nu\mathbf{c} + \frac{1-\nu}{\lambda-\nu}\,\lambda\mathbf{a} = \mathbf{q}, \text{ say.}$$

$$\frac{1-\mu}{\lambda-\mu}\,\lambda\mathbf{a} + \frac{1-\lambda}{\mu-\lambda}\,\mu\mathbf{b} = \mathbf{r}, \text{ say.}$$

We now see that

$$\Sigma\,(1 - \lambda)\,(\mu - \nu t)\,\mathbf{p} = \mathbf{0} \quad \text{where} \quad \Sigma\,(1 - \lambda)\,(\mu - \nu) = 0.$$

Thus, the three points of intersection whose position vectors are **p, q, r** are collinear.

Conversely, let **a, b, c : $\mathbf{a}_1$, $\mathbf{b}_1$, $\mathbf{c}_1$** be the position vectors of the vertices of the triangles ABC, $A_1B_1C_1$ with respect to some origin O. Let P, Q, R be the points of intersection of the pairs of lines

$$BC,\ B_1C_1; \quad CA,\ C_1A_1; \quad AB,\ A_1B_1.$$

These points are given to be collinear.

Let P divide BC in the ratio $z : y$ so that the position vector of the point P is

$$(y\mathbf{b} + z\mathbf{c}) / (y + z).$$

Let Q divide BC in the ratio $x : z$ so that the position vector of the point Q is

$$(z\mathbf{c} + x\mathbf{a}) / (z + x).$$

The position vector of the point R of intersection of PQ and AB may now be easily seen to be

$$(y\mathbf{b} - x\mathbf{a}) / (y - x).$$

Denoting these position vectors by **p, q, r** respectively, we have the relation

$$(y + z)\,\mathbf{p} - (z + x)\,\mathbf{q} - (y - x)\,\mathbf{r} = \mathbf{0} \qquad ...(1)$$

between the position vectors of the collinear points P, Q, R.

Similarly, denoting by $z_1 : y_1$ and $x_1 : y_1$ the ratios in which the points P and Q divide B_1C_1 and C_1A_1 we obtain

$$\mathbf{p} = \frac{y_1\mathbf{b} + z_1\mathbf{c}}{y_1 + z_1}, \mathbf{q} = \frac{z_1\mathbf{c} + x_1\mathbf{a}}{z_1 + x_1}, \ \mathbf{r} = \frac{y_1\mathbf{b} + x_1\mathbf{a}}{y_1 - x_1},$$

and the relation

$$(y_1 + z_1)\,\mathbf{p} - (z_1 + x_1)\,\mathbf{q} - (y_1 - x_1)\,\mathbf{r} = \mathbf{0}. \qquad ...(2)$$

From (1) and (2), we obtain

$$\frac{y+z}{y_1 + z_1} = \frac{z+x}{z_1 + x_1} = \frac{y-x}{y_1 - x_1} = k, \text{ say.}$$

so that

$$y + z = k\,(y_1 + z_1),\ (z + x) = k\,(z_1 + x_1),\ y - x = k\,(y_1 - x_1) \qquad ...(3)$$

Also we have

$$\frac{y\mathbf{b} + z\mathbf{c}}{y+z} = \frac{y_1\mathbf{b}_1 + z_1\mathbf{c}_1}{y_1 + z_1}, \ \frac{z\mathbf{c} + x\mathbf{a}}{z+x} = \frac{z_1\mathbf{c}_1 + x_1\mathbf{a}_1}{z_1 + x_1}$$

$$\frac{y\mathbf{b} - x\mathbf{a}}{y-x} = \frac{y_1\mathbf{b}_1 - x_1\mathbf{a}_1}{y_1 + zx_1} \qquad ...(4)$$

so that with the help of (3), we obtain

$$z\mathbf{c} - kz_1\mathbf{c}_1 = ky_1\mathbf{b}_1 - y\mathbf{b} = kx_1\mathbf{a}_1 - x\mathbf{a}$$

$$\Rightarrow \qquad \frac{z\mathbf{c} - kz_1\mathbf{c}_1}{z - kz_1} = \frac{ky_1\mathbf{b}_1 - y\mathbf{b}}{ky_1 - y} = \frac{kx_1\mathbf{a}_1 - x\mathbf{a}}{kx_1 - x} \qquad ...(5)$$

The three equal vectors in (5) being the position vectors of points on CC_1, BB_1, AA_1, we see that these lines are concurrent.

5. Complete Quadrilateral. (*i*) *The three mid-points of the three diagonals of a complete quadrilateral are collinear.*

(*ii*) *Each diagonal of a complete quadrilateral is cut harmonically by the other two.*

(*i*) Consider any four lines

$$AB,\ BC,\ CA,\ EF.$$

no three of which are concurrent. These lines intersect in pairs in **six** points called the *vertices* of the complete quadrilateral. These six vertices are divided into *three* pairs of opposite vertices, *viz.*, the intersections of the pairs of lines

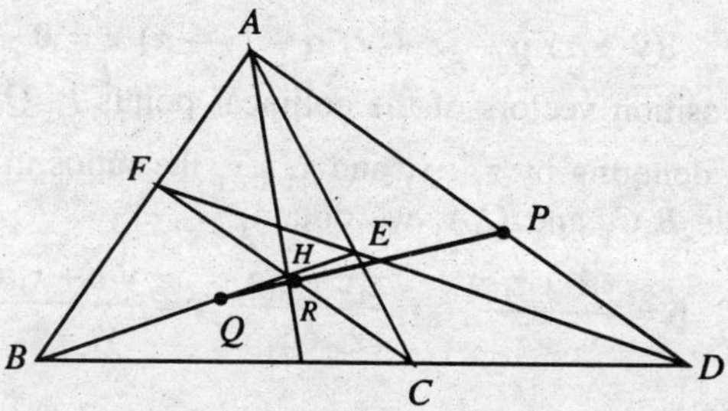

Fig. 5

$$CA,\ AB\ ;\quad BC,\ EF.$$
$$AB,\ BC\ ;\quad CA,\ EF.$$
$$BC,\ CA\ ;\quad AB,\ EF.$$

Thus

$$A,\ D\ ;\quad B\ E\ ;\quad C,\ F$$

are three pairs of opposite vertices. The lines

$$AD,\ BE,\ CF$$

joining the pairs of opposite vertices are called the three diagonals of the complete quadrilateral. We have to prove that the mid-points of *AD, BE, CF* are collinear.

Let

BE, CF meet in *H*.

With any point *O* as the origin of reference, let

$$\mathbf{a},\ \mathbf{b},\ \mathbf{c},\ \mathbf{h}$$

be the position vectors of *A, B, C, H* respectively. These points being coplanar, there exist four scalars *x, y, z, t* such that

$$x\mathbf{a} + y\mathbf{b} + z\mathbf{c} + t\mathbf{h} = \mathbf{0},\quad x + y + z + t = 0. \qquad ...(i)$$

From this we deduce that

$$\frac{z\mathbf{c}+x\mathbf{a}}{z+x}, \quad \frac{x\mathbf{a}+y\mathbf{b}}{x+y}$$

are the position vectors of the points E and F respectively.

We now require the position vector of the point D.

Writing

$$\mathbf{e}=\frac{z\mathbf{c}+x\mathbf{a}}{z+x}, \quad \mathbf{f}=\frac{x\mathbf{a}+y\mathbf{b}}{x+y}$$

and eliminating, **a**, we see that

$$(z+x)\,\mathbf{e}-(x+y)\,\mathbf{f}=z\mathbf{e}-y\mathbf{b}$$

$$\Rightarrow \quad \frac{(z+x)\,\mathbf{e}-(x+y)\,\mathbf{f}}{(z+x)-(x+y)}=\frac{z\mathbf{c}-y\mathbf{b}}{z-y}.$$

This equality shows that

$$\frac{z\mathbf{c}-y\mathbf{b}}{z-y}$$

is the position vector of the point, *D*, of intersection of the lines, *BC, EF.*

The mid-points *P, Q, R* of *AD, BE* and *CF* are

$$\frac{(z-y)\,\mathbf{a}-y\mathbf{b}+z\mathbf{c}}{2(z-y)}, \quad \frac{x\mathbf{a}+(z+x)\,\mathbf{b}+z\mathbf{e}}{2(z+x)}, \quad \frac{x\mathbf{a}+y\mathbf{b}+(x+y)\mathbf{c}}{2(x+y)}.$$

Denoting these by **p, q, r** respectively, we see that

$$x\,(z-y)\,\mathbf{p}+y\,(z+x)\,\mathbf{q}-z\,(x+y)\,\mathbf{r}=\mathbf{0}$$

where

$$x\,(z-y)+y\,(z+x)-z\,(x+y)=0.$$

Thus, the three points *P, Q, R* with position vectors **p, q, r** are collinear.

(*ii*) The position vectors of the points *B, H, E* are

$$\mathbf{b}, \frac{x\mathbf{a}+y\mathbf{b}+z\mathbf{c}}{x+y+z}, \quad \frac{z\mathbf{c}+x\mathbf{a}}{z+x}$$

respectively so that *H* divides *BE* in the ratio $(z+x) : y$. The point *H′* which divides *BE* in the ratio $(z+x) : -y$ is

$$\frac{(x\mathbf{a}+z\mathbf{c})-y\mathbf{b}}{(x+z)-y}=\frac{x\mathbf{a}+(z\mathbf{c}-y\mathbf{b})}{x+(z-y)}$$

which is clearly also a point on the line *AD* dividing the same in the ratio $(z-y) : x$.

Thus, the diagonal *BE* is divided at two points of intersection *H* and *H′* with the other two diagonals *CE, AD* in equal and opposite ratios. Hence, the second part.

Another Proof.

Take A as the origin of reference. Let

$$\overrightarrow{AB} = \mathbf{b}, \ \overrightarrow{AE} = \mathbf{e}.$$

It should be possible to obtain the position vector of every point in the plane of the quadrilateral in terms of **b** and **c.**

Let $\overrightarrow{AF} = \lambda \mathbf{b}$, $\overrightarrow{AC} = \mu \mathbf{e}$; λ, μ being scalars.

The equations of BC and EF are

$$\mathbf{r} = \mathbf{b} + t\,(\mu \mathbf{e} - \mathbf{b}), \quad \mathbf{r} = \mathbf{e} + p\,(\mathbf{e} - \lambda \mathbf{b}).$$

At their point of intersection,

$$\mathbf{b} + t\,(\mu \mathbf{e} - \mathbf{b}) = \mathbf{e} + p\,(\mathbf{e} - \lambda \mathbf{b})$$

$$\Rightarrow \quad (1 - t + p\lambda)\,\mathbf{b} + (\mu t - 1 - p)\,\mathbf{e} = \mathbf{0}$$

$$\Rightarrow \quad 1 - t + p\lambda = 0, \quad \mu t - 1 - p = 0;$$

b, e being non-parallel vectors.

These give

$$t = \frac{1-\lambda}{1-\lambda\mu}.$$

Making substitution, we see that the point D, of intersection of the lines BC and EF is

$$\mathbf{b} + \frac{1-\lambda}{1-\lambda\mu}(\mu \mathbf{e} - \mathbf{b}) = \frac{\lambda(1-\mu)}{1-\lambda\mu}\mathbf{b} + \frac{\mu(1-\lambda)}{1-\lambda\mu}\mathbf{e}.$$

Thus, the mid-points of AD, BE, CF respectively are

$$\frac{1}{2}\left[\frac{\lambda(1-\mu)}{1-\lambda\mu}\mathbf{b} + \frac{\mu(1-\lambda)}{1-\lambda\mu}\mathbf{e}\right], \ \frac{1}{2}(\mathbf{b}+\mathbf{e}), \ \frac{1}{2}(\lambda \mathbf{b} + \mu \mathbf{e}).$$

Denoting these by **p, q, r** respectively, we see that

$$(1 - \lambda\mu)\,\mathbf{p} + \lambda\mu\,\mathbf{q} + (-1)\,\mathbf{r} = \mathbf{0},$$

where

$$(1 - \lambda\mu) + \lambda\mu + (-1) = 0.$$

Thus, the points are collinear.

MISCELLANEOUS EXERCISES I

1. OAB is a given triangle such that $\vec{OA} = \mathbf{a}$, $\vec{OB} = \mathbf{b}$. Also C is a point on AB such that $AB = 2BC$. State which of the following statements are correct ?

(*a*) $\vec{AC} = \frac{2}{3}(\mathbf{b} - \mathbf{a})$, (*b*) $\vec{AC} = \frac{2}{3}(\mathbf{a} - \mathbf{b})$,

(*c*) $\vec{AC} = \frac{2}{3}(\mathbf{b} + \mathbf{a})$, (*d*) $\vec{AC} = \frac{3}{2}(\mathbf{b} - \mathbf{a})$,

(*e*) $\vec{AC} = \frac{3}{2}(\mathbf{a} + \mathbf{b})$.

2. O, A, B, C, D, E are five coplanar points such that

$$\vec{OA} = \mathbf{a},\ \vec{OB} = \mathbf{b},\ \vec{OC} = 2\mathbf{a} + 3\mathbf{b},\ \vec{OD} = \mathbf{a} - 2\mathbf{b},\ \vec{OE} = \mathbf{a} + \mathbf{b}.$$

Show that $\vec{OE}$, $\vec{BC}$ are parallel and $\vec{OB}$, $\vec{AD}$ are parallel.

3. E and F are the mid-points of the diagonals BD and AC of a quadrilateral $ABCD$ respectively. Show that

(*i*) $\vec{AB} + \vec{AD} + \vec{CB} + \vec{CD} = 4\vec{EF}$.

(*ii*) $\vec{AB} + \vec{BC} + \vec{CD} + \vec{AD} = 2\vec{AD}$.

4. Let $\vec{OA} = \mathbf{a}$ and $\vec{OB} = \mathbf{b}$ and $\vec{OC} = \mathbf{a} + \mathbf{b}$.

What is the type of the quadrilateral $OACB$?

5. OAB is a given triangle and P, Q, R are the mid-points of the sides OA, AB and BO respectively of the triangle. Express in terms of the vectors $\vec{OA}$ and $\vec{OB}$ the vectors $\vec{AB}, \vec{PA}, \vec{PQ}$ and $\vec{AR}$.

6. The mid-points of the sides OP and OQ of a triangle $O\ P\ Q$ are S and R respectively. Express the vectors $\vec{PQ}, \vec{SR}, \vec{SQ}$ and $\vec{RP}$ in terms of $\vec{OP}$ and $\vec{OQ}$.

Let SQ and RP meet at the point X and let

$$\frac{SX}{SQ} = m,\ \frac{RX}{RP} = n.$$

Show that

(*i*) $\vec{OX} = \mathbf{a} + m(2\mathbf{b} - \mathbf{a})$ (*ii*) $\vec{OX} = \mathbf{b} + n(2\mathbf{a} - \mathbf{b})$

where $\vec{OP} = 2\mathbf{a}$ and $\vec{OQ} = 2\mathbf{b}$.

Deduce the X divides SQ and RP in the ratio 1 : 2.

7. The diagonals AC, BD of a quadrilateral $ABCD$ meet at a point E. Given that $\overrightarrow{AB} = \mathbf{a}$, $\overrightarrow{BC} = \mathbf{b}$, and that E is a point of trisection of each of AC and BD nearer A as well as B, express in terms of **a** and **b** the vectors $\overrightarrow{BE}$ and $\overrightarrow{DC}$.

8. The point E is a point of trisection of the straight line PQ such that $PE : EQ = 1 : 2$. Also R is any point not on the line PQ and F divides QR internally such that $QF : FR = 2 : 1$. Show that EF is parallel to PR.

9. In the triangle OAB, L is the mid-point of OA and M is a point on OB such that $\dfrac{OM}{MB} = 2$. P is the mid-point of LM, and the line AP is produced to meet OB at Q. Given that $\overrightarrow{OA} = \mathbf{a}$ and $\overrightarrow{OB} = \mathbf{b}$, find in terms of **a** and **b**, the vectors

(*i*) $\overrightarrow{OP}$, (*ii*) $\overrightarrow{AP}$.

If $\overrightarrow{AQ} = \lambda\, \overrightarrow{AP}$ and $\overrightarrow{OQ} = \mu\, \overrightarrow{OB}$, find $\lambda, \mu, \dfrac{AP}{PQ}$ and $\dfrac{OQ}{QB}$.

10. Points X and Y are taken on the sides QR and RS, respectively, of a parallelogram $PQRS$, so that $\overrightarrow{QX} = 4\,\overrightarrow{XR}$ and $\overrightarrow{RY} = 4\,\overrightarrow{YS}$. The line XY cuts the line PR at Z. Prove that $\overrightarrow{PZ} = (21/25)\,\overrightarrow{PR}$.

11. The diagonals of a parallelogram $ABCD$ intersect at E. The position vector with respect to an origin O of A is **a**. Also $\overrightarrow{AB} = \mathbf{p}$ and $\overrightarrow{BC} = \mathbf{q}$. Determine the position vectors of the vertices B, C, D, E with respect to O in terms of **a, p** and **q**.

12. ABC is a triangle and the position vectors of the points A, B, C relative to the origin O are **a, b, c** respectively. The point P is on the side BC such that $BP : PC = 2 : 3$, and the point Q is on the side CA such that $CQ : QA = 1 : 4$. Find the position vector of the common point R of AP and BQ. Find also the ratio in which CR divides AB.

13. In a triangle OAB, $\overrightarrow{OA} = \mathbf{a}$ and $\overrightarrow{OB} = \mathbf{b}$. Point P and Q are taken so that $\overrightarrow{AP} = \mathbf{a}$ and $\overrightarrow{BQ} = 2\mathbf{b}$. Obtain expressions for the vectors $\overrightarrow{AQ}$ and and $\overrightarrow{BP}$ in terms of **a** and **b**. The lines AQ and BP meet at R and the line OR meets AB at S. Find the ratio in which S divides AB.

14. ABC is a triangle, D is the mid-point of AB and E is the point of trisection of BC nearer C. Show that the mid-point F of CD is on AE and find also the ratio $AF : FE$.

Let DE meet AC at L. Find the ratio in which L divides AC.

15. In a triangle OAB, L is a point on the side AB and M is a point on the side OB and the lines OL and AM meet at S.

It is given that $AS = SM$ and $4\,OS = 3\,OL$, and that

$$\frac{OM}{OB} = h \text{ and } \frac{AL}{AB} = k.$$

(*i*) Express the vectors $\vec{AM}$, and $\vec{OS}$ in terms of **a**, **b** and h and the vectors $\vec{OL}$ and $\vec{OS}$ in terms of **a**, **b** and k where $\vec{OA} = \mathbf{a}$ and $\vec{OB} = \mathbf{b}$. Find h and k.

(*ii*) BS meets OA at N. Find the ratios in which L, M and N divide AB, BO and OA respectively.

16. ABC is a triangle and P, Q are the mid-points of AB, AC respectively. If $\vec{AB} = 2\mathbf{a}$ and $\vec{AC} = 2\mathbf{b}$, express in terms of **a** and **b** the vectors (*i*) $\vec{BC}$, (*ii*) $\vec{PQ}$, (*iii*) $\vec{PC}$, (*iv*) $\vec{BQ}$. What can you deduce about the line segments BC and PQ ?

17. OAB is a triangle and X is the point of the line segment AB such that $\vec{OX} = 2\vec{XB}$. If $\vec{OA} = \mathbf{a}$ and $\vec{OB} = \mathbf{b}$, show that $\vec{OX} = \frac{1}{3}\mathbf{a} + \frac{2}{3}\mathbf{b}$. The line OX meets the line through A parallel to OB in Y. Find Y and OX/XY.

18. If $\vec{OA} = \mathbf{a}$, $\vec{OB} = \mathbf{b}$, $\vec{OC} = \mathbf{c}$, $\vec{OD} = \mathbf{d}$ and the points P, Q, R, S are such that

$$\vec{AP} = 2\,\vec{PB},\quad \vec{BQ} = 2\vec{QC},\quad \vec{CR} = 2\,\vec{RD},\quad \vec{DS} = 2\vec{SA}.$$

Express $\vec{PQ}$ and $\vec{SR}$ in terms of **a, b, c, d** and show that the condition for $P\ Q\ R\ S$ to be a parallelogram is $\mathbf{a} + \mathbf{c} = \mathbf{b} + \mathbf{d}$.

19. $OABCD$ is a pentagon in which the sides OA and CB are parallel, and the sides OD and AB are parallel. Also

$$\frac{OA}{CB} = 2 \text{ and } \frac{OD}{AB} = \frac{1}{3}.$$

Given that $\vec{OA} = \mathbf{a}$ and $\vec{OD} = \mathbf{d}$, express each of the vectors $\vec{AD}$, $\vec{OC}$ and $\vec{DC}$ in terms of **a** and **d**. If the diagonals OC and AD meet at X, find $\frac{OX}{XC}$ and $\frac{AX}{XD}$.

20. OAB is given triangle; X is a point on OA such that $\vec{OX} = 2\vec{OA}$ and Y is a point on OB, such that $\vec{OY} = \frac{2}{3}\vec{OB}$. The line XY meets the side AB of the triangle at P. Find the ratio in which P divides AB. Find also the ratio in which P divides YX.

21. In a triangle OAB, L is the point on OA such that $\frac{OL}{LA} = \frac{1}{2}$ and M is the mid-point of AB. N is the point in which the line LM meets the side OB. Find the ratio in which L divides MN.

22. Given a regular hexagon $ABCDEF$ with centre O, show that

(*i*) $\vec{OB} - \vec{OA} = \vec{OC} - \vec{OD}$,

(*ii*) $\vec{OD} + 2\vec{OA} = 2\vec{OB} + \vec{OF}$,

(*iii*) $\vec{AD} + \vec{EB} + \vec{FC} = 4\vec{AB}$.

23. The points O, A, B, C are the vertices of a pyramid and P, Q, R, S are the mid-points of OA, OB, BC, AC respectively.

If $\vec{OA} = \mathbf{a}, \ \vec{OB} = \mathbf{b}, \ \vec{OC} = \mathbf{c}$,

express in terms of **a, b, c** the vectors

(*i*) $\vec{OP}, \ \vec{OQ}, \ \vec{OR}$ and $\vec{OS}$.

(*ii*) $\vec{OG}$ and $\vec{OH}$ where G, H are the mid-points of PR, QS respectively.

State what *can be deduced* about PR and QS.

24. $ABCD$ is parallelogram whose diagonals intersect at E and M is the mid-point of DC. If

$$\vec{AB} = \mathbf{a} \text{ and } \vec{AD} = \mathbf{b},$$

express in terms of **a** and **b** the vectors

(*i*) $\vec{AE}$ (*ii*) $\vec{BD}$ (*iii*) $\vec{MB}$

25. In a quadrilateral $ABCD$, the point P divides DB in the ratio $1 : 2$ and Q is the mid-point of AC. Prove that

$$2\vec{DC} + \vec{BC} - 2\vec{AD} - \vec{AB} = 6\vec{PQ}.$$

26. $PQRS$ is a quadrilateral and

$$\vec{PQ} = \mathbf{a}, \ \vec{QR} = \mathbf{b}, \ \vec{SP} = \mathbf{a} - \mathbf{b},$$

M is the mid-point of QR and X is the point of SM such that

$$\vec{SX} = (4/5)\,\vec{SM}.$$

(*i*) Prove that PR and PX are parallel and find the ratio of their lengths.

(*ii*) Express in terms of **a** and **b** the vectors $\vec{PR}, \vec{SM}, \vec{SX}$ and $\vec{PX}$.

(*iii*) What can be deduced about the line segments PX, XR.

27. Given that O is a point inside the triangle ABC and that D, E and F are the mid-points of BC, CA and AB respectively. Show that

(*i*) $\vec{AD} + \vec{BE} + \vec{CE} = \mathbf{0}$,

(*ii*) $\vec{OD} + \vec{OE} + \vec{OF} = \vec{OA} + \vec{OB} + \vec{OC}$.

28. O, A, B and C are points such that

$$\vec{OA} = \mathbf{a}, \vec{OB} = \mathbf{b}, \vec{OC} = k\mathbf{a} + l\mathbf{b}.$$

where k and l are numbers. Express

$$\vec{AB}, \ \vec{AC} \text{ and } \vec{BC}$$

in terms of k, l, **a** and **b**, show that if $k + l = 1$, then AB, AC and BC are in the same direction. State the conclusion that follows from this, concerning the points A, B and C and express the ratio $AC : CB$ in terms of k and l.

X and Y are points such that

$$\vec{OX} = \frac{2}{3}\mathbf{a} \text{ and } \vec{OY} = 2\mathbf{b}.$$

Express $\vec{XY}$ in terms of **a** and **b** and hence show that the point Z, given by

$$\vec{OZ} = \vec{OX} + \frac{1}{4}\vec{XY}$$

is the mid-point of AB.

29. The triangles OAB, OCD are such that

$$\vec{OA} = \mathbf{a},\ \vec{OB} = \mathbf{b}, \vec{OC} = -3\mathbf{a} \text{ and } \vec{OD} = -3\mathbf{b}.$$

Express in terms of **a** and **b** the vectors $\vec{AB}, \vec{BC}, \vec{AD}$ and $\vec{DC}$ and state which of them are in parallel directions.

30. In a triangle OAB, L is the mid-point of OA and M is the point on OB such that $\dfrac{OM}{MB} = 2$. P is the mid-point of LM and the line AP is produced to meet OB at Q.

Given that $\vec{OA} = \mathbf{a}$ and $\vec{OB} = \mathbf{b}$ find as linear combinations of **a** and **b** the vectors $\vec{OP}$ and $\vec{AP}$.

If $\quad \vec{AQ} = h\vec{AP}$ and $\vec{OQ} = k\vec{OB}$,

find $\quad h, k, AP/PQ$ and OQ/QB.

31. The position vectors of points A, B, C relative to a fixed origin O are **a, b, c** respectively. If D is the mid-point of AB, and if E is the point which divides CB internally in the ratio $1 : 2$, write down the position vectors of D and E in terms of **a, b, c**.

Show that the mid-point F of CD is on AE and find the ratio $AF : FE$.

32. In the triangle OAB, M is the mid-point of AB, C is a point on OM such that $OC = \dfrac{1}{2}CM$ and X is a point on OB such that $OX = 2XB$. The line XC is produced to meet OA at Y. Find

$$\frac{OY}{YA} \text{ and } \frac{XC}{CY}$$

33. $OABC$ is a quadrilateral in which the diagonals OB and AC meet at X. A line drawn through B parallel to CA meets OA produced at D. Show that $\dfrac{BD}{CA} = \dfrac{3}{4}$ and $\dfrac{AD}{OA} = \dfrac{1}{2}$.

34. The vertices P, Q and S of a triangle PQS have position vectors **p, q** and **s** respectively.

(*i*) Find *m*, the position vector of M, the mid-point of PQ, in terms of **p** and **q**.

(*ii*) Find **t**, the position vector of T on SM such that $ST : TM = 2 : 1$, in terms of **p**, **q** and **s**.

(*iii*) If the parallelogram $PQRS$ is now completed. Express **r**, the position vector of the point R in terms of **p**, **q** and **s**.

Prove that P, T and R are collinear.

35. ABC is a triangle and O, any point in the plane of the same; AO, BO and CO meet the sides BC, CA and AB in D, E, F respectively; show that

$$\frac{OD}{AD}+\frac{OE}{BE}+\frac{OF}{CF}=1.$$

36. $ABCD$ is a tetrahedron and O is any point; the lines joining O to the vertices meet the opposite faces in P, Q, R, S respectively. Prove that

$$\frac{OP}{AP}+\frac{OQ}{BQ}+\frac{OR}{CR}+\frac{OS}{DS}=1.$$

Take O, as origin of reference.

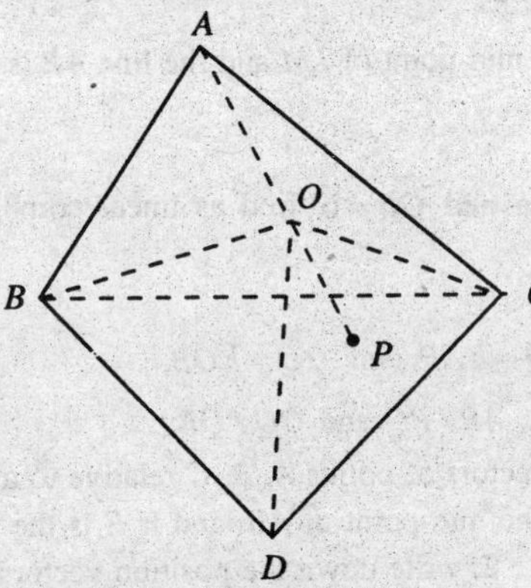

Let $\qquad \vec{OA}=\mathbf{a},\ \vec{OB}=\mathbf{b},\ \vec{OC}=\mathbf{c},\ \vec{OD}=\mathbf{d}.$

As every system of four vectors is linearly dependent, there is a linear relation connecting **a**, **b**, **c**, **d**.

Let this relation be

$$x\mathbf{a}+y\mathbf{b}+z\mathbf{c}+t\mathbf{d}=\mathbf{0}. \qquad ...(1)$$

We re-write (*i*) as

$$\frac{y\mathbf{b}+z\mathbf{c}+t\mathbf{d}}{y+z+t}=\frac{-x\mathbf{a}}{y+z+t}.$$

Now $(y\mathbf{b}+z\mathbf{c}+t\mathbf{d})\ /\ (y+z+t)$ is a point of the plane BCD, and $-x\mathbf{a}\ (y+z+t)$ is a point of the line OA. Thus, this must be the point P of intersection of the line OA with the plane BCD.

$$\therefore \qquad \vec{OP}=-\frac{x}{y+z+t}\mathbf{a} \quad\Rightarrow\quad \mathbf{a}=-\frac{y+z+t}{x}\vec{OP}.$$

Also $\qquad \vec{AP}=\vec{AO}+\vec{OP}$

$$=-\mathbf{a}-\frac{x}{y+z+t}\mathbf{a}=-\frac{\Sigma x}{y+z+t}\mathbf{a}$$

$$= -\frac{\Sigma x}{y+z+t} \cdot \frac{-(y+z+t)}{x} \vec{OP} = \frac{\Sigma x}{x} \vec{OP}$$

$$\Rightarrow \quad \frac{AP}{OP} = \frac{\Sigma x}{x} \Rightarrow \frac{OP}{AP} = \frac{x}{\Sigma x}$$

Similarly

$$\frac{OQ}{BQ} = \frac{y}{\Sigma x}, \frac{OR}{CR} = \frac{z}{\Sigma x}, \frac{OS}{DS} = \frac{t}{\Sigma x}$$

Hence

$$\frac{OP}{AP} + \frac{OQ}{BQ} + \frac{OR}{CR} + \frac{OS}{DS} = 1.$$

37. Any plane cuts the sides *AB, BC, CD, DA* of a skew quadrilateral in *P, Q, R, S* respectively, prove that

$$\frac{AP}{PB} \cdot \frac{BQ}{QC} \cdot \frac{CR}{RD} \cdot \frac{DS}{SA} = 1.$$

Let **b, c, d** be the position vectors of the points *B, C, D* respectively with reference to *A* as origin.

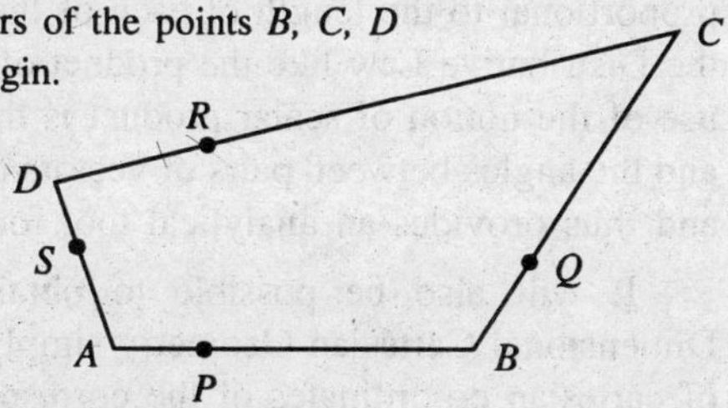

Let

$$\frac{AP}{PB} = \lambda_1, \frac{BQ}{QC} = \lambda_2$$

$$\frac{CR}{DR} = \lambda_3, \frac{DS}{SA} = \lambda_4,$$

so that the position vectors of the points *P, Q, R, S* are

$$\frac{\lambda_1 \mathbf{b}}{\lambda_1 + 1}, \frac{\lambda_2 \mathbf{c} + \mathbf{b}}{\lambda_2 + 1}, \frac{\lambda_2 \mathbf{d} + \mathbf{c}}{\lambda_3 + 1}, \frac{\mathbf{d}}{\lambda_2 + 1}$$

respectively.

Denoting these by **p, q, r, s** respectively, we proceed to find a relation between the same independent of **b, c, d.** Thus

$$(\lambda_2 + 1)\,\mathbf{q} - (\lambda_3 + 1)\,\lambda_2 \mathbf{r} = \mathbf{b} - \lambda_2 \lambda_3 \mathbf{d}$$

$$= \frac{\lambda_1 + 1}{\lambda_1} \mathbf{p} - \lambda_2 \lambda_3 (\lambda_4 + 1)\,\mathbf{s}$$

$$\Rightarrow \quad \frac{\lambda_1 + 1}{\lambda_1} \mathbf{p} - (\lambda_2 + 1)\,\mathbf{q} + (\lambda_3 + 1)\,\lambda_2 \mathbf{r} - \lambda_2 \lambda_3 (\lambda_4 + 1)\,\mathbf{s} = \mathbf{0}.$$

As, however, **p, q, r, s** are the position vectors of four coplanar points, we have

$$\frac{\lambda_1 + 1}{\lambda_1} - (\lambda_2 + 1) + (\lambda_2 + 1)\,\lambda_2 - \lambda_2 \lambda_2 (\lambda_4 + 1) = 0,$$

$$\Rightarrow \quad \lambda_1 \lambda_2 \lambda_3 \lambda_4 = 1.$$

Hence the result.

3

Scalar Product

Introduction. The concept of the *Scalar product* of two vectors, as a result of which a scalar is associated to any given pair of vectors, will be introduced in this chapter. The justification for the use of the word *Product* lies in the fact that the so-called scalar product of two vectors is a scalar proportional to the length of each of the two factor vectors and also obeys the Distributive Law like the product of numbers. Also one very important use of the notion of scalar product is that it enables the lengths of vectors and the angles between pairs of vectors to be expressed in terms of the same and thus provides an analytical tool for the study of *Metric Geometry.*

It will also be possible to obtain various formulae of the three Dimensional Cartesian Geometry simply as a result of translation in terms of cartesian co-ordinates of the corresponding Vector notation formulae.

Some well-known properties of tetrahedra amenable to treatment by scalar products will also be obtained.

3.1. SCALAR PRODUCT

3.1.1. Scalar Product of Two Vectors

Def. *The scalar product of two vectors* **a, b** *is the scalar*

$$|\mathbf{a}|\ |\mathbf{b}|\ \cos\theta$$

where, θ, *is the angle, between the vectors* **a** *and* **b.**

Also the scalar product of any vector with the zero vector is, by definition, the scalar zero.

The scalar product of the vectors **a, b** is denoted by the symbol

$$\mathbf{a} \cdot \mathbf{b}$$

and is, on this account, also called the *Dot product* of the two vectors **a, b.**

Thus, we have, by definition,

$$\mathbf{a} \cdot \mathbf{b} = |\mathbf{a}|\ |\mathbf{b}|\ \cos\theta;$$

θ being the angle between the vectors **a, b.**

Note. It may be easily seen that the scalar product of two vectors remains unaltered when they are replaced by vectors equal to the same so that

$$\mathbf{a} = \mathbf{a}' \text{ and } \mathbf{b} = \mathbf{b}' \Rightarrow \mathbf{a} \cdot \mathbf{b} = \mathbf{a}' \cdot \mathbf{b}'.$$

Ex. D is the mid-point of the side BC of a triangle ABC; show that

$$\overrightarrow{DA} \cdot \overrightarrow{DB} + \overrightarrow{DA} \cdot \overrightarrow{DC} = 0.$$

3.1.2. Sign of the Scalar Product

Def. *If* **a, b** *be two non-zero vectors, then the scalar product*

$$\mathbf{a} \cdot \mathbf{b} = |\mathbf{a}|\ |\mathbf{b}| \cos\theta$$

is positive, negative or zero, according as the angle, θ, between the vectors is acute, obtuse or right.

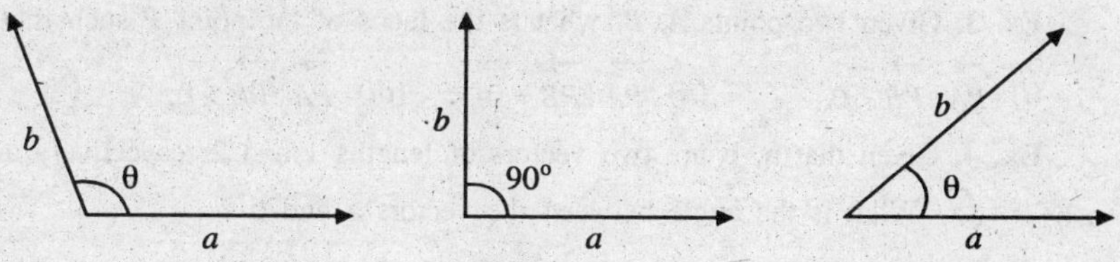

Fig. 3.1

Thus

$$\theta \text{ is acute} \Rightarrow \cos\theta > 0 \Rightarrow \mathbf{a} \cdot \mathbf{b} > 0,$$

$$\theta \text{ is right} \Rightarrow \cos\theta = 0 \Rightarrow \mathbf{a} \cdot \mathbf{b} = 0,$$

$$\theta \text{ is obtuse} \Rightarrow \cos\theta < 0 \Rightarrow \mathbf{a} \cdot \mathbf{b} < 0.$$

We notice that if **a, b** be two vectors, then their scalar product will be zero if and only if, either one at least of the two vectors is the zero vector or the two vectors are at right angles to each other. Thus

$$\mathbf{a} \cdot \mathbf{b} = 0 \Rightarrow \mathbf{a} = \mathbf{0} \text{ or } \mathbf{b} = \mathbf{0} \text{ or } \theta = 90^\circ;$$

θ being the angle between the vectors **a, b.**

It is very important to remember that the *scalar product of two non-zero vectors is zero if and only if they are at right angles to each other.*

3.1.3. Length of a Vector as a Scalar Product

If **a** be any vector, then the scalar product **a.a** of **a** with itself is given by

$$\mathbf{a} \cdot \mathbf{a} = |\mathbf{a}|\ |\mathbf{a}| \cos 0 = |\mathbf{a}|^2.$$

Thus, the length $|\mathbf{a}|$ *of any vector,* **a**, *is the non-negative square root* $\sqrt{(\mathbf{a} \cdot \mathbf{a})}$ *of the scalar product* **a.a.**

A convention. a.a will be denoted as $\mathbf{a}^2$, so that $\mathbf{a}^2$, is a scalar which equals the square of the length of **a.**

3.1.4. Angle between Two Vectors in terms of Scalar Products

If θ be the angle between two non-zero vectors, **a, b,** we have

$$\mathbf{a} \cdot \mathbf{b} = |\mathbf{a}| \; |\mathbf{b}| \cos\theta,$$

$$\Rightarrow \qquad \theta = \cos^{-1}\frac{\mathbf{a}\cdot\mathbf{b}}{|\mathbf{a}|\,|\mathbf{b}|} = \cos^{-1}\frac{\mathbf{a}\cdot\mathbf{b}}{(+\sqrt{\mathbf{a}\cdot\mathbf{a}})(+\sqrt{(\mathbf{b}\cdot\mathbf{b})})}.$$

Ex. 1. Is it true that

(*i*) $\mathbf{a} \cdot \mathbf{b} = \mathbf{a} \cdot \mathbf{c} \;\Rightarrow\; \mathbf{b} = \mathbf{c}$? (*ii*) $\mathbf{a} \cdot \mathbf{b} = 0 \;\Rightarrow\; \mathbf{a} \perp \mathbf{b}$?

Ex. 2. Give two points *O, A*; identify the locus of the point *P* in each of the following cases :

(*i*) $\vec{OP} \cdot \vec{OA} > 0.$ (*ii*) $\vec{OP} \cdot \vec{OA} = 0.$ (*iii*) $\vec{OP} \cdot \vec{OA} < 0.$

Ex. 3. Given two points *A, B,* what is the locus of the point *P* such that

(*i*) $\vec{PA} \cdot \vec{PB} < 0.$ (*ii*) $\vec{PA} \cdot \vec{PB} = 0.$ (*iii*) $\vec{PA} \cdot \vec{PB} > 0.$

Ex. 4. Given that **a, b** are two vectors of lengths 1 and 2 respectively and $\mathbf{a}.\mathbf{b} = -\sqrt{3}$. What is the angle between the vectors **a** and **b** ?

Properties of Scalar Product. We shall now obtain some basic properties of scalar product.

3.1.5. Commutativity

$$\mathbf{a} \cdot \mathbf{b} = \mathbf{b} \cdot \mathbf{a},$$

for every pair of vectors **a, b.**

This property is obvious from the definition.

3.1.6. $\mathbf{a} \cdot (-\mathbf{b}) = -(\mathbf{a} \cdot \mathbf{b})$; $(-\mathbf{a}) \cdot (-\mathbf{b}) = \mathbf{a} \cdot \mathbf{b}$ *for every pair of vectors,* **a, b.**

The proof is simple. [Refer Fig. 3.2]

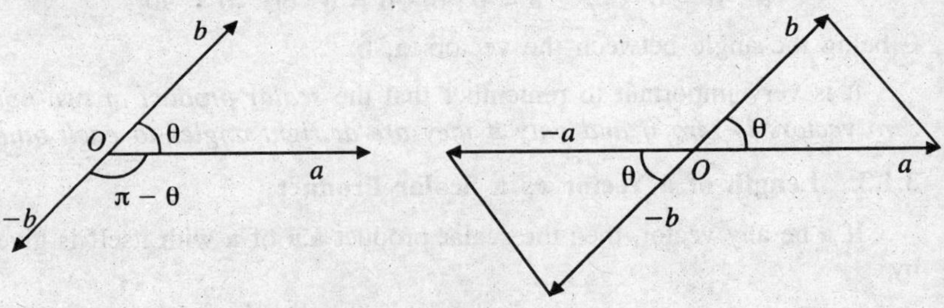

Fig. 3.2

3.1.7. $m\mathbf{a} \cdot n\mathbf{b} = mn\,(\mathbf{a}, \mathbf{b})$ *where* **a . b** *are any vectors and m, n any scalars.*

Let *m, n* be both positive, so that the angle between *m***a** and *n***b** is the same as that between **a** and **b**.

We have

$$ma \cdot nb = |ma|\ |nb|\ \cos\theta$$
$$= |m|\ |a|\ |n|\ |b|\ \cos\theta$$
$$= mn\ |a|\ |b|\cos\theta = mn\ (a \cdot b).$$

Let now m be positive and n be negative so that the angle between m**a** and n**b** is $\pi - \theta$; θ being the angle between **a** and **b.**

We have

$$ma \cdot nb = |ma|\ |nb|\ \cos(\pi - \theta)$$
$$= |m|\ |a|\ |n|\ |b|\ (-\cos\theta)$$
$$= -mn\ |a|\ |b|\ (-\cos\theta)$$
$$= mn\ |a|\ |b|\cos\theta = mn\ (a \cdot b).$$

Other cases may be similarly disposed of.

3.1.8. *If the scalar product of a vector,* **r**, *with each of three* **non-coplanar** *vectors is zero, then,* **r**, *must be the zero vector.*

This follows from the fact that no non-zero vector can be perpendicular to each of three non-coplanar vectors.

In particular, if a vector, **r**, is perpendicular to every vector, then **r** must be the zero vector.

3.1.9. *Let* **b** $= \overrightarrow{PQ}$ *and let L, M be the feet of the perpendiculars for P and Q on the support AB of the vector* **a.**

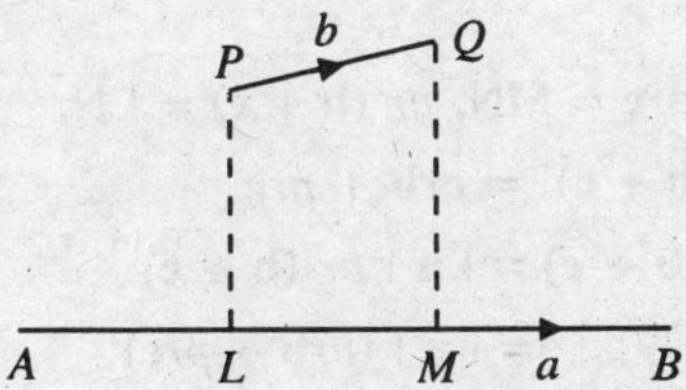

Fig. 3.3

The projection of **b** *upon AB* $= LM = |\mathbf{b}|\cos\theta$.

Thus, the scalar product **a . b** *is the product of the length of* **a**, *with the projection of* **b**, *upon* **a**, *taken with the proper sign.*

3.1.10. Distributivity

Scalar multiplication of vectors distributes the addition of vectors.

We have to prove that

$$\mathbf{a} \cdot (\mathbf{b} + \mathbf{c}) = \mathbf{a} \cdot \mathbf{b} + \mathbf{a} \cdot \mathbf{c}$$

for all vectors **a, b, c.**

This property is an immediate consequence of the fact that *the projection of the sum of two vectors on any line is equal to the sum of their projections on the same line.*

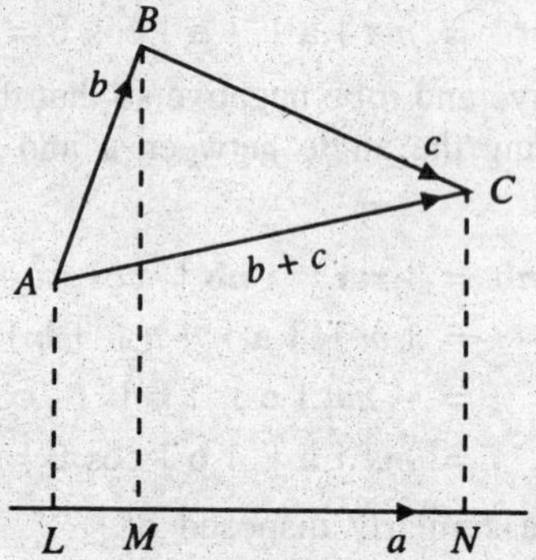

Fig. 3.4

Let

$$\vec{AB} = \mathbf{b},\ \vec{BC} = \mathbf{c} \text{ so that } \vec{AC} = \mathbf{b}+\mathbf{c}.$$

Let *L, M, N* be the feet of the perpendiculars from *A, B, C* on the line of support of the vector **a**.

Let

$$pr\mathbf{b},\ pr\mathbf{c},\ pr\,(\mathbf{b}+\mathbf{c})$$

denote the projections of **b, c** and **b + c** on **a**.

We have

$$pr\mathbf{b} = \text{LM},\ pr\mathbf{c} = \text{MN},\ pr\,(\mathbf{b}+\mathbf{c}) = \text{LN},$$

so that $pr\,(\mathbf{b}+\mathbf{c}) = pr\mathbf{b} + pr\mathbf{c}$

Thus
$$\begin{aligned}\mathbf{a}\,.\,(\mathbf{b}+\mathbf{c}) &= |\,\mathbf{a}\,|\,pr\,(\mathbf{b}+\mathbf{c})\\ &= |\,\mathbf{a}\,|\,(pr\mathbf{b}+pr\mathbf{c})\\ &= |\,\mathbf{a}\,|\,pr\mathbf{b} + |\,\mathbf{a}\,|\,pr\mathbf{c} = \mathbf{a}\,.\,\mathbf{b} + \mathbf{a}\,.\,\mathbf{c}.\end{aligned}$$

It may also be shown that

$$\mathbf{a}\,.\,(\mathbf{b}-\mathbf{c}) = \mathbf{a}\,.\,\mathbf{b} - \mathbf{a}\,.\,\mathbf{c}.$$

We have

$$\begin{aligned}\mathbf{a}\,.\,(\mathbf{b}-\mathbf{c}) &= \mathbf{a}\,.\,[\mathbf{b}+(-\mathbf{c})]\\ &= \mathbf{a}\,.\,\mathbf{b} + \mathbf{a}\,.\,(-\mathbf{c})\\ &= \mathbf{a}\,.\,\mathbf{b} + [-(\mathbf{a}\,.\,\mathbf{c})] = \mathbf{a}\,.\,\mathbf{b} - \mathbf{a}\,.\,\mathbf{c}.\end{aligned}$$

3.1.11. Some Simple Identities

(*i*) $(\mathbf{a}+\mathbf{b})\,.\,(\mathbf{a}-\mathbf{b}) = \mathbf{a}\,.\,\mathbf{a} - \mathbf{b}\,.\,\mathbf{b} = \mathbf{a}^2 - \mathbf{b}^2,$

(*ii*) $(\mathbf{a} + \mathbf{b})^2 = (\mathbf{a} + \mathbf{b}) \cdot (\mathbf{a} + \mathbf{b})$

$= \mathbf{a} \cdot \mathbf{a} + 2\mathbf{a} \cdot \mathbf{b} + \mathbf{b} \cdot \mathbf{b} = \mathbf{a}^2 + 2\mathbf{a} \cdot \mathbf{b} + \mathbf{b}^2,$

(*iii*) $(\mathbf{a} - \mathbf{b})^2 = (\mathbf{a} - \mathbf{b}) \cdot (\mathbf{a} - \mathbf{b})$

$= \mathbf{a} \cdot \mathbf{a} - 2\mathbf{a} \cdot \mathbf{b} + \mathbf{b} \cdot \mathbf{b} = \mathbf{a}^2 - 2\mathbf{a} \cdot \mathbf{b} + \mathbf{b}^2.$

These identities are simple consequences of the fact that the scalar multiplication distributes the addition of vectors and of the other results in § 19.

We may also note that

$$\mathbf{a} \cdot \mathbf{b} = \frac{1}{4}(|\mathbf{a}+\mathbf{b}|^2 - |\mathbf{a}-\mathbf{b}|^2),$$

so that the scalar product of two vectors is expressed in terms of the lengths of their sum and difference.

APPLICATIONS

EXAMPLES

Example 1. Cosine Formula for Triangles. *To prove that for any triangle ABC*

$$c^2 = b^2 + a^2 - 2ab \cos C, \qquad \text{(Kolkata 2005)}$$

in the usual notation of plane Trigonometry.

Solution. We have

$$\vec{AC} + \vec{CB} = \vec{AB}$$

$$\Rightarrow \quad (\vec{AC} + \vec{CB}) \cdot (\vec{AC} + \vec{CB}) = \vec{AB} \cdot \vec{AB}$$

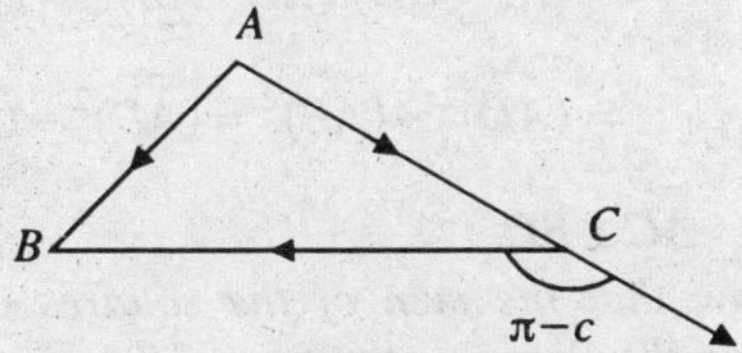

Fig 3.5

$$\Rightarrow \quad AC^2 + CB^2 + 2\vec{AC} \cdot \vec{CB} = AB^2$$

$$\Rightarrow \quad b^2 + a^2 - 2ab \cos C = c^2.$$

Example 2. Projection Formula for Triangles. *To prove that for any triangle ABC*

$$c = b \cos A + a \cos B.$$

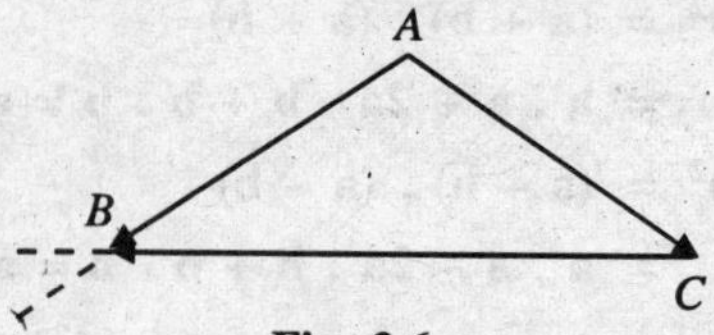

Fig. 3.6

Solution. We have

$$\vec{AC} + \vec{CB} = \vec{AB}$$

$$\Rightarrow \quad \vec{AC} \cdot \vec{AB} + \vec{CB} \cdot \vec{AB} = \vec{AB} \cdot \vec{AB}$$

$$\Rightarrow \quad bc \cos A + ca \cos B = c^2$$

$$\Rightarrow \quad b \cos A + a \cos B = c.$$

Example 3. *Show that the diagonals of a rhombus are at right angles.*

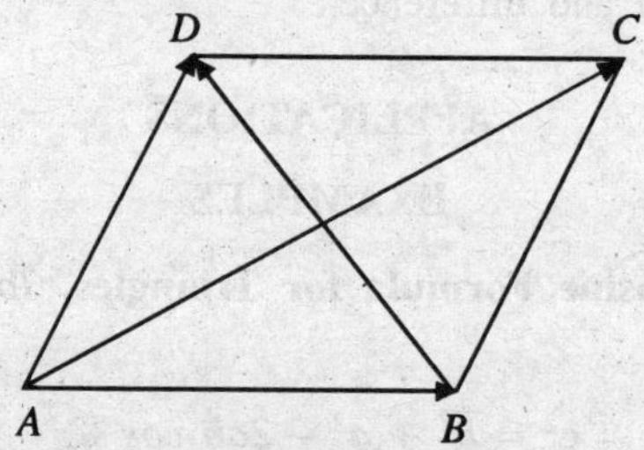

Fig. 3.7

Solution. Let *ABCD* be a rhombus.

We have

$$\vec{AC} \cdot \vec{BD} = (\vec{AB} + \vec{BC}) \cdot (\vec{AD} - \vec{AB})$$

$$= (\vec{AB} + \vec{AD}) \cdot (\vec{AD} - \vec{AB})$$

$$= (\vec{AD})^2 - (\vec{AB})^2 = (\vec{AD})^2 - (\vec{AD})^2 = 0$$

$$\Rightarrow \quad \vec{AC} \perp \vec{BD}.$$

Example 4. *Show that the sum of the squares of the sides of any quadrilateral (plane or skew) equals the sum of the squares of its diagonals together with four times the square of the line joining their middle points.*

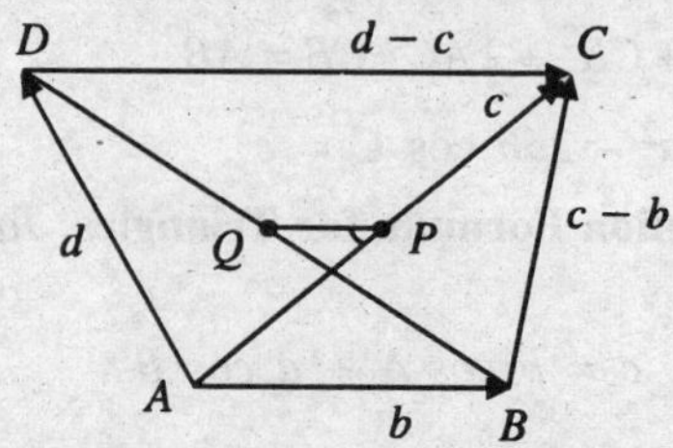

Fig. 3.8

Solution. Take the vertex A as origin.

Let **b, c, d** be the position vectors of the vertices B, C, D.

The position vectors of the mid-points P, Q of the diagonals AC, BD are $\frac{1}{2}\mathbf{c}$, $\frac{1}{2}(\mathbf{b}+\mathbf{d})$ respectively. We have

$$AB^2 + BC^2 + CD^2 + DA^2 = \mathbf{b}^2 + (\mathbf{c} - \mathbf{b})^2 + (\mathbf{d} - \mathbf{c})^2 + \mathbf{d}^2$$
$$= 2(\mathbf{b}^2 + \mathbf{c}^2 + \mathbf{d}^2) - 2(\mathbf{b.c} + \mathbf{c.d})$$

Also $$AC^2 + BD^2 + 4PQ^2 = \mathbf{c}^2 + (\mathbf{d} - \mathbf{b})^2 + 4\left[\frac{1}{2}(\mathbf{b}+\mathbf{d}) - \frac{1}{2}\mathbf{c}\right]^2$$
$$= 2(\mathbf{b}^2 + \mathbf{c}^2 + \mathbf{d}^2) - 2(\mathbf{b.c} + \mathbf{c.d}).$$

Hence the result.

Example 5. *D is the mid-point of the side BC of a triangle ABC; show that*

$$AB^2 + AC^2 = 2(AD^2 + BD^2).$$

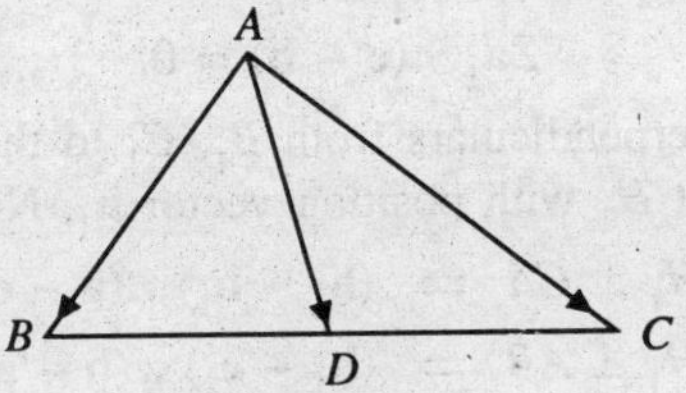

Fig. 3.9

Solution. We have

$$\vec{AB} = \vec{AD} + \vec{DB}$$

$\Rightarrow$ $$AB^2 = (\vec{AD} + \vec{DB})^2$$
$$= AD^2 + DB^2 + 2\vec{AD}.\vec{DB}. \quad ...(i)$$

Also we have

$$\vec{AC} = \vec{AD} = \vec{DC}$$

$\Rightarrow$ $$AC^2 = (\vec{AD} + \vec{DC})^2$$
$$= AD^2 + DC^2 + 2\vec{AD}.\vec{DC}. \quad ...(ii)$$

Adding (*i*) and (*ii*) we get

$$AB^2 + AC^2 = 2AD^2 + 2BD^2 + 2\vec{AD}.(\vec{DB} + \vec{DC})$$
$$= 2(DA^2 + DB^2), \text{ for } \vec{DB} + \vec{DC} = \mathbf{0}.$$

Example 6. *ABC and $A_1B_1C_1$ are two coplanar triangles such that the perpendiculars from A, B, C to the sides B_1C_1, C_1A_1, A_1B_1 of the triangle $A_1B_1C_1$ are concurrent; show that the perpendiculars from A_1, B_1, C_1 to the sides BC, CA, AB of the triangle ABC are also concurrent.*

Solution. Let the position vectors of the vertices of two triangles with respect to any origin *O* of reference be

$$\mathbf{a}, \mathbf{b}, \mathbf{c}; \mathbf{a}_1, \mathbf{b}_1, \mathbf{c}_1.$$

Let the perpendiculars from *A, B, C* to the sides of the second triangle concur at a point *H* with position vector **h**.

Now

$$AH \perp B_1C_1 \Rightarrow (\mathbf{h} - \mathbf{a}) \cdot (\mathbf{c}_1 - \mathbf{b}_1) = 0,$$
$$BH \perp C_1A_1 \Rightarrow (\mathbf{h} - \mathbf{b}) \cdot (\mathbf{a}_1 - \mathbf{c}_1) = 0,$$
$$CH \perp A_1B_1 \Rightarrow (\mathbf{h} - \mathbf{c}) \cdot (\mathbf{b}_1 - \mathbf{a}_1) = 0.$$

Adding we obtain

$$\Sigma (\mathbf{h} - \mathbf{a}) \cdot (\mathbf{c}_1 - \mathbf{b}_1) = 0$$
$$\Rightarrow \quad \Sigma \mathbf{a} \cdot (\mathbf{c}_1 - \mathbf{b}_1) = 0$$
$$\Rightarrow \quad \Sigma \mathbf{a}_1 \cdot (\mathbf{c} - \mathbf{b}) = 0. \qquad ...(i)$$

Again, let the perpendiculars from B_1, C_1 to the sides *CA, AB* of the triangle *ABC* meet at H_1 with position vector $\mathbf{h}_1$. Now

$$B_1H_1 \perp CA \Rightarrow (\mathbf{h}_1 - \mathbf{b}_1) \cdot (\mathbf{a} - \mathbf{c}) = 0 \qquad ...(ii)$$
$$C_1H_1 \perp AB \Rightarrow (\mathbf{h}_1 - \mathbf{c}_1) \cdot (\mathbf{b} - \mathbf{a}) = 0. \qquad ...(iii)$$

Adding (*i*), (*ii*) and (*iii*), we get

$$(\mathbf{h}_1 - \mathbf{a}_1) \cdot (\mathbf{c} - \mathbf{b}) = 0$$
$$\Rightarrow \quad A_1H_1 \perp BC.$$

Hence the result.

Example 7. *Show that the circumcentre, the centroid and the orthocentre of a triangle are collinear and the centroid divides the join of the circumcentre and the orthocentre in the ratio 1 : 2.*

Solution. Let *O, G, H* denote the circumcentre, centroid and orthocentre respectively of a triangle *ABC.*

Let **a, b, c** be the position vectors of the vertices *A, B, C* of the triangle with respect to the circumcentre *O*, as the origin of reference.

Fig. 3.10

We have

$$OA = OB = OC$$
$$\Rightarrow \quad \mathbf{a}^2 = \mathbf{b}^2 = \mathbf{c}^2 \qquad ...(i)$$

Also the position vector of G is

$$\vec{OG} = \frac{1}{3}(\mathbf{a}+\mathbf{b}+\mathbf{c}).$$

We have, by (*i*)

$$[(\mathbf{a} + \mathbf{b} + \mathbf{c}) - \mathbf{a}] \cdot (\mathbf{b} - \mathbf{c}) = 0,$$

$$[(\mathbf{a} + \mathbf{b} + \mathbf{c}) - \mathbf{b}] \cdot (\mathbf{c} - \mathbf{a}) = 0,$$

$$[(\mathbf{a} + \mathbf{b} + \mathbf{c}) - \mathbf{c}] \cdot (\mathbf{a} - \mathbf{b}) = 0.$$

so that if H' denotes the point with position vector

$$\mathbf{a} + \mathbf{b} + \mathbf{c},$$

we see that

$$H'A \perp BC, \quad H'B \perp CA, \quad H'C \perp AB.$$

so that H is the orthocentre of the triangle ABC

and we have $\qquad \vec{OH} = \mathbf{a}+\mathbf{b}+\mathbf{c}.$

Thus, we have

$$\vec{OH} = 3\vec{OG}$$

$\Rightarrow$ G divides OH in the ratio 1 : 2.

Hence the result.

Example 8. *If two pairs of opposite edges of a tetrahedron are at right angles, then show that the third pair is also at right angle. Further show that for such a tetrahedron, the sum of the squares of each pair of opposite edges is the same.*

Solution. Let $OABC$ be a given tetrahedron such that

$$OA \perp BC \text{ and } OB \perp CA.$$

Let

$$\vec{OA} = \mathbf{a}, \quad \vec{OB} = \mathbf{b}, \quad \vec{OC} = \mathbf{c}.$$

Now

$$OA \perp BC \;\Rightarrow\; \vec{OA} \cdot \vec{BC} = 0$$

$\Rightarrow \qquad \mathbf{a} \cdot (\mathbf{c} - \mathbf{b}) = 0 \;\Rightarrow\; \mathbf{a} \cdot \mathbf{c} = \mathbf{a} \cdot \mathbf{b} \qquad$ ...(*i*)

Also

$$OB \perp CA \;\Rightarrow\; \vec{OB} \cdot \vec{CA} = 0$$

$\Rightarrow \qquad \mathbf{b} \cdot (\mathbf{a} - \mathbf{c}) = 0 \;\Rightarrow\; \mathbf{b} \cdot \mathbf{a} = \mathbf{b} \cdot \mathbf{c} \qquad$ ...(*ii*)

From the results (*i*) and (*ii*), we have

$$\mathbf{a} \cdot \mathbf{c} = \mathbf{a} \cdot \mathbf{b} = \mathbf{b} \cdot \mathbf{c}$$

$$\mathbf{a} \cdot \mathbf{c} = \mathbf{b} \cdot \mathbf{c} \quad \Rightarrow \quad (\mathbf{a} - \mathbf{b}) \cdot \mathbf{c} = 0. \qquad ...(iii)$$

$$\Rightarrow \qquad \vec{BA} \cdot \vec{OC} = 0 \Rightarrow OC \perp AB.$$

Again

$$OA^2 + BC^2 = (\vec{OA})^2 + (\vec{BC})^2$$
$$= \mathbf{a}^2 + (\mathbf{c} - \mathbf{b})^2 = \mathbf{a}^2 + \mathbf{b}^2 + \mathbf{c}^2 - 2\mathbf{b} \cdot \mathbf{c}.$$

Similarly, we have

$$OB^2 + CA^2 = \mathbf{a}^2 + \mathbf{b}^2 + \mathbf{c}^2 - 2\mathbf{c} \cdot \mathbf{a},$$
$$OC^2 + AB^2 = \mathbf{a}^2 + \mathbf{b}^2 + \mathbf{c}^2 - 2\mathbf{a} \cdot \mathbf{b}.$$

Thus, we have

$$OA^2 + BC^2 = OB^2 + CA^2 = OC^2 + AB^2.$$

for $\mathbf{a} \cdot \mathbf{b} = \mathbf{b} \cdot \mathbf{c} = \mathbf{c} \cdot \mathbf{a}$.

Example 9. *If the perpendiculars from two vertices B and C to the opposite faces of a tetrahedron ABCD intersect, then BC is perpendicular to AD and the perpendicular from A and D to the opposite faces also intersect.*

Solution. Take A as the origin of reference. Let

$$\vec{AB} = \mathbf{b}, \ \vec{AC} = \mathbf{c}, \ \vec{AD} = \mathbf{d}.$$

Let the perpendiculars from B and C to the opposite faces ACD and ABD meet at a point H whose position vector is $\mathbf{h}$. Thus, we have

$$\vec{BH} \perp \vec{AC} \Rightarrow (\mathbf{h} - \mathbf{b}) \cdot \mathbf{c} = 0; \ \vec{BH} \perp \vec{AD} \Rightarrow (\mathbf{h} - \mathbf{b}) \cdot \mathbf{d} = 0,$$

$$\vec{CH} \perp \vec{AB} \Rightarrow (\mathbf{h} - \mathbf{c}) \cdot \mathbf{b} = 0; \ \vec{CH} \perp \vec{AD} \Rightarrow (\mathbf{h} - \mathbf{c}) \cdot \mathbf{d} = 0.$$

These give

$$\begin{cases} \mathbf{h} \cdot \mathbf{c} = \mathbf{b} \cdot \mathbf{c}, & \mathbf{h} \cdot \mathbf{d} = \mathbf{b} \cdot \mathbf{d} \\ \mathbf{h} \cdot \mathbf{b} = \mathbf{b} \cdot \mathbf{c}, & \mathbf{h} \cdot \mathbf{d} = \mathbf{c} \cdot \mathbf{d} \end{cases}$$

$$\Rightarrow \qquad \mathbf{b} \cdot \mathbf{d} = \mathbf{c} \cdot \mathbf{d} \quad \Rightarrow \quad (\mathbf{b} - \mathbf{c}) \cdot \mathbf{d} = 0 \quad \Rightarrow \quad BC \perp AD.$$

Let K be the foot of the perpendicular from A to the opposite face BCD and let $\mathbf{k}$ be its position vector. As AK is perpendicular to the plane BCD, we have

$$\mathbf{k} \cdot (\mathbf{c} - \mathbf{b}) = 0, \quad \mathbf{k} \cdot (\mathbf{b} - \mathbf{d}) = 0$$

$$\Rightarrow \qquad \mathbf{k} \cdot \mathbf{c} = \mathbf{k} \cdot \mathbf{b} = \mathbf{k} \cdot \mathbf{d}.$$

We shall show that there is a point L on AK such that DL is perpendicular to the face ABC. If $\lambda\mathbf{k}$ be the position vector of this point, the two equations

$$(\lambda\mathbf{k} - \mathbf{d}) \cdot \mathbf{b} = 0, \quad (\lambda\mathbf{k} - \mathbf{d}) \cdot \mathbf{c} = 0$$

must be consistent in relation to λ.

These equations are really the same for

$$\mathbf{k} \cdot \mathbf{b} = \mathbf{k} \cdot \mathbf{c} \quad \text{and} \quad \mathbf{b} \cdot \mathbf{d} = \mathbf{c} \cdot \mathbf{d},$$

so that λ is determined.

Hence the result.

Example 10. *Prove that the angle in a semi-circle is a right angle.*

Solution. Let O be the centre and AB the bounding diameter of the semi-circle. Let P be any point on the circumference. With O as origin, let $\vec{OA} = \mathbf{a}$, $\vec{OB} = -\mathbf{a}$ and $\vec{OP} = \mathbf{r}$.

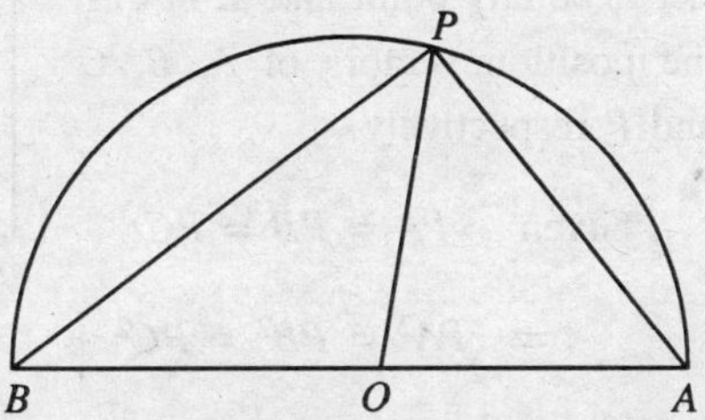

Fig. 3.11

Obviously, $OA = OB = OP$, each being equal to radius of the semi-circle.

$$\vec{AP} = \mathbf{r} - \mathbf{a} \quad \text{and} \quad \vec{BP} = \mathbf{r} - (-\mathbf{a}) = \mathbf{r} + \mathbf{a}$$

$$\therefore \qquad \vec{AP} \cdot \vec{BP} = (\mathbf{r} - \mathbf{a}) \cdot (\mathbf{r} + \mathbf{a}) = \mathbf{r}^2 - \mathbf{a}^2$$

$$= OP^2 - OA^2 = 0$$

$\Rightarrow$ AP and BP are perpendicular to each other, *i.e.*, $\angle APB = 90°$.

Example 11. *The base BC of a ΔABC is divided at D so that $mBD = nCD$. Show that $mAB^2 + nAC^2 = mBD^2 + nCD^2 + (m + n) AD^2$.*

Solution. With A as origin, let the position vectors of B and C be $\mathbf{b}$ and $\mathbf{c}$.

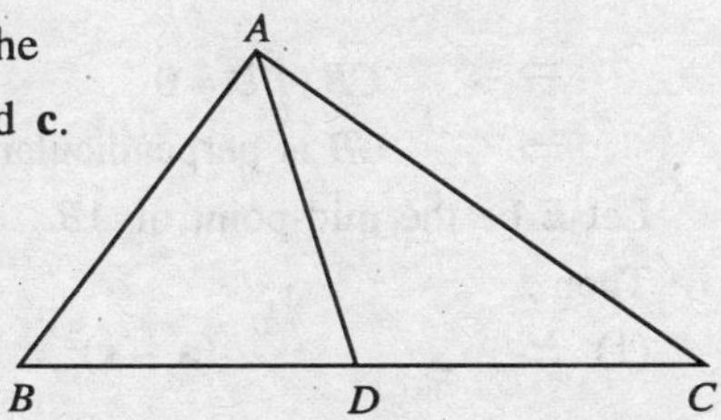

Fig. 3.12

$$m\vec{BD} = n\vec{DC}$$

$$\Rightarrow \qquad m\vec{DB} + n\vec{DC} = 0$$

Now,

$$AB^2 = (\vec{AB})^2 = (\vec{AD} + \vec{DB})^2 = AD^2 + DB^2 + 2\vec{AD} \cdot \vec{DB}$$

$$AC^2 = (\vec{AC})^2 = (\vec{AD} + \vec{DC})^2 = AD^2 + DC^2 + 2\vec{AD} \cdot \vec{DC}$$

Multiplying by m and n respectively and adding, we get

$$mAB^2 + nAC^2 = (m + n) AD^2 + mBD^2 + nDC^2 + 2\vec{AD} \cdot (m\vec{DB} + n\vec{DC})$$

$$= (m + n) AD^2 + mBD^2 + nDC^2.$$

Example 12. *Prove that if a point is equidistant from the vertices of a right angled triangle, its joint to the mid-point of the hypotenuse is perpendicular to the plane of the triangle.*

Solution. ABC be a right angled triangle and let D be the mid-point of its hypotenuse BC. Let P be any point and **a, b, c, r** the position vectors of A, B, C and P respectively.

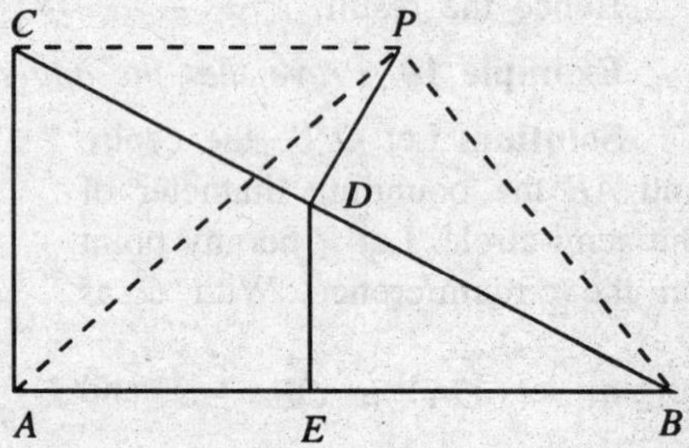

Fig. 3.13

Given $PA = PB = PC$

$\Rightarrow PA^2 = PB^2 = PC^2$

$$\Rightarrow \quad (\mathbf{a} - \mathbf{r})^2 = (\mathbf{b} - \mathbf{r})^2 = (\mathbf{c} - \mathbf{r})^2 \quad \ldots(1)$$

$$(1) \Rightarrow \quad (\mathbf{b} - \mathbf{r}) \cdot (\mathbf{b} - \mathbf{r}) = (\mathbf{c} - \mathbf{r}) \cdot (\mathbf{c} - \mathbf{r})$$

$$\Rightarrow \quad \mathbf{b}^2 + \mathbf{r}^2 - 2\mathbf{b} \cdot \mathbf{r} = \mathbf{c}^2 + \mathbf{r}^2 - 2\mathbf{c} \cdot \mathbf{r}$$

$$\Rightarrow \quad \mathbf{b}^2 - \mathbf{c}^2 - 2(\mathbf{r} \cdot \mathbf{b} - \mathbf{r} \cdot \mathbf{c}) = 0$$

$$\Rightarrow \quad (\mathbf{b} - \mathbf{c}) \cdot (\mathbf{b} + \mathbf{c}) - 2\mathbf{r} \cdot (\mathbf{b} - \mathbf{c}) = 0$$

$$\Rightarrow \quad (\mathbf{b} - \mathbf{c}) \cdot [(\mathbf{b} + \mathbf{c}) - 2\mathbf{r}] = 0$$

$$\Rightarrow \quad 2(\mathbf{b} - \mathbf{c}) \cdot \left[\frac{\mathbf{b} + \mathbf{c}}{2} - \mathbf{r}\right] = 0$$

$$\Rightarrow \quad \vec{CB} \cdot \vec{PD} = 0$$

$\Rightarrow$ CB is perpendicular to PD. ...(1)

Let E be the mid-point of AB.

Then

$$(1) \Rightarrow \quad (\mathbf{a} - \mathbf{r})^2 = (\mathbf{b} - \mathbf{r})^2$$

$$\Rightarrow \quad 2(\mathbf{a} - \mathbf{b}) \cdot \left(\frac{\mathbf{a} + \mathbf{b}}{2} - \mathbf{r}\right) = 0$$

$$\Rightarrow \quad 2\vec{BA} \cdot \vec{PE} = 0$$

$\Rightarrow$ BA is perpendicular to PE. ...(2)

Since D and E are mid-points of BC and AB respectively, DE must be parallel to AC. Again, AC is perpendicular to BA.

Hence, DE is perpendicular to BA, *i.e.*,

$$\vec{AB} \cdot \vec{DE} = 0 \quad \ldots(3)$$

Now,

$$\vec{AD} = \vec{AB} + \vec{BD}$$

$$\Rightarrow \quad \vec{AD} \cdot \vec{DP} = (\vec{AB} + \vec{BD}) \cdot (\vec{DP})$$

$$= \vec{AB} \cdot \vec{DP} + \vec{BD} \cdot \vec{DP}$$

$$= \vec{AB} \cdot \vec{DP} + \frac{1}{2}\vec{BC} \cdot \vec{DP}$$

[$\because$ D is mid-point of BC]

$$= \vec{AB} \cdot \vec{DP} \qquad \text{[by (1)]}$$

$$= \vec{AB} \cdot (\vec{DE} + \vec{EP})$$

$$= \vec{AB} \cdot \vec{DE} + \vec{AB} \cdot \vec{EP} = 0, \qquad \text{[by (2) and (3)]}$$

$\Rightarrow$ DP is perpendicular to AD. ...(4)

From (1) and (4), it follows that PD is perpendicular to both CB and AD and hence, PD must be perpendicular to the plane of ΔABC.

EXERCISES

1. Prove that the altitudes of a triangle are concurrent.
2. Prove that the right bisectors of the sides of a triangle are concurrent.
3. Show that the mid-point of the hypotenuse of a right angled triangle is equidistant from its vertices.
4. Show that a parallelogram whose diagonals are equal is a rectangle.
5. Show that a quadrilateral whose diagonals bisect each other at right angles is a rhombus.
6. Two medians of a triangle are equal, show that the triangle is isosceles.
7. Show that the diagonals of a trapezium having equal non-parallel sides are equal and conversely.
8. Show that the sum of the squares on the diagonals of a parallelogram is equal to the sum of the squares on its sides.
9. Show that the sum of the squares of the six edges of a tetrahedron is equal to
   (*i*) four times the sum of the squares of the lines joining the vertices to its centroid;
   (*ii*) four times the sum of the squares of the lines joining the mid-points of its opposite edges.
10. Show that the sum of the squares of the four diagonals of a parallelopiped is equal to the sum of the squares of its edges.

11. Show that the sides of a trapezium having equal non-parallel sides are equally inclined to the parallel sides.
12. A triangle OAB is right angled at O; squares $OALM$ and $OBPQ$ are constructed on the sides OA and OB externally.

    Show that the lines AP and BL intersect on the altitude through O.
13. Show that a parallelopiped with equal diagonals is a cuboid.

    Also show that a cuboid such that the angle between any two of its diagonals is the same is a cube.
14. If a straight line be perpendicular to each of two intersecting straight lines at their point of intersection, prove that it is perpendicular to every line in the plane determined by the two lines.
15. In a tetrahedron $OABC$, $OA \perp BC$, show that
$$OB^2 + CA^2 = OC^2 + AB^2.$$
16. If two opposite edges of a tetrahedron are equal in length and are at right angles to the line joining their mid-points, show that the remaining pairs of opposite edges have the same property.
17. The line joining the mid-points of two opposite edges of a tetrahedron is perpendicular to the edges; show that the remaining pairs of opposite edges are equal. Also prove the converse.
18. ABCD is a tetrahedron such that the perpendiculars AK, BL, CM and DN to the opposite faces are concurrent. Prove that

    (*i*) any two opposite edges of the tetrahedron are orthogonal.

    (*ii*) K is the orthocentre of the triangle BCD.
19. Two opposite edges AB, CD of a tetrahedron are perpendicular to each other; show that the distance between the mid-points of BD and AC is equal to the distance between the mid-points of BC, AD.
20. $ABCD$ is a tetrahedron and G is the centroid of the base BCD. Prove that
$$AB^2 + AC^2 + AD^2 = GB^2 + GC^2 + GD^2 + 3GA^2.$$

3.2. ORTHOGONAL BASES

It has been seen in Chapter 1 that a linearly independent set of three vectors can be considered as a *Base* in the sense that any given vector can be expressed as a linear combination of its members. Also we have seen that any non-coplanar triad of vectors is a linearly independent system.

In view of the concepts of the length of a vector and angle between vectors as introduced in this chapter, it is possible to consider as Bases systems such that

(*i*) *Length of each member of the system is unity;*

(*ii*) *and any two members of the system are mutually perpendicular.*

Such bases are known as **Orthogonal.**

Let **i, j, k** denote vectors of unit length along three perpendicular lines as shown in the figure. As the length of each of the vectors **i, j, k** is unity and they are mutually perpendicular. We have the relations

$$\mathbf{i} \cdot \mathbf{i} = 1, \quad \mathbf{j} \cdot \mathbf{j} = 1, \quad \mathbf{k} \cdot \mathbf{k} = 1;$$

$$\mathbf{i} \cdot \mathbf{j} = 0 = \mathbf{j} \cdot \mathbf{i}, \quad \mathbf{j} \cdot \mathbf{k} = 0 = \mathbf{k} \cdot \mathbf{j}, \quad \mathbf{k} \cdot \mathbf{i} = 0 = \mathbf{i} \cdot \mathbf{k}. \qquad ...(i)$$

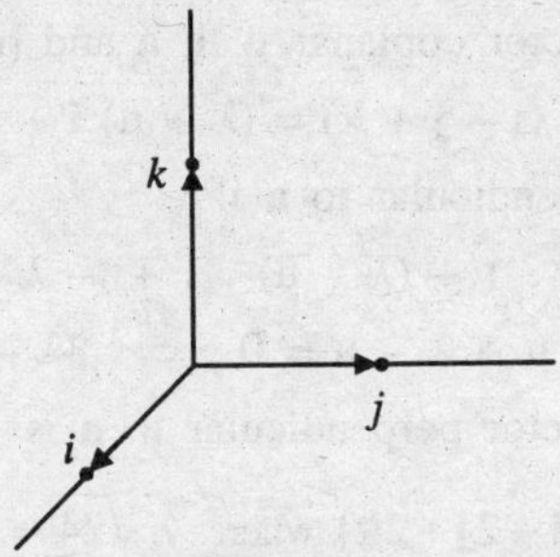

Fig. 3.14

3.2.1. Scalar Product in Terms of Components

Let two given vectors **a, b** expressed as linear combinations of the vectors **i, j, k** of an orthogonal base be

$$\mathbf{a} = a_1\mathbf{i} + a_2\mathbf{j} + a_3\mathbf{k}, \quad \mathbf{b} = b_1\mathbf{i} + b_2\mathbf{j} + b_3\mathbf{k}.$$

Making use of the property that scalar multiplication distributes the sum of vectors and of the relation (*i*), we obtain

$$\mathbf{a.b} = (a_1\mathbf{i} + a_2\mathbf{j} + a_3\mathbf{k}) \cdot (b_1\mathbf{i} + b_2\mathbf{j} + b_3\mathbf{k}) = a_1b_1 + a_2b_2 + a_3b_3.$$

3.2.2. Length of a Vector and Angle between Two given Vectors

We have

$$\mathbf{a} = a_1\mathbf{i} + a_2\mathbf{j} + a_3\mathbf{k} \Rightarrow |a|^2 = \mathbf{a} \cdot \mathbf{a} = a_1^2 + a_2^2 + a_3^2.$$

Also the angle θ between the vectors

$$\mathbf{a} = a_1\mathbf{i} + a_2\mathbf{j} + a_3\mathbf{k}, \quad \mathbf{b} = b_1\mathbf{i} + b_2\mathbf{j} + b_3\mathbf{k}$$

is given by

$$\cos\theta = \frac{\mathbf{a} \cdot \mathbf{b}}{|\mathbf{a}|\,|\mathbf{b}|} = \frac{a_1b_1 + a_2b_2 + a_3b_3}{\sqrt{(a_1^2 + a_2^2 + a_3^2)}\,\sqrt{(b_1^2 + b_2^2 + b_3^2)}}.$$

Cor. For the perpendicularity of the two vectors

$$a_1\mathbf{i} + a_2\mathbf{j} + a_3\mathbf{k}, \quad b_1\mathbf{i} + b_2\mathbf{j} + b_3\mathbf{k},$$

we have

$$(a_1\mathbf{i} + a_2\mathbf{j} + a_3\mathbf{k}) \cdot (b_1\mathbf{i} + b_2\mathbf{j} + b_3\mathbf{k}) = \cos 90^\circ = 0$$

$$\Rightarrow \quad a_1b_1 + a_2b_2 + a_3b_3 = 0.$$

EXAMPLES

Example 1. *Given two vectors*

$$\mathbf{a} = \mathbf{i} + \mathbf{j} - \mathbf{k}; \quad \mathbf{b} = \mathbf{i} - \mathbf{j} + \mathbf{k},$$

find a unit vector **c**, *perpendicular to the vector* **a** *and coplanar with* **a** *and* **b**. *Find also a vector* **d** *perpendicular to both* **a** *and* **c**.

Solution. Any vector coplanar with **a** and **b** is

$\lambda\,(\mathbf{i} + \mathbf{j} - \mathbf{k}) + \mu\,(\mathbf{i} - \mathbf{j} + \mathbf{k}) = (\lambda + \mu)\,\mathbf{i} + (\lambda - \mu)\,\mathbf{j} + (-\lambda + \mu)\,\mathbf{k}.$

This will be perpendicular to **a** if

$$(\lambda + \mu)\,.\,1 + (\lambda - \mu)\,.\,1 + (-\lambda + \mu)\,(-1) = 0$$

$$\Rightarrow \quad \lambda + \mu + \lambda - \mu + \lambda - \mu = 0 \quad \Rightarrow \quad 3\lambda - \mu = 0 \quad \Rightarrow \quad \mu = 3\lambda.$$

Thus, the unit vector perpendicular to **a** is

$$\lambda\left(4\mathbf{i} - 2\mathbf{j} + 2\mathbf{k}\right) \text{ where } \lambda\sqrt{24} = 1 \Rightarrow \lambda = \frac{1}{\sqrt{24}}$$

Thus

$$\mathbf{c} = \frac{1}{\sqrt{24}}(4\mathbf{i} - 2\mathbf{j} + 2\mathbf{k}) = \frac{1}{\sqrt{6}}(2\mathbf{i} - \mathbf{j} + \mathbf{k}).$$

If $p\mathbf{i} + q\mathbf{j} + r\mathbf{k}$ is the required unit vector **d** then, because of its perpendicularity to both **a** and **c**, we have,

$$p + q - r = 0; \quad 2p - q + r = 0$$

which give $p = 0$, $q = r$. [By taking $\mathbf{a}\,.\,\mathbf{d} = 0$ and $\mathbf{c}\,.\,\mathbf{d} = 0$]

Thus, the required vector **d** is given by $(1/\sqrt{2})\,(\mathbf{j} + \mathbf{k})$.

Example 2. *Given two vectors*

$$\mathbf{a} = 2\mathbf{i} - 3\mathbf{j} + \mathbf{k}; \quad \mathbf{b} = -\mathbf{i} + 2\mathbf{j} - \mathbf{k}.$$

Find the projection of **a** *on* **b** *and that of* **b** *on* **a**.

Solution. Let θ denote the angle between the vectors **a** and **b**, so that $|\mathbf{b}|\cos\theta$ is the projection of **b** on **a** and $|\mathbf{a}|\cos\theta$ is the projection of **a** on **b**.

We have

$$|\mathbf{b}|\cos\theta = \frac{\mathbf{a}\,.\,\mathbf{b}}{|\mathbf{a}|} = \frac{-9}{\sqrt{14}},$$

and

$$|\mathbf{a}|\cos\theta = \frac{\mathbf{a}\,.\,\mathbf{b}}{|\mathbf{b}|} = \frac{-9}{\sqrt{6}}$$

Hence the results.

Example 3. *Prove that* $\cos(B - A) = \cos A \cos B + \sin A \sin B$.

Solution. Let $\angle XOL = A, \quad \angle XOM = B.$

Draw $OY \perp OX.$

Let **i**, **j** denote unit vectors along OX and OY.

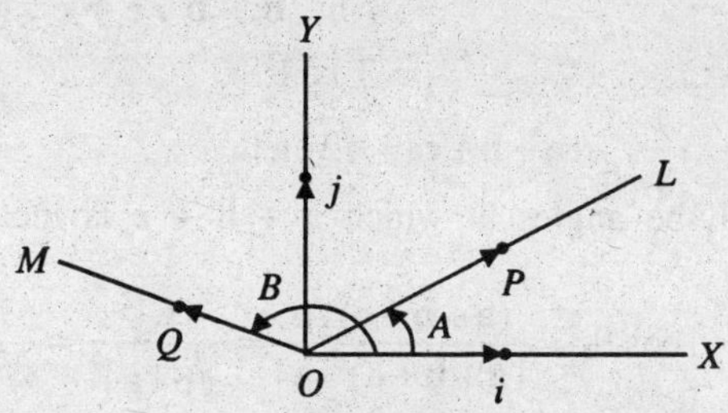

Fig. 3.15

If P and Q be points on OL and OM such that $\vec{OP}$ and $\vec{OQ}$ are unit vectors, we have

$$\vec{OP} = (\cos A)\,\mathbf{i} + (\sin A)\,\mathbf{j},\ \vec{OQ} = (\cos B)\,\mathbf{i} + (\sin B)\,\mathbf{j}.$$

$$\Rightarrow \quad \vec{OP}\cdot\vec{OQ} = \cos A \cos B + \sin A \sin B.$$

Also by def.

$$\vec{OP}\cdot\vec{OQ} = 1.1 \cos(B-A) = \cos(B-A).$$

Hence the result.

Example 4. *If* $\hat{\mathbf{a}}$ and $\hat{\mathbf{b}}$ *are unit vectors and* θ *is the angle between them, show that* $\sin(\theta/2) = \frac{1}{2}|\hat{\mathbf{a}} - \hat{\mathbf{b}}|$.

Solution. $|\hat{\mathbf{a}} - \hat{\mathbf{b}}|^2 = (\hat{\mathbf{a}} - \hat{\mathbf{b}}).(\hat{\mathbf{a}} - \hat{\mathbf{b}})$

$$= \hat{\mathbf{a}}.\hat{\mathbf{a}} - \hat{\mathbf{a}}.\hat{\mathbf{b}} - \hat{\mathbf{b}}.\hat{\mathbf{a}} + \hat{\mathbf{b}}.\hat{\mathbf{b}}$$

$$= 1 - \cos\theta - \cos\theta + 1$$

$$= 2(1-\cos\theta) = 4\sin^2\frac{\theta}{2}$$

$$\therefore \quad \sin\frac{\theta}{2} = \frac{1}{2}|\hat{\mathbf{a}} - \hat{\mathbf{b}}|.$$

Example 5. *If* **a**, **b**, **c** *are mutually perpendicular vectors of equal magnitude, show that* **a** + **b** + **c** *is equally inclined to* **a**, **b**, **c**.

Solution. Given that :

$$\mathbf{b}.\mathbf{a} = \mathbf{a}.\mathbf{b} = \mathbf{b}.\mathbf{c} = \mathbf{c}.\mathbf{b} = \mathbf{a}.\mathbf{c} = \mathbf{c}.\mathbf{a} = 0 \qquad ...(1)$$

and

$$|\mathbf{a}| = |\mathbf{b}| = |\mathbf{c}| \qquad ...(2)$$

Now,

$$|\mathbf{a}+\mathbf{b}+\mathbf{c}|^2 = (\mathbf{a}+\mathbf{b}+\mathbf{c}).(\mathbf{a}+\mathbf{b}+\mathbf{c})$$

$$= \mathbf{a} \cdot \mathbf{a} + \mathbf{a} \cdot \mathbf{b} + \mathbf{a} \cdot \mathbf{c} + \mathbf{b} \cdot \mathbf{a}$$
$$+ \mathbf{b} \cdot \mathbf{b} + \mathbf{b} \cdot \mathbf{c} + \mathbf{c} \cdot \mathbf{a} + \mathbf{c} \cdot \mathbf{b} + \mathbf{c} \cdot \mathbf{c}$$
$$= 3 \,|\, \mathbf{a} \,|^2$$

$\therefore$ $$|\mathbf{a}+\mathbf{b}+\mathbf{c}| = \sqrt{3}\,|\mathbf{a}|.$$

Let θ_1, θ_2, θ_3 be angles at which **a** + **b** + **c** is inclined to **a, b, c** respectively, then

$$\cos\theta_1 = \frac{(\mathbf{a}+\mathbf{b}+\mathbf{c}) \cdot \mathbf{a}}{|\mathbf{a}+\mathbf{b}+\mathbf{c}|\;|\mathbf{a}|} = \frac{|\mathbf{a}|^2}{\sqrt{3}\,|\mathbf{a}|^2} = \frac{1}{\sqrt{3}}.$$

Similarly $\cos\theta_2 = \cos\theta_3 = \dfrac{1}{\sqrt{3}}.$

$\Rightarrow$ $\theta_1 = \theta_2 = \theta_3.$

Example 6. *If* **a**, **b** *are vectors and a, b their lengths, show that*

$$\left(\frac{\mathbf{a}}{a^2} - \frac{\mathbf{b}}{b^2}\right)^2 = \left(\frac{\mathbf{a}-\mathbf{b}}{ab}\right)^2$$

Solution. $$\left(\frac{\mathbf{a}}{a^2} - \frac{\mathbf{b}}{b^2}\right)^2 = \left(\frac{\mathbf{a}}{a^2} - \frac{\mathbf{b}}{b^2}\right) \cdot \left(\frac{\mathbf{a}}{a^2} - \frac{\mathbf{b}}{b^2}\right)$$

$$= \frac{\mathbf{a} \cdot \mathbf{a}}{a^4} - \frac{2\mathbf{a} \cdot \mathbf{b}}{a^2 b^2} + \frac{\mathbf{b} \cdot \mathbf{b}}{b^4} = \frac{a^2}{a^4} - \frac{2\mathbf{a} \cdot \mathbf{b}}{a^2 b^2} + \frac{b^2}{b^4}$$

$$= \frac{1}{a^2} - \frac{2\mathbf{a} \cdot \mathbf{b}}{a^2 b^2} + \frac{1}{b^2} = \frac{b^2 - 2\mathbf{a} \cdot \mathbf{b} + a^2}{a^2 b^2}$$

$$= \frac{(\mathbf{a}-\mathbf{b}) \cdot (\mathbf{a}-\mathbf{b})}{a^2 b^2} = \frac{(\mathbf{a}-\mathbf{b})^2}{a^2 b^2}.$$

EXERCISES

1. Find the lengths of the following vectors :

$$3\mathbf{i} + 2\mathbf{j} - \mathbf{k}, \quad \sqrt{2}\mathbf{i} - \sqrt{3}\mathbf{k}, \quad 5\mathbf{i} + 4\mathbf{j} - 2\mathbf{k}.$$

2. Find the angles between the following pairs of vectors :

   (*i*) **i** – **j** + **k**, – **i** + **j** + 2**k**, (*ii*) 3**i** + 4**j**, 2**j** – 5**k**,

   (*iii*) 2**i** – 3**k**, **i** + **j** + **k**.

3. Find the unit vectors perpendicular to each of the following pairs of vectors :

   (*i*) **i** – **j** + **k**, **i** + 2**j** – **k**, (*ii*) 2**i** + **k**, **i** + **j** + **k**

   (*iii*) **i** + **j**, **i** – **j** + **k**, (*iv*) 2**i** + **j** + **k**, **i** – 2**j** + **k**.

4. Determine the unit vectors which make an angle of 60° with **i** – **j** and an angle of 60° with **i** + **k**.

5. Given that **i, j, k** is an orthogonal base, show that

$$\mathbf{i},\ \sqrt{\frac{1}{2}}(\mathbf{j}+\mathbf{k}),\ \sqrt{\frac{1}{2}}(\mathbf{j}-\mathbf{k})$$

is also an orthonormal base.

6. Given that **i – j, i + 2j** are two vectors. Find a unit vector coplanar with these vectors and perpendicular to the first vector **i – j**. Find also the unit vector which is perpendicular to the plane of the two given vectors. Do you thus obtain an orthonormal triad ?

7. Let

(*i*) $\mathbf{a} = \mathbf{i} - \mathbf{j},\ \mathbf{b} = \mathbf{i} + \mathbf{k},$

(*ii*) $\mathbf{a} = \mathbf{i}+\mathbf{j}+\mathbf{k},\ \ \mathbf{b} = \mathbf{i}+\sqrt{2}\mathbf{j}-\sqrt{6}\mathbf{k}.$

Find in each case a unit vector **c** perpendicular to **a** and coplanar with **a** and **b**. Also find a unit vector **d** perpendicular to both **a** and **c**.

8. Given that

$$\mathbf{a} = \mathbf{i} + \mathbf{j} + \mathbf{k};\ \ \mathbf{b} = \mathbf{i} - \mathbf{j} + \mathbf{k};\quad \mathbf{c} = \mathbf{i} + \mathbf{j} - \mathbf{k},$$

evaluate

(*i*) $(\mathbf{a} \cdot \mathbf{b}) + (\mathbf{b} \cdot \mathbf{c}) \div (\mathbf{c} \cdot \mathbf{a});$

(*ii*) $(\mathbf{a} \cdot \mathbf{c})\, \mathbf{c} + (\mathbf{c} \cdot \mathbf{b})\, \mathbf{a};$

(*iii*) $(\mathbf{a} + 2\mathbf{b}) \cdot [\mathbf{a} + (\mathbf{a} \cdot \mathbf{c})\, \mathbf{b}].$

9. Given the vectors **a** and **b** as follows :

(*i*) $\mathbf{a} = 2\mathbf{i} - \frac{3}{2}\mathbf{j} + \frac{4}{5}\mathbf{k};\ \ \mathbf{b} = \mathbf{i} + \sqrt{2}\mathbf{j} + \sqrt{6}\mathbf{k}.$

(*ii*) $\mathbf{a} = \mathbf{i}+\mathbf{j}+\mathbf{k};\ \ \mathbf{b} = \sqrt{3\mathbf{i}+3\mathbf{j}-2\mathbf{k}}.$

Find in each case the projections of **a** on **b** and of **b** on **a**.

3.3. APPLICATIONS TO CARTESIAN GEOMETRY

Rectangular Cartesian Coordinates and Position Vectors. We shall now consider some preliminary applications to Cartesian Geometry. The major results will be obtained in the following chapter with the help of scalar triple products.

Take any point *O* and three mutually perpendicular straight lines *OX*, *OY*, *OZ* through *O*. Let *A*, *B*, *C* be three points on these lines such that

$$\overrightarrow{OA} = \mathbf{i},\ \ \overrightarrow{OB} = \mathbf{j},\ \ \overrightarrow{OC} = \mathbf{k},$$

where **i, j, k** denote mutually perpendicular unit vectors.

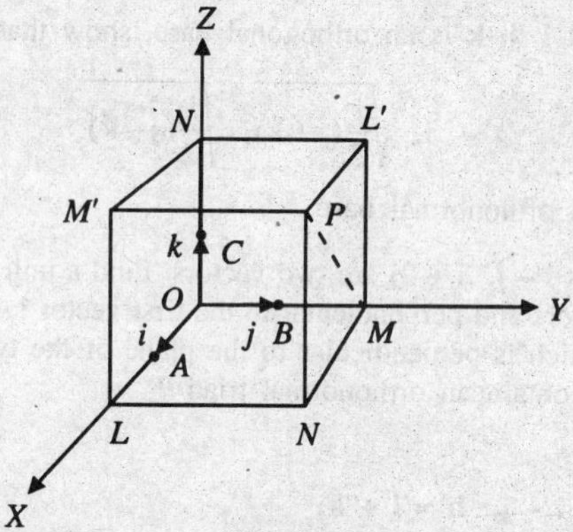

Fig. 3.16

Let P be any given point. Through P draw planes parallel to the three planes BOC, COA, AOB meeting OA, OB, OC in L, M, N respectively so that we obtain a rectangular parallelopiped having OP as a diagonal. We have

$$\mathbf{r} = \vec{OP} = \vec{OM} + \vec{MP}$$
$$= \vec{OM} + \vec{ML'} + \vec{L'P}$$
$$= \vec{OM} + \vec{ON} + \vec{OL}$$
$$= \vec{OL} + \vec{OM} + \vec{ON}.$$

There exist 3 scalars, x, y, z such that

$$\vec{OL} = x\vec{OA} = x\mathbf{i}, \quad \vec{OM} = y\vec{OB} = y\mathbf{j}, \quad \vec{ON} = z\vec{OC} = z\mathbf{k}.$$

Thus, we have

$$\mathbf{r} = x\mathbf{i} + y\mathbf{j} + z\mathbf{k}.$$

Here $\vec{OP} = \mathbf{r}$ has been expressed as a linear combination of three mutually perpendicular unit vectors **i, j, k.**

The numbers x, y, z are *Rectangular Cartesian Coordinates* of the point P with respect to OX, OY, OZ respectively as coordinate axes (compare 10.2 Page 16 Chapter 1).

Thus, with reference to the coordinate axes chosen as above, we see that the statement that

$x\mathbf{i} + y\mathbf{j} + z\mathbf{k}$ *is the position vector of a point* $\Leftrightarrow$ *x, y, z are the rectangular cartesian coordinates of the point.*

3.4. DISTANCE BETWEEN POINTS WHOSE RECTANGULAR CARTESIAN COORDINATES ARE

$$P\ (x_1, y_1, z_1), \quad Q\ (x_2, y_2, z_2).$$

The position vectors of the points P, Q are

$$\vec{OP} = x_1\mathbf{i} + y_1\mathbf{j} + z_1\mathbf{k},\ \vec{OQ} = x_2\mathbf{i} + y_2\mathbf{j} + z_2\mathbf{k} \text{ respectively.}$$

We have

$$\vec{PQ} = \vec{OQ} - \vec{OP} = (x_2 - x_1)\,\mathbf{i} + (y_2 - y_1)\,\mathbf{j} + (z_2 - z_1)\,\mathbf{k}$$

so that

$$PQ^2 = \vec{PQ}\,.\,\vec{PQ} = [(x_2 - x_1)\,\mathbf{i} + (y_2 - y_1)\,\mathbf{j} + (z_2 - z_1)\,\mathbf{k}].$$
$$[(x_2 - x_1)\,\mathbf{i} + (y_2 - y_1)\,\mathbf{j} + (z_2 - z_1)\,\mathbf{k}]$$
$$= (x_2 - x_1)^2 + (y_2 - y_1)^2 + (z_2 - z_1)^2.$$

3.5. DIRECTION COSINES OF A LINE

Consider any directed line and let $\vec{OA}$ denote the vector of unit length parallel to this line.

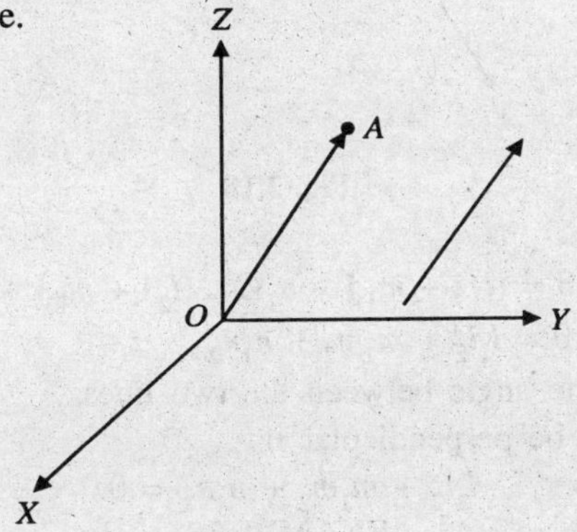

Fig. 3.17

Let $\qquad \vec{OA} = l\mathbf{i} + m\mathbf{j} + n\mathbf{k}.$

We have

$$\vec{OA}\,.\,\mathbf{i} = (l\mathbf{i} + m\mathbf{j} + n\mathbf{k})\,.\,\mathbf{i} = l$$

$$\Rightarrow \qquad 1.1\,.\,\cos\alpha = l \quad \Rightarrow \quad l = \cos\alpha$$

where α denotes the angle which the given line makes with x-axis.

Similarly we may show $m = \cos\beta$, $n = \cos\gamma$ where β and γ denote the angles which the given line makes with y-axis and z-axis respectively.

Thus, if $l\mathbf{i} + m\mathbf{j} + n\mathbf{k}$

denotes a vector of unit length parallel to a given line, then l, m, n are the *direction cosines of the line.*

Cor. *The sum of the squares of the direction cosines of any line is unity.*

We have $\qquad \vec{OA}\,.\,\vec{OA} = (l\mathbf{i} + m\mathbf{j} + n\mathbf{k})\,.\,(l\mathbf{i} + m\mathbf{j} + n\mathbf{k})$

$$\Rightarrow \qquad 1 = l^2 + m^2 + n^2.$$

3.6. ANGLE BETWEEN TWO LINES

Let $l_1, m_1, n_1; l_2, m_2, n_2$ be the direction cosines of two given lines. The vectors of unit length along the given lines being

$$l_1\mathbf{i} + m_1\mathbf{j} + n_1\mathbf{k}, \quad l_2\mathbf{i} + m_2\mathbf{j} + n_2\mathbf{k},$$

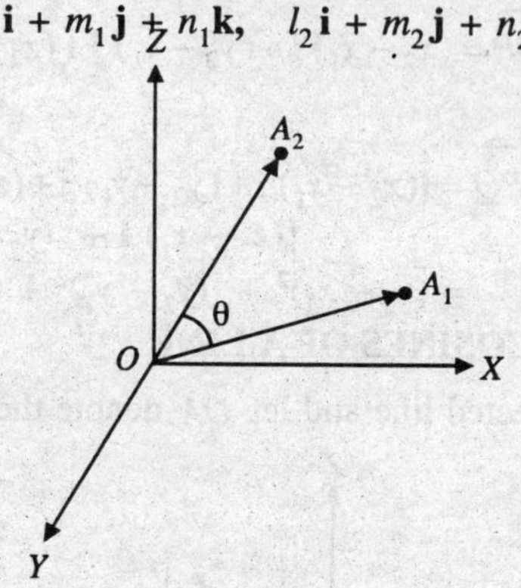

Fig. 3.18

we have

$$1.1 \,.\, \cos\theta = (l_1\mathbf{i} + m_1\mathbf{j} + n_1\mathbf{k}) \,.\, (l_2\mathbf{i} + m_2\mathbf{j} + n_2\mathbf{k})$$

$$\Rightarrow \quad \cos\theta = l_1 l_2 + m_1 m_2 + n_1 n_2$$

where θ denotes the angle between the two lines.

The lines will be perpendicular if

$$l_1 l_2 + m_1 m_2 + n_1 n_2 = 0.$$

EXAMPLES

Example 1. *Find the angle between the lines AB, AC where A, B, C are the three points with rectangular cartesian coordinates*

$$(1, 2, -1), \quad (2, 0, 3), \quad (3, -1, 2)$$

respectively.

Solution. In terms of usual notation, the position vectors of the given points are

$$\vec{OA} = \mathbf{i}+2\mathbf{j}+\mathbf{k}, \quad \vec{OB} = 2\mathbf{i}+3\mathbf{k}, \quad \vec{OC} = 3\mathbf{i}-\mathbf{j}+2\mathbf{k},$$

O being the origin. Thus, we have

$$\vec{AB} = \vec{OB} - \vec{OA} = \mathbf{i}-2\mathbf{j}+4\mathbf{k}, \quad \vec{AC} = \vec{OC} - \vec{OA} = 2\mathbf{i}-3\mathbf{j}+3\mathbf{k}.$$

$$\therefore \quad \vec{AB} \,.\, \vec{AC} = (\mathbf{i}-2\mathbf{j}+4\mathbf{k}) \,.\, (2\mathbf{i}-3\mathbf{j}+3\mathbf{k}) = 20.$$

Also

$$\vec{AB}^2 = (\mathbf{i}-2\mathbf{j}+4\mathbf{k}) \,.\, (\mathbf{i}-2\mathbf{j}+4\mathbf{k}) = 21.$$

$$\vec{AC}^2 = (2\mathbf{i}-3\mathbf{j}+3\mathbf{k}) \,.\, (2\mathbf{i}-3\mathbf{j}+3\mathbf{k}) = 22.$$

If θ denote the angle between AB and AC, we have

$$\cos\theta = \frac{\vec{AB} \,.\, \vec{AC}}{|\vec{AB}|\,|\vec{AC}|} = \frac{20}{\sqrt{21}\sqrt{22}}$$

Example 2. *A line makes angles α, β, γ, δ with the diagonals of a cube; show that*

$$\cos^2\alpha + \cos^2\beta + \cos^2\gamma + \cos^2\delta = \frac{4}{3}.$$

Solution.

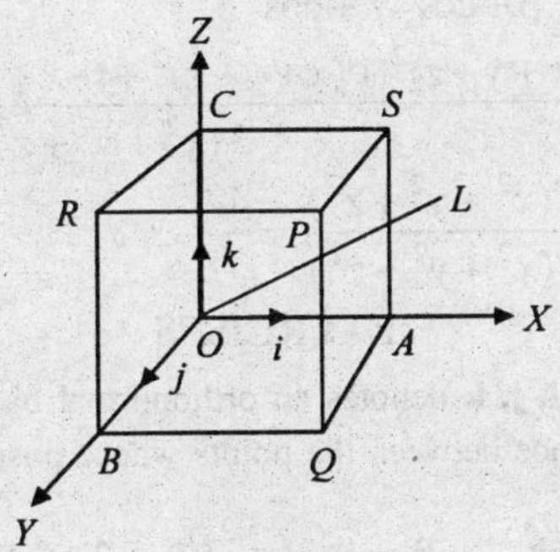

Fig. 3.19

Since the angle will remain unchanged for any size of cube, consider a unit cube. Represent the coterminous edges $\vec{OA}$, $\vec{OB}$, $\vec{OC}$ by unit vectors **i, j, k.**

Then $\qquad \vec{OA} = \mathbf{i},\ \vec{OB} = \mathbf{j},\ \vec{OC} = \mathbf{k},$

$$\vec{OP} = \vec{OA} + \vec{AQ} + \vec{QP} = \mathbf{i} + \mathbf{j} + \mathbf{k}$$

$$\vec{CQ} = -\mathbf{k} + \mathbf{i} + \mathbf{j},\ \vec{AR} = -\mathbf{i} + \mathbf{k} + \mathbf{j}$$

and $\qquad \vec{BS} = \mathbf{i} - \mathbf{j} + \mathbf{k}.$

$\therefore \qquad OP = CQ = AR = BS = \sqrt{3}.$

Let any line OL be given by

$$\vec{OL} = x\mathbf{i} + y\mathbf{j} + z\mathbf{k}.$$

If α be the angle between OL and OP, then

$$\vec{OL} \,.\, \vec{OP} = OL \times OP \times \cos\alpha$$

$$\Rightarrow \quad (x\mathbf{i} + y\mathbf{j} + z\mathbf{k}) \,.\, (\mathbf{i} + \mathbf{j} + \mathbf{k}) = \sqrt{(x^2 + y^2 + z^2)}.\sqrt{3}\cos\alpha$$

$$\Rightarrow \quad \cos\alpha = \frac{(x + y + z)}{\sqrt{3}\,\sqrt{(x^2 + y^2 + z^2)}}$$

Similarly,

$$\cos\beta = \frac{(x + y - z)}{\sqrt{3}\,\sqrt{(x^2 + y^2 + z^2)}}$$

$$\cos \gamma = \frac{(-x+y+z)}{\sqrt{3}\sqrt{(x^2+y^2+z^2)}}$$

and $$\cos \delta = \frac{(x-y+z)}{\sqrt{3}\sqrt{(x^2+y^2+z^2)}}$$

$$\therefore \quad \cos^2 \alpha + \cos^2 \beta + \cos^2 \gamma + \cos^2 \delta$$

$$= \frac{(x+y+z)^2 + (x+y-z)^2 + (-x+y+z)^2 + (x-y+z)^2}{3(x^2+y^2+z^2)}$$

$$= \frac{4(x^2+y^2+z^2)}{3(x^2+y^2+z^2)} = \frac{4}{3}.$$

EXERCISES

(In the following **i, j, k** denotes an orthonormal base.)

1. Find the distance between the points whose position vectors are given as follows :
   (*i*) 4**i** + 3**j** – 6**k**, – 2**i** + **j** – **k**; (*ii*) – 2**i** + 3**j** + 5**k**, 7**i** – **k**.
2. Find the distance between the pairs of points whose cartesian coordinates are :
   (*i*) (– 1, 1, 3), (0, 5, 6); (*ii*) (2, 3, – 1), (2, 6, 2).
3. Find the angle between the lines parallel to the following pairs of vectors :
   (*i*) 6**i** + 2**j** + 3**k**, 3**i** + 12 **j** + 4**k**; (*ii*) **i** + 2**j** + 3**k**, – **i** + **j** + 2**k**.
4. What are the direction cosines of the joins of the following pairs of points :
   (*i*) (6, 3, 2), (5, 1, 4), (*ii*) (3, – 4, 7), (0, 2, 5).
5. Given that *P, Q, R, S* are the points (6, – 6, 0), (– 1, – 7, 6), (3, – 4, 4), (2, – 9, 2) respectively, show that $PQ \perp RS$.
6. Find the angle between the lines joining the following pairs of points :
   (*i*) (0, 4, 1), (2, 3, – 1); (*ii*) (4, 5, 0), (2, 6, 2).
7. Show that the points
   (5, – 1, 1), (7, – 4, 7), (1, – 6, 10), (– 1, – 3, 4)
   are the vertices of a rhombus.
8. Show that the join of points (1, 2, 3), (4, 5, 7) is parallel to the join of the points (– 4, 3, – 6), (2, 9, 2).
9. Show that the points
   (4, 7, 8), (2, 3, 4), (– 1, – 2, 1), (1, 2, 5)
   are the vertices of a parallelogram.
10. Show that the line *AB* is perpendicular to the line *CD* if *A, B, C, D* are the points (2, 3, 4), (5, 4, – 1), (3, 6, 2), (1, 2, 0) respectively.
11. Show that the three angles of the triangle with vertices
    (1, – 1, 1), (2, 3, – 1), (3, 0, 2)
    are
    $$\cos^{-1}\frac{2}{\sqrt{114}},\ \cos^{-1}\frac{4}{\sqrt{126}},\ \cos^{-1}\frac{-17}{\sqrt{399}}$$
    respectively.

12. The points P, Q, R have rectangular cartesian coordinates $(1, 1, -1)$, $(4, 1, 2)$ and $(-2, 1, 2)$ respectively. Which of the following is true for the triangle PQR ?
    A. All the sides have the same length.
    B. $QR = 2\ PQ$.
    C. Angle PQR is a right angle.
    D. The area of the triangle is 18 square units.
13. Find the angle between two diagonals of a cube.
14. If the edges of a rectangular parallelopiped are α, β, γ, show that the angles between the four diagonals are
$$\cos^{-1} \frac{\alpha^2 \pm \beta^2 \pm \gamma^2}{\alpha^2 + \beta^2 + \gamma^2}.$$
15. With reference to mutually perpendicular axes A, B, C and D are the points $(-6, 1, 6)$, $(6, -2, 3)$, $(-2, -3, -1)$ and $(-5, -9, -7)$ respectively.
    (*i*) Show that the points A, B and D lie on the surface of a sphere with centre C.
    (*ii*) If DCE is a diameter of the sphere, find the coordinates of E.
    (*iii*) Prove that the angle BCA is a right angle and that angle BCE equals angle ECA.
    (*iv*) Show that the diameter DCE does not lie in the plane of the triangle ABC.

SUMMARY

1. The scalar product of two vectors is the product of the length of one vector and projection of other on it.
2. The square of the length of the vector **a** is
$$\mathbf{a} \cdot \mathbf{a} \text{ denoted by } |\mathbf{a}|^2.$$
3. The angle between two vectors **a** and **b** is
$$\cos^{-1} \frac{\mathbf{a} \cdot \mathbf{b}}{|\mathbf{a}|\,|\mathbf{b}|}.$$
4. If **a, b** are two non-zero vectors, then
$$\mathbf{a} \cdot \mathbf{b} = 0 \iff \mathbf{a} \text{ and } \mathbf{b} \text{ are perpendicular.}$$
5. Scalar multiplication distributes the addition of vectors.
6. If **i, j, k** denote mutually perpendicular vectors, each of unit length, then
$$\mathbf{i} \cdot \mathbf{i} = 1, \quad \mathbf{j} \cdot \mathbf{j} = 1, \quad \mathbf{k} \cdot \mathbf{k} = 1;$$
$$\mathbf{i} \cdot \mathbf{j} = 0, \quad \mathbf{j} \cdot \mathbf{k} = 0, \quad \mathbf{k} \cdot \mathbf{i} = 0,$$
7. A base consisting of three mutually perpendicular vectors is called *orthogonal*. In case each vector is of unit length, the base is called *orthonormal*.
8. $(a_1\mathbf{i} + a_2\mathbf{j} + a_3\mathbf{k}) \cdot (b_1\mathbf{i} + b_2\mathbf{j} + b_3\mathbf{k}) = a_1 b_1 + a_2 b_2 + a_3 b_3$.
9. (*i*) Length of a vector $a_1\mathbf{i} + a_2\mathbf{j} + a_3\mathbf{k}$ is $\sqrt{(\mathbf{a}_1^2 + \mathbf{a}_2^2 + \mathbf{a}_3^2)}$.
   (*ii*) Angle between two vectors

$a_1\mathbf{i} + a_2\mathbf{j} + a_3\mathbf{k}$ and $b_1\mathbf{i} + b_2\mathbf{j} + b_3\mathbf{k}$ is

$$\cos^{-1} \frac{a_1b_1 + a_2b_2 + a_3b_3}{\sqrt{(a_1^2 + a_2^2 + a_3^2)}\ \sqrt{(b_1^2 + b_2^2 + b_3^2)}}.$$

(*iii*) The vectors $a_1\mathbf{i} + a_2\mathbf{j} + a_3\mathbf{k}$, $b_1\mathbf{i} + b_2\mathbf{j} + b_3\mathbf{k}$ are perpendicular if and only if

$$a_1b_1 + a_2b_2 + a_3b_3 = 0.$$

10. If OX, OY, OZ be three mutually perpendicular axes and **i, j, k** are unit vectors along OX, OY, OZ respectively, then

(x, y, z) are the coordinates of P

$\Leftrightarrow$ $x\mathbf{i} + y\mathbf{j} + z\mathbf{k}$ is the position vector $\overrightarrow{OP}$ of P.

OBJECTIVE QUESTIONS

For each of the following questions, four alternatives are given for the answer. Only one of them is correct. Choose the correct alternative.

1. $a_1\mathbf{i} + a_2\mathbf{j}$ is a unit vector perpendicular to $4\mathbf{i} - 3\mathbf{j}$ if

(*a*) $a_1 = .6,\ a_2 = .8$ (*b*) $a_1 = 3,\ a_2 = 4$

(*c*) $a_1 = .8,\ a_2 = .6$ (*d*) $a_1 = 4,\ a_2 = 3$

2. If $\mathbf{a} = 2\mathbf{i} - 3\mathbf{j}$, $\mathbf{b} = 2\mathbf{j} + 3\mathbf{k}$, then $(\mathbf{a} + \mathbf{b}) \cdot (\mathbf{a} - \mathbf{b}) =$

(*a*) 0 (*b*) -8 (*c*) 9 (*d*) -10

3. If $\overrightarrow{PO} + \overrightarrow{OQ} = \overrightarrow{QO} + \overrightarrow{OR}$, then P, Q, R are

(*a*) the vertices of an equilateral triangle

(*b*) the vertices of an isosceles triangle

(*c*) collinear (*d*) None of these

4. The triangle ABC is defined by the vertices $A\ (1, -2, 2)$, $B\ (1, 4, 0)$ and $C\ (-4, 1, 1)$. Let M be the foot of the altitude drawn from the vertex B to side AC. Then $\overrightarrow{BM} =$

(*a*) $(-20/7, -30/7, 10/7)$ (*b*) $-20, -30, 10)$

(*c*) $(2, 3, -1)$ (*d*) None of these

5. The vector **b**, which is collinear with the vector $\mathbf{a} = (2, 1, -1)$ and satisfies the condition $\mathbf{a} \cdot \mathbf{b} = 3$, is

(*a*) $(1, 1/2, -1/2)$ (*b*) $(2/3, 1/3, -1/3)$

(*c*) $(1/2, 1/4, -1/4)$ (*d*) $(1, 1, 0)$

6. The angle between **a** and **b** is $\pi/6$, then angle between 2**a** and 3**b** is

(*a*) $\pi/3$ (*b*) $\pi/2$ (*c*) $\pi/6$ (*d*) None of these

7. If θ be the angle between the vectors $\mathbf{i} + \mathbf{j}$ and $\mathbf{j} + \mathbf{k}$, then θ is (*a*) 0 (*b*) $\pi/4$ (*c*) $\pi/2$ (*d*) $\pi/3$

8. If **a**, **b**, **c** are three vectors such that each is inclined at an angle $\pi/3$ with the other two and $|\mathbf{a}| = 1$, $|\mathbf{b}| = 2$, $|\mathbf{c}| = 3$, then the scalar product of the vectors $2\mathbf{a} + 3\mathbf{b} - 5\mathbf{c}$ and $4\mathbf{a} - 6\mathbf{b} + 10\mathbf{c}$ is equal to

(*a*) – 334 (*b*) 188 (*c*) – 522 (*d*) – 514

9. If $\mathbf{a} + \mathbf{b} + \mathbf{c} = 0$, $|\mathbf{a}| = 3$, $|\mathbf{b}| = 5$, $|\mathbf{c}| = 7$, then the angle between **a** and **b** is

(*a*) $\pi/6$ (*b*) $2\pi/3$ (*c*) $5\pi/3$ (*d*) $\pi/3$

10. Projection of the vector $2\mathbf{i} + 3\mathbf{j} - 2\mathbf{k}$ on the vector $\mathbf{i} - 2\mathbf{j} + 3\mathbf{k}$ is

(*a*) $2/\sqrt{(14)}$ (*b*) $1/\sqrt{(14)}$ (*c*) $3/\sqrt{(14)}$ (*d*) None of these

11. The points *A* (1, 1, 2), *B* (3, 4, 2) and *C* (5, 6, 4). The exterior angle of the triangle *B* is

(*a*) $\cos^{-1}\left[-5/\sqrt{(39)}\right]$ (*b*) $\cos^{-1}\left[5/\sqrt{(39)}\right]$

(*c*) $\cos^{-1}(5/6)$ (*d*) None of these

12. If $3\mathbf{i} + 2\mathbf{j} + 8\mathbf{k}$ and $2\mathbf{i} + x\mathbf{j} + \mathbf{k}$ are at right angles then $x =$

(*a*) 7 (*b*) – 7 (*c*) 5 (*d*) – 4

13. The vectors $2\mathbf{i} + 3\mathbf{j} - 4\mathbf{k}$ and $a\mathbf{i} + b\mathbf{j} + c\mathbf{k}$ are perpendicular, when

(*a*) $a = 2, b = 3, c = -4$ (*b*) $a = 4, b = 4, c = 5$

(*c*) $a = 4, b = 4, c = -5$ (*d*) None of these

14. If $A = 2\mathbf{i} + 2\mathbf{j} + 4\mathbf{k}$, $B = -\mathbf{i} + 2\mathbf{j} + \mathbf{k}$ and $C = 3\mathbf{i} + \mathbf{j}$, then $A + tB$ is perpendicular to *c* if *t* is equal to

(*a*) 8 (*b*) 4 (*c*) 6 (*d*) 2

15. Let **a**, **b**, **c** and **d** be position vectors of four points *A*, *B*, *C* and *D* lying in a plane. If $(\mathbf{a} - \mathbf{d}) \cdot (\mathbf{b} - \mathbf{c}) = 0 = (\mathbf{b} - \mathbf{d}) \cdot (\mathbf{c} - \mathbf{a})$, then ΔABC has Δ as

(*a*) in-centre (*b*) circum-centre

(*c*) ortho-centre (*d*) centroid

16. A unit vector in *xy*-plane that makes an angle of 45° with the vector $\mathbf{i} + \mathbf{j}$ and an angle of 60° with the vector $3\mathbf{i} - 4\mathbf{j}$ is

(*a*) *i* (*b*) $(\mathbf{a} + \mathbf{j})/\sqrt{2}$

(*c*) $(\mathbf{i} - \mathbf{j})/\sqrt{2}$ (*d*) None of these

17. $(\mathbf{a} - \mathbf{i})\,\mathbf{i} + (\mathbf{a} \cdot \mathbf{j})\,\mathbf{j} + (\mathbf{a} \cdot \mathbf{k})\,\mathbf{k} =$

(*a*) **0** (*b*) **a** (*c*) 3**a** (*d*) None of these

18. Let α, β, γ be distinct real numbers. The points with position vectors $\alpha\mathbf{i} + \beta\mathbf{j} + \gamma\mathbf{k}$, $\beta\mathbf{i} + \gamma\mathbf{j} + \alpha\mathbf{k}$, $\gamma\mathbf{i} + \alpha\mathbf{j} + \beta\mathbf{k}$.

(*a*) are collinear
(*b*) form an equilateral triangle
(*c*) form an scalene triangle
(*d*) form a right angled triangle

19. If (**A** + **B**) is perpendicular to **B** and (**A** + 2**B**) is perpendicular to **A**, then

(*a*) $\mathbf{A} = \sqrt{2}\mathbf{B}$ (*b*) $\mathbf{A} = 2\mathbf{B}$ (*c*) $2\mathbf{A} = \mathbf{B}$ (*d*) $\mathbf{A} = \mathbf{B}$

20. If **a, b, c** are three non-zero vectors such that $\mathbf{a} + \mathbf{b} + \mathbf{c} = 0$ then the value of $\mathbf{a} \cdot \mathbf{b} + \mathbf{b} \cdot \mathbf{c} + \mathbf{c} \cdot \mathbf{a}$ is

(*a*) less than zero (*b*) equal to zero
(*c*) greater than zero (*d*) 3

21. If $|\mathbf{a} + \mathbf{b}| = |\mathbf{a} - \mathbf{b}|$, $\mathbf{a}, \mathbf{b} \neq 0$, then

(*a*) **a** is $\parallel$ to **b** (*b*) **a** is $\perp$ to **b**
(*c*) $|\mathbf{a}| = |\mathbf{b}|$ (*d*) None of these

22. If **x** and **y** are two unit vectors and ϕ is the angle between them then $\frac{1}{2}|\mathbf{x} - \mathbf{y}|$ is equal to

(*a*) 0 (*b*) $\pi/2$ (*c*) $\left|\sin \frac{1}{2}\phi\right|$ (*d*) $\left|\cos \frac{1}{2}\phi\right|$

23. The value of *b* such that the scalar product of the vector $\mathbf{i} + \mathbf{j} + \mathbf{k}$ with the unit vector parallel to the sum of the vector $2\mathbf{i} + 4\mathbf{j} - 5\mathbf{k}$, and $b\mathbf{i} + 2\mathbf{j} + 3\mathbf{k}$ is one is

(*a*) – 2 (*b*) – 1 (*c*) 0 (*d*) 1

24. The projection of $\mathbf{a} = 3\mathbf{i} - \mathbf{j} + 5\mathbf{k}$ on $\mathbf{b} = 2\mathbf{i} + 3\mathbf{j} + \mathbf{k}$ is

(*a*) $8/\sqrt{(35)}$ (*b*) $8/\sqrt{(39)}$ (*c*) $8/\sqrt{(14)}$ (*d*) $\sqrt{(14)}$

25. Given two vectors $\mathbf{a} = 2\mathbf{i} - 3\mathbf{j} + 6\mathbf{k}$ on $\mathbf{b} = -2\mathbf{i} + 2\mathbf{j} - \mathbf{k}$ and

$$\lambda = \frac{\text{the projection of } \mathbf{a} \text{ on } \mathbf{b}}{\text{the projection of } \mathbf{b} \text{ on } \mathbf{a}},$$

then the value of λ is

(*a*) 3/7 (*b*) 7/3 (*c*) 3 (*d*) 7

ANSWERS

1. (*a*)	**2.** (*a*)	**3.** (*c*)	**4.** (*a*)	**5.** (*a*)
6. (*c*)	**7.** (*d*)	**8.** (*a*)	**9.** (*d*)	**10.** (*a*)
11. (*b*)	**12.** (*b*)	**13.** (*b*)	**14.** (*a*)	**15.** (*c*)
16. (*d*)	**17.** (*b*)	**18.** (*b*)	**19.** (*a*)	**20.** (*a*)
21. (*b*)	**22.** (*c*)	**23.** (*d*)	**24.** (*c*)	**25.** (*b*)

4

Applications to Metric Geometry

Introduction. While in the Chapter 2, we obtained the parametric vectorial equations of planes, we shall in this chapter obtain the vectorial equations of planes in **non-parametric form.** A study of *geometry of sphere* with the help of scalar product of vectors will also be taken into account.

4.1. NORMAL FORM OF THE VECTOR EQUATION OF A PLANE.

Non-Parametric Form. Let, **n**, be the unit vector normal to given plane and let, p, be the length of the perpendicular from the origin of reference O, to the plane; p, will be always considered as positive.

Draw OK perpendicular to the plane; K being the foot of the perpendicular, We have

$$\vec{OK} = p\mathbf{n}.$$

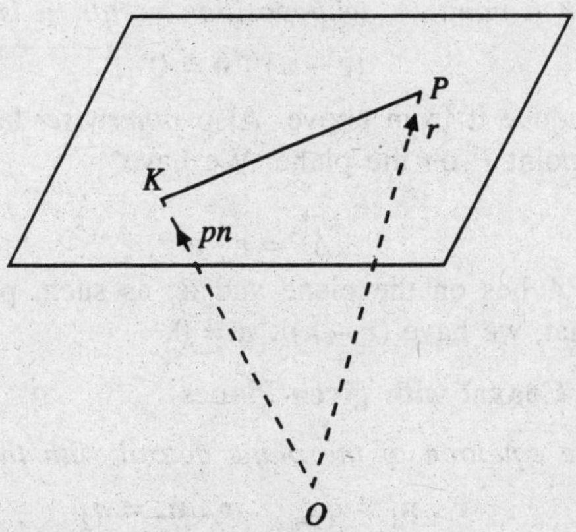

Fig. 4.1

Let, **r**, be the position vector $\vec{OP}$ of any point P on the plane. As KP lies in the plane to which OK is normal, we have

$$KP \perp OK \Rightarrow \vec{OK} \,.\, \vec{KP} = 0.$$

Also we have

$$\vec{KP} = \vec{OP} - \vec{OK} = \mathbf{r} - p\mathbf{n}.$$

Thus

$$p\mathbf{n} \,.\, (r - p\mathbf{n}) = 0 \qquad ...(i)$$

$$\Rightarrow \qquad \mathbf{r} \,.\, \mathbf{n} = p\mathbf{n}^2 = p; \quad \mathbf{n} \text{ being a unit vector.}$$

This equation (*i*) is satisfied by the position vector of every point on the plane. Also, conversely, any point P whose position vector, **r**, satisfies (*i*), is a point of the plane. Thus

$$\mathbf{r} \,.\, \mathbf{n} = p$$

is the vector equation of the plane, such that **n** is the unit vector normal to the plane and p is the length of the perpendicular from the origin to the plane.

Cor. 1. The equation $\mathbf{r} \,.\, \mathbf{n} = q$ *is the equation of a plane normal to the vector* **n,** even if, **n,** be not a unit vector, for we may rewrite it as

$$\mathbf{r} \,.\, \frac{\mathbf{n}}{|\mathbf{n}|} = \frac{q}{|\mathbf{n}|}.$$

Hence, $q / |\mathbf{n}|$ is the length of the perpendicular from the origin of reference to the plane $\mathbf{r} \,.\, \mathbf{n} = q$.

Cor 2. *The vectorial equation of the plane normal to the vector,* **n,** *and passing through a point, A, with position vector,* **a** *is*

$$(\mathbf{r} - \mathbf{a}) \,.\, \mathbf{n} = 0$$

We can deduce it from above. Also *otherwise,* let **r** be the position vector of any point P on the plane. We have

$$\vec{AP} = \mathbf{r} - \mathbf{a}.$$

The line PA lies on the plane and is, as such, perpendicular to the vector, **n**, so that, we have $(\mathbf{r} - \mathbf{a}) \,.\, \mathbf{n} = 0$.

4.1.1. Planes Coaxal with given Planes

To find the equation of the plane coaxal with the planes

$$\mathbf{r} \,.\, \mathbf{n}_1 = q_1, \qquad \mathbf{r} \,.\, \mathbf{n}_2 = q_2 \qquad ...(1)$$

and passing through the point with position vector, **a.**

The equation

$$\mathbf{r} \,.\, \mathbf{n}_1 - q_1 + k\,(\mathbf{r} \,.\, \mathbf{n}_2 - q_2) = 0$$

$$\Rightarrow \qquad \mathbf{r} \,.\, (\mathbf{n}_1 + k\mathbf{n}_2) = q_1 + kq_2 \qquad ...(2)$$

represents a plane through the line of intersection of the two given planes, whatever value k may have. The plane (2) will pass through the point, **a**, for the value of k, given by

$$\mathbf{a} \cdot (\mathbf{n}_1 + k\mathbf{n}_2) = q_1 + kq_2 \qquad \text{...(3)}$$

$$\Rightarrow \qquad k = \frac{\mathbf{a} \cdot \mathbf{n}_1 - q_1}{q_2 - \mathbf{a} \cdot \mathbf{n}_2}.$$

Hence, the required equation is

$$(\mathbf{r} \cdot \mathbf{n}_1 - q_1) + \frac{\mathbf{a} \cdot \mathbf{n}_1 - q_1}{q_2 - \mathbf{a} \cdot \mathbf{n}_2}(\mathbf{r} \cdot \mathbf{n}_2 - q_2) = 0$$

$$\Leftrightarrow \qquad (\mathbf{r} \cdot \mathbf{n}_1 - q_1) \cdot (\mathbf{a} \cdot \mathbf{n}_2 - q_2) = (\mathbf{r} \cdot \mathbf{n}_2 - q_2) \cdot (\mathbf{a} \cdot \mathbf{n}_1 - q_1).$$

4.1.2. Angle between Two Planes

The angle θ, between the two planes

$$\mathbf{r} \cdot \mathbf{n}_1 = q_1, \qquad \mathbf{r} \cdot \mathbf{n}_2 = q_2,$$

being equal to the angle between the vectors $\mathbf{n}_1$ and $\mathbf{n}_2$ which are normal to the planes, we have

$$\theta = \cos^{-1} \frac{\mathbf{n}_1 . \mathbf{n}_2}{|\mathbf{n}_1|\ |\mathbf{n}_2|}.$$

4.1.3. Angle between a Line and a Plane

The angle θ, between any line

$$\mathbf{r} = \mathbf{a} + \mathbf{b}t$$

and any plane

$$\mathbf{r} \cdot \mathbf{n} = q,$$

being equal to the complement of the angle between the normal vector, **n**, of the plane and the direction vector, **b**, of the line, we have

$$\theta = \sin^{-1} \frac{\mathbf{n} \cdot \mathbf{b}}{|\mathbf{n}|\ |\mathbf{b}|}.$$

4.1.4. Perpendicular Distance of a Point from a Plane

Let **a** be the position vector of a given point A and

$$\mathbf{r} \cdot \mathbf{n} = q$$

be the equation of the given plane.

The equation of the line through A, normal to the plane is

$$\mathbf{r} = \mathbf{a} + t\mathbf{n}.$$

At the point of intersection of this line with the plane, we have

$$(\mathbf{a} + t\mathbf{n}) \cdot \mathbf{n} = q$$

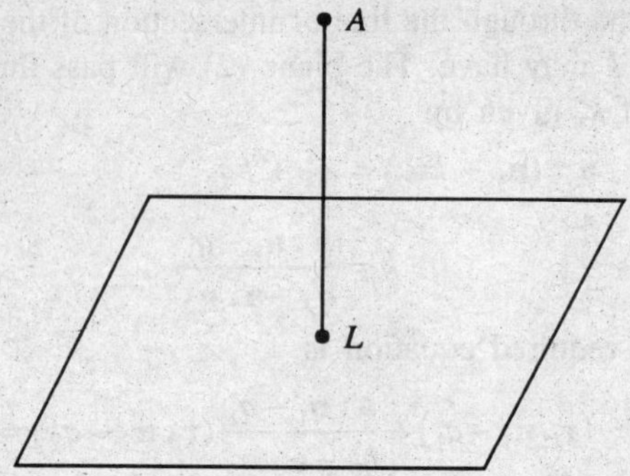

Fig. 4.2

$$\Rightarrow \qquad t = \frac{(q - \mathbf{a} \cdot \mathbf{n})}{\mathbf{n}^2}.$$

Thus, the foot of the perpendicular L from the point with position vector **a** *to the given place* **r** . **n** = *q is*

$$\mathbf{a} + \frac{q - \mathbf{a} \cdot \mathbf{n}}{\mathbf{n}^2} \mathbf{n}.$$

$\therefore$ the length $AL = |\overrightarrow{AL}| = \left| \mathbf{a} + \frac{q - \mathbf{a} \cdot \mathbf{n}}{\mathbf{n}^2} \mathbf{n} - \mathbf{a} \right|$

$$= \left| \frac{q - \mathbf{a} \cdot \mathbf{n}}{\mathbf{n}^2} \mathbf{n} \right| = \left| \frac{q - \mathbf{a} \cdot \mathbf{n}}{\mathbf{n}^2} \right| |\mathbf{n}|$$

$$= \frac{|q - \mathbf{a} \cdot \mathbf{n}|}{|\mathbf{n}|} \text{ for } \mathbf{n}^2 = |\mathbf{n}|^2.$$

4.1.5. Planes Bisecting the Angles between Two given Planes

Let

$$\mathbf{r} \cdot \mathbf{n}_1 = q_1 \text{ and } \mathbf{r} \cdot \mathbf{n}_2 = q_2$$

be the given planes. The perpendicular distance of any point, **r**, on either bisecting plane from the two planes being equal, we have

$$\frac{|\mathbf{r} \cdot \mathbf{n}_1 - q_1|}{|\mathbf{n}_1|} = \frac{|\mathbf{r} \cdot \mathbf{n}_2 - q_2|}{|\mathbf{n}_2|}$$

$$\Rightarrow \qquad \frac{\mathbf{r} \cdot \mathbf{n}_1 - q_1}{|\mathbf{n}_1|} = \pm \frac{\mathbf{r} \cdot \mathbf{n}_2 - q_2}{|\mathbf{n}_2|} \qquad ...(i)$$

Thus, (*i*) gives the equations of the two required bisecting planes. In the standard form, these may be re-written as

$$\mathbf{r} \cdot \left(\frac{\mathbf{n}_1}{|\mathbf{n}_1|} \pm \frac{\mathbf{n}_2}{|\mathbf{n}_2|} \right) = \left(\frac{q_1}{|\mathbf{n}_1|} \pm \frac{q_2}{|\mathbf{n}_2|} \right).$$

4.1.6. Perpendicular Distance of a Point from a Line

Foot of the Perpendicular. Let, **a** be the position vector of the given point P and let

$$\mathbf{r} = \mathbf{b} + \mathbf{c}t$$

be the equation of the given line AB. If, $\mathbf{b} + \mathbf{c}t$ be the position vector of the foot L of the perpendicular, we have

$$PL \perp AB$$

$$\Rightarrow \quad (\mathbf{b} + \mathbf{c}t - \mathbf{a}) \cdot \mathbf{c} = 0$$

$$\Rightarrow \quad (\mathbf{b} - \mathbf{a}) \cdot \mathbf{c} = -\mathbf{c}^2 t$$

$$\Rightarrow \quad t = \frac{(\mathbf{a} - \mathbf{b}) \cdot \mathbf{c}}{\mathbf{c}^2}$$

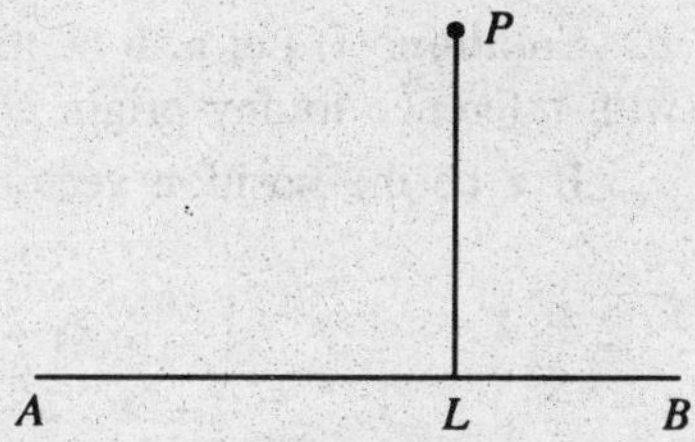

Fig. 4.3

Thus, the foot of the perpendicular is

$$\mathbf{b} + \frac{(\mathbf{a} - \mathbf{b}) \cdot \mathbf{c}}{\mathbf{c}^2} \mathbf{c}.$$

Also the perpendicular distance PL

$$= |\overrightarrow{PL}| = \left| \mathbf{b} + \frac{(\mathbf{a} - \mathbf{b}) \cdot \mathbf{c}}{\mathbf{c}^2} \mathbf{c} - \mathbf{a} \right|.$$

EXAMPLES

Example 1. *Find the projection of the line*

$$\mathbf{r} = \mathbf{a} + \mathbf{b}t$$

on the plane

$$\mathbf{r} \cdot \mathbf{n} = q.$$

Solution. The equation of the line through any point, $\mathbf{a} + \mathbf{b}t$, of the given line normal to the given plane is

$$\mathbf{r} = \mathbf{a} + \mathbf{b}t + p\mathbf{n},$$

so that the projection of the point, $\mathbf{a} + \mathbf{b}t$, on the given plane is given for the value of p, satisfying

$$(\mathbf{a} + \mathbf{b}t + p\mathbf{n}) \cdot \mathbf{n} = q$$

$$\Rightarrow \quad p = \frac{q - (\mathbf{a} + \mathbf{b}t) \cdot \mathbf{n}}{\mathbf{n}^2}$$

Thus, the projection of the point $\mathbf{a} + \mathbf{b}t$ is the point

$$\mathbf{a} + \mathbf{b}t + \frac{q - (\mathbf{a} + \mathbf{b}t) \cdot \mathbf{n}}{\mathbf{n}^2} \mathbf{n} = \left(\mathbf{a} + \frac{q - \mathbf{a} \cdot \mathbf{n}}{\mathbf{n}^2} \mathbf{n} \right) + t \left(\mathbf{b} - \frac{\mathbf{b} \cdot \mathbf{n}}{\mathbf{n}^2} \mathbf{n} \right).$$

Hence, the required projection is the line

$$\mathbf{r} = \left(\mathbf{a} + \frac{q - \mathbf{a}\,.\,\mathbf{n}}{\mathbf{n}^2}\mathbf{n}\right) + t\left(\mathbf{b} - \frac{\mathbf{b}\,.\,\mathbf{n}}{\mathbf{n}^2}\mathbf{n}\right).$$

Example 2. *Find the locus of a point equidistant from*
(i) two given points (ii) three given points.

Solution. (*i*) Let **a, b** be the position vectors of the given points *A, B,* with reference to any origin *O.*

If **r** be the position vector of any point *P* on the locus, we have

$$PA^2 = PB^2$$

$$\Rightarrow \quad (\mathbf{r} - \mathbf{a})^2 = (\mathbf{r} - \mathbf{b})^2$$

$$\Rightarrow \quad -2\mathbf{r}\,.\,\mathbf{a} + \mathbf{a}^2 = -2\mathbf{r}\,.\,\mathbf{b} + \mathbf{b}^2$$

$$\Rightarrow \quad \mathbf{r}\,.\,(\mathbf{a} - \mathbf{b}) = \frac{1}{2}(\mathbf{a}^2 - \mathbf{b}^2) = \frac{1}{2}(\mathbf{a} + \mathbf{b})\,(\mathbf{a} - \mathbf{b})$$

$$\Rightarrow \quad \left[\mathbf{r} - \frac{1}{2}(\mathbf{a} + \mathbf{b})\right] \cdot (\mathbf{a} - \mathbf{b}) = 0.$$

Thus, the required locus is the plane bisecting the line *AB* normally.

(*ii*) If **a, b, c** be the position vectors of the given points, then the required locus is the line of intersection of the planes

$$\left[\mathbf{r} - \frac{1}{2}(\mathbf{a} + \mathbf{b})\right] \cdot (\mathbf{a} - \mathbf{b}) = 0, \quad \left[\mathbf{r} - \frac{1}{2}(\mathbf{b} + \mathbf{c})\right] \cdot (\mathbf{b} - \mathbf{c}) = 0.$$

This line passes through the circumcentre of the triangle *A, B, C* and is perpendicular to the plane of the same.

Example 3. *Show that the centre of the sphere passing through the four points with position vectors* **a, b, c, d** *is the point common to the three planes*

$$\left[\mathbf{r} - \frac{1}{2}(\mathbf{a} + \mathbf{b})\right] \cdot (\mathbf{a} - \mathbf{b}) = 0, \quad \left[\mathbf{r} - \frac{1}{2}(\mathbf{b} + \mathbf{c})\right] \cdot (\mathbf{b} - \mathbf{c}) = 0,$$

$$\left[\mathbf{r} - \frac{1}{2}(\mathbf{c} + \mathbf{a})\right] \cdot (\mathbf{c} - \mathbf{a}) = 0$$

Solution. This is an immediate consequence of the preceding result.

Example 4. *Show that the six planes through the middle point of each edge of a tetrahedron and perpendicular to the opposite edge meet in a point.*

Solution. Taking the centroid of the tetrahedron as the origin of reference, let **a, b, c, d** be the position vectors of the four vertices. We have

$$\frac{1}{4}(\mathbf{a} + \mathbf{b} + \mathbf{c} + \mathbf{d}) = \mathbf{0} \Rightarrow \mathbf{a} + \mathbf{b} + \mathbf{c} + \mathbf{d} = \mathbf{0}.$$

The equation of the plane through the mid-point of the edge AB and the perpendicular to the opposite edge CD is

$$\left[\mathbf{r}-\frac{1}{2}(\mathbf{a}+\mathbf{b})\right]\cdot(\mathbf{d}-\mathbf{c})=0$$

$$\Leftrightarrow \qquad \left[\mathbf{r}+\frac{1}{2}(\mathbf{c}+\mathbf{d})\right]\cdot(\mathbf{d}-\mathbf{c})=\mathbf{0} \text{ for } \mathbf{a}+\mathbf{b}+\mathbf{c}+\mathbf{d}=\mathbf{0}.$$

$$\Leftrightarrow \qquad (\mathbf{r}+\mathbf{c})^2=(\mathbf{r}+\mathbf{d})^2.$$

Thus, every point on this plane is equidistant from the points $-\mathbf{c}$, $-\mathbf{d}$.

Considering other planes, we see that the six planes, in question, pass through the point given by

$$(\mathbf{r}+\mathbf{a})^2=(\mathbf{r}+\mathbf{b})^2=(\mathbf{r}+\mathbf{c})^2=(\mathbf{r}+\mathbf{d})^2$$

which is the centre of the sphere through the four points

$$-\mathbf{a},\ -\mathbf{b},\ -\mathbf{c},\ -\mathbf{d}.$$

This point is the centre of the circumsphere of the tetrahedron $A'B'C'D'$ where A', B', C', D' are the points such that AA', BB', CC', and DD' are bisected at the centroid of the tetrahedron $ABCD$.

Note. The tetrahedron $A'B'C'D'$ obtained above is said to be the Associate of the tetrahedron $ABCD$. Both these tetrahedra have the same centroid.

It is easy to see that the planes through the mid-point of each edge perpendicular to the opposite edge of the tetrahedron $A'B'C'D'$ pass through the circumcentre of the tetrahedron $ABCD$.

EXERCISES

1. Show that six planes bisecting the six edges of a tetrahedron perpendicularly meet in a point.
2. Find the equation of the plane through the line of intersection of the planes $\mathbf{r}\cdot\mathbf{n}_1=1$, $\mathbf{r}\cdot\mathbf{n}_2=1$ and perpendicular to the plane $\mathbf{r}\cdot\mathbf{n}_3=1$.
3. Find the reflection of the point, $\mathbf{a}$, in the plane

   $$\mathbf{r}\cdot\mathbf{n}=q.$$

   Also find the reflection of the line $\mathbf{r}=\mathbf{a}+t\mathbf{b}$ in the same plane.
4. Find the reflection of the point, $\mathbf{a}$, in the line

   $$\mathbf{r}=\mathbf{b}+t\mathbf{c}.$$
5. Show that the square of any straight line is equal to the sum of the squares of its projections on three mutually perpendicular straight lines.
6. Prove that the square on any straight line is equal to half the sum of the squares of its projections on three mutually perpendicular planes.
7. Find the distance of the point, $\mathbf{a}$, from the plane $\mathbf{r}\cdot\mathbf{n}=q$ measured parallel to the line $\mathbf{r}=\mathbf{b}+t\mathbf{c}$.

4.2. EQUATION OF A SPHERE

To find the equation of the sphere, whose centre is **c** *and radius* **a.**

Let O be the origin and C the centre of the sphere with position vector **c.** Let the position vector of any point P on the sphere be **r.** If **a** be the radius of the sphere, then

$$CP^2 = \mathbf{a}^2 \quad \text{or} \quad (\mathbf{r} - \mathbf{c})^2 = a^2$$

$$\Rightarrow \quad \mathbf{r}^2 - 2\mathbf{r} \cdot \mathbf{c} + \mathbf{c}^2 - \mathbf{a}^2 = 0$$

$$\Rightarrow \quad \mathbf{r}^2 - 2\mathbf{r} \cdot \mathbf{c} + k = 0 \quad ...(1)$$

where $k = \mathbf{c}^2 - \mathbf{a}^2$.

Since (1) is satisfied by the position vector of every point on the surface of the sphere hence it represents the equation of the sphere.

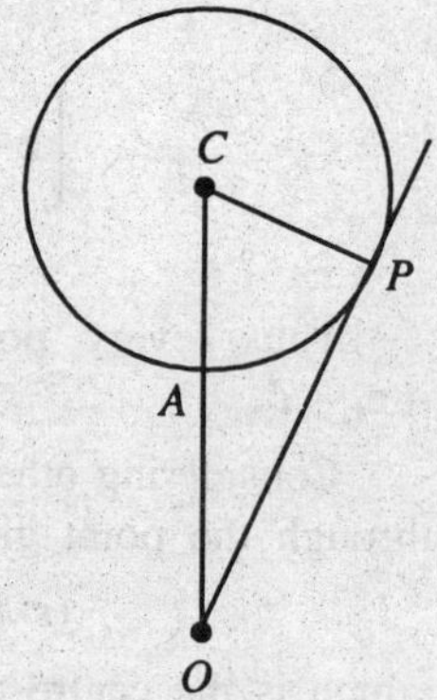

Fig. 4.4

Particular Cases :

(*i*) When the origin is the centre of the sphere, then $\mathbf{c} = 0$ and $k = \mathbf{c}^2 - \mathbf{a}^2 = -\mathbf{a}^2$. Hence, the equation of the sphere is

$$\mathbf{r}^2 = a^2.$$

(*ii*) When the origin lies on the surface of the sphere, then $|\mathbf{c}| = c = a$ and hence $k = \mathbf{c}^2 - a^2 = 0$. Therefore, the equation of the required sphere is

$$\mathbf{r}^2 - 2\mathbf{r} \cdot \mathbf{c} = 0.$$

4.2.1. Diameter Form of Equation of the Sphere

Let the position vectors of the extremities A and B of the diameter AB be **a** and **b** respectively. Let **r** be the position vector of any point P on the surface of the sphere. Then

$$\vec{AP} \cdot \vec{BP} = 0$$

$$\Rightarrow \quad (\mathbf{r} - \mathbf{a}) \cdot (\mathbf{r} - \mathbf{b}) = 0.$$

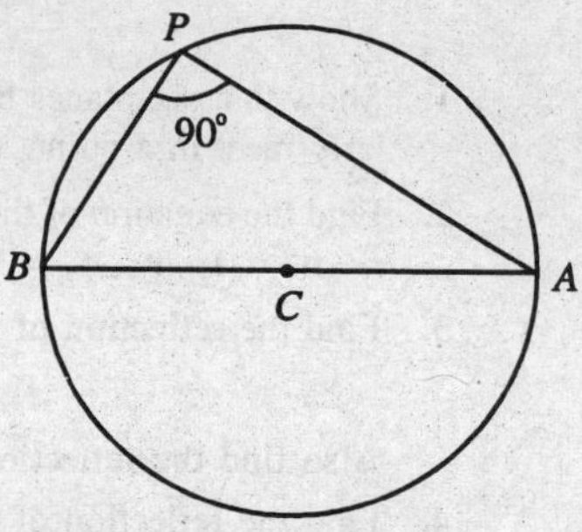

Fig. 4.5

4.2.2. Intersection of a Straight Line and a Sphere

Let

$$F(\mathbf{r}) \equiv \mathbf{r}^2 - 2\mathbf{r} \cdot \mathbf{c} + k = 0. \quad ...(1)$$

be the equation of the sphere, where $k = \mathbf{c}^2 - a^2$. Let the equation of the straight line be

$$\mathbf{r} = \mathbf{d} + t\,\hat{\mathbf{b}} \qquad ...(2)$$

which passes through the point D whose position vector is $\mathbf{d}$ and is parallel to a unit vector $\hat{\mathbf{b}}$.

For the points of intersection the values of $\mathbf{r}$ satisfy (1) and (2) both. On eliminating $\mathbf{r}$ between (1) and (2), we get

$$(\mathbf{d} + t\hat{\mathbf{b}})^2 - 2\,(\mathbf{d} + t\hat{\mathbf{b}})\,.\,\mathbf{c} + k = 0$$

$$\Rightarrow \quad t^2 + 2t\hat{\mathbf{b}}\,.(\mathbf{d} - \mathbf{c}) + \mathbf{d}^2 - 2\mathbf{d}\,.\,\mathbf{c} + k = 0$$

$$\Rightarrow \quad t^2 + 2t\hat{\mathbf{b}}\,.(\mathbf{d} - \mathbf{c}) + F(\mathbf{d}) = 0 \qquad ...(3)$$

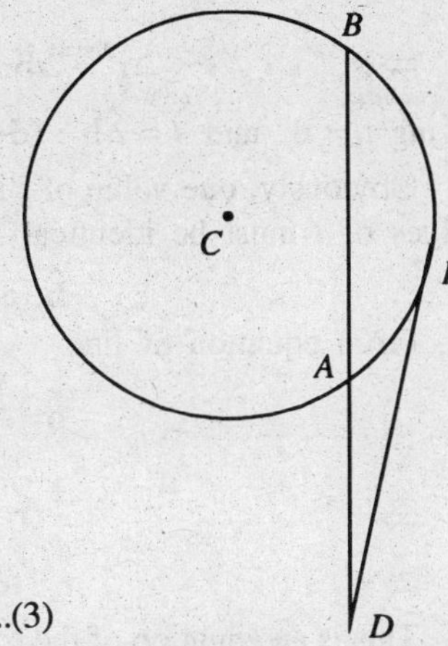

Fig. 4.6

This equation is quadratic in t and hence it gives two values t_1 and t_2 of t, when substituted in (2) give the position vectors of the two points A and B on the sphere.

Cor. 1. The points of intersection are real, coincident or imaginary according as

$$4\,[\hat{\mathbf{b}}\,.(\mathbf{d} - \mathbf{c})]^2 - 4F(\mathbf{d}) > 0 = 0 \text{ or } < 0.$$

Also $\quad t_1 = DA$ and $t_2 = DB$.

$\therefore \quad t_1 t_2 = DA\,.\,DB = F\,(\mathbf{d}) = \mathbf{d}^2 - 2\mathbf{d}\,.\,\mathbf{c} + k$

which is independent of $\hat{\mathbf{b}}$, *i.e.*, the direction of the line through the point D. Hence, for all lines through D, the product of the segment DA and DB remains unaltered.

Cor 2. If A and B tend to coincide at T, then the line becomes a tangent at T. Then

$$DT^2 = DA\,.\,DB = F\,(\mathbf{d})$$
$$= \mathbf{d}^2 - 2\mathbf{d}\,.\,\mathbf{c} + k.$$

This gives the *square of the length of the tangent from any point to the sphere.* It is also called the *power of the point D* with respect to the sphere.

4.2.3. Tangent Plane at a given Point

As in § 4.2.2, for the point of intersection of sphere (1) and line (2), we have

$$t^2 - 2t\hat{\mathbf{b}}\,.\,(\mathbf{d} - \mathbf{c}) + F\,(\mathbf{d}) = 0$$

since $\mathbf{d}$ lies on the sphere, hence

$$F\,(\mathbf{d}) = 0$$

$$\Rightarrow \qquad t^2 - 2t\hat{\mathbf{b}} \cdot (\mathbf{d}-\mathbf{c}) = 0$$

giving $t = 0$ and $t = 2\,\hat{\mathbf{b}} \cdot (\mathbf{d}-\mathbf{c})$

Obviously, one value of t is zero. If (2) touches (1) [§ 4.2.2], both the values of t must be identical

$$\therefore \qquad \hat{\mathbf{b}}.(\mathbf{d}-\mathbf{c}) = 0 \qquad \text{...(3)}$$

From equation of line

$$\hat{\mathbf{b}} = \frac{1}{t}(\mathbf{r}-\mathbf{d})$$

$$\Rightarrow \qquad \frac{1}{t}(\mathbf{r}-\mathbf{d})\cdot(\mathbf{d}-\mathbf{c}) = 0$$

$$\Rightarrow \qquad \mathbf{r} \cdot (\mathbf{d} - \mathbf{c}) = \mathbf{d} \cdot (\mathbf{d} - \mathbf{c}) \qquad \text{...(4)}$$

This is an equation of the form $\mathbf{r} \cdot \mathbf{n} = q$. So (4) represents a plane which is perpendicular to the radius through the point $\mathbf{d}$.

From (4),

$$\mathbf{r} \cdot \mathbf{d} - \mathbf{r} \cdot \mathbf{c} - \mathbf{d}^2 + \mathbf{d} \cdot \mathbf{c} = 0$$

or $$\mathbf{r} \cdot \mathbf{d} - \mathbf{r} \cdot \mathbf{c} - \mathbf{d}^2 + \mathbf{d} \cdot \mathbf{c} + (\mathbf{d}^2 - 2\mathbf{d} \cdot \mathbf{c} + k) = 0$$

or $$\mathbf{r} \cdot \mathbf{d} - \mathbf{c} \cdot (\mathbf{r} + \mathbf{d}) + k = 0 \qquad \text{...(5)}$$

This is the required equation to the tangent plane to the sphere (1) at the given point $\mathbf{d}$.

4.2.4. Condition for Tangency of any Plane to a Sphere

Let

$$\mathbf{r}^2 - 2\mathbf{r} \cdot \mathbf{c} + k = 0 \qquad \text{...(1)}$$

be the given sphere whose centre is $\mathbf{c}$ and radius a.

Let the given plane be

$$\mathbf{r} \cdot \mathbf{n} = q \qquad \text{...(2)}$$

Since the tangent plane at any point is perpendicular to the radius through that point, the square of the perpendicular from the centre to the tangent plane must be equal to the square of the radius. Hence, the required condition will be

$$\left(\frac{q - \mathbf{c} \cdot \mathbf{n}}{|\mathbf{n}|}\right)^2 = a^2 = \mathbf{c}^2 - k.$$

4.2.5. Condition of Orthogonality

Let the two spheres be

$$\mathbf{r}^2 - 2\mathbf{r} \cdot \mathbf{c}_1 + k_1 = 0; \quad k_1 = \mathbf{c}_1{}^2 - a_1{}^2 \qquad \text{...(1)}$$

$$\mathbf{r}^2 - 2\mathbf{r} \cdot \mathbf{c}_2 + k_2 = 0; \quad k_2 = \mathbf{c}_2{}^2 - a_2{}^2 \qquad \text{...(2)}$$

These spheres will intersect orthogonally if the tangent plane to any of them at the point of intersection passes through the centre of the other. So the square of the distance between the centres must be equal to the sum of the squares at their radii. Thus,

$$(\mathbf{c}_1 - \mathbf{c}_2)^2 = a_1^2 + a_2^2 = \mathbf{c}_1^2 - k_1 + \mathbf{c}_2^2 - k_2$$

$$\Rightarrow \quad 2\mathbf{c}_1 \cdot \mathbf{c}_2 = k_1 + k_2.$$

4.2.6. Polar Plane

Def. The polar plane of a given point with respect to a sphere is the locus of the points the tangent planes at which pass through the given point.

4.2.6.1. To find the polar plane of a point

Let

$$\mathbf{r}^2 - 2\mathbf{r} \cdot \mathbf{c} + k = 0$$

be the the given sphere and **a** the given point. Let **d** be any point on the sphere the tangent plane at which pass through **a**. Tangent plane at **d** to the sphere is

$$\mathbf{r} \cdot \mathbf{d} - \mathbf{c} \cdot (\mathbf{r} + \mathbf{d}) + k = 0$$

It passes through **a**, hence

$$\mathbf{a} \cdot \mathbf{d} - \mathbf{c} \cdot (\mathbf{a} + \mathbf{d}) + k = 0$$

∴ Locus of **d** is

$$\mathbf{a} \cdot \mathbf{r} - \mathbf{c} \cdot (\mathbf{a} + \mathbf{r}) + k = 0$$

or

$$\mathbf{r} \cdot \mathbf{a} - \mathbf{c} \cdot (\mathbf{r} + \mathbf{a}) + k = 0.$$

4.2.7. Diametral Plane

Def. The locus of points which bisect a system of parallel chords of a given sphere is called the diametral plane, all chords being parallel to a given line.

4.2.7.1. To find the diametral plane of a sphere

Let

$$F(\mathbf{r}) = \mathbf{r}^2 - 2\mathbf{r} \cdot \mathbf{c} + k = 0 \qquad \text{...(1)}$$

be the given sphere and let a chord parallel to a given unit vector $\hat{\mathbf{b}}$ and passing through the point **d** be

$$\mathbf{r} = \mathbf{d} + t\hat{\mathbf{b}} \qquad \text{...(2)}$$

As in § 4.2.2, the points of intersection of (1) and (2) are given by

$$t^2 - 2t\mathbf{b} \cdot (\hat{\mathbf{d}} - \mathbf{c}) + F(\mathbf{d}) = 0 \qquad \text{...(3)}$$

If *d* be the middle point of the chord (2), then the two values of *t* must be equal and opposite in sign, *i.e.*, the sum of the roots of (3) must be zero.

Thus, $\hat{\mathbf{b}} \cdot (\mathbf{d} - \mathbf{c}) = 0$

∴ the locus of the point **d** is

$$\hat{\mathbf{b}}.(\mathbf{r}-\mathbf{c})=0.$$

This is the required diametral plane.

4.2.8. Radical Plane

Def. The locus of a point, which moves so that the squares of the lengths of the tangents drawn from it to the given spheres are equal (or, powers of the point w.r.t. two spheres are equal), is called the radical plane of the two spheres.

4.2.8.1. To obtain radical plane of two spheres

Let $\quad \mathbf{r}^2 - 2\mathbf{r} \cdot \mathbf{c}_1 + k_1 = 0$

and $\quad \mathbf{r}^2 - 2\mathbf{r} \cdot \mathbf{c}_2 + k_2 = 0$

be the two given spheres. Let **d** be the position vector of a point which moves so that the square of the tangents drawn from it to spheres (1) are equal. Then

$$\mathbf{d}^2 - 2\mathbf{d} \cdot \mathbf{c}_1 + k_1 = \mathbf{d}^2 - 2\mathbf{d} \cdot \mathbf{c}_2 + k_2$$

or $\quad 2\mathbf{d} \cdot (\mathbf{c}_1 - \mathbf{c}_2) = k_1 - k_2$

$\therefore$ Locus of **d** is

$$2\mathbf{r} \cdot (\mathbf{c}_1 - \mathbf{c}_2) = k_1 - k_2.$$

Remark. The radical plane is perpendicular to $\mathbf{c}_1 - \mathbf{c}_2$. Thus, the radical plane of two spheres is perpendicular to the line joining the centres of the given spheres.

4.2.8.2. Coaxal System of Spheres

Def. A system of spheres every pair of which has the same radical plane is called a coaxal system of spheres.

4.2.8.3. Equation of a System of Coaxal Spheres

To show that the equation $\mathbf{r}^2 - 2\lambda\, \mathbf{r} \cdot \mathbf{c} + k = 0$, where λ is a parameter and k is constant represents a system of coaxal spheres.

Consider two members of the system of spheres corresponding to values λ_1 and λ_2 of the parameter λ,

$$\mathbf{r}^2 - 2\lambda_1\, \mathbf{r} \cdot \mathbf{c} + k = 0$$

and $\quad \mathbf{r}^2 - 2\lambda_2\, \mathbf{r} \cdot \mathbf{c} + k = 0.$

The radical plane of these spheres is

$$2(\lambda_1 - \lambda_2)\, \mathbf{r} \cdot \mathbf{c} = 0$$

$\Rightarrow \quad \mathbf{r} \cdot \mathbf{c} = 0, \qquad (\because \lambda_1 \neq \lambda_2)$

This is independent of λ. Hence, whatever be the value of λ, the radical plane of every two members will be the same. Thus, $\mathbf{r}^2 - 2\lambda\, \mathbf{r} \cdot \mathbf{c} + d = 0$, represents a coaxal system of spheres.

Cor. The member of coaxal system of zero radius are called *limiting points.* The radius of the sphere of the system is $\sqrt{\{(\lambda c)^2 - k\}}$. Equating to zero, we get $\lambda = \pm\sqrt{(k/c)}$. Hence, the limiting points are $\pm\sqrt{(k/c)}\,\mathbf{c}$.

EXAMPLES

Example 1. *Show that the diameter subtends a right angle at any point on the surface of the sphere.*

Solution. Let ACB be the diameter of a sphere whose centre is the origin C and whose radius is a.

Let P be any point on the surface of the sphere. Equation of the sphere is

$$\mathbf{r}^2 = \mathbf{a}^2$$

$$\Rightarrow \quad (\mathbf{r} - \mathbf{a})(\mathbf{r} + \mathbf{a}) = 0$$

$$\Rightarrow \quad \vec{AP} \cdot \vec{BP} = 0$$

showing that $\angle APB = 90^\circ$.

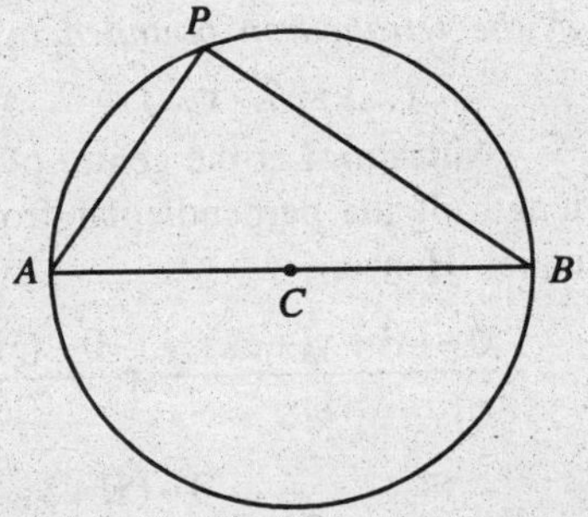

Fig. 4.7

Example 2. *A straight line is drawn from a point O to meet a fixed sphere in P. A point Q taken in OP such that OP : OQ is a fixed ratio. Show that the locus of Q is a sphere.*

Solution. With O as origin, let the fixed sphere be

$$\mathbf{r}^2 - 2\mathbf{r} \cdot \mathbf{c} + k = 0$$

Let $\vec{OQ} = \mathbf{r}'$ be the position vector Q and let $OP : OQ = n$.

$$\Rightarrow \quad \vec{OP} = n\vec{OQ} \;\Rightarrow\; \mathbf{r} = n\mathbf{r}'$$

Since P lies on the sphere, we must have

$$n^2 \mathbf{r}'^2 - 2n\mathbf{r}' \cdot \mathbf{c} + k = 0$$

or

$$\mathbf{r}'^2 - 2\mathbf{r}' \cdot (1/n)\,\mathbf{c} + k/n^2 = 0$$

$\therefore$ Locus of $\mathbf{r}'$ is

$$\mathbf{r}^2 - 2\mathbf{r} \cdot (1/n)\,\mathbf{c} + k/n^2 = 0,$$

which is a sphere.

Example 3. *Show that the sphere which cuts the spheres $F_1(\mathbf{r}) = 0$ and $F_2(\mathbf{r}) = 0$ orthogonally, also cuts $F_1(\mathbf{r}) - \lambda F_2(\mathbf{r}) = 0$ orthogonally.*

Solution. Let

$$F_1(\mathbf{r}) = \mathbf{r}^2 - 2\mathbf{r} \cdot \mathbf{c}_1 + k_1 = 0 \qquad ...(1)$$

and

$$F_2(\mathbf{r}) = \mathbf{r}^2 - 2\mathbf{r} \cdot \mathbf{c}_2 + k_2 = 0 \qquad ...(2)$$

be the given spheres. Then the sphere $F_1(\mathbf{r}) - F_2(\mathbf{r}) = 0$ is

$$\mathbf{r}^2 - 2\mathbf{r} \cdot \mathbf{c}_1 + k_1 - \lambda(\mathbf{r}^2 - 2\mathbf{r} \cdot \mathbf{c}_2 + k_2) = 0$$

or $$\mathbf{r}^2 - 2\mathbf{r}\cdot\frac{\mathbf{c}_1 - \lambda\mathbf{c}_2}{1-\lambda} + \frac{k_1 - \lambda k_2}{1-\lambda} = 0 \qquad ...(3)$$

Let the sphere $\mathbf{r}^2 - 2\mathbf{r} \cdot \mathbf{c} + k = 0$ cuts (1) and (2) orthogonally, then

$$2\mathbf{c} \cdot \mathbf{c}_1 = k + k_1 \qquad ...(4)$$

and $$2\mathbf{c} \cdot \mathbf{c}_2 = k + k_2 \qquad ...(5)$$

Multiplying (5) by λ and subtracting from (4), we get

$$2\mathbf{c} \cdot (\mathbf{c}_1 - \lambda\mathbf{c}_2) = k(1-\lambda) + k_1 - \lambda k_2$$

or $$2\mathbf{c}\cdot\frac{\mathbf{c}_1 - \lambda\mathbf{c}_2}{1-\lambda} = k + \frac{k_1 - \lambda k_2}{1-\lambda}$$

showing that the given sphere cuts (3) orthogonally.

Example 4. *Find the co-ordinates of the centre of the sphere inscribed in the tetrahedron bounded by the planes*

$$\mathbf{r} \cdot \mathbf{i} = 0,\ \ \mathbf{r} \cdot \mathbf{j} = 0,\ \ \mathbf{r} \cdot \mathbf{k} = 0 \ \text{ and } \mathbf{r} \cdot (\mathbf{i} + \mathbf{j} + \mathbf{k}) = a.$$

Solution. Let the centre of the given sphere be $x\mathbf{i} + y\mathbf{j} + z\mathbf{k}$. Then the length of the perpendicular from the centre on the given planes must be equal. Hence,

$$\frac{0-(x\mathbf{i}+y\mathbf{j}+z\mathbf{k})\cdot\mathbf{i}}{|\mathbf{i}|} = \frac{0-(x\mathbf{i}+y\mathbf{j}+z\mathbf{k})\cdot\mathbf{j}}{|\mathbf{j}|}$$

$$= \frac{0-(x\mathbf{i}+y\mathbf{j}+z\mathbf{k})\cdot\mathbf{k}}{|\mathbf{k}|} = \frac{a-(x\mathbf{i}+y\mathbf{j}+z\mathbf{k}),(\mathbf{i}+\mathbf{j}+\mathbf{k})}{|\mathbf{i}+\mathbf{j}+\mathbf{k}|}$$

$$\Rightarrow \quad \frac{x}{1} = \frac{y}{1} = \frac{z}{1} = \frac{a-(x+y+z)}{\sqrt{3}} = \frac{x+y+z+a-(x+y+z)}{1+1+1+\sqrt{3}} = \frac{a}{3+\sqrt{3}}$$

$$\therefore \quad x = y = z = \frac{a}{3+\sqrt{3}} = \frac{a}{6}\left(3-\sqrt{3}\right).$$

Example 5. *Prove that any straight line drawn from a point O to intersect a sphere is cut harmonically by the surface and the polar plane of O.*

Solution. Let

$$\mathbf{r}^2 - 2\mathbf{r} \cdot \mathbf{c} + k = 0$$

be the given sphere with O as origin. Polar plane of O w.r.t. the sphere is

$$O^2 - (\mathbf{r} + O) \cdot \mathbf{c} + k = 0$$

or $$\mathbf{r} \cdot \mathbf{c} = k.$$

Let the equation of the given line through O is

$$\mathbf{r} = t\hat{\mathbf{b}}$$

For the points of intersection of sphere and line, we have

$$t^2 - 2t(\hat{\mathbf{b}} \cdot \mathbf{c}) + k = 0$$

This is quadratic in t and so it gives two values t_1, t_2 which represent the distances OP and OQ respectively.

$$\therefore \quad \frac{1}{OP} + \frac{1}{OQ} = \frac{t_1 + t_2}{t_1 t_2} = \frac{2\hat{\mathbf{b}} \cdot \mathbf{c}}{k}.$$

Again, for the points of intersection of polar plane and the sphere, we have

$$t\hat{\mathbf{b}} \cdot \mathbf{c} = k, \text{ so that } t = OR$$

$$= \frac{k}{\hat{\mathbf{b}} \cdot \mathbf{c}} \quad \text{or} \quad \frac{1}{OR} = \frac{\hat{\mathbf{b}} \cdot \mathbf{c}}{k}$$

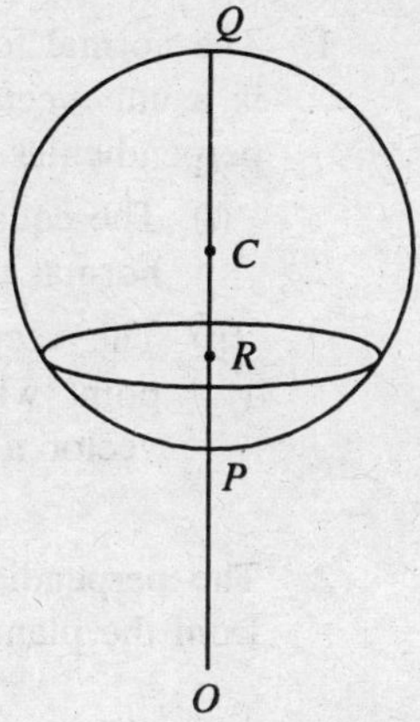

Fig. 4.8

Hence, $$\frac{1}{OP} + \frac{1}{OQ} = \frac{2}{OR}.$$

$\Rightarrow$ OP, OR, OQ are in $H.P.$

Hence, the result.

EXERCISES

1. Prove that the locus of a point, the sum of the squares of whose distances from n given points is constant, is a sphere.
2. Prove that the equation of the sphere circumscribing the tetrahedron bounded by the planes
$$\mathbf{r} \cdot \mathbf{i} = 0, \quad \mathbf{r} \cdot \mathbf{j} = 0, \quad \mathbf{r} \cdot \mathbf{k} = 0$$
and $$\mathbf{r} \cdot (\mathbf{i} + \mathbf{j} + \mathbf{k}) = a, \text{ is}$$
$$\mathbf{r} \cdot \{\mathbf{r} - a\,(\mathbf{i} + \mathbf{j} + \mathbf{k})\} = 0.$$
3. If the plane of the point H passes through G, then show that the polar plane of G passes through H.
4. If from any point on the surface of a sphere, straight lines are drawn to the extremities of any diameter of a concentric sphere, the sum of the squares on these lines is constant.
5. If the line joining the centre O of a sphere to any point P meets the polar plane of P in Q, prove that $OP \cdot OQ = a^2$, where a is the radius of the sphere.
6. Find the locus of a point whose powers with respect to two spheres are equal.
7. Prove by vector method that the triangle inscribed in a semi-circle is a right angle.
8. Show that the locus of the straight lines which intersect the sphere $\mathbf{r}^2 - 2\mathbf{r} \cdot \mathbf{c} + k = 0$ and bisected at a given point $\mathbf{d}$ is $(\mathbf{r} - \mathbf{d}) \cdot (\mathbf{d} - \mathbf{c}) = 0$.

SUMMARY

1. The normal form of the equation of a plane is $\mathbf{r} \cdot \mathbf{n} = p$. Here $\mathbf{n}$ is a unit vector normal to the plane and p is the length of the perpendicular from the origin to the plane.
   (*i*) The equation $\mathbf{r} \cdot \mathbf{n} = q$, represents a plane, $\mathbf{n}$ being a vector normal to the plane.
   (*ii*) The vector equation of the plane which passes through a point with position vector $\mathbf{a}$ and which is normal to the vector $\mathbf{n}$ is
$$(\mathbf{r} - \mathbf{a}) \cdot \mathbf{n} = 0.$$
2. The perpendicular distance of the point with position vector a from the plane $\mathbf{r} \cdot \mathbf{n} = q$ is
$$\frac{|q - \mathbf{a} \cdot \mathbf{n}|}{|\mathbf{n}|}.$$
3. Equation of the sphere, whose centre is $\mathbf{c}$ and radius $\mathbf{a}$ is
$$\mathbf{r}^2 - 2\mathbf{r} \cdot \mathbf{c} + k = 0, \quad \text{where} \quad k = \mathbf{c}^2 - a^2.$$
   (*i*) when the origin is the centre, the equation of sphere is
$$\mathbf{r}^2 = a^2$$
   (*ii*) when the origin lies on the surface of the sphere, the equation of the sphere is
$$\mathbf{r}^2 - 2\mathbf{r} \cdot \mathbf{c} = 0.$$
   (*iii*) Equation of the sphere on the join of two given points $\mathbf{a}$ and $\mathbf{b}$ as diameter is
$$(\mathbf{r} - \mathbf{a}) \cdot (\mathbf{r} - \mathbf{b}) = 0.$$
4. Equation of the tangent plane to the sphere $\mathbf{r}^2 - 2\mathbf{r} \cdot \mathbf{c} + k = 0$ at the point $\mathbf{d}$ is
$$\mathbf{r} \cdot \mathbf{d} - \mathbf{c} \cdot (\mathbf{r} + \mathbf{d}) + k = 0.$$
5. Condition that a plane $\mathbf{r} \cdot \mathbf{n} = q$ will touch the sphere $\mathbf{r}^2 - 2\mathbf{r} \cdot \mathbf{c} + k = 0$ is
$$\left(\frac{q - \mathbf{c} \cdot \mathbf{n}}{|\mathbf{n}|}\right)^2 = a^2 = \mathbf{c}^2 - k.$$
6. Two spheres $\mathbf{r}^2 - 2\mathbf{r} \cdot \mathbf{c}_1 + k_1 = 0$ and $\mathbf{r}^2 - 2\mathbf{r} \cdot \mathbf{c}_2 + k_2 = 0$ will intersect orthogonally if $2\mathbf{c}_1 \cdot \mathbf{c}_2 = k_1 + k_2$.
7. Polar plane of a point $\mathbf{a}$ with respect to the sphere $\mathbf{r}^2 - 2\mathbf{r} \cdot \mathbf{c} + k = 0$ is
$$\mathbf{r} \cdot \mathbf{a} - c \cdot (\mathbf{r} + \mathbf{a}) + k = 0.$$
8. Diameteral plane of a sphere $\mathbf{r}^2 - 2\mathbf{r} \cdot \mathbf{c} + k = 0$, parallel to a given unit vector $\hat{\mathbf{b}}$ is

$$\hat{\mathbf{b}} . (\mathbf{r} - \mathbf{c}) = 0.$$

9. The radical plane of two spheres can be obtained by subtracting the equation of one sphere from the other after making the coefficients of $\mathbf{r}^2$ in each equation unity.

10. Equation $\mathbf{r}^2 - 2\lambda\mathbf{r} . \mathbf{c} + k = 0$, where λ is a parameter and k is constant represents a system of coaxal spheres.

The members of coaxal system of zero radius are called limiting points. For the coaxal system $\mathbf{r}^2 - 2\lambda\mathbf{r} . \mathbf{c} + k = 0$, the limiting points are $\pm \sqrt{(k/c)}\, \mathbf{c}$.

OBJECTIVE QUESTIONS

For each of the following questions, four alternatives are given for the answer. Only one of them is correct. Choose the correct alternative.

1. The angle between two planes $\mathbf{r} . \mathbf{n} = q$ and $\mathbf{r} . \mathbf{n}' = q'$ is

(*a*) $\sin^{-1}\left(\frac{\mathbf{n} . \mathbf{n}'}{nn'}\right)$ (*b*) $\cos^{-1}\left(\frac{\mathbf{n} . \mathbf{n}'}{nn'}\right)$

(*c*) $\tan^{-1}\left(\frac{\mathbf{n} . \mathbf{n}'}{nn'}\right)$ (*d*) None of these

2. The intercept made by the plane $\mathbf{r} . \mathbf{n} = q$ on the x-axis is equal to

(*a*) $q / (\mathbf{i} . \mathbf{n})$ (*b*) $(\mathbf{i} . \mathbf{n}) / q$

(*c*) $q\,(\mathbf{i} . \mathbf{n})$ (*d*) None of these

3. Angle between a line $\mathbf{r} = \mathbf{a} + \mathbf{b}t$ and the plane $\mathbf{r} . \mathbf{n} = q$ is

(*a*) $\sin^{-1} \frac{\mathbf{n} . \mathbf{b}}{|\mathbf{n}|\,|\mathbf{b}|}$ (*b*) $\cos^{-1} \frac{\mathbf{n} . \mathbf{b}}{|\mathbf{n}|\,|\mathbf{b}|}$

(*c*) $\tan^{-1} \frac{\mathbf{n} . \mathbf{b}}{|\mathbf{n}|\,|\mathbf{b}|}$ (*d*) None of these

4. The equation of a sphere whose centre is the origin and radius **a** is

(*a*) $\mathbf{r}^2 - 2\mathbf{r} . \mathbf{a} = 0$ (*b*) $\mathbf{r} = \mathbf{a}$

(*c*) $\mathbf{r}^2 = \mathbf{a}^2$ (*d*) $\mathbf{r} . \mathbf{a} = k$

5. The two spheres $\mathbf{r}^2 - 2\mathbf{r} . \mathbf{c}_1 + k_1 = 0$ and $\mathbf{r}^2 - 2\mathbf{r} . \mathbf{c}_2 + k_2 = 0$ will cut orthogonally if

(*a*) $\mathbf{c}_1 . \mathbf{c}_2 = k_1 + k_2$ (*b*) $2\mathbf{c}_1 . \mathbf{c}_2 = k_1 + k_2$

(*c*) $\mathbf{c}_1 + \mathbf{c}_2 = k_1 k_2$ (*d*) $\mathbf{c}_1 + \mathbf{c}_2 = 2k_1 k_2$

ANSWERS

1. (*b*) **2.** (*a*) **3.** (*a*) **4.** (*c*) **5.** (*b*)

5

Vector Product and Scalar Triple Product

Introduction. The concept of the *Vector product* of two vectors, as a result of which a well-defined vector is associated to a given ordered pair of vectors, will be introduced in this chapter. Like Scalar products, Vector products are also proportional to the lengths of the factor vectors and obey Distributive Laws. It should be, however, remembered that quite a number of the properties of Scalar and Vector products are different from those of ordinary products of numbers.

The notion of vector products greatly facilitates the study of a type of metrical relations such as those involving areas and volumes. Also in those contexts where we require to deal with vectors perpendicular to two given vectors, vector products play an important part.

It will also be seen in the last chapter that characterisation of forces by vectors is possible with the help of Vector products.

In addition to Scalar and Vector products, *Mixed product*

$$\mathbf{a} \times \mathbf{b} \cdot \mathbf{c},$$

known as *Scalar triple product,* also plays an important part. We shall also consider this in this chapter.

The volume of a parallelopiped is given as a scalar triple product and the vanishing of a scalar triple product provides a very neat form of condition for three vectors to be coplanar and four points to be coplanar.

The part of the Algebra of vectors developed in this chapter will also provide important tools for the solution of Vector Equations.

5.1. RIGHT-HANDED AND LEFT-HANDED VECTOR TRIADS

A screw experiences a motion of translation when it is rotated and, as such, we can distinguish a screw as being right-handed or left-handed by the direction of its translation when it is rotated in any given manner. Also we shall now associate a screw with any given ordered vector triad and thus be enabled to distinguish any given vector triad also as being right-handed or left-handed.

Let **a, b, c** be three vectors, given in this order. We shall suppose that they are coinitial and no two are collinear.

Suppose, that, **a**, points towards the right of the reader and, **b**, towards the upper part of the paper.

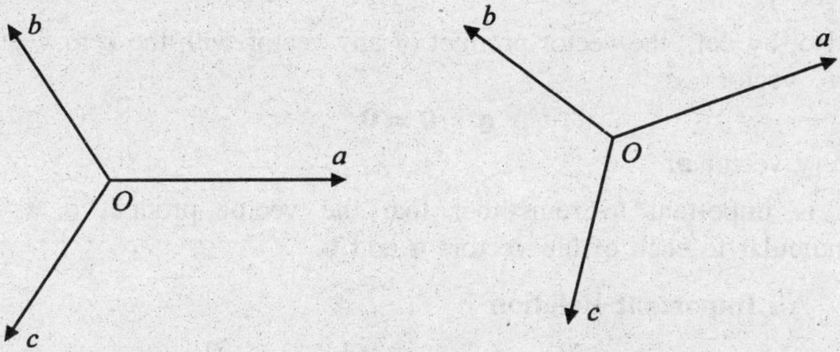

Left-hand Right-hand

Fig. 5.1

If now a right-handed (left-handed) screw be rotated from **a**, towards **b**, through an angle < 180°, then it undergoes translation in a direction pointing towards the reader (away from the reader). On this account, *an ordered vector triad*

a, b, c

is said to be right-handed or left-handed according as the right-handed screw is translated along **c** *or opposite to* **c**, *when it is rotated from* **a** *towards* **b**, *through an angle less than 180°.*

It is easily seen that a cyclic permutation of the vectors of a triad does not change the character of the triad from the point of view of its right-handedness and left-handedness. Thus, if **a, b, c** is a right-handed (left-handed) triad, then

b, c, a and **c, a, b**

are also right-handed (left-handed), but

a, c, b; b, a, c; c, b, a

are left-handed (right-handed).

5.2. VECTOR PRODUCT OR CROSS PRODUCT

Def. *The vector product, denoted by*

$$\mathbf{a} \times \mathbf{b}$$

of two vectors, **a, b** *taken in this order, is the vector* **c**, *where*

(*i*) $|\mathbf{c}| = |\mathbf{a}|\ |\mathbf{b}| \sin\theta,$

θ *being the angle between the vectors,* **a, b** *and* $0 \leq \theta \leq 180°$.

(*ii*) *the support of* **c** *is perpendicular to the supports of* **a** *and* **b**.

(*iii*) *the sense of* **c** *is such that the triad*

$$\mathbf{a}, \mathbf{b}, \mathbf{c}$$

forms a right-handed system.

As $0 \leq \theta \leq 180°$, $\sin \theta$ cannot be negative so that $|\,\mathbf{c}\,|$ will not be negative.

Also, by def., the vector product of any vector with the zero vector is the zero vector, *i.e.*,

$$\mathbf{a} \times \mathbf{0} = \mathbf{0}$$

for every vector **a**.

It is important to remember that the vector product $\mathbf{a} \times \mathbf{b}$ is perpendicular to each of the vectors **a** and **b**.

5.2.1. An Important Relation

$$(\mathbf{a} \times \mathbf{b})^2 = \mathbf{a}^2\mathbf{b}^2 - (\mathbf{a} \cdot \mathbf{b})^2.$$

We have

$$\begin{aligned}(\mathbf{a} \times \mathbf{b})^2 &= (\mathbf{a} \times \mathbf{b}) \cdot (\mathbf{a} \times \mathbf{b}) \\ &= (|\,\mathbf{a}\,|\ |\,\mathbf{b}\,| \sin \theta)^2 \\ &= \mathbf{a}^2\mathbf{b}^2 \sin^2 \theta \\ &= \mathbf{a}^2\mathbf{b}^2 (1 - \cos^2 \theta) \\ &= \mathbf{a}^2\mathbf{b}^2 - (|\,\mathbf{a}\,|\ |\,\mathbf{b}\,| \cos \theta)^2 \\ &= \mathbf{a}^2\mathbf{b}^2 - (\mathbf{a} \cdot \mathbf{b})^2\end{aligned}$$

This relation expresses the length of the vector $\mathbf{a} \times \mathbf{b}$ in terms of the scalar products

$$\mathbf{a} \cdot \mathbf{a}, \quad \mathbf{b} \cdot \mathbf{b}, \quad \mathbf{a} \cdot \mathbf{b}.$$

5.2.2. Formulation of Vector Product in Terms of Scalar Products

The vector product $\mathbf{a} \times \mathbf{b}$ is the vector **c**, such that

(*i*) $|\,\mathbf{c}\,| = \sqrt{[\mathbf{a}^2\mathbf{b}^2 - (\mathbf{a}.\mathbf{b})^2]}$

(*ii*) $\mathbf{c} \cdot \mathbf{a} = 0, \quad \mathbf{c} \cdot \mathbf{b} = 0.$

(*iii*) **a, b, c** form a right-handed system.

5.3. SOME PROPERTIES OF VECTOR PRODUCTS

5.3.1. (*i*) *The vector product of two parallel vectors is the zero vector*, for, in this case, $\theta = 0$ or $180°$, so that $\sin \theta = 0$ and as such $|\,\mathbf{c}\,| = 0$. Thus, $\mathbf{c} = \mathbf{0}$.

5.3.2. *The vector multiplication is not commutative.* In fact

$$\mathbf{a} \times \mathbf{b} = -\,\mathbf{b} \times \mathbf{a}.$$

This follows from the fact that while the magnitude and the support of $\mathbf{b} \times \mathbf{a}$ is the same as those of $\mathbf{a} \times \mathbf{b}$, their senses are different.

5.3.3.

$$-\mathbf{a} \times \mathbf{b} = -(\mathbf{a} \times \mathbf{b}), \quad \mathbf{a} \times (-\mathbf{b}) = -(\mathbf{a} \times \mathbf{b}),$$
$$(-\mathbf{a}) \times (-\mathbf{b}) = \mathbf{a} \times \mathbf{b}.$$

Generally,

$$m\mathbf{a} \times n\mathbf{b} = mn(\mathbf{a} \times \mathbf{b}) = \mathbf{a} \times mn\mathbf{b},$$

where m and n are any scalars, positive or negative.

These results are immediate consequences of the definition of Vector product.

5.3.4. Relations between the mutually perpendicular unit vectors i, j, k.

If **i, j, k** are mutually perpendicular unit vectors forming a right-handed system, then, as may be easily seen, from the definition of vector products, we have

$$\mathbf{i} \times \mathbf{i} = \mathbf{0}, \quad \mathbf{j} \times \mathbf{j} = \mathbf{0}, \quad \mathbf{k} \times \mathbf{k} = \mathbf{0},$$
$$\mathbf{i} \times \mathbf{j} = \mathbf{k}, \quad \mathbf{j} \times \mathbf{i} = -\mathbf{k};$$
$$\mathbf{j} \times \mathbf{k} = \mathbf{i}, \quad \mathbf{k} \times \mathbf{j} = -\mathbf{i};$$
$$\mathbf{k} \times \mathbf{i} = \mathbf{j}, \quad \mathbf{i} \times \mathbf{k} = -\mathbf{j}.$$

Note. Later on, we shall show that the Distributive Law also holds for vector multiplications, *i.e.*,

$$\mathbf{a} \times (\mathbf{b} + \mathbf{c}) = (\mathbf{a} \times \mathbf{b}) + (\mathbf{a} \times \mathbf{c}) \text{ for all vectors } \mathbf{a}, \mathbf{b}, \mathbf{c}.$$

After establishing the Distributive Law, we shall be able to express

$$(a_1\mathbf{i} + a_2\mathbf{j} + a_3\mathbf{k}) \times (b_1\mathbf{i} + b_2\mathbf{j} + b_3\mathbf{k})$$

as a linear combination of the vectors, **i, j, k.**

5.3.5. Vector Product as a Determinant

Let $\mathbf{a} = a_1\mathbf{i} + a_2\mathbf{j} + a_3\mathbf{k}, \ \mathbf{b} = b_1\mathbf{i} + b_2\mathbf{j} + b_3\mathbf{k}$

$$\mathbf{a} \times \mathbf{b} = a_1b_1\, \mathbf{i} \times \mathbf{i} + a_1b_2\, \mathbf{i} \times \mathbf{j} + a_1b_3\, \mathbf{i} \times \mathbf{k}$$
$$+ a_2b_1\, \mathbf{j} \times \mathbf{i} + a_2b_2\, \mathbf{j} \times \mathbf{j} + a_2b_3\, \mathbf{j} \times \mathbf{k}$$
$$+ a_3b_1\, \mathbf{k} \times \mathbf{i} + a_3b_2\, \mathbf{k} \times \mathbf{j} + a_3b_3\, \mathbf{k} \times \mathbf{k}$$

By § 5.3.4, it becomes

$$\mathbf{a} \times \mathbf{b} = (a_2b_3 - a_3b_2)\,\mathbf{i} + (a_3b_1 - a_1b_3)\,\mathbf{j} + (a_1b_2 - a_2b_1)\,\mathbf{k}$$

This expression can be re-written in the determinant form as

$$\mathbf{a} \times \mathbf{b} = \begin{vmatrix} \mathbf{i} & \mathbf{j} & \mathbf{k} \\ a_1 & a_2 & a_3 \\ b_1 & b_2 & b_3 \end{vmatrix}$$

5.3.6. Angle between Two Vectors a and b

From § 5.2, we have

$$|\mathbf{a} \times \mathbf{b}| = |\mathbf{a}|\ |\mathbf{b}| \sin\theta$$

$$\Rightarrow \qquad \sin\theta = \frac{|\mathbf{a}\times\mathbf{b}|}{|\mathbf{a}|\ |\mathbf{b}|}$$

$$= \left|\hat{\mathbf{a}}\times\hat{\mathbf{b}}\right|$$

In terms of components, it can easily be shown that

$$\sin\theta = \frac{\sqrt{\{(a_2b_3-a_3b_2)^2+(a_3b_1-a_1b_3)^2+(a_1b_2-a_2b_1)^2\}}}{\sqrt{(a_1^2+a_2^2+a_3^2)}+\sqrt{(b_1^2+b_2^2+b_3^2)}}$$

5.3.7. General Form of Vector Product

Let **a** and **b** be any two vectors and *l, m, n* be three non-coplanar vectors.

Let $\mathbf{a} = a_1\mathbf{l} + a_2\mathbf{m} + a_3\mathbf{n}$ and $\mathbf{b} = b_1\mathbf{l} + b_2\mathbf{m} + b_3\mathbf{n}$. Then

$$\mathbf{a}\times\mathbf{b} = (a_1\mathbf{l} + a_2\mathbf{m} + a_3\mathbf{n}) \times (b_1\mathbf{l} + b_2\mathbf{m} + b_3\mathbf{n})$$

$$= (a_1b_2 - a_2b_1)\,\mathbf{l}\times\mathbf{m} - (a_1b_3 - a_3b_1)$$

$$\mathbf{n}\times\mathbf{l} + (a_2b_3 - a_3b_2)\,\mathbf{m}\times\mathbf{n}$$

$$= \begin{vmatrix} \mathbf{m}\times\mathbf{n} & \mathbf{n}\times\mathbf{l} & \mathbf{l}\times\mathbf{m} \\ a_1 & a_2 & a_3 \\ b_1 & b_2 & b_3 \end{vmatrix}$$

EXAMPLES

Example 1. *Determine a unit vector perpendicular to the plane of* **a** *and* **b**, *where* $\mathbf{a} = 4\mathbf{i} + 3\mathbf{j} - \mathbf{k}$ *and* $\mathbf{b} = 2\mathbf{i} - 6\mathbf{j} - 3\mathbf{k}$. *Also obtain sine of the angle between* **a** *and* **b.**

Solution. $\mathbf{a}\times\mathbf{b} = \begin{vmatrix} \mathbf{i} & \mathbf{j} & \mathbf{k} \\ 4 & 3 & -1 \\ 2 & -6 & -3 \end{vmatrix}$

$$= -15\mathbf{i} + 10\mathbf{j} + 30\mathbf{k}.$$

As **a** × **b** is the vector perpendicular to the plane of **a** and **b**, hence a unit vector perpendicular to the plane of **a** and **b** is

$$= \frac{\mathbf{a}\times\mathbf{b}}{|\mathbf{a}\times\mathbf{b}|} = \frac{-15\mathbf{i}+10\mathbf{j}-30\mathbf{k}}{\sqrt{(-15)^2+(10)^2+(-30)^2}}$$

$$-\frac{3}{7}\mathbf{i}+\frac{2}{7}\mathbf{j}+\frac{6}{7}\mathbf{k}.$$

If θ be the angle between the vectors, then

$$\sin\theta = \frac{|\mathbf{a}\times\mathbf{b}|}{|\mathbf{a}|\,|\mathbf{b}|} = \frac{\sqrt{\{(-15)^2+10^2+(-30)^2\}}}{\sqrt{\{4^2+3^2+(-1)^2\}}\,\sqrt{\{2^2+(-6)^2+(-3)^2\}}}$$

$$= \frac{35}{\sqrt{26}\,\sqrt{(49)}} = \frac{5\sqrt{26}}{26}.$$

Example 2. *If* $\mathbf{a}\times\mathbf{b} = \mathbf{c}\times\mathbf{d}$ *and* $\mathbf{a}\times\mathbf{c} = \mathbf{b}\times\mathbf{d}$, *show that* $\mathbf{a}-\mathbf{d}$ *and* $\mathbf{b}-\mathbf{c}$ *are parallel vectors.*

Solution. $(\mathbf{a}-\mathbf{d})\times(\mathbf{b}-\mathbf{c}) = \mathbf{a}\times\mathbf{b} - \mathbf{a}\times\mathbf{c} - \mathbf{d}\times\mathbf{b} + \mathbf{d}\times\mathbf{c}$

$$= (\mathbf{a}\times\mathbf{b} - \mathbf{c}\times\mathbf{d}) - (\mathbf{a}\times\mathbf{c} - \mathbf{b}\times\mathbf{d})$$

$$= 0 \qquad \text{[as given]}$$

$\therefore$ $(\mathbf{a}-\mathbf{d})$ and $(\mathbf{b}-\mathbf{c})$ are parallel vectors.

Example 3. *Prove by vector method that*

$$\sin(\alpha-\beta) = \sin\alpha\cos\beta - \cos\alpha\sin\beta.$$

Solution. Let OX and OY be a set of rectangular co-ordinate axes with O as origin. Let OA and OB subtend angles **a** and **b** with OX. Let **k** be the usual unit vector perpendicular to the plane of the paper such that, **i, j, k** form a right-handed system.

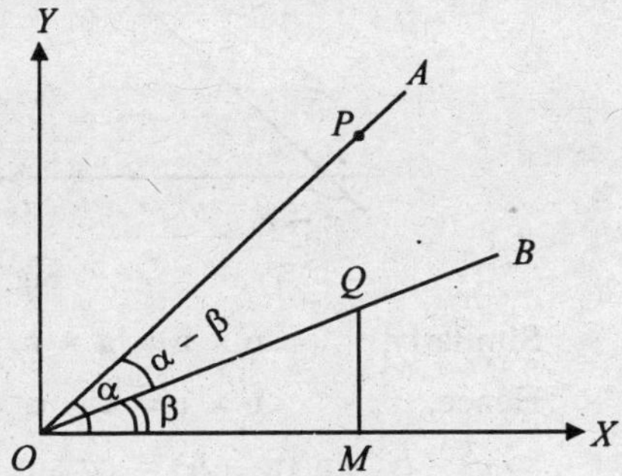

Fig. 5.2.

Let $OP = \hat{\mathbf{p}}$ and $OQ = \hat{\mathbf{q}}$ be the unit vectors along OA and OB respectively. Then, we have

$$\hat{\mathbf{q}}\times\hat{\mathbf{p}} = 1.1\,.\sin(\alpha-\beta)\,k$$

$$= \sin(\alpha-\beta)\,k. \qquad ...(1)$$

Let QM be perpendicular to OX. We have

$$OM = OQ\cos\beta = \cos\beta \text{ and } MQ = OQ\sin\beta = \sin\beta$$

$$\left[\because OQ = \left|\vec{OQ}\right| = |\hat{\mathbf{q}}| = 1\right]$$

$\therefore$ $\hat{\mathbf{q}} = (\cos\beta)\,\mathbf{i} + (\sin\beta)\,\mathbf{j}$

Similarly, $\hat{\mathbf{p}} = (\cos\alpha)\,\mathbf{i} + (\sin\alpha)\,\mathbf{j}$

$\therefore$ $\hat{\mathbf{q}}\times\hat{\mathbf{p}} = [(\cos\beta)\,\mathbf{i} + (\sin\beta)\,\mathbf{j}]\times[(\cos\alpha)\,\mathbf{i} + (\sin\alpha)\,\mathbf{j}]$

$$= (\sin\alpha\cos\beta - \cos\alpha\sin\beta)\,\hat{\mathbf{k}}.$$

Hence, from (1), we get

$$\sin(\alpha - \beta) = \sin\alpha\cos\beta - \cos\alpha\sin\beta.$$

Example 4. *If* **a, b, c** *be three vectors such that* $\mathbf{a} + \mathbf{b} + \mathbf{c} = 0$, *prove that* $\mathbf{a} \times \mathbf{b} = \mathbf{b} \times \mathbf{c} = \mathbf{c} \times \mathbf{a}$ *and deduce the sine rule*

$$\frac{\sin A}{a} = \frac{\sin B}{b} = \frac{\sin C}{c}. \qquad \text{(Kolkata, 99)}$$

Solution. Let $\overrightarrow{BC}, \overrightarrow{CA}, \overrightarrow{AB}$ represent the vectors **a, b, c** respectively. Then, we have

$$\mathbf{a} + \mathbf{b} + \mathbf{c} = 0,$$

$$\Rightarrow \quad \mathbf{c} = -(\mathbf{a} + \mathbf{b})$$

$$\Rightarrow \quad \mathbf{b} \times \mathbf{c} = \mathbf{b} \times (-\mathbf{a} - \mathbf{b})$$

$$= -\mathbf{b} \times \mathbf{a} = \mathbf{a} \times \mathbf{b}$$

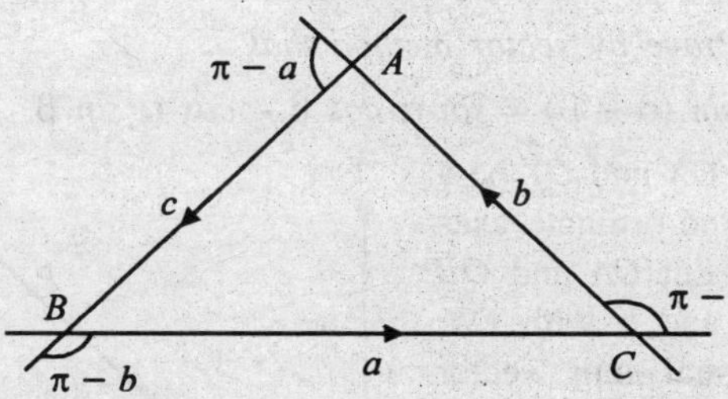

Fig. 5.3.

Similarly, $\quad \mathbf{c} \times \mathbf{a} = \mathbf{a} \times \mathbf{b}.$

Hence, $\quad \mathbf{b} \times \mathbf{c} = \mathbf{c} \times \mathbf{a} = \mathbf{a} \times \mathbf{b}$

$$\Rightarrow \quad bc\sin(\pi - A) = ca\sin(\pi - B)$$

$$= ab\sin(\pi - C)$$

$$\Rightarrow \quad bc\sin A = ca\sin B = ab\sin C$$

$$\Rightarrow \quad \frac{\sin A}{a} = \frac{\sin B}{b} = \frac{\sin C}{c}.$$

Example 5. *Given that* $\mathbf{a} \cdot \mathbf{b} = \mathbf{a} \cdot \mathbf{c}$, $\mathbf{a} \times \mathbf{b} = \mathbf{a} \times \mathbf{c}$ *and* **a** *is a non-zero vector. Show that* $\mathbf{b} = \mathbf{c}$.

Solution. $\quad \mathbf{a} \cdot \mathbf{b} = \mathbf{a} \cdot \mathbf{c}$ and $\mathbf{a} \neq \mathbf{0}$

$\Rightarrow \quad \mathbf{a} \cdot (\mathbf{b} - \mathbf{c}) = 0$

$\Rightarrow \quad \mathbf{b} - \mathbf{c} = 0$ or **a** is perpendicular to $(\mathbf{b} - \mathbf{c})$ ...(1)

Again $\quad \mathbf{a} \times \mathbf{b} = \mathbf{a} \times \mathbf{c}$ and $\mathbf{a} \neq \mathbf{0}$

$\Rightarrow \quad \mathbf{a} \times (\mathbf{b} - \mathbf{c}) = \mathbf{0}$

$\Rightarrow \quad \mathbf{b} - \mathbf{c} = 0$ or **a** is parallel to $(\mathbf{b} - \mathbf{c})$ ...(2)

Since the two relations hold simultaneously. Both **a** is perpendicular to (**b** – **c**) and **a** is parallel to (**b** – **c**) is an absurd result. Hence, the only possibility is **b** – **c** = 0 so that **b** = **c**.

EXERCISES

1. Prove that the unit vector perpendicular to each of the vectors 3**i** + **j** + 2**k** and 2**i** – 2**j** + 4**k** is (**i** – **j** – **k**) / $\sqrt{(3)}$ and the sine of the angle between them in $2/\sqrt{(7)}$.
2. If $\hat{\mathbf{a}}, \hat{\mathbf{b}}, \hat{\mathbf{c}}$ be unit vectors such that $\hat{\mathbf{a}}$ and $\hat{\mathbf{b}}$ are perpendicular and $\hat{\mathbf{c}}$ is inclined at an angle θ to both $\hat{\mathbf{a}}$ and $\hat{\mathbf{b}}$.

   Show that $\mathbf{c} = \alpha\,(\mathbf{a} + \mathbf{b}) + \beta\,(\mathbf{a} \times \mathbf{b})$.

   where $\alpha = -\cos\theta$ and $\beta = -\cos 2\theta$.
3. If **a** . **b** = 0 and **a** × **b** = 0 simultaneously, prove that atleast one of the vectors **a** and **b** must be a null vector.
4. Show that the vector **a** × (**b** × **a**) is perpendicular to **a** and coplanar with **a** and **b**.
5. Show that **a** × (**b** × **c**) is coplanar with **b** and **c**.
6. **a**, **b**, **c** are three vectors such that

   $$\mathbf{a} \times \mathbf{b} = \mathbf{c}, \quad \mathbf{b} \times \mathbf{c} = \mathbf{a};$$

   show that the three vectors **a**, **b**, **c** are orthogonal in pairs and

   $$|\mathbf{b}| = 1, \quad |\mathbf{c}| = |\mathbf{a}|.$$
7. **a**, **b** are two vectors. Also **c**, **d** are two vectors coplanar with **a** and **b** and perpendicular to **a** and **b** respectively. Show that

   $(\mathbf{a} \times \mathbf{b}) \times \mathbf{c} = (\mathbf{a} \cdot \mathbf{c})\,\mathbf{b} - (\mathbf{b} \cdot \mathbf{c})\,\mathbf{a}$, $(\mathbf{a} \times \mathbf{b}) \times \mathbf{d} = (\mathbf{a} \cdot \mathbf{d})\,\mathbf{b} - (\mathbf{b} \cdot \mathbf{d})\,\mathbf{a}$
8. Show that **a** . (**b** × **c**) = 0 if and only if the vectors **a**, **b**, **c** are coplanar.

5.4. INTERPRETATION OF VECTOR PRODUCT AS VECTOR AREA

We shall now show how it is possible to represent a plane area bounded by a closed curve which does not cross itself by a vector. For this purpose it is necessary to distinguish between the two senses in which the curve may be described.

A plane area bounded by a closed curve which does not cross itself is represented by a vector, **c**, defined as follows :

(*i*) The number of units of the length of **c** is equal to the number of units of the given area.

(*ii*) The support of **c** is perpendicular to the plane of the area.

(*iii*) The sense of **c** is such that the direction of description of the boundary of the curve and the sense of **c** correspond to a right-handed screw.

The sense of **c** will be reversed, if we reverse the direction of description of the boundary of the area.

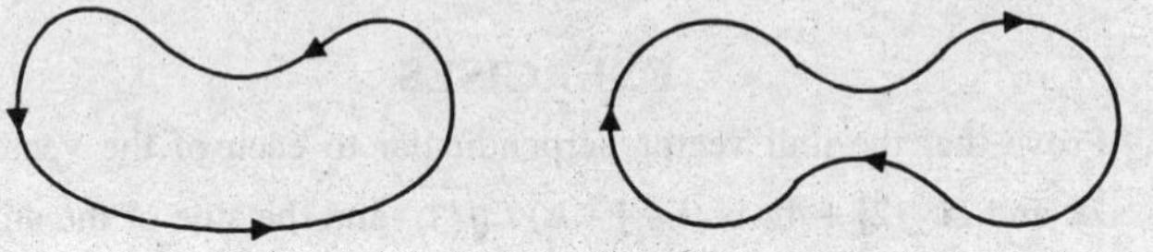

Fig. 5.4.

5.4.1. Area of a Triangle as Vector Product

Let $\overrightarrow{OA} = \mathbf{a},\ \overrightarrow{OB} = \mathbf{b}.$

Consider the triangle OAB and the vector representing the vector area.

$$\overrightarrow{\Delta OAB}.$$

We shall see that

$$\overrightarrow{\Delta OAB} = \frac{1}{2}\mathbf{a} \times \mathbf{b}. \qquad ...(i)$$

The magnitude of each side of (*i*) is $\frac{1}{2} OA \,.\, OB \sin\theta$ and the support of each is also the same. Also according to our conventions the senses of the vectors on the two sides of (*i*) are the same.

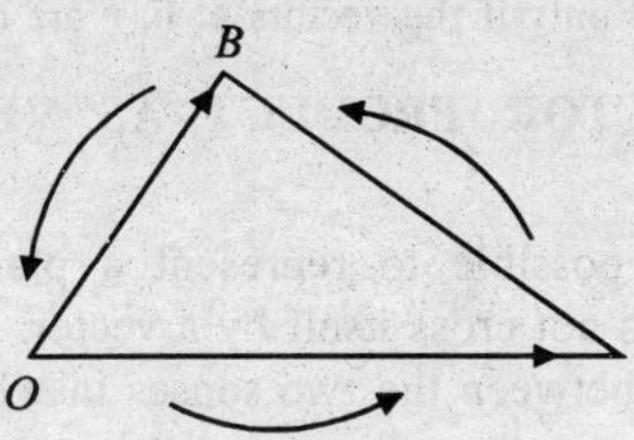

Fig. 5.5.

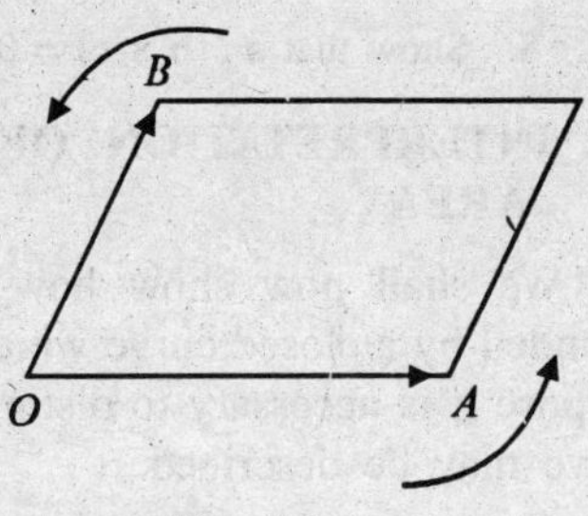

Fig. 5.6.

Hence, the equality (*i*). We also have

$$\Delta OAB = \frac{1}{2}\mathbf{a} \times \mathbf{b} = \frac{1}{2}\ \overrightarrow{OA} \times \overrightarrow{OB}.$$

It may also be now easily seen that **a** × **b** is the vector area of the parallelogram constructed with $\overrightarrow{OA}$ and $\overrightarrow{OB}$ as adjacent sides (Fig. 5.6).

5.5. SCALAR TRIPLE PRODUCT (*Patna 2003, Avadh 2005*)

The use of scalar triple product will enable us to prove that vector product distributes the sum of vectors. Also the concept of scalar triple product plays a useful part in the applications of vectors of Geometry.

Let **a, b, c** be any three vectors. Consider the expression

$$\mathbf{a} \times \mathbf{b} \cdot \mathbf{c}$$

which is the scalar product of the vectors **a** × **b** and **c**

Let

$$\overrightarrow{OA} = \mathbf{a},\ \overrightarrow{OB} = \mathbf{b},\ \overrightarrow{OC} = \mathbf{c}.$$

The figures show only the three faces of the parallelopiped through the point *O*.

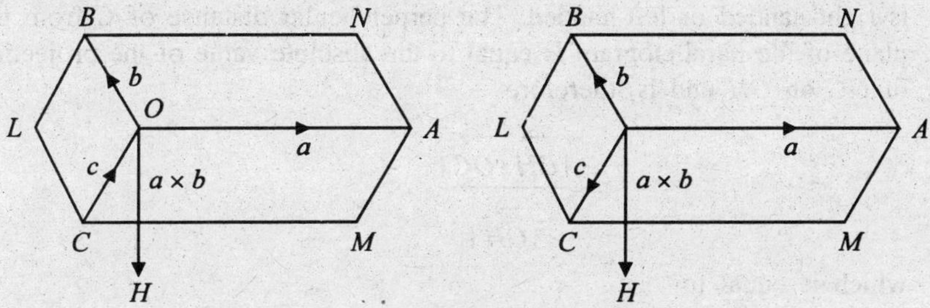

Fig. 5.7.

Firstly suppose that the three vectors are coplanar. The vector **a** × **b,** being perpendicular to the plane of the vectors, is perpendicular to the vector, **c** and, as such

$$\mathbf{a} \times \mathbf{b} \cdot \mathbf{c} = 0.$$

Next suppose that the three vectors are not coplanar so that the lines *OA, OB* and *OC* do not lie in the same plane.

Complete the parallelopiped with *OA, OB* and *OC* as coterminous edges. Let *V*, be the volume of this parallelopiped. We shall regard, *V*, as necessarily positive.

We shall now prove the following :

$$V = [\mathbf{a} \times \mathbf{b} \cdot \mathbf{c}] \quad \text{or} \quad V = -[\mathbf{a} \times \mathbf{b} \cdot \mathbf{c}]$$

according as the ordered vector triad **a, b, c** *is right-handed or left-handed.*

The vector **a** × **b** denotes the vector area of the parallelogram with *OA* and *OB* as adjacent sides.

Let

$$\overrightarrow{OH} = \mathbf{a} \times \mathbf{b}.$$

The volume of the parallelopiped is the product of the area of this parallelogram with the perpendicular distance of C from its plane.

The vector triad

$$\mathbf{a},\ \mathbf{b},\ \mathbf{a}\times\mathbf{b}$$

is, by definition, right-handed.

The angle between

$$\mathbf{a}\times\mathbf{b} \text{ and } \mathbf{c}$$

i.e., between

$$\overrightarrow{OH} \text{ and } \overrightarrow{OC}$$

is acute or obtuse according as the vector triad

$$\mathbf{a},\ \mathbf{b},\ \mathbf{c}$$

is right-handed or left-handed. The perpendicular distance of C from the plane of the parallelogram is equal to the absolute value of the projection of OC on OH and is, therefore

$$\frac{|\overrightarrow{OH}\cdot\overrightarrow{OC}|}{|\overrightarrow{OH}|}$$

which is equal to

$$\frac{\overrightarrow{OH}\cdot\overrightarrow{OC}}{|\overrightarrow{OH}|} \text{ or } \frac{-\overrightarrow{OH}\cdot\overrightarrow{OC}}{|\overrightarrow{OH}|}$$

according as the angle between $\overrightarrow{OH}$ and $\overrightarrow{OC}$ is acute or obtuse.

Thus, if **a, b, c** is a right-handed system, we have

$$V=\frac{\overrightarrow{OH}\cdot\overrightarrow{OC}}{|\overrightarrow{OH}|}\,|\overrightarrow{OH}|=\overrightarrow{OH}\cdot\overrightarrow{OC}=\mathbf{a}\times\mathbf{b}\,.\,\mathbf{c}$$

and if **a, b, c** is a left-handed system, then

$$V=\frac{\overrightarrow{OH}\cdot\overrightarrow{OC}}{|\overrightarrow{OH}|}\,|\overrightarrow{OH}|=-\overrightarrow{OH}\cdot\overrightarrow{OC}=-\,\mathbf{a}\times\mathbf{b}\cdot\mathbf{c}.$$

The foregoing investigation also shows that

$$\mathbf{a}\times\mathbf{b}\ .\ \mathbf{c}$$

is positive or negative according as **a, b, c** is a right or left-handed system.

Def. *The expression* $\mathbf{a} \times \mathbf{b} \cdot \mathbf{c}$ *is called a scalar triple or mixed product of the three vectors* **a, b, c** *taken in this order.*

5.5.1. Different Scalar Triple Products with the same Three Vectors

Let **a, b, c** be three given vectors. We can permute the three given vectors in six different ways. Also each manner of writing down the three vectors gives rise to two scalar triple products depending upon the positions of dot and cross. Thus, we have the following *twelve* scalar triple products :

$$\mathbf{a} \times \mathbf{b} \cdot \mathbf{c} \quad \mathbf{b} \times \mathbf{c} \cdot \mathbf{a} \quad \mathbf{c} \times \mathbf{a} \cdot \mathbf{b} \quad \mathbf{a} \cdot \mathbf{b} \times \mathbf{a} \quad \mathbf{b} \cdot \mathbf{c} \times \mathbf{a} \quad \mathbf{c} \cdot \mathbf{a} \times \mathbf{b}$$

$$\mathbf{a} \times \mathbf{c} \cdot \mathbf{b} \quad \mathbf{b} \times \mathbf{a} \cdot \mathbf{c} \quad \mathbf{c} \times \mathbf{b} \cdot \mathbf{a} \quad \mathbf{a} \cdot \mathbf{c} \times \mathbf{b} \quad \mathbf{b} \cdot \mathbf{a} \times \mathbf{c} \quad \mathbf{c} \cdot \mathbf{b} \times \mathbf{a}.$$

We shall now prove the following two important results :

I. *A cyclic permutation of three vectors does not change the value of the scalar triple product and an anti-cyclic permutation changes the value in sign but not in magnitude.*

II. *The positions of dot and cross can be interchanged without any change in the value of the scalar triple product.* (*Patna* 2003)

Firstly suppose that **a, b, c** is a right-handed system so have

$$V = \mathbf{a} \times \mathbf{b} \cdot \mathbf{c}.$$

The vector triads **b, c, a** and **c, a, b** are also right-handed and the parallelopiped with *OA, OB, OC* as coterminous edges is the same as that with *OB, OC, OA* or with *OC, OA, OB* as coterminous edges. Thus,

$$V = \mathbf{b} \times \mathbf{c} \cdot \mathbf{a} \quad \text{and} \quad V = \mathbf{c} \times \mathbf{a} \cdot \mathbf{b}.$$

Also since scalar multiplication is commutative, we shall

$$\mathbf{a} \cdot \mathbf{b} \times \mathbf{c} = \mathbf{b} \times \mathbf{c} \cdot \mathbf{a} = V,$$

$$\mathbf{b} \cdot \mathbf{c} \times \mathbf{a} = \mathbf{c} \times \mathbf{a} \cdot \mathbf{b} = V,$$

$$\mathbf{c} \cdot \mathbf{a} \times \mathbf{b} = \mathbf{a} \times \mathbf{b} \cdot \mathbf{c} = V.$$

Thus, six out of the twelve scalar triple products are equal to *V* and in each of these cases, we have the same cyclic order of the vectors and interchange of dot and cross has not mattered.

Again **a, c, b; b, a, c; c, b, a** being left-handed systems, we have

$$\mathbf{a} \times \mathbf{c} \cdot \mathbf{b} = -V, \quad \mathbf{b} \times \mathbf{a} \cdot \mathbf{c} = -V, \quad \mathbf{c} \times \mathbf{b} \cdot \mathbf{a} = -V$$

and

$$\mathbf{a} \cdot \mathbf{c} \times \mathbf{b} = \mathbf{c} \times \mathbf{b} \cdot \mathbf{a} = -V,$$

$$\mathbf{b} \cdot \mathbf{a} \times \mathbf{c} = \mathbf{a} \times \mathbf{c} \cdot \mathbf{b} = -V,$$

$$\mathbf{c} \cdot \mathbf{b} \times \mathbf{a} = \mathbf{b} \times \mathbf{a} \cdot \mathbf{c} = -V.$$

Each of these latter six scalar products involves an anti-cyclic permutation of the vectors in $\mathbf{a} \times \mathbf{b} \cdot \mathbf{c}$.

Notation. In view of the properties of scalar triple product obtained above, we write

$$\mathbf{a} \times \mathbf{b} \cdot \mathbf{c} = [\mathbf{a}\ \mathbf{b}\ \mathbf{c}]$$

This notation takes note of the cyclic order of the vectors and disregards the positions of dot and cross which are really not important. We have

$$[\mathbf{a}\ \mathbf{b}\ \mathbf{c}] = [\mathbf{b}\ \mathbf{c}\ \mathbf{a}] = [\mathbf{c}\ \mathbf{a}\ \mathbf{b}]$$
$$= -[\mathbf{a}\ \mathbf{c}\ \mathbf{b}] = -[\mathbf{b}\ \mathbf{a}\ \mathbf{c}] = -[\mathbf{c}\ \mathbf{b}\ \mathbf{a}].$$

Note. It is very important to notice that the scalar triple product

$$[\mathbf{a}\ \mathbf{b}\ \mathbf{c}]$$

is zero if, and only if, the three vectors **a, b, c** *are coplanar.* In particular, the scalar triple product is zero if any two of the three vectors are the same.

Note. If **i, j, k** constitute an orthonormal right-handed triad, then

$$[\mathbf{i}\ \mathbf{j}\ \mathbf{k}] = \mathbf{i} \times \mathbf{j} \cdot \mathbf{k} = \mathbf{k} \cdot \mathbf{k} = 1.$$

5.5.2. Volume of a Tetrahedron — Theorem

The volume of a tetrahedron ABCD is

$$\frac{1}{6} | \overrightarrow{AB} \times \overrightarrow{AC} \cdot \overrightarrow{AD} |.$$

We know that the volume of a tetrahedron $ABCD$ is $\frac{1}{3} \times$ area of ΔABC $\times$ height of D from the plane ABC. Also the volume, V, of the parallelopiped with AB, AC, AD as its coterminous edges is equal to the area of the parallelogram with AB, AC as adjacent sides $\times$ height of D from the plane ABC.

$\therefore$ volume of the tetrahedron

$$= \frac{1}{6} | \overrightarrow{AB} \times \overrightarrow{AC} \cdot \overrightarrow{AD} |.$$

Cor. Volume of the tetrahedron, the position vectors of whose vertices A, B, C, D are **a, b, c, d** is

$$= \frac{1}{6} | \overrightarrow{AB} \times \overrightarrow{AC} \cdot \overrightarrow{AD} |$$
$$= \frac{1}{6} | (\mathbf{b} - \mathbf{a}) \times (\mathbf{c} - \mathbf{a}) \cdot (\mathbf{d} - \mathbf{a}) |.$$

Ex. 1. Show that each of the four faces of a tetrahedron subtends the same volume at the centroid.

Ex. 2. Compare the volume of a tetrahedron with that of the tetrahedron formed by the centroids of its faces.

Ex. 3. G_1, G_2, G_3 are the centroids of the triangular faces OBC, OCA, OAB of a tetrahedron $OABC$, compare the volume of the tetrahedron $OABC$ with that of the parallelopiped constructed with OG_1, OG_2, OG_3 as coterminous edges.

5.6. DISTRIBUTIVE LAW

Theorem. $\mathbf{a} \times (\mathbf{b} + \mathbf{c}) = \mathbf{a} \times \mathbf{b} + \mathbf{a} \times \mathbf{c}$,
where **a, b, c,** *are any three vectors.*

We shall now prove that the vector product distributes the sum of vectors, *i.e.*,

$$\mathbf{a} \times (\mathbf{b} + \mathbf{c}) = \mathbf{a} \times \mathbf{b} + \mathbf{a} \times \mathbf{c}.$$

This result will be proved by showing that the scalar product of the vector $\mathbf{a} \times (\mathbf{b} + \mathbf{c}) - \mathbf{a} \times \mathbf{b} - \mathbf{a} \times \mathbf{c}$ with *every* vector is zero.

Let, **r**, be *any* vector whatsoever. We have

$$\mathbf{r} \cdot [\mathbf{a} \times (\mathbf{b} + \mathbf{c}) - \mathbf{a} \times \mathbf{b} - \mathbf{a} \times \mathbf{c}]$$
$$= \mathbf{r} \cdot \mathbf{a} \times (\mathbf{b} + \mathbf{c}) - \mathbf{r} \cdot \mathbf{a} \times \mathbf{b} - \mathbf{r} \cdot \mathbf{a} \times \mathbf{c},$$

(for scalar multiplication distributes vector addition)

$$\mathbf{r} \times \mathbf{a} \cdot (\mathbf{b} + \mathbf{c}) - \mathbf{r} \times \mathbf{a} \cdot \mathbf{b} - \mathbf{r} \times \mathbf{a} \cdot \mathbf{c},$$

(for the positions of dot and cross are interchangeable)

$$= \mathbf{r} \times \mathbf{a} \cdot [(\mathbf{b} + \mathbf{c}) - \mathbf{b} - \mathbf{c}]$$

(for scalar multiplication distributes vector addition)

$$= \mathbf{r} \times \mathbf{a} \cdot \mathbf{0} = 0.$$

Thus,

$$\mathbf{r} \cdot [\mathbf{a} \times (\mathbf{b} + \mathbf{c}) - \mathbf{a} \times \mathbf{b} - \mathbf{a} \times \mathbf{c}] = 0,$$

for *every* vector **r**. It follows that

$$\mathbf{a} \times (\mathbf{b} + \mathbf{c}) - \mathbf{a} \times \mathbf{b} - \mathbf{a} \times \mathbf{c} = 0$$

$$\Rightarrow \quad \mathbf{a} \times (\mathbf{b} + \mathbf{c}) = \mathbf{a} \times \mathbf{b} + \mathbf{a} \times \mathbf{c}$$

Cor. 1. $(\mathbf{b} + \mathbf{c}) \times \mathbf{a} = -\mathbf{a} \times (\mathbf{b} + \mathbf{c})$

$$= -[\mathbf{a} \times \mathbf{b} + \mathbf{a} \times \mathbf{c}]$$
$$= -\mathbf{a} \times \mathbf{b} - \mathbf{a} \times \mathbf{c} = \mathbf{b} \times \mathbf{a} + \mathbf{c} \times \mathbf{a}.$$

Cor. 2. $\mathbf{a} \times (\mathbf{b} - \mathbf{c}) = \mathbf{a} \times [\mathbf{b} + (-\mathbf{c})]$

$$= \mathbf{a} \times \mathbf{b} + \mathbf{a} \times (-\mathbf{c}) = \mathbf{a} \times \mathbf{b} - \mathbf{a} \times \mathbf{c}.$$

Similarly

$$(\mathbf{b} - \mathbf{c}) \times \mathbf{a} = \mathbf{b} \times \mathbf{a} - \mathbf{c} \times \mathbf{a}.$$

Cor. 3. Scalar product in terms of an orthonormal base. If

$$\mathbf{a} = a_1\mathbf{i} + a_2\mathbf{j} + a_3\mathbf{k}, \quad \mathbf{b} = b_1\mathbf{i} + b_2\mathbf{j} + b_3\mathbf{k}, \quad \mathbf{c} = c_1\mathbf{i} + c_2\mathbf{j} + c_3\mathbf{k},$$

we have

$$\mathbf{a} \times \mathbf{b} \cdot \mathbf{c} = [(a_2b_3 - a_3b_2)\,\mathbf{i} + (a_3b_1 - a_1b_3)\,\mathbf{j} + (a_1b_2 - a_2b_1)\,\mathbf{k}] \cdot (c_1\mathbf{i} + c_2\mathbf{j} + c_3\mathbf{k})$$

$$= \begin{vmatrix} a_1 & a_2 & a_3 \\ b_1 & b_2 & b_3 \\ c_1 & c_2 & c_3 \end{vmatrix}$$

5.7. SOME PROPERTIES OF SCALAR TRIPLE PRODUCT

5.7.1. *To express the scalar triple product* **[a b c]** *in terms of any three non-coplanar vectors* **l, m, n.**

Let $\mathbf{a} = a_1\mathbf{l} + a_2\mathbf{m} + a_3\mathbf{n}$; $\mathbf{b} = b_1\mathbf{l} + b_2\mathbf{m} + b_3\mathbf{n}$; $\mathbf{c} = c_1\mathbf{l} + c_2\mathbf{m} + c_3\mathbf{n}$

$$\therefore \quad \mathbf{a} \times \mathbf{b} = (a_1\mathbf{l} + a_2\mathbf{m} + a_3\mathbf{n}) \times (b_1 l + h_2\mathbf{m} + b_3\mathbf{n})$$

$$= (a_2b_3 - a_3b_2)(\mathbf{m} \times \mathbf{n}) + (a_3b_1 - a_1b_3)(\mathbf{n} \times \mathbf{l}) + (a_1b_2 - a_2b_1)(\mathbf{l} \times \mathbf{m})$$

$$\therefore \quad (\mathbf{a} \times \mathbf{b}) \cdot \mathbf{c} = [(a_2b_3 - a_3b_2)(\mathbf{m} \times \mathbf{n}) + (a_3b_1 - a_1b_3)(\mathbf{n} \times \mathbf{l}) + (a_1b_2 - a_2b_1)(\mathbf{l} \times \mathbf{m})] \cdot (c_1\mathbf{l} + c_2\mathbf{m} + c_3\mathbf{n})$$

$$= c_1(a_2b_3 - a_3b_2)[\mathbf{m}\ \mathbf{n}\ \mathbf{l}] + c_2(a_3b_1 - a_1b_3)[\mathbf{n}\ \mathbf{l}\ \mathbf{m}] + c_3(a_1b_2 - a_2b_1)[\mathbf{l}\ \mathbf{m}\ \mathbf{n}]$$

But, as $[\mathbf{m}\ \mathbf{n}\ \mathbf{l}] = [\mathbf{n}\ \mathbf{l}\ \mathbf{m}] = [\mathbf{l}\ \mathbf{m}\ \mathbf{n}]$

we have

$$(\mathbf{a} \times \mathbf{b}) \cdot \mathbf{c} = [a_1(b_2c_3 - b_2c_3) - a_2(b_1c_3 - b_3c_1) + a_3(b_1c_2 - b_2c_1)[\mathbf{l}\ \mathbf{m}\ \mathbf{n}]$$

$$\Rightarrow \quad [\mathbf{a}\ \mathbf{b}\ \mathbf{c}] = \begin{vmatrix} a_1 & a_2 & a_3 \\ b_1 & b_2 & b_3 \\ c_1 & c_2 & c_3 \end{vmatrix} = [\mathbf{l}\ \mathbf{m}\ \mathbf{n}]$$

5.7.2. *To express any vector* **r** *as a linear combination of three non coplanar vectors* **a, b, c.** (Kolkata, 97

Let x, y, z be three scalars. Then

$$\mathbf{r} = x\mathbf{a} + y\mathbf{b} + z\mathbf{c}$$

$$\therefore \quad \mathbf{r} \cdot (\mathbf{b} \times \mathbf{c}) = x[\mathbf{a}\ \mathbf{b}\ \mathbf{c}] + y[\mathbf{b}\ \mathbf{b}\ \mathbf{c}] + z[\mathbf{c}\ \mathbf{b}\ \mathbf{c}]$$

$$\Rightarrow \quad [\mathbf{r}\ \mathbf{b}\ \mathbf{c}] = x[\mathbf{a}\ \mathbf{b}\ \mathbf{c}]$$

Similarly $\quad [\mathbf{r}\ \mathbf{c}\ \mathbf{a}] = y[\mathbf{a}\ \mathbf{b}\ \mathbf{c}]$

and $\quad [\mathbf{r}\ \mathbf{a}\ \mathbf{b}] = z[\mathbf{a}\ \mathbf{b}\ \mathbf{c}]$

$$\Rightarrow \quad \mathbf{r} = \frac{[\mathbf{r}\,\mathbf{b}\,\mathbf{c}]}{[\mathbf{a}\,\mathbf{b}\,\mathbf{c}]}\mathbf{a} + \frac{[\mathbf{r}\,\mathbf{c}\,\mathbf{a}]}{[\mathbf{a}\,\mathbf{b}\,\mathbf{c}]}\mathbf{b} + \frac{[\mathbf{r}\,\mathbf{a}\,\mathbf{b}]}{[\mathbf{a}\,\mathbf{b}\,\mathbf{c}]}\mathbf{c}.$$

5.7.3. Distributive Law for Scalar Products

(*i*) $[\mathbf{a}, \mathbf{b} + \mathbf{c}, \mathbf{d}] = [\mathbf{a}\ \mathbf{b}\ \mathbf{d}] + [\mathbf{a}\ \mathbf{c}\ \mathbf{d}]$

(*ii*) $[\mathbf{a}, \mathbf{b} + \mathbf{c}, \mathbf{d} + \mathbf{e}] = [\mathbf{a}\ \mathbf{b}\ \mathbf{d}] + [\mathbf{a}\ \mathbf{b}\ \mathbf{e}] + [\mathbf{a}\ \mathbf{c}\ \mathbf{d}] + [\mathbf{a}\ \mathbf{c}\ \mathbf{e}]$

Proof : (*i*) $[\mathbf{a}, \mathbf{b} + \mathbf{c}, \mathbf{d}] = [\mathbf{a}\ ´\ (\mathbf{b}\ ´\ \mathbf{c})] \cdot \mathbf{d}$

$$= [\mathbf{a} \times \mathbf{b} + \mathbf{a} \times \mathbf{c}] \cdot \mathbf{d}$$

$$= (\mathbf{a} \times \mathbf{b}) \cdot \mathbf{d} + (\mathbf{a} \times \mathbf{c}) \cdot \mathbf{d}$$

$$= [\mathbf{a}\ \mathbf{b}\ \mathbf{d}] + [\mathbf{a}\ \mathbf{c}\ \mathbf{d}]$$

Similarly part (*ii*) can be proved.

EXAMPLES

Example 1. *Find the volume of the parallelopiped whose coterminous edges are* $2\mathbf{i} - 3\mathbf{j} + \mathbf{k}$, $\mathbf{i} - \mathbf{j} + 2\mathbf{k}$ *and* $2\mathbf{i} + \mathbf{j} - \mathbf{k}$.

Solution. Volume of the parallelopiped

$$= \begin{vmatrix} 2 & -3 & 1 \\ 1 & -1 & 2 \\ 2 & 1 & -1 \end{vmatrix}$$

$$= 2(1-2) + 3(-1-4) + 1(1+2)$$

$$= -14 \text{ or to say } 14 \text{ units.}$$

Example 2. *Show that the vectors* $\mathbf{a} - 2\mathbf{b} + 3\mathbf{c}$, $-2\mathbf{a} + 3\mathbf{b} - 4\mathbf{c}$ *and* $\mathbf{a} - 3\mathbf{b} + 5\mathbf{c}$ *are coplanar.*

Solution. Let $\mathbf{p} = \mathbf{a} - 2\mathbf{b} + 3\mathbf{c}$, $\mathbf{q} = -2\mathbf{a} + 3\mathbf{b} - 4\mathbf{c}$, $\mathbf{r} = \mathbf{a} - 3\mathbf{b} + 5\mathbf{c}$. By § 5.7.1.

$$[\mathbf{p}\,\mathbf{q}\,\mathbf{r}] = \begin{vmatrix} 1 & -2 & 3 \\ -2 & 3 & -4 \\ 1 & -3 & 5 \end{vmatrix} = [\mathbf{a}\,\mathbf{b}\,\mathbf{c}]$$

$$= 0 \times [\mathbf{a}\,\mathbf{b}\,\mathbf{c}] = 0$$

showing that $\mathbf{p}, \mathbf{q}, \mathbf{r}$ are coplanar.

Example 3. *Prove that if* $\mathbf{l}, \mathbf{m}, \mathbf{n}$ *be three non-coplanar vectors, then*

$$[\mathbf{l}\,\mathbf{m}\,\mathbf{n}]\,[\mathbf{a} \times \mathbf{b}] = \begin{vmatrix} \mathbf{l}.\mathbf{a} & \mathbf{l}.\mathbf{b} & \mathbf{l} \\ \mathbf{m}.\mathbf{a} & \mathbf{m}.\mathbf{b} & \mathbf{m} \\ \mathbf{n}.\mathbf{a} & \mathbf{n}.\mathbf{b} & \mathbf{n} \end{vmatrix}$$

(Awadh 99, 2005, Rohilkhand 98, 2004, Garhwal 94, 99, 2006)

Solution. Let $\mathbf{a} = a_1\mathbf{i} + a_2\mathbf{j} + a_3\mathbf{k}$, $\mathbf{b} = b_1\mathbf{i} + b_2\mathbf{j} + b_3\mathbf{k}$,

$\mathbf{l} = l_1\mathbf{i} + l_2\mathbf{j} + l_3\mathbf{k}$, $\mathbf{m} = m_1\mathbf{i} + m_2\mathbf{j} + m_3\mathbf{k}$,

$\mathbf{n} = n_1\mathbf{i} + n_2\mathbf{j} + n_3\mathbf{k}$. Then

$$[\mathbf{l}\,\mathbf{m}\,\mathbf{n}]\,[\mathbf{a} \times \mathbf{b}] = \begin{vmatrix} l_1 & l_2 & l_3 \\ m_1 & m_2 & m_3 \\ n_1 & n_2 & n_3 \end{vmatrix} \begin{vmatrix} \mathbf{i} & \mathbf{j} & \mathbf{k} \\ a_1 & a_2 & a_3 \\ b_1 & b_2 & b_3 \end{vmatrix}$$

$$= \begin{vmatrix} l_1\mathbf{i} + l_2\mathbf{j} + l_3\mathbf{k} & l_1a_1 + l_2a_2 + l_3a_3 & l_1b_1 + l_2b_2 + l_3b_3 \\ m_1\mathbf{i} + m_2\mathbf{j} + m_3\mathbf{k} & m_1a_1 + m_2a_2 + m_3a_3 & m_1b_1 + m_2b_2 + m_3b_3 \\ n_1\mathbf{i} + n_2\mathbf{j} + n_3\mathbf{k} & n_1a_1 + n_2a_2 + n_3a_3 & n_1b_1 + n_2b_2 + n_3b_3 \end{vmatrix}$$

$$= \begin{vmatrix} \mathbf{l} & \mathbf{l}.\mathbf{a} & \mathbf{l}.\mathbf{b} \\ \mathbf{m} & \mathbf{m}.\mathbf{a} & \mathbf{m}.\mathbf{b} \\ \mathbf{n} & \mathbf{n}.\mathbf{a} & \mathbf{n}.\mathbf{b} \end{vmatrix} = \begin{vmatrix} \mathbf{l}.\mathbf{a} & \mathbf{l}.\mathbf{b} & \mathbf{l} \\ \mathbf{m}.\mathbf{a} & \mathbf{m}.\mathbf{b} & \mathbf{m} \\ \mathbf{n}.\mathbf{a} & \mathbf{n}.\mathbf{b} & \mathbf{n} \end{vmatrix}$$

Example 4. *Find the value of p so that the vectors* **2i − j + k**, **i** + 2**j** − 3**k** *and* 3**i** + *p***j** + 5**k** *are coplanar.* (*Avadh 2000*)

Solution. Given vectors will be coplanar if

$$\begin{vmatrix} 2 & -1 & 1 \\ 1 & 2 & -3 \\ 3 & p & 5 \end{vmatrix} = 0$$

$\Rightarrow$ $2(10 + 3p) + 1(5 + 9) + 1(p - 6) = 0$

$\Rightarrow$ $7p + 28 = 0 \quad \Rightarrow \quad p = -4.$

Example 5. *Prove that*

$$\mathbf{a} \times \mathbf{b} = [(\mathbf{i} \times \mathbf{a}) \,.\, \mathbf{b}]\, \mathbf{i} + [(\mathbf{j} \times \mathbf{a}) \,.\, \mathbf{b}]\, \mathbf{j} + [(\mathbf{k} \times \mathbf{a}) \,.\, \mathbf{k}]\, \mathbf{k}.$$

Solution. For any vector **r**, we have

$$\mathbf{r} = (\mathbf{i} \,.\, \mathbf{r})\, \mathbf{i} + (\mathbf{j} \,.\, \mathbf{r})\, \mathbf{j} + (\mathbf{k} \,.\, \mathbf{r})\, \mathbf{k}$$

Replacing **r** by (**a** × **b**), we get

$$(\mathbf{a} \times \mathbf{b}) = [\mathbf{i} \,.\, (\mathbf{a} \times \mathbf{b})]\, \mathbf{i} + [\mathbf{j} \,.\, (\mathbf{a} \times \mathbf{b})]\, \mathbf{j} + [\mathbf{k} \,.\, (\mathbf{a} \times \mathbf{b})]\, \mathbf{k}$$

$$= [(\mathbf{i} \times \mathbf{a}) \,.\, \mathbf{b}]\, \mathbf{i} + [(\mathbf{j} \times \mathbf{a}) \,.\, \mathbf{b}]\, \mathbf{j} + [(\mathbf{k} \times \mathbf{a}) \,.\, \mathbf{b}]\, \mathbf{k}.$$

since the position of the dot and cross can be interchanged in a scalar triple product.

EXERCISES

1. If **i, j, k** are three mutually perpendicular unit vectors, compute

(*a*) (2**i** − **j** + **k**) × (3**i** + **k**), (*b*) (**i** − 2**j** + 3**k**) × (2**j** − 3**k**)

(*c*) (3**i** + 4**j**) × (**i** − **j** + **k**),

and verify in each case with the help of scalar product that the vector product is perpendicular to the given vectors.

Find also the lengths of these vector products.

2. Compute the following scalar triple products :

(*a*) (**i** − 2**j** + 3**k**) × (2**i** + **j** − **k**) . (**j** + **k**).

(*b*) (2**i** − 3**j** + **k**) . (**i** − **j** + 2**k**) × (2**i** + **j** − **k**).

3. Compute the following vector products :

(*a*) [(**i** − **j** + **k**) × (2**i** − 3**j** − **k**)] × [(− 3**i** + **j** + **k**) × (2**j** + **k**)]

(*b*) [(3**i** − 2**j** − 2**k**) × (**i** − **k**)] × [(**i** + **j** + **k**) × (**i** − 2**j** + 3**k**)].

4. Show that the vectors

(*i*) **j** − 2**k**, **i** − **j** + **k**, − 2**i** + 3**j** − 4**k**;

(*ii*) 2**i** − **j** + 2**k**, 4**i** + **j**, 3**i** + **k**,

are coplanar.

5. Show that the vectors **i + j**, **i + k**, **k + j** are not coplanar.

6. Find the volume of the parallelopiped whose coterminous edges are represented by **a** = 2**i** – 3**j** + 4**k**, **b** = **i** + 2**j** – **k**, **c** = 3**i** – **j** + 2**k**.

7. If **a**, **b**, **c** are three vectors such that **a** × **b** = **c** and **b** × **c** = **a**, show that three vectors **a**, **b**, **c** are orthogonal in pairs and | **b** | = 1, | **c** | = | **a** |.

8. Prove that

$$[\mathbf{l\,m\,n}]\,[\mathbf{a\,b\,c}] = \begin{vmatrix} \mathbf{l.a} & \mathbf{l.b} & \mathbf{l.c} \\ \mathbf{m.a} & \mathbf{m.b} & \mathbf{m.c} \\ \mathbf{n.a} & \mathbf{n.b} & \mathbf{n.c} \end{vmatrix}$$

(Kumaon 2006, Nagpur 2005, Avadh, 2005)

Also deduce that

$$[\mathbf{a\,b\,c}]^2 = \begin{vmatrix} \mathbf{a.a} & \mathbf{a.b} & \mathbf{a.c} \\ \mathbf{b.a} & \mathbf{b.b} & \mathbf{b.c} \\ \mathbf{c.a} & \mathbf{c.b} & \mathbf{c.c} \end{vmatrix}$$

9. Show that the four points **a**, **b**, **c**, **d** are coplanar if

$$[\mathbf{b\ c\ d}] + [\mathbf{c\ a\ d}] + [\mathbf{a\ b\ d}] = [\mathbf{a\ b\ c}].$$

EXAMPLES

Example 1. *If one diagonal of a quadrilateral bisects the other, then it also bisects the quadrilateral.*

Solution. Let *OABC* be the given quadrilateral such that its diagonal $\overrightarrow{OB}$ bisects the diagonal $\overrightarrow{AC}$.

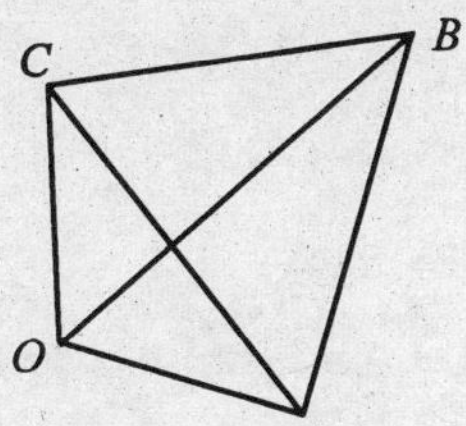

Fig. 5.8.

Let $\overrightarrow{OA} = \mathbf{a},\ \overrightarrow{OB} = \mathbf{b},\ \overrightarrow{OC} = \mathbf{c}.$

Since the mid-points $\frac{\mathbf{a}+\mathbf{c}}{2}$ of $\overrightarrow{AC}$ lies on $\overrightarrow{OB}$, there exists a scalar *t* such that

$$\frac{\mathbf{a}+\mathbf{c}}{2} = t\mathbf{b} \Rightarrow \mathbf{a}+\mathbf{c} = 2t\mathbf{b}.$$

Multiplying both sides vectorially with **b**, we have

$$(\mathbf{a} + \mathbf{c}) \times \mathbf{b} = 2t\, \mathbf{b} \times \mathbf{b} = \mathbf{0}$$

$$\Rightarrow \quad \mathbf{a} \times \mathbf{b} = \mathbf{b} \times \mathbf{c}$$

$$\Rightarrow \quad \frac{1}{2}\mathbf{a} \times \mathbf{b} = \frac{1}{2}\mathbf{b} \times \mathbf{c}$$

$$\Rightarrow \quad \frac{1}{2}|\mathbf{a} \times \mathbf{b}| = \frac{1}{2}|\mathbf{b} \times \mathbf{c}|$$

$\Rightarrow \quad \Delta\, OAB = \Delta\, OBC$

for $OABC$ is a plane quadrilateral.

Hence, the diagonal OB bisects the quadrilateral.

Example 2. *P, Q are the mid-points of the non-parallel sides BC and AD of a trapezium ABCD. Show that* $\Delta\, APD = \Delta\, CQB$.

Solution. Let $\overrightarrow{AB} = \mathbf{b}$ and $\overrightarrow{AD} = \mathbf{d}$.

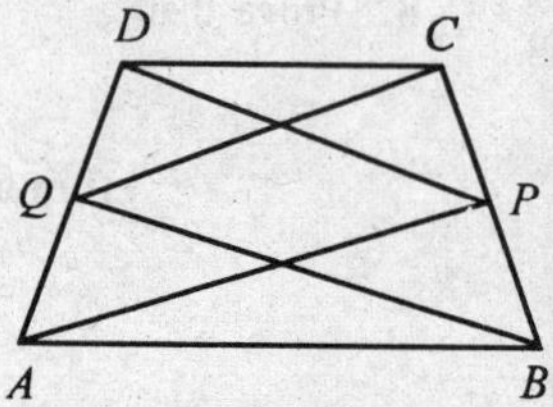

Fig. 5.9.

Now DC is parallel to AB

$\Rightarrow$ there exists a scalar t such that

$$\overrightarrow{DC} = t\, \overrightarrow{AB} = t\, \mathbf{b}.$$

$$\therefore \quad \overrightarrow{AC} = \overrightarrow{AD} + \overrightarrow{DC} = \mathbf{d} + t\, \mathbf{b}.$$

The position vectors of P and Q are $\frac{1}{2}(\mathbf{b} + \mathbf{d} + t\, \mathbf{b})$ and $\frac{1}{2}\mathbf{d}$ respectively.

Now
$$2\, \Delta\, APD = \overrightarrow{AP} \times \overrightarrow{AD}$$
$$= \frac{1}{2}(\mathbf{b} + \mathbf{d} + t\mathbf{b}) \times \mathbf{d} = \frac{1}{2}(1+t)\,(\mathbf{b} \times \mathbf{d}).$$

Also
$$2\, \Delta\, CQB = \overrightarrow{CQ} \times \overrightarrow{CB} = \left[\frac{1}{2}\mathbf{d} - (\mathbf{d} + t\mathbf{b})\right] \times [\mathbf{b} - (\mathbf{d} + t\, \mathbf{b})]$$
$$= \left[-\frac{1}{2}\mathbf{d} - t\mathbf{b}\right] \times [-\mathbf{d} + (1-t)\, \mathbf{b}]$$
$$= -\frac{1}{2}(1-t)\, \mathbf{d} \times \mathbf{b} + t\mathbf{b} \times \mathbf{d}$$
$$= \frac{1}{2}(1-t)\, \mathbf{b} \times \mathbf{d} + t\mathbf{b} \times \mathbf{d} = \frac{1}{2}(1 - t + 2t)\, \mathbf{b} \times \mathbf{d}$$
$$= \frac{1}{2}(1+t)\, \mathbf{b} \times \mathbf{d} = 2\, \Delta\, \overrightarrow{APD}.$$

Hence, the result.

Example 3. *Given that* **a, b, c** *are the position vectors of the vertices of a* $\Delta\, ABC$*, find the vector area of the triangle.*

Solution. The required vector area

$$= \frac{1}{2}(\overrightarrow{AB} \times \overrightarrow{AC})$$
$$= \frac{1}{2}[(\mathbf{b} - \mathbf{a}) \times (\mathbf{c} - \mathbf{a})]$$
$$= \frac{1}{2}[\mathbf{a} \times \mathbf{b} + \mathbf{b} \times \mathbf{c} + \mathbf{c} \times \mathbf{a}].$$

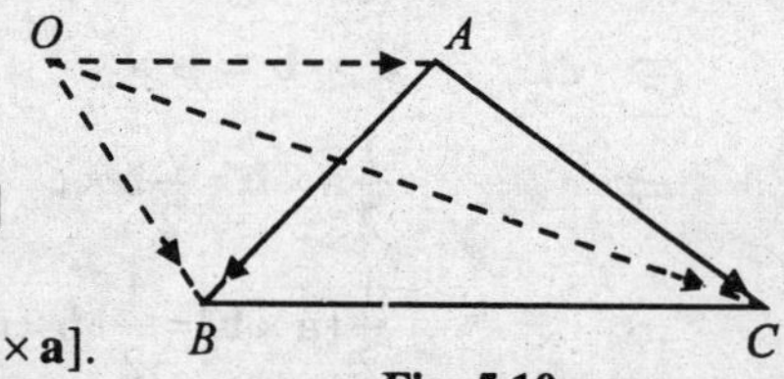

Fig. 5.10.

Example 4. *Find the area of the triangle whose vertices are the points with rectangular cartesian co-ordinates*

$$(1,\ 2,\ 3),\quad (-2,\ 1,\ -4),\quad (3,\ 4,\ -2).$$

Solution. Let **i, j, k** denote unit vectors along the three rectangular axes. If O be the origin and A, B, C denote the given vertices, we have

$$\vec{OA} = \mathbf{i} + 2\mathbf{j} + 3\mathbf{k},\ \vec{OB} = -2\mathbf{i} + \mathbf{j} - 4\mathbf{k},\ \vec{OC} = 3\mathbf{i} + 4\mathbf{j} - 2\mathbf{k}.$$

$$\therefore \quad \vec{AB} = \vec{OB} - \vec{OA} = -3\mathbf{i} - \mathbf{j} - 7\mathbf{k},$$

$$\vec{AC} = \vec{OC} - \vec{OA} = 2\mathbf{i} + 2\mathbf{j} - 5\mathbf{k}.$$

$$\vec{AB} \times \vec{AC} = (-3\mathbf{i} - \mathbf{j} - 7\mathbf{k}) \times (2\mathbf{i} + 2\mathbf{j} - 5\mathbf{k})$$

$$= 19\mathbf{i} - 29\mathbf{j} - 4\mathbf{k}.$$

$$\therefore \quad \text{the required area} = \frac{1}{2} |\vec{AB} \times \vec{AC}| = \frac{1}{2} |19\mathbf{i} - 29\mathbf{j} - 4\mathbf{k}|$$

$$= \frac{1}{2}\sqrt{(19^2 + 29^2 + 4^2)} = \frac{\sqrt{1218}}{2}.$$

Example 5. *Find the volume of the tetrahedron the rectangular cartesian co-ordinates of whose vertices are*

$$(0,\ 1,\ 2),\quad (3,\ 0,\ 1),\quad (4,\ 3,\ 6),\quad (2,\ 3,\ 2).$$

Solution. Let, as usual, **i, j, k** denote unit vectors along the three rectangular axes. If A, B, C, D denote the given vertices, we have

$$\vec{OA} = \mathbf{j} + 2\mathbf{k}, \qquad \vec{OB} = 3\mathbf{i} + \mathbf{k},$$

$$\vec{OC} = 4\mathbf{i} + 3\mathbf{j} + 6\mathbf{k}, \qquad \vec{OD} = 2\mathbf{i} + 3\mathbf{j} + 2\mathbf{k}.$$

The required volume is

$$\frac{1}{6} | \vec{AB} \times \vec{AC} \cdot \vec{AD} |.$$

We have

$$\vec{AB} = \vec{OB} - \vec{OA} = 3\mathbf{i} - \mathbf{j} - \mathbf{k}.$$

$$\vec{AC} = \vec{OC} - \vec{OA} = 4\mathbf{i} + 2\mathbf{j} + 4\mathbf{k}.$$

$$\vec{AD} = \vec{OD} - \vec{OA} = 2\mathbf{i} + 2\mathbf{j}.$$

$$\therefore \quad \vec{AB} \times \vec{AC} = (3\mathbf{i} - \mathbf{j} - \mathbf{k}) \times (4\mathbf{i} + 2\mathbf{j} + 4\mathbf{k})$$

$$= -2\mathbf{i} - 16\mathbf{j} + 10\mathbf{k}.$$

$$\therefore \quad \vec{AB} \times \vec{AC} \cdot \vec{AD} = (-2\mathbf{i} - 16\mathbf{j} + 10\mathbf{k}) \cdot (2\mathbf{i} + 2\mathbf{j}) = -36.$$

Thus, the required volume = 36.

Example 6. *Show that the perpendicular distance of a point whose position vector is* **a** *from the plane through three points with position vectors* **b, c, d,** *is*

$$[\mathbf{b}\ \mathbf{c}\ \mathbf{d}] + [\mathbf{c}\ \mathbf{a}\ \mathbf{d}] + [\mathbf{a}\ \mathbf{b}\ \mathbf{d}] - [\mathbf{a}\ \mathbf{b}\ \mathbf{c}] \,/\, |\,\mathbf{b} \times \mathbf{c} + \mathbf{c} \times \mathbf{d} + \mathbf{d} \times \mathbf{b}\,|$$

Solution. Let *ABCD* be the tetrahedron whose vertices *A, B, C, D* have position vectors **a, b, c, d** respectively.

Volume of tetrahedron *ABCD*

$$= \frac{1}{3}(\text{area } \Delta\, BCD) \times h,$$

where *h* is the perpendicular distance of *A* from the plane *BCD*

$$\therefore \quad h = \frac{3 \times \text{volume of tetrahedron } ABCD}{\text{Area of } \Delta\ BCD}$$

$$= 3 \cdot \frac{\frac{1}{6}[\vec{AB}, \vec{AC}, \vec{AD}]}{\left(\frac{1}{2}\right) |\,\vec{BC} \times \vec{BD}\,|}$$

$$= \frac{[\mathbf{b}-\mathbf{a},\ \mathbf{c}-\mathbf{a},\ \mathbf{d}-\mathbf{a}]}{|\,(\mathbf{c}-\mathbf{b}) \times (\mathbf{d}-\mathbf{b})\,|}$$

$$= \frac{(\mathbf{b}-\mathbf{a}) \cdot [(\mathbf{c}-\mathbf{a}) \times (\mathbf{d}-\mathbf{a})]}{|\,\mathbf{c} \times \mathbf{d} - \mathbf{c} \times \mathbf{b} - \mathbf{b} \times \mathbf{d}\,|}$$

$$= \frac{(\mathbf{b}-\mathbf{a}) \cdot [\mathbf{c} \times \mathbf{d} - \mathbf{c} \times \mathbf{a} - \mathbf{a} \times \mathbf{d}]}{|\,\mathbf{b} \times \mathbf{c} + \mathbf{c} \times \mathbf{d} + \mathbf{d} \times \mathbf{b}\,|}$$

$$= \frac{[\mathbf{b}\,\mathbf{c}\,\mathbf{d}] - [\mathbf{b}\,\mathbf{c}\,\mathbf{a}] - [\mathbf{b}\,\mathbf{a}\,\mathbf{d}] - [\mathbf{a}\,\mathbf{c}\,\mathbf{d}]}{|\,\mathbf{b} \times \mathbf{c} + \mathbf{c} \times \mathbf{d} + \mathbf{d} \times \mathbf{b}\,|}$$

$$= \frac{[\mathbf{b}\,\mathbf{c}\,\mathbf{d}] + [\mathbf{c}\,\mathbf{a}\,\mathbf{d}] + [\mathbf{a}\,\mathbf{b}\,\mathbf{d}] - [\mathbf{a}\,\mathbf{b}\,\mathbf{c}]}{|\,\mathbf{b} \times \mathbf{c} + \mathbf{c} \times \mathbf{d} + \mathbf{d} \times \mathbf{b}\,|}$$

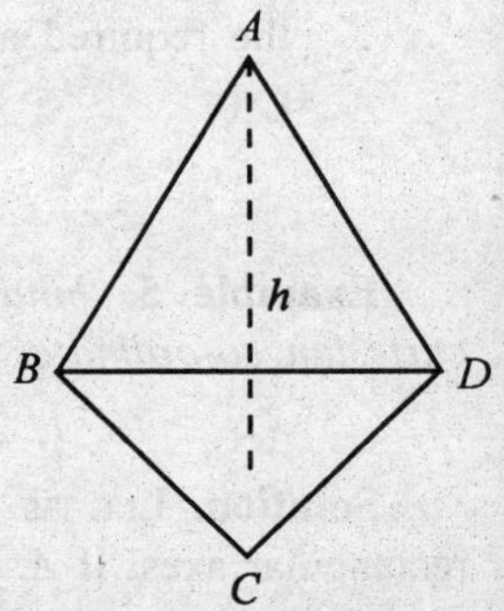

Fig. 5.11.

Example 7. *Prove that the formula for the volume of a tetrahedron in terms of the lengths of three concurrent edges and their mutual inclinations is*

$$V^2 = \frac{a^2b^2c^2}{36} \begin{vmatrix} 1 & \cos\phi & \cos\psi \\ \cos\phi & 1 & \cos\theta \\ \cos\psi & \cos\theta & 1 \end{vmatrix}$$

Solution. Let $OABC$ be the tetrahedron with O as origin. Let **a, b, c** be the position vectors of A, B, C and **a, b, c** be their magnitudes.

Let $\mathbf{a} = a_1\mathbf{i} + a_2\mathbf{j} + a_3\mathbf{k}$, $\mathbf{b} = b_1\mathbf{i} + b_2\mathbf{j} + b_3\mathbf{k}$, $\mathbf{c} = c_1\mathbf{i} + c_2\mathbf{j} + c_3\mathbf{k}$

Then

$$V = \frac{1}{6}[\mathbf{a}\ \mathbf{b}\ \mathbf{c}] = \frac{1}{6}\begin{vmatrix} a_1 & a_2 & a_3 \\ b_1 & b_2 & b_3 \\ c_1 & c_2 & c_3 \end{vmatrix}$$

$$\therefore \qquad V^2 = \frac{1}{36}\begin{vmatrix} a_1 & a_2 & a_3 \\ b_1 & b_2 & b_3 \\ c_1 & c_2 & c_3 \end{vmatrix} \times \begin{vmatrix} a_1 & a_2 & a_3 \\ b_1 & b_2 & b_3 \\ c_1 & c_2 & c_3 \end{vmatrix}$$

$$= \frac{1}{36}\begin{vmatrix} a_1^2 + a_2^2 + a_3^2 & a_1b_1 + a_2b_2 + a_3b_3 & a_1c_1 + a_2c_2 + a_3c_3 \\ a_1b_1 + a_2b_2 + a_3b_3 & b_1^2 + b_2^2 + b_3^2 & b_1c_1 + b_2c_2 + b_3c_3 \\ a_1c_1 + a_2c_2 + a_3c_3 & b_1c_1 + b_2c_2 + b_3c_3 & c_1^2 + c_2^2 + c_3^2 \end{vmatrix}$$

$$= \frac{1}{36}\begin{vmatrix} |\mathbf{a}|^2 & \mathbf{a}.\mathbf{b} & \mathbf{a}.\mathbf{c} \\ \mathbf{a}.\mathbf{b} & |\mathbf{b}|^2 & \mathbf{b}.\mathbf{c} \\ \mathbf{a}.\mathbf{c} & \mathbf{b}.\mathbf{c} & |\mathbf{c}|^2 \end{vmatrix}$$

$$= \frac{1}{36}\begin{vmatrix} a^2 & ab\cos\phi & ca\cos\psi \\ ab\cos\phi & b^2 & bc\cos\theta \\ ca\cos\psi & bc\cos\theta & c^2 \end{vmatrix}$$

$$= \frac{a^2b^2c^2}{36}\begin{vmatrix} 1 & \cos\phi & \cos\psi \\ \cos\phi & 1 & \cos\theta \\ \cos\psi & \cos\theta & 1 \end{vmatrix}$$

Example 8. *Prove that each of the four faces of a tetrahedron subtends the same volume at the centroid.*

Solution. $ABCD$ be a tetrahedron with G as centroid. With G as origin, let the position vectors of A, B, C, D be **a, b, c** and **d** respectively. Then

$$\frac{\mathbf{a}+\mathbf{b}+\mathbf{c}+\mathbf{d}}{4} = 0 \Rightarrow \mathbf{a}+\mathbf{b}+\mathbf{c}+\mathbf{d} = 0.$$

Volume of tetrahedron $GABC = \frac{1}{6}[\mathbf{a}\ \mathbf{b}\ \mathbf{c}]$

Volume of tetrahedron $GBCD = \frac{1}{6}[\mathbf{b}\ \mathbf{c}\ \mathbf{d}]$

$$= \frac{1}{6}(\mathbf{b} \times \mathbf{c}) \cdot (-\mathbf{a} - \mathbf{b} - \mathbf{c})$$

$$= -\frac{1}{6}(\mathbf{b} \times \mathbf{c}) \cdot \mathbf{a} = \frac{1}{6}[\mathbf{a}\ \mathbf{b}\ \mathbf{c}]$$

Similarly other volumes can be shown.

Example 9. *Compare the volume of a tetrahedron with that of the tetrahedron formed by the centroids of its faces.*

Solution. Let **a, b, c, d** be the position vectors of the vertices *A, B, C, D* respectively of given tetrahedron. Then

$$V = \frac{1}{6}[\mathbf{b} - \mathbf{a},\ \mathbf{c} - \mathbf{a},\ \mathbf{d} - \mathbf{a}]$$

$$= \frac{1}{6}[\mathbf{a} - \mathbf{b},\ \mathbf{a} - \mathbf{c},\ \mathbf{a} - \mathbf{d}] \text{ (numerically)}$$

Now, the position vectors $\mathbf{g}_1, \mathbf{g}_2, \mathbf{g}_3, \mathbf{g}_4$ of the centroids G_1, G_2, G_3, G_4 of the four faces *BCD, ACD, ABD* are *ABC* respectively are given by

$$\mathbf{g}_1 = \frac{\mathbf{b} + \mathbf{c} + \mathbf{d}}{3}, \qquad \mathbf{g}_2 = \frac{\mathbf{a} + \mathbf{c} + \mathbf{d}}{3},$$

$$\mathbf{g}_3 = \frac{\mathbf{a} + \mathbf{b} + \mathbf{d}}{3}, \qquad \mathbf{g}_4 = \frac{\mathbf{a} + \mathbf{b} + \mathbf{c}}{3}.$$

$$\therefore \qquad V' = \frac{1}{6}[\mathbf{g}_2 - \mathbf{g}_1,\ \mathbf{g}_3 - \mathbf{g}_1,\ \mathbf{g}_4 - \mathbf{g}_1]$$

$$= \frac{1}{6}\left[\frac{1}{3}(\mathbf{a} - \mathbf{b}),\ \frac{1}{3}(\mathbf{a} - \mathbf{c}),\ \frac{1}{3}(\mathbf{a} - \mathbf{d})\right]$$

$$= \frac{1}{27} \cdot \frac{1}{6}\left[(\mathbf{a} - \mathbf{b}),\ (\mathbf{a} - \mathbf{c}),\ (\mathbf{a} - \mathbf{d})\right]$$

$$= \frac{1}{27} V.$$

EXERCISES

1. *ABC* is a triangle, *E* and *F* are the mid-points of *AC* and *AB* respectively. *CP* is drawn parallel to *AB* to meet BE produced in *P*, show that

$$\Delta\ FEP = \Delta\ FCE = \frac{1}{4} \Delta\ ABC.$$

2. Show that $\overrightarrow{AC} \times \overrightarrow{BD}$ represents twice the vector area of a plane quadrilateral *ABCD*.

3. ABC is a triangle and EF is any straight line parallel to BC meeting AC, AB in E, F respectively. If BR and CQ be drawn parallel to AC, AB respectively to meet EF in R and Q respectively, prove that

$$\Delta\, ARB = \Delta\, ACQ.$$

4. Two triangles of equal area are on the opposite sides of the same base; prove that the straight line joining their vertices is bisected by the base.

5. $ABCD$ is a quadrilateral such that

$$\overrightarrow{AB} = \mathbf{b}, \quad \overrightarrow{AD} = \mathbf{d}, \quad \overrightarrow{AC} = m\mathbf{b} + p\mathbf{d};$$

show that the area of the quadrilateral $ABCD$ is

$$\frac{1}{2} | m + p | \, | \mathbf{b} \times \mathbf{d} |.$$

6. $ABCD$ is a plane quadrilateral with $\overrightarrow{AB} = a$, $\overrightarrow{BC} = b$, $\overrightarrow{CD} = c$. Prove that the area of the quadrilateral is

$$\frac{1}{2} | \mathbf{a} \times \mathbf{b} + \mathbf{b} \times \mathbf{c} - \mathbf{c} \times \mathbf{a} |.$$

7. Find the areas of the triangles with the following vertices :

(*a*) (0, 0, 0), (1, 2, 3), (2, – 1, 4);

(*b*) (1, 0, 0), (0, 1, 0), (1, 1, 1);

(*c*) (– 1, 2, 3), (2, – 1, – 1), (1, 1, – 1);

(*d*) (a, 0, 0), (0, b, 0), (0, 0, c).

8. Find the volumes of the tetrahedron with the following vertices :

(*a*) (0, 0, 0), (1, 1, – 1), (1, – 1, 1), (– 1, 1, 1).

(*b*) (– 1, 0, 1), (2, – 1, 0), (3, 2, 5), (1, 2, 1).

9. Show that the volume of the tetrahedron, the co-ordinates of whose vertices are (x_1, y_1, z_1), (x_2, y_2, z_2), (x_3, y_3, z_3) and (x_4, y_4, z_4) is the absolute value of

$$\frac{1}{6}\begin{vmatrix} x_2 - x_1 & y_2 - y_1 & z_2 - z_1 \\ x_3 - x_1 & y_3 - y_1 & z_3 - z_1 \\ x_4 - x_1 & y_4 - y_1 & z_4 - z_1 \end{vmatrix}$$

10. Show that the volume of the tetrahedron, the position vectors of whose vertices are **a, b, c, d** is

$$\frac{1}{6}\big| [\mathbf{b\,c\,d}] + [\mathbf{c\,a\,d}] + [\mathbf{a\,b\,d}] - [\mathbf{a\,b\,c}] \big|$$

Deduce the condition for the points with position vectors **a, b, c, d** to be coplanar.

11. Show that the volume of a tetrahedron bounded by the four planes $\mathbf{r} \cdot (m\mathbf{j} + n\mathbf{k}) = 0$, $\mathbf{r} \cdot (n\mathbf{k} + l\mathbf{i}) = 0$, $\mathbf{r} \cdot (l\mathbf{i} + m\mathbf{j}) = 0$ and $\mathbf{r} \cdot (l\mathbf{i} + m\mathbf{j} + n\mathbf{k}) = p$ is $2p^3 / 3lmn$. (*Avadh, 1998*)

5.8. VECTOR TRIPLE PRODUCTS

Let **a, b, c** be any three vectors. Then an expression of the form

$$(\mathbf{a} \times \mathbf{b}) \times \mathbf{c}$$

which is the vector product of **a** × **b** with **c**, is known as vector triple product.

It will be shown that this product is a linear combination of **a** and **b** and precisely that

$$(\mathbf{a} \times \mathbf{b}) \times \mathbf{c} = (\mathbf{a} \cdot \mathbf{c})\, \mathbf{b} - (\mathbf{b} \cdot \mathbf{c})\, \mathbf{a}.$$

The vector (**a** × **b**) × **c** being perpendicular to the vector **a** × **b**, is coplanar with **a** and **b** and, as such, it is a linear combination of **a** and **b.** Let

$$(\mathbf{a} \times \mathbf{b}) \times \mathbf{c} = l\mathbf{a} + m\mathbf{b}. \qquad ...(i)$$

Also the vector on the left is perpendicular to the vector **c** so that its scalar product with, **c**, is zero.

Multiplying both sides of (*i*) scalarly with, **c**, we obtain

$$l\, \mathbf{a} \cdot \mathbf{c} + m\, \mathbf{b} \cdot \mathbf{c} = 0. \qquad ...(ii)$$

From (*i*) and (*ii*), we obtain an equality of the form

$$(\mathbf{a} \times \mathbf{b}) \times \mathbf{c} = k\, [(\mathbf{a} \cdot \mathbf{c})\, \mathbf{b} - (\mathbf{b} \cdot \mathbf{c})\, \mathbf{a}] \qquad ...(iii)$$

This equality has been obtained by using the fact that *the vector* (**a** × **b**) × **c** *is coplanar with* **a** *and* **b** *and is perpendicular to* **c.**

It will here below be shown that $k = 1$. The proof of this fact involves an examination of the *magnitudes* of the vectors whereas the equality (*ii*) has been obtained by considering *directions* only. A proof of the result will now be given.

Note. The vector multiplication is not associative, *i.e.*, (**a** × **b**) × **c** and **a** × (**b** × **c**) are not equal for all vectors **a, b, c.** In fact, as seen above, (**a** × **b**) × **c** is a linear combination of **a** and **b** and **a** × (**b** × **c**) is a linear combination of **b** and **c.**

The student may compute

$$\{(\mathbf{i} - \mathbf{j}) \times (\mathbf{i} + \mathbf{j} + \mathbf{k})\} \times (\mathbf{j} - \mathbf{k})$$

and $$(\mathbf{i} - \mathbf{j}) \times \{(\mathbf{i} + \mathbf{j} + \mathbf{k})\} \times (\mathbf{j} - \mathbf{k})\}$$

and see that they are *not* equal.

Theorem. *To prove that*

$$(\mathbf{a} \times \mathbf{b}) \times \mathbf{c} = (\mathbf{a} \cdot \mathbf{c})\, \mathbf{b} - (\mathbf{b} \cdot \mathbf{c})\, \mathbf{a}$$ (*Kolkata, 99, 2005 Garhwal 98*)

where **a, b, c** *are three vectors.*

The proof will be given by suitably selecting an orthonormal base consisting of vectors **i, j, k** related to **a, b, c.**

i, is a unit vector along **a**;

j, is a unit vector perpendicular to **a** and in the plane of **a** and **b**.

k, is a unit vector perpendicular to **i** and **j** such that **i, j, k** is a right-handed triad.

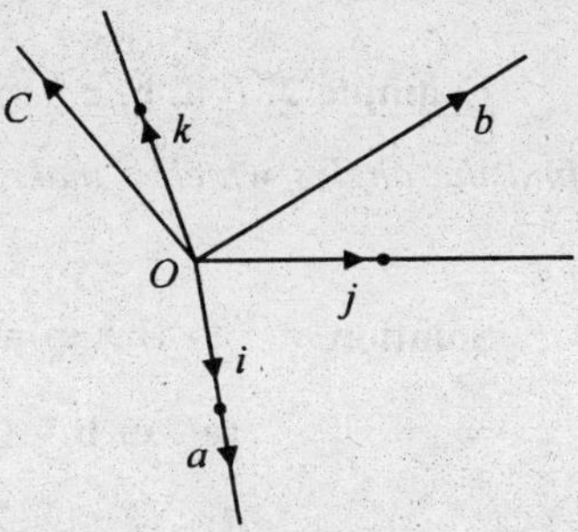

Fig. 5.11.

Thus, we have expressions of the form

$$\mathbf{a} = a_1\mathbf{i},$$

$$\mathbf{b} = b_1\mathbf{i} + b_2\mathbf{j},$$

$$\mathbf{c} = c_1\mathbf{i} + c_2\mathbf{j} + c_3\mathbf{k},$$

Now $\mathbf{a} \times \mathbf{b} = a_1\mathbf{i} \times (b_1\mathbf{i} + b_2\mathbf{j})$

$$= a_1b_2\mathbf{k}.$$

$$(\mathbf{a} \times \mathbf{b}) \times \mathbf{c} = a_1b_2\mathbf{k} \times (c_1\mathbf{i} + c_2\mathbf{j} + c_3\mathbf{k})$$

$$= -a_1b_2c_2\mathbf{i} + a_1b_2c_1\mathbf{j}.$$

Also $(\mathbf{a} \cdot \mathbf{c})\,\mathbf{b} - (\mathbf{b} \cdot \mathbf{c})\,\mathbf{a} = a_1c_1(b_1\mathbf{i} + b_2\mathbf{j}) - (b_1c_1 + b_2c_2)\,a_1\mathbf{i}$

$$= -a_1b_2c_2\mathbf{i} + a_1b_2c_1\mathbf{j}.$$

Hence $(\mathbf{a} \times \mathbf{b}) \times \mathbf{c} = (\mathbf{a} \cdot \mathbf{c})\,\mathbf{b} - (\mathbf{b} \cdot \mathbf{c})\,\mathbf{a}$ ...(*ii*)

Cor. $\mathbf{a} \times (\mathbf{b} \times \mathbf{c}) = -(\mathbf{b} \times \mathbf{c}) \times \mathbf{a}$

$$= -[(\mathbf{b} \cdot \mathbf{a})\,\mathbf{c} - (\mathbf{c} \cdot \mathbf{a})\,\mathbf{b}]$$

$$= (\mathbf{c} \cdot \mathbf{a})\,\mathbf{b} - (\mathbf{b} \cdot \mathbf{a})\,\mathbf{c} = (\mathbf{a} \cdot \mathbf{c})\,\mathbf{b} - (\mathbf{a} \cdot \mathbf{b})\,\mathbf{c}.$$

Rule to Remember. As a help to memory, we notice that each of the two scalar products on the right of (*ii*) involves the outer vector **c**, and of the two other vectors, **a, b** in the bracket; **a** is remote from **c** and **b** is adjacent **c**. Thus, we have the rule :

(*remote* × *adjacent*) × *outer* = (*outer* . *remote*) *adjacent* −
(*outer* . *adjacent*) *remote*.

The formula in the cor. is also written by the same rule.

EXAMPLES

Example 1. *Prove* $\mathbf{i} \times (\mathbf{a} \times \mathbf{i}) + \mathbf{j} \times (\mathbf{a} \times \mathbf{j}) + \mathbf{k} \times (\mathbf{a} \times \mathbf{k}) = 2\mathbf{a}.$

(*U.P.P.C.S. 2004, Kumaon 2002 Awadh 2000, 2001, Rohilkhand 2000*)

Solution. L.H.S. $= (\mathbf{i} \cdot \mathbf{i})\,\mathbf{a} - (\mathbf{i} \cdot \mathbf{a})\,\mathbf{i} + (\mathbf{j} \cdot \mathbf{j})\,\mathbf{a}$

$$- (\mathbf{j} \cdot \mathbf{a})\,\mathbf{j} + (\mathbf{k} \cdot \mathbf{k})\,\mathbf{a} - (\mathbf{k} \cdot \mathbf{a})\,\mathbf{k}$$

$$= 3\mathbf{a} - [(\mathbf{i} \cdot \mathbf{a})\,\mathbf{i} + (\mathbf{j} \cdot \mathbf{a})\,\mathbf{j} + (\mathbf{k} \cdot \mathbf{a})\,\mathbf{k}]$$

$$= 3\mathbf{a} - \mathbf{a} = 2\mathbf{a}.$$

Example 2. *If* **a, b, c** *be three unit vectors such that* $\mathbf{a}\times(\mathbf{b}\times\mathbf{c})=\frac{1}{2}\mathbf{b}$, *find the angles which* **a** *makes with* **b** *and* **c**, **b** *and* **c** *being non-parallel.*

(*Awadh 99, 2001, 2003; Kumaon 98*)

Solution. $\mathbf{a}\times(\mathbf{b}\times\mathbf{c})=\frac{1}{2}\mathbf{b}$

$\Rightarrow$ $(\mathbf{a}\,.\,\mathbf{c})\,\mathbf{b}-(\mathbf{a}\,.\,\mathbf{b})\,\mathbf{c}=\frac{1}{2}\mathbf{b}$

$\Rightarrow$ $\left(\mathbf{a}\,.\,\mathbf{c}-\frac{1}{2}\right)\mathbf{b}-(\mathbf{a}\,.\,\mathbf{b})\,\mathbf{c}=0$

Since **b** and **c** are non-parallel, hence coefficients of **b** and **c** vanish separately.

$\therefore$ $\mathbf{a}\,.\,\mathbf{c}-\frac{1}{2}=0$ and $\mathbf{a}\,.\,\mathbf{b}=0$

$\Rightarrow$ Angle between **a** and $\mathbf{c}=\cos^{-1}\left(\frac{1}{2}\right)=\frac{\pi}{3}$ and Angle between **a** and $\mathbf{b}=\pi/2$.

Example 3. *Show that* $\mathbf{a}\times(\mathbf{b}\times\mathbf{c})=(\mathbf{a}\times\mathbf{b})\times\mathbf{c}$ *if and only if either* $\mathbf{b}=0$, *or* **c** *is collinear with* **a**, *or* **b** *is perpendicular to both* **a** *and* **c**.

(*Purvanchal 2004, Avadh 2002*)

Solution. $\mathbf{a}\times(\mathbf{b}\times\mathbf{c})=(\mathbf{a}\times\mathbf{b})\times\mathbf{c}$

$\Leftrightarrow$ $(\mathbf{a}\,.\,\mathbf{c})\,\mathbf{b}-(\mathbf{a}\,.\,\mathbf{b})\,\mathbf{c}=(\mathbf{a}\,.\,\mathbf{c})\,\mathbf{b}-(\mathbf{b}\,.\,\mathbf{c})\,\mathbf{a}$

$\Leftrightarrow$ $(\mathbf{a}\,.\,\mathbf{b})\,\mathbf{c}=(\mathbf{b}\,.\,\mathbf{c})\,\mathbf{a}$ ...(*i*)

The condition is necessary. If **c** and **a** are collinear then theorem is proved. But if **c** and **a** are not collinear, then (1) implies that

$$\mathbf{a}\,.\,\mathbf{b}=0 \text{ and } \mathbf{b}\,.\,\mathbf{c}=0$$

It is true if $\mathbf{b}=0$ or **b** is perpendicular to both **a** and **c**.

The condition is sufficient. Let $\mathbf{b}=0$, or **b** be perpendicular to both **a** and **c**. Then both sides of (1) are equal and hence the given relation holds. Again, if **c** is collinear with **a**, then there exists a non-zero scalar x such that $\mathbf{c}=x\mathbf{a}$. Then, we have

$$(\mathbf{a}\,.\,\mathbf{b})\,\mathbf{c}=(\mathbf{a}\,.\,\mathbf{b})\,x\mathbf{a}=x\,(\mathbf{a}\,.\,\mathbf{b})\,\mathbf{a}$$

$$=(x\mathbf{a}\,.\,\mathbf{b})\,.\,\mathbf{a}=(\mathbf{c}\,.\,\mathbf{b})\,\mathbf{a}=(\mathbf{b}\,.\,\mathbf{c})\,\mathbf{a}.$$

Hence the proposition.

Example 4. *Prove that* $[\mathbf{a}\times\mathbf{b},\ \mathbf{b}\times\mathbf{c},\ \mathbf{c}\times\mathbf{a}]=[\mathbf{a}\ \mathbf{b}\ \mathbf{c}]^2$.

(*Patna 2003, Avadh 2005, Kumaon 2002, 2005, Rohilkhand 2005, 2006*)

Solution. Let $\mathbf{b}\times\mathbf{c}=\mathbf{d}$

Then $(\mathbf{b}\times\mathbf{c})\times(\mathbf{c}\times\mathbf{a})=\mathbf{d}\times(\mathbf{c}\times\mathbf{a})=(\mathbf{d}\,.\,\mathbf{a})\,\mathbf{c}-(\mathbf{d}\,.\,\mathbf{c})\,\mathbf{a}$

$=[\mathbf{b}\ \mathbf{c}\ \mathbf{a}]\,\mathbf{c}-[\mathbf{b}\ \mathbf{c}\ \mathbf{c}]\,\mathbf{a}$

$=[\mathbf{b}\ \mathbf{c}\ \mathbf{a}]\,\mathbf{c}$ $(\because [\mathbf{b}\,\mathbf{c}\,\mathbf{c}]=0)$

$$\therefore \quad [\mathbf{a} \times \mathbf{b},\ \mathbf{b} \times \mathbf{c},\ \mathbf{c} \times \mathbf{a}] = (\mathbf{a} \times \mathbf{b}) \cdot [(\mathbf{b} \times \mathbf{c}) \times (\mathbf{c} \times \mathbf{a})]$$
$$= (\mathbf{a} \times \mathbf{b}) \cdot [\mathbf{b}\ \mathbf{c}\ \mathbf{a}]\ \mathbf{c}$$
$$= [\mathbf{b}\ \mathbf{c}\ \mathbf{a}]\ [\mathbf{a}\ \mathbf{b}\ \mathbf{c}] = [\mathbf{a}\ \mathbf{b}\ \mathbf{c}]^2.$$

EXERCISES

1. Show that

$$(\mathbf{a} \times \mathbf{b}) \times \mathbf{c} = \mathbf{a} \times (\mathbf{b} \times \mathbf{c})$$

if, and only if, the vectors **a** and **c** are collinear.

2. Show that

(*i*) $[\mathbf{a} + \mathbf{b}, \mathbf{b} + \mathbf{c}, \mathbf{c} + \mathbf{a}] = 2\ [\mathbf{a}\ \mathbf{b}\ \mathbf{c}]$.

(*Kumaon 2000, 2005, Avadh 97, Purvanchal 2004, Rohilkhand 99, Garhwal 2000*)

(*ii*) $(\mathbf{a} \times \mathbf{b}) \times (\mathbf{a} \times \mathbf{c}) \cdot \mathbf{d} = \mathbf{a} \cdot \mathbf{d}\ [\mathbf{a}\ \mathbf{b}\ \mathbf{c}]$.

3. Prove that

(*i*) $\mathbf{a} \times (\mathbf{b} + \mathbf{c}) + \mathbf{b} \times (\mathbf{c} + \mathbf{a}) + \mathbf{c} \times (\mathbf{a} + \mathbf{b}) = \mathbf{0}$.

(*Awadh 99, Rohilkhand 98*)

(*ii*) $\mathbf{a} \times (\mathbf{b} \times \mathbf{c}) + \mathbf{b} \times (\mathbf{c} \times \mathbf{a}) + \mathbf{c} \times (\mathbf{a} \times \mathbf{b}) = \mathbf{0}$.

(*Rohilkhand 98, U.P.P.C.S. 2000, Garhwal 99*)

4. Show that

$$(\mathbf{b} \times \mathbf{c}) \cdot (\mathbf{a} \times \mathbf{d}) + (\mathbf{c} \times \mathbf{a}) \cdot (\mathbf{b} \times \mathbf{d}) + (\mathbf{a} \times \mathbf{b}) \cdot (\mathbf{c} \times \mathbf{d}) = \mathbf{0}.$$

5. Prove that

$$\mathbf{d} \cdot [\mathbf{a} \times \{\mathbf{b} \times (\mathbf{c} \times \mathbf{d})\}] = (\mathbf{b} \cdot \mathbf{d})\ [\mathbf{a}\ \mathbf{c}\ \mathbf{d}].$$

6. **p, q, r** are three vectors defined by the relations

$$\mathbf{p} = \frac{\mathbf{b} \times \mathbf{c}}{[\mathbf{a}\ \mathbf{b}\ \mathbf{c}]},\ \mathbf{q} = \frac{\mathbf{c} \times \mathbf{a}}{[\mathbf{a}\ \mathbf{b}\ \mathbf{c}]},\ \mathbf{r} = \frac{\mathbf{a} \times \mathbf{b}}{[\mathbf{a}\ \mathbf{b}\ \mathbf{c}]},$$

and $\quad [\mathbf{a}\ \mathbf{b}\ \mathbf{c}] \neq \mathbf{0}$.

Prove that

(*i*) $\mathbf{a} \times \mathbf{p} + \mathbf{b} \times \mathbf{q} + \mathbf{c} \times \mathbf{r} = 0$,

(*ii*) $\mathbf{a} \cdot \mathbf{p} + \mathbf{b},\ \mathbf{q} + \mathbf{c} \cdot \mathbf{r} = 3$,

(*iii*) $[\mathbf{a}\ \mathbf{b}\ \mathbf{c}]\ [\mathbf{p}\ \mathbf{q}\ \mathbf{r}] = 1$,

(*iv*) $\mathbf{a} = \dfrac{\mathbf{q} \times \mathbf{r}}{[\mathbf{p}\ \mathbf{q}\ \mathbf{r}]},\ \mathbf{b} = \dfrac{\mathbf{r} \times \mathbf{p}}{[\mathbf{p}\ \mathbf{q}\ \mathbf{r}]},\ \mathbf{a} = \dfrac{\mathbf{p}\mathbf{q}}{[\mathbf{p}\ \mathbf{q}\ \mathbf{r}]}$

7. **a, b, c; a′, b′, c′** are two systems of non-coplanar vectors and

$$\mathbf{p} = \mathbf{a} \times (\mathbf{b}' \times \mathbf{c}'),\ \mathbf{q} = \mathbf{b} \times (\mathbf{c}' \times \mathbf{a}'),\ \mathbf{r} = \mathbf{c} \times (\mathbf{a}' \times \mathbf{b}');$$
$$\mathbf{p}' = \mathbf{a}' \times (\mathbf{b} \times \mathbf{c}),\ \mathbf{q}' = \mathbf{b}' \times (\mathbf{c} \times \mathbf{a}),\ \mathbf{r}' = \mathbf{c}' \times (\mathbf{a} \times \mathbf{b}).$$

Show that

$$[\mathbf{p}\ \mathbf{q}\ \mathbf{r}] \neq 0, \quad \Leftrightarrow \quad [\mathbf{p}', \mathbf{q}', \mathbf{r}'] \neq 0.$$

8. **a, b, c** are three non-coplanar vectors; express $\mathbf{b} \times \mathbf{c}, \mathbf{c} \times \mathbf{a}, \mathbf{a} \times \mathbf{b}$ as linear combination of **a, b, c** and show that if

$$\mathbf{b} \times \mathbf{c} = l_{11}\mathbf{a} + l_{12}\mathbf{b} + l_{13}\mathbf{c}, \quad \mathbf{c} \times \mathbf{a} = l_{21}\mathbf{a} + l_{22}\mathbf{b} + l_{23}\mathbf{c},$$
$$\mathbf{a} \times \mathbf{b} = l_{31}\mathbf{a} + l_{32}\mathbf{b} + l_{33}\mathbf{c},$$

then

$$l_{ij} = A_{ij} / [\mathbf{a}\ \mathbf{b}\ \mathbf{c}],$$

where A_{ij} is the cofactor of the element in the ith row and jth column in the determinantal expression for $[\mathbf{a}\ \mathbf{b}\ \mathbf{c}]^2$.

SUMMARY

1. The vector multiplication is neither commutative nor associative.
2. If **i, j, k** are mutually perpendicular unit vectors, then

$$\mathbf{i} \times \mathbf{i} = 0, \quad \mathbf{j} \times \mathbf{j} = 0, \quad \mathbf{k} \times \mathbf{k} = 0,$$
$$\mathbf{i} \times \mathbf{j} = \mathbf{k}, \quad \mathbf{j} \times \mathbf{k} = \mathbf{i}, \quad \mathbf{k} \times \mathbf{i} = \mathbf{j},$$
$$\mathbf{j} \times \mathbf{i} = -\mathbf{k}, \quad \mathbf{k} \times \mathbf{j} = -\mathbf{i}, \quad \mathbf{i} \times \mathbf{k} = -\mathbf{j},$$

3. The area of a triangle OAB where $\overrightarrow{OA} = \mathbf{a}$ and $\overrightarrow{OB} = \mathbf{b}$ is

$$\frac{1}{2} |\mathbf{a} \times \mathbf{b}|.$$

4. The scalar triple product $\mathbf{a} \cdot \mathbf{b} \times \mathbf{c}$ is 0 if and only if the three vectors **a, b, c** are coplanar.
5. A cyclic permutation of the three vectors does not change the value of the scalar triple product and an anti-cyclic permutation changes the value in sign but not in magnitude.
6. The positions of dot and cross can be interchanged without any change in the value of a scalar triple product.
7. The expression $\frac{1}{6} \mathbf{a} \cdot \mathbf{b} \times \mathbf{c}$ denotes the volume of the tetrahedron with coterminous edges

$$\overrightarrow{OA} = \mathbf{a}, \quad \overrightarrow{OB} = \mathbf{b}, \quad \overrightarrow{OC} = \mathbf{c}$$

if the vectors **a, b, c** form a right-handed system.

8. Vector multiplication distributes the sum of vectors, *i.e.*,

$$\mathbf{a} \times (\mathbf{b} + \mathbf{c}) = \mathbf{a} \times \mathbf{b} + \mathbf{a} \times \mathbf{c}$$

for all vectors **a, b, c.**

9. $(a_1\mathbf{i} + a_2\mathbf{j} + a_3\mathbf{k}) \times (b_1\mathbf{i} + b_2\mathbf{j} + b_3\mathbf{k})$

$$= (a_2b_3 - a_3b_2)\,\mathbf{i} + (a_3b_1 - a_1b_3)\,\mathbf{j} + (a_1b_2 - a_2b_1)\,\mathbf{k}$$

which is usually denoted by

$$\begin{vmatrix} \mathbf{i} & \mathbf{j} & \mathbf{k} \\ a_1 & a_2 & a_3 \\ b_1 & b_2 & b_3 \end{vmatrix}$$

10. If $\mathbf{a} = a_1\mathbf{i} + a_2\mathbf{j} + a_3\mathbf{k}$, $\mathbf{b} = b_1\mathbf{i} + b_2\mathbf{j} + b_3\mathbf{k}$,

and $\mathbf{c} = c_1\mathbf{i} + c_2\mathbf{j} + c_3\mathbf{k}$,

then $$\mathbf{a} \cdot \mathbf{b} \times \mathbf{c} = \begin{vmatrix} a_1 & a_2 & a_3 \\ b_1 & b_2 & b_3 \\ c_1 & c_2 & c_3 \end{vmatrix}.$$

11. $(\mathbf{a} \times \mathbf{b}) \times \mathbf{c} = (\mathbf{a} \cdot \mathbf{c})\, \mathbf{b} - (\mathbf{b} \cdot \mathbf{c})\, \mathbf{a}$.

APPLICATION OF SCALAR TRIPLE PRODUCTS TO GEOMETRY

5.9. EQUATIONS OF PLANES PARALLEL TO GIVEN VECTORS AND PASSING THROUGH GIVEN POINTS

5.9.1. *Equation of the plane through three points A, B, C with position vector* **a, b, c.**

We have

$$\vec{AB} = \mathbf{b} - \mathbf{a},$$

$$\vec{AC} = \mathbf{c} - \mathbf{a}.$$

The vector

$$\vec{AB} \times \vec{AC} = (\mathbf{b} - \mathbf{a}) \times (\mathbf{c} - \mathbf{a})$$

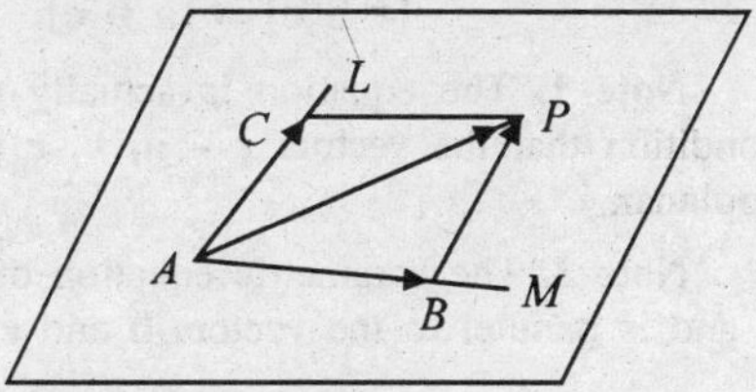

Fig. 5.12.

is normal to the plane. Thus, the equation of the plane is

$$(\mathbf{r} - \mathbf{a}) \cdot (\mathbf{b} - \mathbf{a}) \times (\mathbf{c} - \mathbf{a}) = 0$$

$$\Rightarrow \quad \mathbf{r} \cdot (\mathbf{b} \times \mathbf{c} + \mathbf{c} \times \mathbf{a} + \mathbf{a} \times \mathbf{b}) = [\mathbf{a}\ \mathbf{b}\ \mathbf{c}]$$

which is the required equation.

Note. The equation can also be deduced from the parametric equation $\mathbf{r} = \mathbf{a} + t(\mathbf{b} - \mathbf{a}) + p(\mathbf{c} - \mathbf{a})$ on multiplying both sides of

$\mathbf{r} - \mathbf{a} = t(\mathbf{b} - \mathbf{a}) + p(\mathbf{c} - \mathbf{a})$ scalarly with $(\mathbf{b} - \mathbf{a}) \times (\mathbf{c} - \mathbf{a})$.

Cor. Condition for coplanarity of four points. Four points **a, b, c, d** will be coplanar if, and only if, the point **d**, lies on the plane through the three points **a, b, c**, *i.e.*, **d** satisfies the equation

$$\mathbf{r} \cdot (\mathbf{b} \times \mathbf{c} + \mathbf{c} \times \mathbf{a} + \mathbf{a} \times \mathbf{b}) = [\mathbf{a}\ \mathbf{b}\ \mathbf{c}].$$

Thus, the required condition is

$$[\mathbf{d}\ \mathbf{b}\ \mathbf{c}] + [\mathbf{d}\ \mathbf{c}\ \mathbf{a}] + [\mathbf{d}\ \mathbf{a}\ \mathbf{b}] = [\mathbf{a}\ \mathbf{b}\ \mathbf{c}].$$

Note. As seen in chapter 2 four points with position vectors **a, b, c, d** are coplanar, if and only if, there exist four scalars x, y, z, t, not all zero, such that

$$x\mathbf{a} + y\mathbf{b} + z\mathbf{c} + t\mathbf{d} = 0, \qquad x + y + z + t = 0.$$

From these eliminating t, we obtain

$$x(\mathbf{a} - \mathbf{d}) + y(\mathbf{b} - \mathbf{d}) + z(\mathbf{c} - \mathbf{d}) = 0$$

Now supposing that $x \neq 0$ and multiplying scalarly with

$$(\mathbf{b} - \mathbf{d}) \times (\mathbf{c} - \mathbf{d}),$$

we obtain

$$x(\mathbf{a} - \mathbf{d}) \cdot (\mathbf{b} - \mathbf{d}) \times (\mathbf{c} - \mathbf{d}) = 0$$

$$\Rightarrow \quad [\mathbf{a}\ \mathbf{b}\ \mathbf{c}] = [\mathbf{d}\ \mathbf{b}\ \mathbf{c}] + [\mathbf{d}\ \mathbf{c}\ \mathbf{a}] + [\mathbf{d}\ \mathbf{a}\ \mathbf{b}],$$

which is the condition as obtained above.

5.9.2. *Equation of the plane which passes through a given point A with position vector* **a** *and is parallel to the given vectors* **b** *and* **c.**

The vector **b** × **c** is normal to the plane and the point with position vector **a** lies on the plane. The equation of the plane, therefore, is

$$(\mathbf{r} - \mathbf{a}) \cdot \mathbf{b} \times \mathbf{c} = 0$$

$$\Leftrightarrow \quad [\mathbf{r}\ \mathbf{b}\ \mathbf{c}] = [\mathbf{a}\ \mathbf{b}\ \mathbf{c}].$$

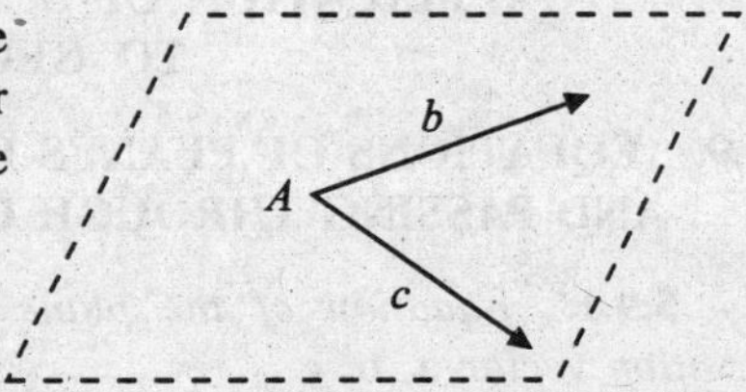

Fig. 5.13.

Note 1. The equation is actually the condition that the vectors **r** − **a, b, c** are coplanar.

Note 2. The parametric equation of the plane which passes through a point **a** and is parallel to the vectors **b** and **c** is

$$\mathbf{r} = \mathbf{a} + t\mathbf{b} + p\mathbf{c}.$$

$$\Rightarrow \quad \mathbf{r} - \mathbf{a} = t\mathbf{b} + p\mathbf{c}.$$

and on multiplying both sides scalarly with **b** × **c,** we get

$$(\mathbf{r} - \mathbf{a}) \cdot (\mathbf{b} \times \mathbf{c}) = t\mathbf{b}, \qquad \mathbf{b} \times \mathbf{c} + p\mathbf{c} \cdot \mathbf{b} \times \mathbf{c} = 0$$

5.9.3. *Equation of the plane through two given points A, B with position vectors* **a, b** *and parallel to a given vector* **c.**

The vector (**a** − **b**) × **c** is normal to the plane. Thus, we require the equation of the plane which passes through a point **a** and which is normal to the vector

$$(\mathbf{a} - \mathbf{b}) \times \mathbf{c}.$$

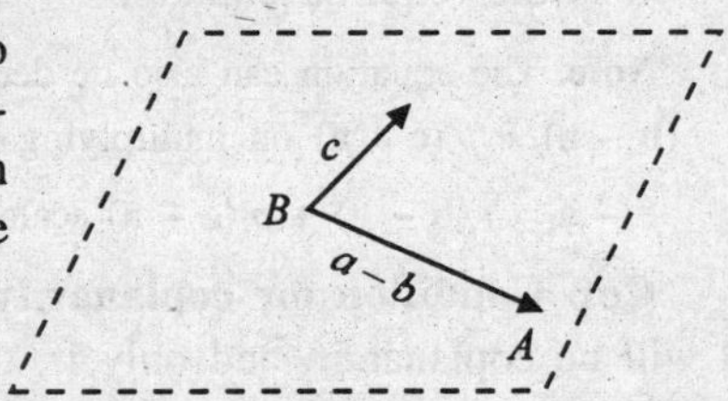

Fig. 5.14.

The required equation therefore is

$$(\mathbf{r} - \mathbf{a}) \cdot (\mathbf{a} - \mathbf{b}) \times \mathbf{c} = 0$$

$$\Leftrightarrow \quad \mathbf{r} \cdot [\mathbf{b} \times \mathbf{c} + \mathbf{c} \times \mathbf{a}] = [\mathbf{a}\ \mathbf{b}\ \mathbf{c}]. \qquad ...(i)$$

Note 1. We may also obtain this result on taking the parametric equation

$$\mathbf{r} = \mathbf{a} + t(\mathbf{b} - \mathbf{a}) + p\mathbf{c} \quad \Leftrightarrow \quad \mathbf{r} - \mathbf{a} = t(\mathbf{b} - \mathbf{a}) + p\mathbf{c}$$

and multiplying both sides scalarly with $(\mathbf{b} - \mathbf{a}) \times \mathbf{c}$.

Note 2. The vector $\mathbf{b} \times \mathbf{c} + \mathbf{c} \times \mathbf{a}$ is normal to the plane which passes through points with position vectors **a**, **b** and which is parallel to the vector **c**.

5.10. COPLANARITY OF TWO LINES

To find the condition for the lines

$$\mathbf{r} = \mathbf{a} + t\mathbf{b}, \qquad \mathbf{r} = \mathbf{c} + p\mathbf{d}$$

to be coplanar.

The equation of the plane which contains the line $\mathbf{r} = \mathbf{a} + t\mathbf{b}$ and which is parallel to the vector **d** is

$$(\mathbf{r} - \mathbf{a}) \cdot (\mathbf{b} \times \mathbf{d}) = 0$$

The two lines will be coplanar if this plane passes through the point with position vector **c**, *i.e.*, if $[\mathbf{c}\ \mathbf{b}\ \mathbf{d}] = [\mathbf{a}\ \mathbf{b}\ \mathbf{d}]$. ...(*i*)

Cor. 1. Plane through two coplanar lines. Assuming the condition for coplanarity to be satisfied, we now find the equation of the plane through the two lines.

The plane which passes through the point with position vector **a** and which is parallel to the vectors **b** and **d** is the plane through the two coplanar lines. The equation of the required plane therefore is

$$(\mathbf{r} - \mathbf{a}) \cdot (\mathbf{b} \times \mathbf{d}) = 0 \quad \Leftrightarrow \quad \mathbf{r} \cdot \mathbf{b} \times \mathbf{d} = [\mathbf{a}\ \mathbf{b}\ \mathbf{d}] \qquad ...(ii)$$

The condition (*i*) of coplanarity is actually the condition for the plane (*ii*) to contain the point **c** of the line $\mathbf{r} = \mathbf{c} + p\mathbf{d}$.

5.11. SHORTEST DISTANCE BETWEEN TWO LINES

To find the shortest distance between the lines

$$\mathbf{r} = \mathbf{a} + t\mathbf{b}, \quad \mathbf{r} = \mathbf{c} + p\mathbf{d}. \qquad (Kolkata, 99)$$

The line of shortest distance between two given lines is the line which meets the two lines perpendicularly.

Being perpendicular to each of the two given lines, the line of shortest distance LM is parallel to the vector

$$\mathbf{b} \times \mathbf{d}.$$

Also the length LM is the projection of $\overrightarrow{AC}$ upon $\overrightarrow{LM}$.

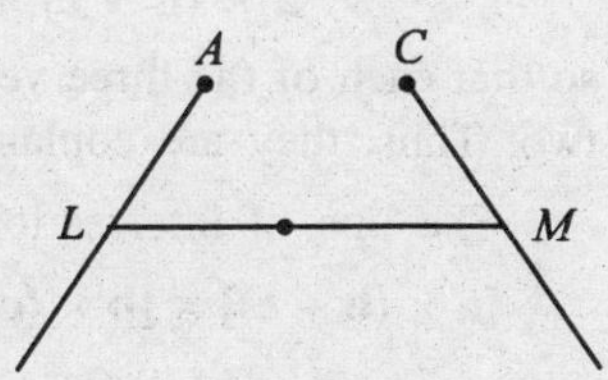

Fig. 5.15.

Thus

$$LM = \frac{|(\mathbf{c}-\mathbf{a}).(\mathbf{b}\times\mathbf{d})|}{|\mathbf{b}\times\mathbf{d}|} = \frac{|[\mathbf{c}\ \mathbf{b}\ \mathbf{d}]-[\mathbf{a}\ \mathbf{b}\ \mathbf{d}]|}{|\mathbf{b}\times\mathbf{d}|}.$$

Also the line of shortest distance is the line of intersection of the two planes through the given lines and the line LM. The equations of these planes are

$$(\mathbf{r}-\mathbf{a})\,.\,\mathbf{b}\times(\mathbf{b}\times\mathbf{d}) = 0, \quad (\mathbf{r}-\mathbf{c})\,.\,\mathbf{d}\times(\mathbf{b}\times\mathbf{d}) = 0.$$

The line of intersection of these planes is the required line of shortest distance.

EXAMPLES

Example 1. *Show that the planes through each of the three edges of a trihedral angle bisecting the internal angle between the other two are coaxal.*

Find also the common line in terms of the vectors along the three edges.

Solution. Suppose that OA, OB, OC are three edges of a trihedral angle at O. We suppose that

$$\overrightarrow{OA} = \mathbf{a},\ \overrightarrow{OB} = \mathbf{b},\ \overrightarrow{OC} = \mathbf{c}$$

are unit vectors.

Let OD, OE, OF bisect the internal angles between OA, OB and OC taken in pairs. Then the bisectors OD, OE and OF lie along the vectors

$$\mathbf{b}+\mathbf{c},\quad \mathbf{c}+\mathbf{a},\quad \mathbf{a}+\mathbf{b}$$

respectively. (Refer Ex. 5, Page 38-39). The equations of the planes

$$OAD,\ OBE,\ OCF$$

are

$$\mathbf{r}\,.\,\mathbf{a}\times(\mathbf{b}+\mathbf{c}) = 0,\quad \mathbf{r}\,.\,\mathbf{b}\times(\mathbf{c}+\mathbf{a}) = 0,\quad \mathbf{r}\,.\,\mathbf{c}\times(\mathbf{a}+\mathbf{b}) = 0.$$

These planes have a point, O, in common and will thus be coaxal if the normals to them are coplanar, *i.e.*, the vectors

$$\mathbf{a}\times(\mathbf{b}+\mathbf{c}),\ \mathbf{b}\times(\mathbf{c}+\mathbf{a}),\ \mathbf{c}\times(\mathbf{a}+\mathbf{b}) \quad \ldots(i)$$

are coplanar. Now we have the identity

$$\mathbf{a}\times(\mathbf{b}+\mathbf{c}) + \mathbf{b}\times(\mathbf{c}+\mathbf{a}) + \mathbf{c}\times(\mathbf{a}+\mathbf{b}) = 0,$$

so that each of the three vectors in (*i*) is a linear combination of the other two. Thus, they are coplanar.

The line of intersection of these planes is parallel to the vector

$$[\mathbf{a}\times(\mathbf{b}+\mathbf{c})]\times[\mathbf{b}\times(\mathbf{c}+\mathbf{a})] = (\mathbf{a}\times\mathbf{b}+\mathbf{a}\times\mathbf{c})\times(\mathbf{b}\times\mathbf{c}+\mathbf{b}\times\mathbf{a})$$

$$= (\mathbf{a}\times\mathbf{b})\times(\mathbf{b}\times\mathbf{c}) + (\mathbf{a}\times\mathbf{c})\times(\mathbf{b}\times\mathbf{c}) + (\mathbf{a}\times\mathbf{c})\times(\mathbf{b}\times\mathbf{a})$$

$$= [\mathbf{a}\ \mathbf{b}\ \mathbf{c}]\,\mathbf{b} - [\mathbf{a}\ \mathbf{c}\ \mathbf{b}]\,\mathbf{c} - [\mathbf{a}\ \mathbf{c}\ \mathbf{b}]\,\mathbf{a}$$

$$= [\mathbf{a}\ \mathbf{b}\ \mathbf{c}]\ [\mathbf{a} + \mathbf{b} + \mathbf{c}].$$

Thus, the line of intersection is parallel to the vector

$$\mathbf{a} + \mathbf{b} + \mathbf{c}$$

where **a, b, c** are unit vectors along the three edges.

Example 2. *In each of the three planes determined by two of the lines*

$$OA,\ OB,\ OC$$

a straight line is drawn through O perpendicular to the third line, prove that the three lines so determined are coplanar.

Solution. Let $\vec{OA} = \mathbf{a},\ \vec{OB} = \mathbf{b},\ \vec{OC} = \mathbf{c}.$

The plane containing OA and OB is

$$\mathbf{r} \cdot \mathbf{a} \times \mathbf{b} = 0.$$

Any line in this plane is perpendicular to the vector

$$\mathbf{a} \times \mathbf{b}.$$

Thus, the line in this plane perpendicular to OC is parallel to the vector

$$(\mathbf{a} \times \mathbf{b}) \times \mathbf{c}$$

so that the three lines in question are

$$\mathbf{r} = t\ (\mathbf{a} \times \mathbf{b}) \times \mathbf{c},\quad \mathbf{r} = p\ (\mathbf{b} \times \mathbf{c}) \times \mathbf{a},\quad \mathbf{r} = k\ (\mathbf{c} \times \mathbf{a}) \times \mathbf{b}.$$

We now have the identity

$$(\mathbf{a} \times \mathbf{b}) \times \mathbf{c} + (\mathbf{b} \times \mathbf{c}) \times \mathbf{a} + (\mathbf{c} \times \mathbf{a}) \times \mathbf{b} = 0.$$

so that the three vectors

$$(\mathbf{a} \times \mathbf{b}) \times \mathbf{c},\quad (\mathbf{b} \times \mathbf{c}) \times \mathbf{a},\quad (\mathbf{c} \times \mathbf{a}) \times \mathbf{b}$$

are coplanar.

Thus, the three lines are coplanar.

Example 3. *l_1, l_2, l_3 and m_1, m_2, m_3 are two sets of lines such that the lines of shortest distances between l_1, l_2, l_3 taken in pairs are separately parallel to m_1, m_2, m_3; show that the lines of shortest distances between m_1, m_2, m_3 taken in pairs are also, separately parallel to l_1, l_2, l_3.*

Solution. Let the given lines be parallel to the vectors

$$\mathbf{a}_1, \mathbf{a}_2, \mathbf{a}_3 \text{ and } \mathbf{b}_1, \mathbf{b}_2, \mathbf{b}_3.$$

The given conditions are equivalent to

$$k_1\ \mathbf{b}_1 = \mathbf{a}_2 \times \mathbf{a}_3,\quad k_2\ \mathbf{b}_2 = \mathbf{a}_3 \times \mathbf{a}_1,\quad k_3\ \mathbf{b}_3 = \mathbf{a}_1 \times \mathbf{a}_2,$$

where k_1, k_2, k_3 are some scalars.

$$\therefore \qquad k_2\ \mathbf{b}_2 \times k_3\ \mathbf{b}_3 = (\mathbf{a}_3 \times \mathbf{a}_1) \times (\mathbf{a}_1 \times \mathbf{a}_2)$$

$$= [\mathbf{a}_3\ \mathbf{a}_1\ \mathbf{a}_2]\ \mathbf{a}_1 = [\mathbf{a}_1\ \mathbf{a}_2\ \mathbf{a}_3]\ \mathbf{a}_1$$

so that the vector $\mathbf{a}_1$ is parallel to the line of shortest distance between the lines m_2 and m_3.

Hence the result.

Example 4. *A square PQRS is folded along the diagonal PR so that the planes PRQ and PRS are perpendicular to one another. Determine the shortest distance between PQ and RS in terms of p, the length of the side of the square.*

Solution. Let the new position of Q be Q'. Let **i, j, k** denote unit vector along PQ', $Q'R$, PS so that they form a right-handed orthonormal system.

Taking P as the origin of reference, the position vectors of Q', R and S are

$$p\mathbf{i},\quad p\mathbf{i} + p\mathbf{j},\quad p\mathbf{k}$$

respectively.

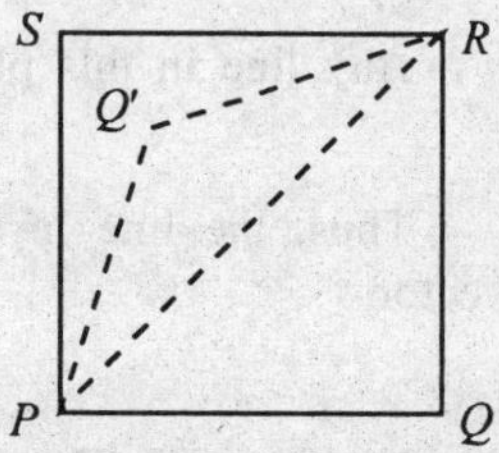

Fig. 5.16.

The equations of PQ' and RS are

$$\mathbf{r} = tp\mathbf{i},$$

$$\mathbf{r} = p\mathbf{i} + p\mathbf{j} + t\,(p\mathbf{i} + p\mathbf{j} - p\mathbf{k}).$$

$\therefore$ the required shortest distance

$$= \frac{|(p\mathbf{i}+p\mathbf{j})\,.\,p\mathbf{i}\times(p\mathbf{i}+p\mathbf{j}-p\mathbf{k})|}{|\,p\mathbf{i}\times(p\mathbf{i}+p\mathbf{j}-p\mathbf{k})|}$$

$$= \frac{|(p\mathbf{i}+p\mathbf{j})\,.\,(p^2\mathbf{k}+p^2\mathbf{j})|}{|\,p^2\mathbf{k}+p^2\mathbf{j}\,|} = \frac{p^3}{\sqrt{(p^4+p^4)}} = \frac{p}{\sqrt{2}}.$$

Example 5. *OA, OB, OC are three coterminous edges of a rectangular parallelopiped and OA = x, OB = y, OC = z. Show that the shortest distance between OA and either diagonal skew to it is*

$$yz/\sqrt{(y^2+z^2)}$$

and the shortest distances divide OA in the ratios

$$z^2/y^2 \text{ and } y^2/z^2.$$

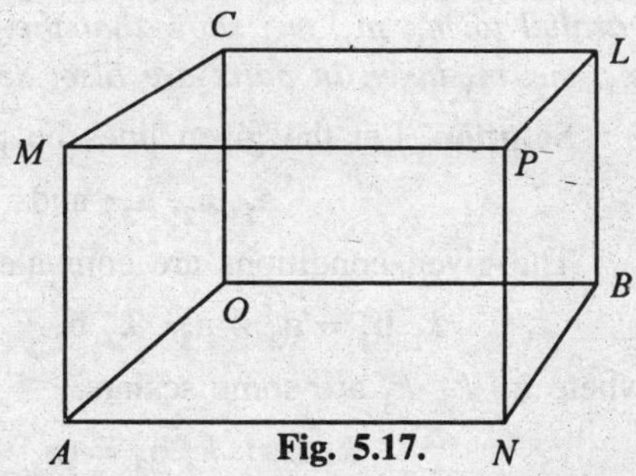

Fig. 5.17.

Solution. Let **i, j, k** denote unit vectors OA, OB, OC so that they form an orthonormal system.

The diagonals *CN* and *BM* are both skew to the edge *OA*.

We shall determine the shortest distance between *OA* and the diagonal *CN*.

The position vectors of *A, C, N* are

$$x\mathbf{i},\ z\mathbf{k},\ \ x\mathbf{i} + y\mathbf{j}$$

so that the equations of *OA* and *CN* are

$$\mathbf{r} = t\mathbf{i},\ \ \mathbf{r} = z\mathbf{k} + p\ (x\mathbf{i} + y\mathbf{j} - z\mathbf{k}).$$

The length of the shortest distance between these lines is

$$= \frac{|(z\mathbf{k}-0)\,.\,\mathbf{i}\times(x\mathbf{i}+y\mathbf{j}-z\mathbf{k})|}{|\mathbf{i}\times(x\mathbf{i}+y\mathbf{j}-z\mathbf{k})|}$$

$$= \frac{|z\mathbf{k}\,.\,(y\mathbf{k}+z\mathbf{j})|}{|y\mathbf{k}+z\mathbf{j}|} = \frac{yz}{\sqrt{(y^2+z^2)}}.$$

Suppose now that the line of shortest distance meets the two lines in the points

$$t\mathbf{i},\ \ z\mathbf{k} + p\ (x\mathbf{i} + y\mathbf{j} - z\mathbf{k})$$

so that their join is perpendicular to both the lines. Thus, we have

$$\begin{cases} [z\mathbf{k}+p(x\mathbf{i}+y\mathbf{j}-z\mathbf{k}-t\mathbf{i}]\,.\,\mathbf{i}=0 \\ [z\mathbf{k}+p(x\mathbf{i}+y\mathbf{j}-z\mathbf{k})-t\mathbf{i}]\,.\,(x\mathbf{i}+y\mathbf{j}-z\mathbf{k})=0 \end{cases}$$

$$\Rightarrow \quad \begin{cases} px-t=0 \\ (px-t)\,x+py^2-z\,(z-pz)=0 \end{cases}$$

$$\Rightarrow \quad p=\frac{z^2}{y^2+z^2},\quad t=\frac{xz^2}{y^2+z^2}.$$

If the line of shortest distance meets *OA* in *L'*, the position vector $\overrightarrow{OL'}$ of *L'* is

$$\frac{xz^2}{y^2+z^2}\,\mathbf{i},$$

$$\Rightarrow \quad \overrightarrow{OL'} = xz^{2''}/(y^2+z^2).$$

$$\therefore \quad LA = x-\frac{xz^2}{y^2+z^2}=\frac{xy^2}{y^2+z^2}$$

$$\Rightarrow \quad \frac{OL'}{L'A}=\frac{z^2}{y^2}.$$

We may also now consider the other diagonal *BM* and prove the corresponding result.

EXERCISES

1. Find the line through the point **a** parallel to the line of intersection of the planes
$$\mathbf{r} \cdot \mathbf{n}_1 = 1, \quad \mathbf{r} \cdot \mathbf{n}_2 = 1.$$
2. Find the line through the point **a** parallel to the planes
$$\mathbf{r} \cdot \mathbf{n}_1 = q_1, \quad \mathbf{r} \cdot \mathbf{n}_2 = q_2.$$
3. Find the line through the point **c** parallel to the plane $\mathbf{r} \cdot \mathbf{n} = 1$ and perpendicular to the line, $\mathbf{r} = \mathbf{a} + t\mathbf{b}$.
4. What is the equation of the line which passes through the point **a**, meets the line $\mathbf{r} = \mathbf{b} + t\mathbf{c}$ and is parallel to the plane $\mathbf{r} \cdot \mathbf{n} = 1$.
5. Find the plane which passes through a points **a** and is perpendicular to the two planes
$$\mathbf{r} \cdot \mathbf{n}_1 = q_1, \quad \mathbf{r} \cdot \mathbf{n}_2 = q_2.$$
6. Find the plane which passes through a point, **a**, is perpendicular to the plane $\mathbf{r} \cdot \mathbf{n} = q$ and is parallel to the line $\mathbf{r} = \mathbf{b} + t\mathbf{c}$.
7. Find the plane which passes through the two points **a**, **b** and is perpendicular to the plane $\mathbf{r} \cdot \mathbf{n} = q$.
8. Find the plane containing the line $\mathbf{r} = \mathbf{a} + t\mathbf{b}$ and perpendicular to the plane $\mathbf{r} \cdot \mathbf{n} = q$.
9. What is the equation of the plane which passes through the line of intersection of the planes
$$\mathbf{r} \cdot \mathbf{n}_1 = q_1, \quad \mathbf{r} \cdot \mathbf{n}_2 = q_2,$$
and is parallel to the line of intersection of the planes
$$\mathbf{r} \cdot \mathbf{n}_3 = q_3, \quad \mathbf{r} \cdot \mathbf{n}_4 = q_4.$$
10. What is the equation of the plane containing the parallel lines
$$\mathbf{r} = \mathbf{a} + t\mathbf{c}, \qquad \mathbf{r} = \mathbf{b} + p\mathbf{c}.$$
11. What is the condition for the lines
$$\mathbf{r} = \mathbf{a} + t\,(\mathbf{b} \times \mathbf{c}), \qquad \mathbf{r} = \mathbf{b} + p\,(\mathbf{c} \times \mathbf{a}).$$
to intersect. Assuming the condition satisfied, express the position vector of the point of intersection in terms of the non-coplanar vectors **a, b, c.**
12. Show that the lines
$$\mathbf{r} \times \mathbf{a} = \mathbf{b} \times \mathbf{a}, \quad \mathbf{r} \times \mathbf{b} = \mathbf{a} \times \mathbf{b}$$
intersect and find the point of intersection.
13. Find the equation of the line which passes through the point **a** and intersect the lines
$$\mathbf{r} = \mathbf{b} + t\mathbf{c}, \qquad \mathbf{r} = \mathbf{d} + p\mathbf{c}.$$
14. Find the equation of the plane through the line
$$\mathbf{r} = \mathbf{a} + t\mathbf{b},$$
and parallel to the line
$$\mathbf{r} = \mathbf{c} + p\mathbf{d},$$
and hence obtain the shortest distance between the two lines.
15. Show that the line of shortest distance between the lines
$$\mathbf{r} = \mathbf{a} + t\mathbf{b}, \qquad \mathbf{r} = \mathbf{c} + p\mathbf{d}$$
meet the lines at points given by
$$t = \frac{[\mathbf{d} \times (\mathbf{a} - \mathbf{c})] \cdot (\mathbf{b} \times \mathbf{d})}{(\mathbf{b} \times \mathbf{d})^2}, \quad p = \frac{[\mathbf{b} \times (\mathbf{a} - \mathbf{c})] \cdot (\mathbf{b} \times \mathbf{d})}{(\mathbf{b} \times \mathbf{d})^2}$$
16. The position vector of four points A, B, C, D relative to any origin O, are

denoted by *a, b, c, d*. Interpret geometrically the equations
(*i*) $(\mathbf{a} - \mathbf{b}) \times (\mathbf{c} \times \mathbf{d}) = 0$, (*ii*) $(\mathbf{a} - \mathbf{b}) \cdot (\mathbf{c} \times \mathbf{d}) = 0$.

17. *OA, OB* and *OC* are three concurrent lines and the lines OA_1, OB_1, OC_1 are normal to the planes BOC, COA and AOB respectively; show that the lines OA, OB, OC are also normal to the planes B_1OC_1, C_1OA_1 and A_1OB_1 respectively.

18. *AA′* is the common perpendicular to two skew lines PQA and *P′Q′A′*; *P, Q* being any two points on the first line and *P′, Q′*, any two points on the second. Prove that the common perpendicular of *AA′* and the line joining the mid-points of *PP′, QQ′* bisects *AA′*.

19. If *a, b, c, d* are four lines in space and (*ad*) represents any plane parallel to *a* and *d* and so on and if (*ad*) is perpendicular to (*bc*) and (*bd*) is perpendicular to (*ca*), prove that (*cd*) is perpendicular to (*ab*).

20. *OA, OB, OC* are coterminous edges of a parallelopiped and *P* is the vertex opposite to *O*; show that the distance of *O* from the plane *ABC* is twice that of *P* from the same plane.

21. *OA, OB* and *OC* are three coterminous vectors. T_1 and T_2 are two parallelopipeds such that *OA, OB, OC* are coterminous edges of T_1 and are the altitudes through *O* of T_2; show that the product of the volumes of the two tetrahedron is

$$\alpha^2 \beta^2 \gamma^2$$

where α, β, γ are the lengths of *OA, OB, OC*.

22. *OA, OB* and *OC* are three mutually perpendicular straight lines and *p* is the length of the perpendicular from *O* to the plane *ABC*; show that

$$p^{-2} = a^{-2} + b^{-2} + c^{-2}$$

and the area of the Δ *ABC* is

$$\frac{1}{2}\sqrt{(b^2c^2 + c^2a^2 + a^2b^2)},$$

a, b, c being the lengths of *OA, OB* and *OC* respectively.

23. Two skew lines *AP, BQ* are met by the shortest distance between them at *A, B* and *P, Q* are points on them such that $AP = p$, $BQ = q$. If the planes *APQ* and *BPQ* are perpendicular, show that *pq* is constant.

24. Two skew lines *AP, BQ* inclined to one another at an angle of 60° are intersected by the shortest distance between them at *A, B* respectively and *P, Q* are points on the lines such that *AQ* is perpendicular to *BP*, prove that $AP \cdot BQ = 2AB^2$.

25. Points *P* and *Q* are taken on two skew lines so that *PQ* is always parallel to a given plane, show that the locus of a point *R* which divides *PQ* in a given ratio is a straight line.

26. A straight line intersects two skew lines. Show that the locus of a point which divides the intercept in a given ratio is a plane perpendicular to the line of shortest distance between the given lines. Find also the condition for the plane to bisect the line of shortest distance.

27. *ABCD* is a skew quadrilateral. Points *E, F, G* and *H* are taken on *AB, BC, CD* and *DA* respectively such that

$$AE : EB :: DG : GC \text{ and } BF : FC :: AH : HD.$$

Show that the points *EF* and *GH* are coplanar.

28. l_1, l_2, l_3 are three given mutually skew lines m_1, m_2, m_3 are the lines of shortest distance between $l_2, l_3 : l_3, l_1$, and l_1, l_2 respectively. Show that the planes through any point P parallel to the pairs of lines l_1, m_1; l_2, m_2; l_3, m_3 are coaxal.

OBJECTIVE QUESTIONS

For each of the following questions, four alternatives are given for the answer. Only one of them is correct. Choose the correct alternative.

1. A vector that is perpendicular to both the vectors $\mathbf{a} = \mathbf{i} - 2\mathbf{j} + \mathbf{k}$ and $\mathbf{b} = \mathbf{i} - \mathbf{j} + \mathbf{k}$ is
   (*a*) $-\mathbf{i} + \mathbf{k}$ (*b*) $-\mathbf{i} - 2\mathbf{j} + \mathbf{k}$
   (*c*) $\mathbf{i} - 2\mathbf{j} + \mathbf{k}$ (*d*) $\mathbf{i} + \mathbf{k}$
2. Given $\mathbf{a} = \mathbf{i} + \mathbf{j} - \mathbf{k}$, $\mathbf{b} = -\mathbf{i} + 2\mathbf{j} + \mathbf{k}$ and $\mathbf{c} = -\mathbf{i} + 2\mathbf{j} - \mathbf{k}$, a unit vector perpendicular to both $\mathbf{a} + \mathbf{b}$ and $\mathbf{b} + \mathbf{c}$ is
   (*a*) $\mathbf{i}$ (*b*) $\mathbf{j}$ (*c*) $\mathbf{k}$ (*d*) $(\mathbf{i} + \mathbf{j} + \mathbf{k})/\sqrt{3}$
3. If $\mathbf{a}$ is perpendicular to $\mathbf{b}$ and $\mathbf{c}$, then
   (*a*) $\mathbf{a} \times (\mathbf{b} \times \mathbf{c}) = 1$ (*b*) $\mathbf{a} \times (\mathbf{b} \times \mathbf{c}) = \mathbf{0}$
   (*c*) $\mathbf{a} \times (\mathbf{b} \times \mathbf{c}) = -1$ (*d*) None of these
4. $\mathbf{a} \times (\mathbf{b} + \mathbf{c}) + \mathbf{b} \times (\mathbf{c} + \mathbf{a}) + \mathbf{c} \times (\mathbf{a} + \mathbf{b})$ is equal to
   (*a*) $2\,[\mathbf{a}\ \mathbf{b}\ \mathbf{c}]$ (*b*) $\mathbf{0}$
   (*c*) 3 (*d*) None of these
5. If $\mathbf{a} \times \mathbf{b} = \mathbf{a} \times \mathbf{c}$, $\mathbf{a} \neq 0$, then
   (*a*) $\mathbf{b} = \mathbf{c} + \lambda\mathbf{a}$ (*b*) $\mathbf{c} = \mathbf{a} + \lambda\mathbf{b}$
   (*c*) $\mathbf{a} = \mathbf{b} + \lambda\mathbf{c}$ (*d*) None of these
6. If $\mathbf{a} \cdot \mathbf{b} = \mathbf{a} \cdot \mathbf{c}$ and $\mathbf{a} \times \mathbf{b} = \mathbf{a} \times \mathbf{c}$, then
   (*a*) either $\mathbf{a} = 0$ or $\mathbf{b} = \mathbf{c}$ (*b*) $\mathbf{a}$ is parallel to $(\mathbf{b} - \mathbf{c})$
   (*c*) $\mathbf{a}$ is perpendicular to $(\mathbf{b} - \mathbf{c})$ (*d*) None of these
7. If $\mathbf{a} \times \mathbf{b} = \mathbf{c}$ and $\mathbf{b} \times \mathbf{c} = \mathbf{a}$, then
   (*a*) $\mathbf{a}, \mathbf{b}, \mathbf{c}$ are orthogonal in pairs and $|\mathbf{a}| = |\mathbf{c}|$ and $|\mathbf{b}| = 1$
   (*b*) $\mathbf{a}, \mathbf{b}, \mathbf{c}$ are orthogonal to each other
   (*c*) $\mathbf{a}, \mathbf{b}, \mathbf{c}$ are orthogonal in pairs but $|\mathbf{a}| \neq |\mathbf{c}|$
   (*d*) $\mathbf{a}, \mathbf{b}, \mathbf{c}$ are orthogonal but $|\mathbf{b}| \neq 1$
8. The vector $\mathbf{a} \times (\mathbf{b} \times \mathbf{a})$ is :
   (*a*) perpendicular to $\mathbf{a}$ (*b*) perpendicular to $\mathbf{b}$
   (*c*) null vector (*d*) perpendicular to both $\mathbf{a}$ and $\mathbf{b}$
9. The triple product $(\mathbf{a} + \mathbf{c}) \cdot (\mathbf{b} + \mathbf{c}) \times (\mathbf{a} + \mathbf{b} + \mathbf{c})$ equals to
   (*a*) $[\mathbf{a}\ \mathbf{b}\ \mathbf{c}]$ (*b*) $2\,[\mathbf{a}\ \mathbf{b}\ \mathbf{c}]$
   (*c*) 0 (*d*) None of these
10. $(\mathbf{a} + 2\mathbf{b} - \mathbf{c}) \cdot (\mathbf{a} - \mathbf{b}) \times (\mathbf{a} - \mathbf{b} - \mathbf{c})$ equals to
    (*a*) $[\mathbf{a}\ \mathbf{b}\ \mathbf{c}]$ (*b*) $-[\mathbf{a}\ \mathbf{b}\ \mathbf{c}]$
    (*c*) $3\,[\mathbf{a}\ \mathbf{b}\ \mathbf{c}]$ (*d*) $-3\,[\mathbf{a}\ \mathbf{b}\ \mathbf{c}]$

11. $\mathbf{a} \cdot (\mathbf{b} + \mathbf{c}) \times (\mathbf{a} + \mathbf{b} + \mathbf{c}) =$

(*a*) 0 (*b*) 2 [**a b c**]

(*c*) [**a b c**] (*d*) None of these

12. Let **a** and 2**b** denote the diagonals of a parallelogram. Then area of the parallelogram is given by

(*a*) $|\mathbf{a} \times \mathbf{b}|$ (*b*) $\frac{1}{2}|\mathbf{a} \times \mathbf{b}|$

(*c*) $2|\mathbf{a} \times \mathbf{b}|$ (*d*) None of these

13. If **a, b, c** are three unit vectors, $\mathbf{b} \times \mathbf{c}$ and $\mathbf{a} \times (\mathbf{b} \times \mathbf{c}) = \frac{1}{2}\mathbf{b}$, then angle between **a** and **c** is

(*a*) $\pi/6$ (*b*) $\pi/2$

(*c*) $\pi/3$ (*d*) None of these

14. The value of $\left\{|\mathbf{a} \times \mathbf{b}|^2 + (\mathbf{a} \cdot \mathbf{b})^2\right\} \div \left(\mathbf{a}^2\mathbf{b}^2\right)$ is

(*a*) unity (*b*) zero

(*c*) 2 (*d*) None of these

15. $(\mathbf{a} - \mathbf{b}) \times (\mathbf{a} + \mathbf{b})$ **is equal to**

(*a*) 0 (*b*) $\mathbf{a} \times \mathbf{b}$

(*c*) $2(\mathbf{a} \times \mathbf{b})$ (*d*) $|\mathbf{a}|^2 + |\mathbf{b}|^2$

16. The unit vector perpendicular to each of the vectors 2**i** – **j** + **k** and 3**i** + 4**j** – **k** is

(*a*) – 3**i** + 5**j** + 11**k** (*b*) (– 3**i** + 5**j** + 11**k**) / 155

(*c*) (– 3**i** + 5**j** + 11**k**) / $\sqrt{(155)}$ (*d*) None of these

17. If θ is the angle between vectors **a** and **b**, and $|\mathbf{a} \times \mathbf{b}| = |\mathbf{a} \cdot \mathbf{b}|$, then θ is equal to

(*a*) 0° (*b*) 180° (*c*) 135° (*d*) 45°

18. If the vector **c**, **a** = *x***i** + *y***j** + *z***k** and **b** = **j** are such that **a, c** and **b** form a right handed system then **c** is

(*a*) *z***i** – *x***k** (*b*) **0**

(*c*) *y***j** (*d*) – *z***i** + *x***k**

19. The value of *x* such that the vectors **a** = 2**i** – **j** + **k**, **b** = *x***i** + 2**j** – 3**k** and **c** = 3**i** – 4**j** + 5**k** are coplanar is

(*a*) 1 (*b*) 3 (*c*) – 2 (*d*) 0

20. The value of *p* such that the vectors **i** + 3**j** – 2**k**, 2**i** – **j** + 4**k** and 3**i** + 2**j** + *p***k** are coplanar is

(*a*) 4 (*b*) 2 (*c*) 8 (*d*) 10

21. If **r** satisfies the equation **r** × (**i** + 2**j** + **k**) = **i** – **k**, then for any scalar *t*, **r** is equal to

(*a*) **i** + *t* (**i** + 2**j** + **k**) (*b*) **j** + *t* (**i** + 2**j** + **k**)

(*c*) **k** + *t* (**i** + 2**j** + **k**) (*d*) None of these

22. The area of the triangle whose two sides are given by $4\mathbf{i} - \mathbf{j} + \mathbf{k}$ and $3\mathbf{i} + \mathbf{j} - \mathbf{k}$ is

(*a*) $7\sqrt{2}$ (*b*) $14\sqrt{2}$ (*c*) $14/\sqrt{2}$ (*d*) $7/\sqrt{2}$

23. The area of the parallelogram whose adjacent sides are $2\mathbf{i} - 3\mathbf{k}$ and $4\mathbf{j} + 2\mathbf{k}$ is

(*a*) $2\sqrt{(14)}$ (*b*) $4\sqrt{(14)}$ (*c*) $16\sqrt{(14)}$ (*d*) $\sqrt{(14)}$

24. The area of the parallelogram having diagonals $\mathbf{a} = 3\mathbf{i} + \mathbf{j} - 2\mathbf{k}$ and $\mathbf{b} = \mathbf{i} - 3\mathbf{j} + 4\mathbf{k}$ is

(*a*) $10\sqrt{3}$ (*b*) $5\sqrt{3}$ (*c*) 8 (*d*) 4

25. If the non-zero vectors **a** and **b** are perpendicular to each other, then the solution of the equation $\mathbf{r} \times \mathbf{a} = \mathbf{b}$, is given by

(*a*) $\mathbf{r} = x\mathbf{a} + \frac{1}{\mathbf{a}\,.\,\mathbf{a}}(\mathbf{a} \times \mathbf{b});\ x \in R$

(*b*) $\mathbf{r} = x\mathbf{b} - \frac{1}{\mathbf{b}\,.\,\mathbf{b}}(\mathbf{a} \times \mathbf{b});\ x \in R$

(*c*) $\mathbf{r} = x\mathbf{a} \times \mathbf{b},\ x \in R$ (*d*) None of these

26. $[\mathbf{a} + \mathbf{b}, \mathbf{b} + \mathbf{c}, \mathbf{c} + \mathbf{a}] =$

(*a*) $[\mathbf{a}\ \mathbf{b}\ \mathbf{c}]$ (*b*) $\Sigma\,(\mathbf{a}\,.\,\mathbf{b})\,\mathbf{c}$

(*c*) $2\,[\mathbf{a}\ \mathbf{b}\ \mathbf{c}]$ (*d*) $|\,\mathbf{a}\,|\,|\,\mathbf{b}\,|\,|\,\mathbf{c}\,|$

27. The value of $[\mathbf{a} - \mathbf{b}, \mathbf{b} - \mathbf{c}, \mathbf{c} - \mathbf{a}]$ where $|\,\mathbf{a}\,| = 1$, $|\,\mathbf{b}\,| = 5$, $|\,\mathbf{c}\,| = 3$ is

(*a*) 0 (*b*) 1 (*c*) 6 (*d*) None of these

28. $[\mathbf{a}, \mathbf{b}, \mathbf{a} \times \mathbf{b}]$ is equal to

(*a*) $|\,\mathbf{a} \times \mathbf{b}\,|$ (*b*) $|\,\mathbf{a} \times \mathbf{b}\,|^2$

(*c*) $|\,\mathbf{a}\,.\,\mathbf{b}\,|$ (*d*) $|\,\mathbf{a}\,|\,|\,\mathbf{b}\,|$

29. $\mathbf{i} \times (\mathbf{a} \times \mathbf{i}) + \mathbf{j} \times (\mathbf{a} \times \mathbf{j}) + \mathbf{k} \times (\mathbf{a} \times \mathbf{k}) =$

(*a*) **a** (*b*) 2**a** (*c*) 3**a** (*d*) **0** (*U.P.P.C.S. 2C*

30. If $\mathbf{u} = \mathbf{i} \times (\mathbf{a} \times \mathbf{i}) + \mathbf{j} \times (\mathbf{a} \times \mathbf{j}) + \mathbf{k} \times (\mathbf{a} \times \mathbf{k})$, then

(*a*) **u** is a unit vector (*b*) $\mathbf{u} = \mathbf{a} + \mathbf{i} + \mathbf{j} + \mathbf{k}$

(*c*) $\mathbf{u} = 2\mathbf{a}$ (*d*) $\mathbf{u} = 8\,(\mathbf{i} + \mathbf{j} + \mathbf{k})$

31. If $\mathbf{u} = \mathbf{a} - \mathbf{b}$, $\mathbf{v} = \mathbf{a} + \mathbf{b}$ and $|\,\mathbf{a}\,| = |\,\mathbf{b}\,| = 2$, then $|\,\mathbf{u} \times \mathbf{v}\,|$ is equal to

(*a*) $2\sqrt{\left[16 - (\mathbf{a} \cdot \mathbf{b})^2\right]}$ (*b*) $\sqrt{\left[16 - (\mathbf{a} \cdot \mathbf{b})^2\right]}$

(*c*) $2\sqrt{\left[4 - (\mathbf{a} \cdot \mathbf{b})^2\right]}$ (*d*) $\left[4 - (\mathbf{a} \cdot \mathbf{b})^2\right]$

32. For non-zero vectors **a, b, c**, $|\,(\mathbf{a} \times \mathbf{b})\,.\,\mathbf{c}\,| = |\,\mathbf{a}\,|\,|\,\mathbf{b}\,|\,|\,\mathbf{c}\,|$ holds if and only if (*a*) $\mathbf{a} \cdot \mathbf{b} = 0$, $\mathbf{b} \cdot \mathbf{c} = 0$ (*b*) $\mathbf{b} \cdot \mathbf{c} = 0$, $\mathbf{c} \cdot \mathbf{a} = 0$

(*c*) $\mathbf{c} \cdot \mathbf{a} = 0$, $\mathbf{a} \cdot \mathbf{b} = 0$ (*d*) $\mathbf{a} \cdot \mathbf{b} = \mathbf{b} \cdot \mathbf{c} = \mathbf{c} \cdot \mathbf{a} = 0$

33. If $\mathbf{u} = \mathbf{a} \times (\mathbf{b} \times \mathbf{c}) + \mathbf{b} \times (\mathbf{c} \times \mathbf{a}) + \mathbf{c} \times (\mathbf{a} \times \mathbf{b})$ then

(*a*) $\mathbf{u}$ is a unit vector (*b*) $\mathbf{u} = \mathbf{a} + \mathbf{b} + \mathbf{c}$

(*c*) $\mathbf{u} = \mathbf{0}$ (*d*) $\mathbf{u} \neq \mathbf{0}$

34. The vector **a** lies in the plane of vectors **b** and **c**, which of the following is correct :

(*a*) $\mathbf{a} \cdot (\mathbf{b} \times \mathbf{c}) = 0$ (*b*) $\mathbf{a} \cdot (\mathbf{b} \times \mathbf{c}) = 1$

(*c*) $\mathbf{a} \cdot (\mathbf{b} \times \mathbf{c}) = -1$ (*d*) $\mathbf{a} \cdot (\mathbf{b} \times \mathbf{c}) = 3$

35. Let *a, b, c* be distinct non-negative numbers. If the vectors $a\mathbf{i} + a\mathbf{j} + c\mathbf{k}$, $\mathbf{i} + \mathbf{k}$ and $c\mathbf{i} + c\mathbf{j} + b\mathbf{k}$ lie in a plane, then **c** is

(*a*) the A.M. of *a* and *b* (*b*) the G.M. of *a* and *b*

(*c*) the H.M. of *a* and *b* (*d*) equal to zero

36. Let $\mathbf{a} = \mathbf{i} + \mathbf{j}$ and $\mathbf{b} = 2\mathbf{i} - \mathbf{k}$, the point of intersection of the lines $\mathbf{r} \times \mathbf{a} = \mathbf{b} \times \mathbf{a}$ and $\mathbf{r} \times \mathbf{b} = \mathbf{a} \times \mathbf{b}$ is

(*a*) $-\mathbf{i} + \mathbf{j} + \mathbf{k}$ (*b*) $3\mathbf{i} - \mathbf{j} + \mathbf{k}$

(*c*) $3\mathbf{i} + \mathbf{j} + \mathbf{k}$ (*d*) $\mathbf{i} - \mathbf{j} - \mathbf{k}$

37. If the vectors $(-bc, b^2 + bc, c^2 + bc)$, $(a^2 + ac, -ac, c^2 + ac)$ and $(a^2 + ab, b^2 + ab, -ab)$ are coplanar, where none of *a, b* or *c* is zero, then

(*a*) $a^2 + b^2 + c^2 = 1$ (*b*) $bc + ca + ab = 0$

(*c*) $a + b + c = 0$ (*d*) $a^2 + b^2 + c^2 = bc + ca + ab$

ANSWERS

1. (*a*)	**2.** (*c*)	**3.** (*b*)	**4.** (*b*)	**5.** (*a*)
6. (*a*)	**7.** (*a*)	**8.** (*a*)	**9.** (*a*)	**10.** (*c*)
11. (*a*)	**12.** (*a*)	**13.** (*c*)	**14.** (*a*)	**15.** (*c*)
16. (*c*)	**17.** (*d*)	**18.** (*a*)	**19.** (*a*)	**20.** (*b*)
21. (*b*)	**22.** (*d*)	**23.** (*b*)	**24.** (*b*)	**25.** (*a*)
26. (*c*)	**27.** (*a*)	**28.** (*b*)	**29.** (*b*)	**30.** (*c*)
31. (*a*)	**32.** (*d*)	**33.** (*c*)	**34.** (*a*)	**35.** (*b*)
36. (*c*)	**37.** (*b*)			

6

Geometry with Cartesian Co-ordinates

Cartesian Equations. We shall now see how the properties of scalar products enable us to obtain various results in Cartesian Geometry, where points, lines and planes are given in terms of cartesian co-ordinates determined with reference to three mutually perpendicular concurrent lines taken as cartesian axes.

6.1. NORMAL FORM OF THE CARTESIAN EQUATION OF A PLANE

We have already shown in § 4.1, page 100 that if **n** be a unit vector normal to a given plane and p be the length of the perpendicular from the origin to the plane, then the vector equation of the plane is

$$\mathbf{r} \cdot \mathbf{n} = p.$$

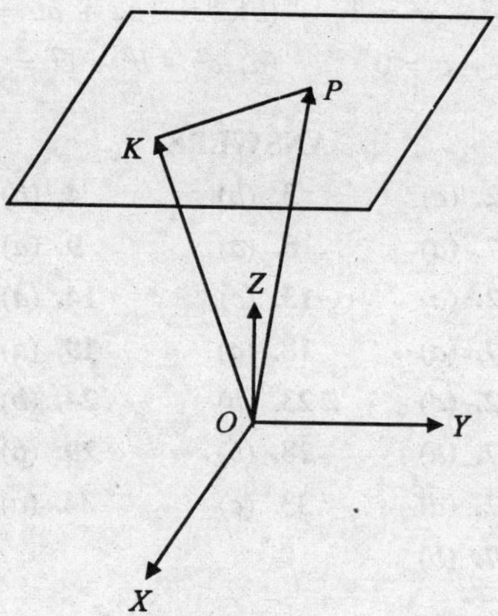

Fig. 6.1.

Substituting

$$\mathbf{r} = x\mathbf{i} + y\mathbf{j} + z\mathbf{k}; \quad \mathbf{n} = a\mathbf{i} + b\mathbf{j} + c\mathbf{k}$$

in the equation, we see that $ax + by + cz = p$ is the cartesian equation of the plane such that a, b, c are the direction cosines of the normal to the plane and p is the length of the perpendicular from the origin to the plane.

Note 1. We may see that $ax + by + cz = d$ is also the equation of a plane. Here, however, a, b, c are the direction ratios of the normal to the plane.

The coefficients a, b, c are proportional to the direction cosines of the normal to the plane whose equation is

$$ax + by + cz + d = 0.$$

The actual direction cosines are

$$a \Big/ \sqrt{\Sigma a^2},\ b \Big/ \sqrt{\Sigma a^2},\ c \Big/ \sqrt{\Sigma a^2}.$$

Note 2. Equation of a plane is of the first degree.

6.1.1. *To find the equation of the plane passing through the three points*

$$P_1(x_1, y_1, z_1),\ \ P_2(x_2, y_2, z_2),\ \ P_3(x_3, y_3, z_3).$$

We have

$$\overrightarrow{OP_1} = x_1\mathbf{i} + y_1\mathbf{j} + z_1\mathbf{k},\ \text{etc.},$$

where O is the origin and **i, j, k** are unit vectors along the coordinate axes. If $P(x, y, z)$ be any point of the plane, we have

$$\overrightarrow{P_1P} \times \overrightarrow{P_1P_2} \cdot \overrightarrow{P_1P_3} = 0. \qquad ...(i)$$

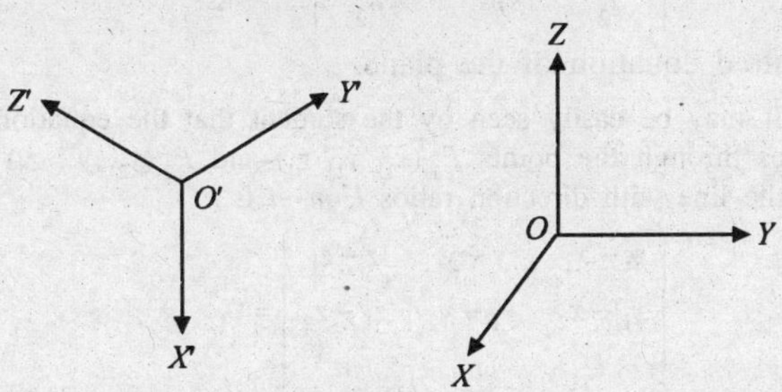

Fig 6.2.

Now

$$\overrightarrow{P_1P} = (x - x_1)\,\mathbf{i} + (y - y_1)\,\mathbf{j} + (z - z_1)\,\mathbf{k},$$

$$\overrightarrow{P_1P_2} = (x_2 - x_1)\,\mathbf{i} + (y_2 - y_1)\,\mathbf{j} + (z_2 - z_1)\,\mathbf{k},$$

$$\overrightarrow{P_1P_3} = (x_3 - x_1)\,\mathbf{i} + (y_3 - y_1)\,\mathbf{j} + (z_3 - z_1)\,\mathbf{k}.$$

By Cor. 3, Page 130 (i) gives the following required equation :

$$\begin{vmatrix} x - x_1 & y - y_1 & z - z_1 \\ x_2 - x_1 & y_2 - y_1 & z_2 - z_1 \\ x_3 - x_1 & y_3 - y_1 & z_3 - z_1 \end{vmatrix} = 0$$

6.1.2. *To find the equation of the plane which passes through $P_1(x_1, y_1, z_1)$ and which is parallel to two lines with direction ratios l_1, m_1, n_1 and l_2, m_2, n_2.*

Let $P(x, y, z)$ be any point of the plane so that

$$\vec{OP} = x\mathbf{i} + y\mathbf{j} + z\mathbf{k}.$$

We have

$$\vec{OP_1} = x_1\mathbf{i} + y_1\mathbf{j} + z_1\mathbf{k}$$

so that

$$\vec{P_1P} = (x - x_1)\,\mathbf{i} + (y - y_1)\,\mathbf{j} + (z - z_1)\,\mathbf{k}.$$

Also we have

$$\mathbf{a} = l_1\mathbf{i} + m_1\,\mathbf{j} + n_1\mathbf{k}, \quad \mathbf{b} = l_2\mathbf{i} + m_2\,\mathbf{j} + n_2\mathbf{k}.$$

The vectors $\vec{P_1P}$, **a** and **b** being coplanar, the scalar product

$$\vec{P_1P} \times \mathbf{a} \,.\, \mathbf{b} = 0$$

so that

$$\begin{vmatrix} x - x_1 & y - y_1 & z - z_1 \\ l_1 & m_1 & n_1 \\ l_2 & m_2 & n_2 \end{vmatrix} = 0$$

is the required equation of the plane.

Note. It may be easily seen by the student that the equation of the plane which passes through the points $P_1(x_1, y_1, z_1)$ and $P_2(x_2, y_2, z_2)$ and which is parallel to the line with direction ratios l, m, n is

$$\begin{vmatrix} x - x_1 & y - y_1 & z - z_1 \\ x_1 - x_2 & y_1 - y_2 & z_1 - z_2 \\ l & m & n \end{vmatrix} = 0.$$

6.2. CARTESIAN EQUATIONS OF A LINE

To find the line which passes through a point $A(x_1, y_1, z_1)$ having direction ratios $p, q. r.$

The position vector of the point A is $x_1\mathbf{i} + y_1\,\mathbf{j} + z_1\mathbf{k}$ and the vector $p\mathbf{i} + q\mathbf{j} + r\mathbf{k}$ is parallel to the given line. Replacing the vectors **a** and **b** by $x_1\mathbf{i} + y_1\,\mathbf{j} + z_1\mathbf{k}$ and $p\mathbf{i} + q\mathbf{j} + r\mathbf{k}$ respectively in the parametric equation

$$\mathbf{r} = \mathbf{a} + t\mathbf{b}$$

we obtain

$$x\mathbf{i} + y\mathbf{j} + z\mathbf{k} = (x_1\mathbf{i} + x_2\,\mathbf{j} + x_3\mathbf{k}) + t\,(p\mathbf{i} + q\mathbf{j} + r\mathbf{k})$$

$$\Rightarrow \ (x - x_1 - tp)\,\mathbf{i} + (y - y_1 - tq)\,\mathbf{j} + (z - z_1 - tr)\,\mathbf{k} = 0$$

$$\Rightarrow \ x - x_1 - tp = 0, \ y - y_1 - tq = 0, \ z - z_1 - tr = 0$$

$$\Rightarrow \ x = x_1 - tp, \ y = y_1 - tq, \ z = z_1 + tr$$

$$\Rightarrow \ \frac{x-x_1}{p} = \frac{y-y_1}{q} = \frac{z-z_1}{r}$$

which are the required equations of the line.

Cor. *To find the line through two points $P_1 (x_1, y_1, z_1)$ and $P_2 (x_2, y_2, z_2)$.*

We write

$$\overrightarrow{OP_1} = \mathbf{a} = x_1\mathbf{i} + y_1\mathbf{j} + z_1\mathbf{k}$$

$$\mathbf{b} = (x_2 - x_1)\,\mathbf{i} + (y_2 - y_1)\,\mathbf{j} + (z_2 - z_1)\,\mathbf{k}$$

so that **a** is the position vector of a point on the line and **b** is a vector parallel to the line. The required equations therefore are

$$\frac{x-x_1}{x_2-x_1} = \frac{y-y_1}{y_2-y_1} = \frac{z-z_1}{z_2-z_1},$$

6.3. COPLANARITY OF LINES

To find the condition for the lines

$$\frac{x-x_1}{l_1} = \frac{y-y_1}{m_1} = \frac{z-z_1}{n_1},$$

$$\frac{x-x_2}{l_2} = \frac{y-y_2}{m_2} = \frac{z-z_2}{n_2},$$

to be coplanar.

In vector notation, we can rewrite these equations as

$$\mathbf{r} = \mathbf{a} + t\mathbf{b}, \quad \mathbf{r} = \mathbf{c} + t\mathbf{d}$$

where

$$\mathbf{a} = x_1\mathbf{i} + y_1\mathbf{j} + z_1\mathbf{k}, \quad \mathbf{b} = l_1\mathbf{i} + m_1\mathbf{j} + n_1\mathbf{k}.$$

$$\mathbf{c} = x_2\mathbf{i} + y_2\mathbf{j} + z_2\mathbf{k}, \quad \mathbf{d} = l_2\mathbf{i} + m_2\mathbf{j} + n_2\mathbf{k}.$$

The condition for coplanarity is

$$(\mathbf{a} - \mathbf{c}) \cdot \mathbf{b} \times \mathbf{d} = 0$$

$$\begin{vmatrix} x_2-x_1 & y_2-y_1 & z_2-z_1 \\ l_1 & m_1 & n_1 \\ l_2 & m_2 & n_2 \end{vmatrix} = 0.$$

Assuming the condition to be satisfied, the equation of the plane containing the two lines is

$$(\mathbf{a} - \mathbf{a}) \cdot \mathbf{b} \times \mathbf{d} = 0$$

$$\Leftrightarrow \quad \begin{vmatrix} x-x_1 & y-y_1 & z-z_1 \\ l_1 & m_1 & n_1 \\ l_2 & m_2 & n_2 \end{vmatrix} = 0.$$

6.4. SHORTEST DISTANCE

Adopting the preceding notation, the length of shortest distance is

$$\frac{(\mathbf{c}-\mathbf{a}) \cdot \mathbf{b} \times \mathbf{d}}{|\mathbf{b} \times \mathbf{d}|}$$

where
$$(\mathbf{c}-\mathbf{a}) \cdot \mathbf{b} \times \mathbf{d} = \begin{vmatrix} x_2-x_1 & y_2-y_1 & z_2-z_1 \\ l_1 & m_1 & n_1 \\ l_2 & m_2 & n_2 \end{vmatrix}$$

and
$$\mathbf{b} \times \mathbf{d} = \Sigma\,(m_1 n_2 - m_2 n_1)\,\mathbf{i}$$

$$\Rightarrow \quad |\mathbf{b} \times \mathbf{d}| = \sqrt{\Sigma (m_1 n_2 - m_2 n_1)^2}\,.$$

Of course the required shortest distance is the absolute value of

$$\frac{(\mathbf{c}-\mathbf{a}) \cdot \mathbf{b} \times \mathbf{d}}{|\mathbf{b} \times \mathbf{d}|}.$$

6.5. CHANGE OF AXES

Let

$$OX,\ OY,\ OZ \text{ and } O'X',\ O'Y',\ O'Z'$$

be two sets of rectangular co-ordinate axes.

Let (f, g, h) be the co-ordinates of O' and let

$$l_1, m_1, n_1; \quad l_2, m_2, n_2; \quad l_3, m_3, n_3$$

be the direction ratios of OX', OY', OZ', with respect to OX, OY, OZ.

Let (x, y, z), (x', y', z') be the co-ordinates of a point P with respect to the two sets of axes.

If, $\mathbf{i}, \mathbf{j}, \mathbf{k} : \mathbf{i}', \mathbf{j}', \mathbf{k}'$ be unit vectors along the two sets of axes, we have

$$\vec{OP} = x\mathbf{i} + y\mathbf{j} + z\mathbf{k}.$$

$$\vec{O'P} = x'\mathbf{i}' + y'\mathbf{j}' + z'\mathbf{k}', \quad \vec{OO'} = f\,\mathbf{i} + g\mathbf{j} + h\mathbf{k},$$

$$\mathbf{i}' = l_1\mathbf{i} + m_1\mathbf{j} + n_1\mathbf{k},$$

$$\mathbf{j}' = l_2\mathbf{i} + m_2\mathbf{j} + n_2\mathbf{k},$$
$$\mathbf{k}' = l_3\mathbf{i} + m_3\mathbf{j} + n_3\mathbf{k}.$$

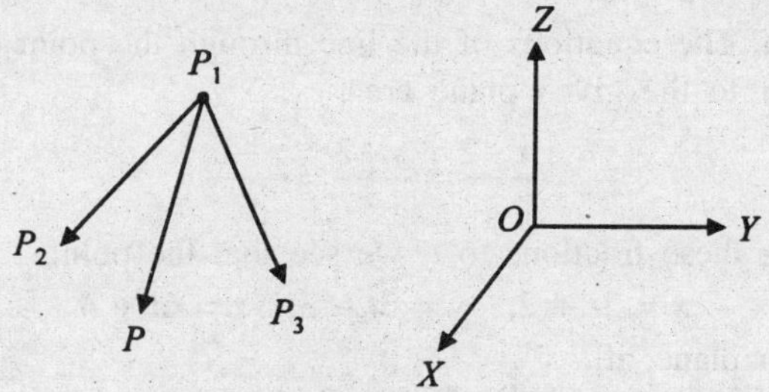

Fig. 6.3.

Now $\quad \vec{OP} = \vec{OO'} + \vec{O'P}$

$\Rightarrow \quad x\mathbf{i} + y\mathbf{j} + z\mathbf{k} = f\mathbf{i} + g\mathbf{j} + h\mathbf{k} + x'\mathbf{i}' + y'\mathbf{j}' + z'\mathbf{k}'$

$\Rightarrow \quad (x - f)\,\mathbf{i} + (y - g)\,\mathbf{j} + (z - h)\,\mathbf{k} = x'\,(l_1\mathbf{i} + m_1\mathbf{j} + n_1\mathbf{k}) + y'\,(l_2\mathbf{i} + m_2\mathbf{j} + n_2\mathbf{k}) + z'\,(l_3\mathbf{i} + m_3\mathbf{j} + n_3\mathbf{k})$

$$\Rightarrow \quad \begin{cases} x = f + x'l_1 + y'l_2 + z'l_3, \\ y = g + x'm_1 + y'm_2 + z'm_3, \\ z = h + x'n_1 + y'n_2 + z'n_3, \end{cases}$$

which are the required formulae for the transformation of co-ordinates from one set of axes to another.

EXAMPLES

Example 1. *If OX, OY, OZ and O′ X′, O′ Y′, O′ Z′ are two sets of rectangular co-ordinate axes and*

$$l_1, m_1, n_1; \quad l_2, m_2, n_2; \quad l_3, m_3, n_3$$

denote the direction cosines of the members of either set with respect to other, then

$$\begin{vmatrix} l_1 & m_1 & n_1 \\ l_2 & m_2 & n_2 \\ l_3 & m_3 & n_3 \end{vmatrix} = 1.$$

Solution. If **i**, **j**, **k**; **i′**, **j′**, **k′** denote unit vectors along the axes, we have

$$\mathbf{i}' = l_1\mathbf{i} + m_1\mathbf{j} + n_1\mathbf{k},$$
$$\mathbf{j}' = l_2\mathbf{i} + m_2\mathbf{j} + n_2\mathbf{k},$$
$$\mathbf{k}' = l_3\mathbf{i} + m_3\mathbf{j} + n_3\mathbf{k}.$$

Also we have

$$[\mathbf{i}'\,\mathbf{j}'\,\mathbf{k}'] = \begin{vmatrix} l_1 & m_1 & n_1 \\ l_2 & m_2 & n_2 \\ l_3 & m_3 & n_3 \end{vmatrix} = [\mathbf{i}\,\mathbf{j}\,\mathbf{k}]$$

As [**i j k**] = [**i**′, **j**′, **k**′], we have the required result.

Example 2. *Find the reflection of the point (2, – 3, 4) in the plane* $2x + 3y + 5z + 2 = 0$.

Solution. The equations of the line through the point (2, – 3, 4) and perpendicular to the given plane are

$$\frac{x-2}{2}=\frac{y+3}{3}=\frac{z-6}{6}$$

Equating these fractions to t, we see that the point,

$$x = 2t + 2, \quad y = 3t - 3, \quad z = 6t + 4 \qquad ...(i)$$

lies on given plane, if

$$2\,(2t + 2) + 3\,(3t - 3) + 6\,(6t + 4) + 2 = 0$$

$$4t + 4 + 9t - 9 + 36t + 24 + 2 = 0$$

$$\Rightarrow \qquad 49t = -21$$

$$t=\frac{-3}{7}.$$

Putting this value of t in (i) we see that the foot of the perpendicular is

$$2\left(\frac{-3}{7}\right)+2,\ 3\left(\frac{-3}{7}\right)-3,\ 6\left(\frac{-3}{7}\right)+4,\ i.e.,\ \left(\frac{8}{7},\ \frac{-30}{7},\ \frac{10}{7}\right)$$

i.e.,

$$\left(\frac{8}{7},\frac{-30}{7},\frac{10}{7}\right).$$

If (a, b, c) be the reflection P of the given point in the given plane, then the mid-point of AP, *viz.*,

$$\frac{a+2}{2},\ \frac{b-3}{2},\ \frac{c+4}{2}$$

must be the foot of the perpendicular.

$$\therefore \qquad \frac{a+2}{2}=\frac{8}{7},\ \frac{b-3}{2}=\frac{-30}{7},\ \frac{c+4}{2}=\frac{10}{7}$$

$$\Rightarrow \qquad a=\frac{16}{7}-2,\ b=-\frac{60}{7}+3,\ c=\frac{20}{7}-4$$

$$\Rightarrow \qquad a=\frac{2}{7},\ b=\frac{-39}{7},\ c=\frac{-8}{7}.$$

Thus

$$\left(\frac{2}{7},\frac{-39}{7},\frac{-8}{7}\right)$$

is the required reflection of the given point in the given plane.

Example 3. *Find the reflection of the point A (1, 0, 0) in the line*

$$\frac{x-1}{2}=\frac{y+1}{-3}=\frac{z+10}{8}.$$

Solution. Any point P on the line is

$$2t + 1, \quad -3t - 1, \quad 8t - 10. \qquad ...(i)$$

The line AP with direction ratios $2t, -3t - 1, 8t - 10$ will be perpendicular to the given line, if

$$2(2t) - 3(-3t - 1) + 8(8t - 10) = 0$$

$$\Rightarrow \quad 4t + 9t + 3 + 64t - 80 = 0$$

$$\Rightarrow \quad 77t - 77 = 0 \Rightarrow t = 1.$$

Putting this value of t in the co-ordinates (i) r of the point P we see that the co-ordinates of the foot of the perpendicular, P are $(3, -4, -2)$.

If $B(a, b, c)$ be the reflection of the point $A(1, 0, 0)$, then the mid-point $\left(\frac{a+1}{2}, \frac{b}{2}, \frac{c}{2}\right)$ of AB must be the foot P of the perpendicular from the point on A the given line so that we have

$$\frac{a+1}{2} = 3, \ \frac{b}{2} = -4, \ \frac{c}{2} = -2$$

$$\Rightarrow \quad a = 5, \quad b = -8, \quad c = -4.$$

Thus, $(5, -8, -4)$ is the reflection of the given point in the given line.

Example 4. *Find the equation of the plane through the points A, B, C whose rectangular cartesian co-ordinates are*

$$(1, 1, 1), \quad (1, -1, 1), \quad (-1, -3, -5).$$

Solution. The position vectors of these points are

$$\mathbf{i} + \mathbf{j} + \mathbf{k}, \quad \mathbf{i} - \mathbf{j} + \mathbf{k}, \quad -\mathbf{i} - 3\mathbf{j} + 5\mathbf{k}.$$

respectively.

Let P

$$x\mathbf{i} + y\mathbf{j} + z\mathbf{k}$$

be any point on the plane. We have

$$(\vec{AB} \times \vec{AC}) \cdot \vec{AP} = 0$$

$$\Leftrightarrow \quad [(-2\mathbf{j} \times (-2\mathbf{i} - 4\mathbf{j} - 6\mathbf{k})] \cdot [(x - 1)\mathbf{i} + (y - 1)\mathbf{j} + (z - 1)\mathbf{k}] = 0$$

$$\Leftrightarrow \quad (12\mathbf{i} - 4\mathbf{k}) \cdot [(x - 1)\mathbf{i} + (y - 1)\mathbf{j} + (z - 1)\mathbf{k}] = 0$$

$$\Leftrightarrow \quad 12(x - 1) - 4(z - 1) = 0$$

$$\Leftrightarrow \quad 3x - z - 2 = 0.$$

which is the required equation.

Example 5. *Show that the points whose rectangular cartesian co-ordinates are*

$$(-6, 3, 2), \quad (3, -2, 4), \quad (5, 7, 3), \quad (-13, 17, -1)$$

are coplanar.

Solution. In the usual notation we have

$$\vec{OA} = -6\mathbf{i} + 3\mathbf{j} + 2\mathbf{k}, \ \vec{OB} = 3\mathbf{i} - 2\mathbf{j} + 4\mathbf{k},$$

$$\vec{OC} = 5\mathbf{i}+7\mathbf{j}+3\mathbf{k},\ \vec{OD} = -13\mathbf{i}+17\mathbf{j}-\mathbf{k},$$

Thus,

$$\vec{AB} = 9\mathbf{i}-5\mathbf{j}+2\mathbf{k},\ \vec{AC} = 11\mathbf{i}+4\mathbf{j}+\mathbf{k},\ \vec{AD} = -7\mathbf{i}+14\mathbf{j}-3\mathbf{k}.$$

Finally we have

$$\vec{AB} \times \vec{AC} \cdot \vec{AD} = (9\mathbf{i}-5\mathbf{j}+2\mathbf{k}) \times (11\mathbf{i}+4\mathbf{j}+\mathbf{k}) \cdot (-7\mathbf{i}+14\mathbf{j}-3\mathbf{k}) = 0$$

$\Rightarrow$ $\vec{AB}, \vec{AC}, \vec{AD}$ are coplanar vectors.

$\Rightarrow$ A, B, C, D are coplanar points.

Example 6. *Find the length of the shortest distance between the lines*

$$\frac{x+3}{-4} = \frac{y-6}{3} = \frac{z}{2};\ \frac{x+2}{-4} = \frac{y}{1} = \frac{z-7}{1}.$$

Solution. The line $\dfrac{x+3}{-4} = \dfrac{y-6}{3} = \dfrac{z}{2}$ passes through the point (– 3, 6, 0), *i.e.*, through the point with position vector – 3**i** + 6**j** and is parallel to the line with direction ratios – 4, 3, 2, *i.e.*, parallel to the vector – 4**i** + 3**j** + 2**k**.

The second line $\dfrac{x+2}{-4} = \dfrac{y}{1} = \dfrac{z-7}{1}$ passes through the point (– 2, 0, 7), *i.e.*, through the point that position vector – 2**i** + 7**k** and is parallel to the line with direction ratios – 4, 1, 1, *i.e.*, to the vector – 4**i** + **j** + **k**.

The line of shortest distance is parallel to the line perpendicular to the given lines, *i.e.*, it is parallel to the vector

$$(-4\mathbf{i} + 3\mathbf{j} + 2\mathbf{k}) \times (-4\mathbf{i} + \mathbf{j} + \mathbf{k})$$

$$= \begin{vmatrix} \mathbf{i} & \mathbf{j} & \mathbf{k} \\ -4 & 3 & 2 \\ -4 & 1 & 1 \end{vmatrix} = \mathbf{i}-4\mathbf{j}+8\mathbf{k}.$$

The length of the shortest distance is the projection on this line of the join of the points on the two lines, *i.e.*, of the vector

$$= (-3\mathbf{i} + 6\mathbf{j}) - (-2\mathbf{i} + 7\mathbf{k}) = -\mathbf{i} + 6\mathbf{j} - 7\mathbf{k}.$$

Thus, the required length of the shortest distance is

$$\frac{(-\mathbf{i}+6\mathbf{j}-7\mathbf{k})\cdot(\mathbf{i}-4\mathbf{j}+8\mathbf{k})}{|\mathbf{i}-4\mathbf{j}+8\mathbf{k}|} = \frac{-81}{9} = -9.$$

Thus, the shortest distance between the given two lines is 9.

EXERCISES

1. Find the reflections of the points

   (*i*) (– 5, 11, 6) in the plane $5x - 6y - 2z - 7 = 0$;

(*ii*) (3, 5, 8) in the plane $3x - y - z + 26 = 0$;

(*iii*) (5, – 9, – 6) in the plane $2x - y + 3z - 8 = 0$.

2. Find the feet of the perpendiculars from the origin to the lines

(*i*) $\dfrac{x-3}{3} = \dfrac{y-8}{-4} = \dfrac{z-3}{-2}$

(*ii*) $\dfrac{x+1}{2} = \dfrac{y}{1} = \dfrac{z-4}{-1}$

(*iii*) $\dfrac{x-3}{1} = \dfrac{1-y}{2} = \dfrac{z-5}{1}$.

Find also the reflections in each case.

3. Find the reflection of the point (7, – 1, 2) in the line

$$\frac{x-9}{1} = \frac{y-5}{3} = \frac{z-5}{5}.$$

4. Given a triangle with vertices (1, 2, 3), (2, 1, 4), (5, 1, 3). Show that the orthocentre of this triangle is

$$\frac{1}{13}(23\mathbf{i} + \mathbf{j} + 9\mathbf{k}).$$

5. Examine if the following sets of points are coplanar :

(*a*) (3, 2, – 5), (– 3, 8, – 5), (– 3, 2, 1), (– 1, 4, – 3).

(*b*) (6, – 4, 4), (0, 0, – 4), (– 1, – 2, – 3), (1, 2, – 5).

(*c*) (– 2, – 1, 0), (1, – 2, – 1), (2, 1, 4), (0, 1, 0).

6. Find the equations of the planes through the points :

(*i*) (– 2, – 2, 2), (1, 1, 1), (1, – 1, 2)

(*ii*) (– 6, 3, 2), (3, – 2, 4), (5, 7, 3).

7. By vector method prove that the shortest distance between the straight line

$$y + z = 0, \quad z + x = 0$$

and the straight line

$$x + y = 0, \quad x + y + z - a = 0$$

is $\sqrt{6a/3}$.

8. A tetrahedron has vertices

$$A\ (0, 0, 0),\ B\ (1, 1, 0),\ C\ (0, 1, -1),\ D\ (1, 0, -1)$$

Show that the line joining the origin to the point (1, 1, 1) is perpendicular to the face *BCD*.

9. $A = (0, 1, 2)$, $B = (3, 0, 1)$, $C = (4, 3, 6)$, $D = (2, 3, 2)$ are the rectangular cartesian co-ordinates of four points. Find

(*i*) the area of the triangle *ABC*.

(*ii*) the perpendicular distance from *A* to the line *BC*.

(*iii*) the volume of the tetrahedron *ABCD*.

(*iv*) the perpendicular distance from *D* to the plane *ABC*.

(*v*) the shortest distance between the lines *AB* and *CD*.

10. Find the volume of the tetrahedron formed by the planes

$$my + nz = 0,\quad nz + lx = 0,\quad lx + my = 0,$$
$$lx + my + nz = p.$$

11. Find the equation of the plane which contains the line

$$x = 2, \quad y - z = 0$$

and is perpendicular to the plane $x + z = 3$.

Find the point where this plane meets the line

$$\frac{x}{2} = \frac{y}{3} = \frac{z}{1}.$$

12. Prove that a plane normally bisecting a diagonal of a cube meets the edges of the cube in points which are the vertices of a regular hexagon.

13. Prove that the lines

$$\frac{x+1}{3} = \frac{y+3}{5} = \frac{z+5}{7}; \quad \frac{x-2}{1} = \frac{y-4}{3} = \frac{z-6}{5}.$$

intersect. Find their point of intersection and the plane in which they lie.

14. Prove that the lines

$$\frac{x-a}{a'} = \frac{y-b}{b'} = \frac{z-c}{c'} \text{ and } \frac{x-a'}{a} = \frac{y-b'}{b} = \frac{z-c'}{c}$$

intersect and find the co-ordinates of the point of intersection and the equation of the plane in which they lie.

15. Prove that the lines

$$\frac{x-1}{2} = \frac{y+1}{-3} = \frac{z+10}{8}; \quad \frac{x-4}{1} = \frac{y+3}{-4} = \frac{z+1}{7}$$

intersect. Find also their point of intersection and the plane through them.

16. Find the magnitude and the equations of the line of shortest distance between the two lines :

(*i*) $\dfrac{x-3}{2} = \dfrac{y+15}{-7} = \dfrac{z-9}{5}; \quad \dfrac{x+1}{2} = \dfrac{y-1}{1} = \dfrac{z-9}{-3}.$

(*ii*) $\dfrac{x-3}{-1} = \dfrac{y-4}{2} = \dfrac{z+2}{1}; \quad \dfrac{x-1}{1} = \dfrac{y+7}{3} = \dfrac{z+2}{2}.$

17. Obtain the co-ordinates of the points where the shortest distance between the lines

$$\frac{x-23}{-6} = \frac{y-19}{-4} = \frac{z-25}{3}; \quad \frac{x-12}{-9} = \frac{y-1}{4} = \frac{z-5}{2}.$$

meets them.

18. Find the shortest distance between the lines

$$\frac{x-1}{2} = \frac{y-2}{3} = \frac{z-3}{4}; \quad \frac{x-2}{3} = \frac{y-3}{4} = \frac{z-4}{5}.$$

Show also that the lines are coplanar.

MISCELLANEOUS EXERCISES II

1. Show that the perpendicular distance of any point **a** from the line $\mathbf{r} = \mathbf{b} + t\mathbf{c}$ is
$$|(\mathbf{b} - \mathbf{a}) \times \mathbf{c}| / |\mathbf{c}|.$$

2. Two triangles of equal area are on the opposite sides of the same base. Prove that the straight line joining their vertices is bisected by the base.

3. If **a, b, c** are three vectors such that $\mathbf{a} + \mathbf{b} + \mathbf{c} = 0$, prove that
$$\mathbf{a} \times \mathbf{b} = \mathbf{b} \times \mathbf{c} = \mathbf{c} \times \mathbf{a}$$
and interpret the result geometrically.

4. A', B', C', D' are the projections of the four coplanar points A, B, C, D on any given plane; show that the volumes of the tetrahedron $AB'C'D'$ and $A'BCD$ are equal.

5. Lines are drawn in a given direction through the vertices O, A, B, C of a tetrahedron $OABC$ to meet the opposite faces in O', A', B', C'. Prove that
$$\text{Volume } O', A', B', C' = 3 \text{ Volume } OABC.$$

6. Show that the volume of a tetrahedron $ABCD$ is
$$\frac{1}{6}\,|\overrightarrow{AB} \times \overrightarrow{CD} \cdot \overrightarrow{AC}|$$
and deduce that if the lengths of two opposite edges of a tetrahedron are a, b and the angle between them is α and the shortest distance between the same is $2c$, then the volume of the tetrahedron is $\frac{1}{3} abc \sin \alpha$.

7. Two lines AB and CD of fixed lengths move along given straight lines; show that the volume of the tetrahedron remains constant.

Deduce that the volume of a tetrahedron $ABCD$ is
$$\frac{1}{6}\,|\overrightarrow{AB} \times \overrightarrow{CD} \cdot \overrightarrow{LM}|,$$
where LM is the line of shortest distance between the edges AB and CD.

8. A, B, C, D are any four points, show that
$$\overrightarrow{AB} \cdot \overrightarrow{CD} + \overrightarrow{BC} \cdot \overrightarrow{AD} + \overrightarrow{CA} \cdot \overrightarrow{BD} = 0.$$

9. $ABCD$ is a quadrilateral, show that
$$AC^2 + BD^2 = AB^2 + CD^2 + 2\,\overrightarrow{BC} \cdot \overrightarrow{AD}.$$

10. $OABC$ is a tetrahedron and the position vectors of A, B, C with respect to O as origin are **a, b, c** respectively.

Write down the position vectors of the mid-points P, Q, R, L, M, N of OA, OB, OC, BC, CA, AB respectively.

Determine the vector equations of the lines PL and QM and show that these lines intersect and find the position vector of the point of intersection.

Deduce that the lines PL, QM and RN are concurrent.

If a tetrahedron OABC is regular, that is all six edges have the same lengths, show that

(*i*) $\mathbf{a} \cdot \mathbf{b} = \mathbf{b} \cdot \mathbf{c} = \mathbf{c} \cdot \mathbf{a}$.

(*ii*) PL and BC are perpendicular.

11. Given that $\mathbf{p} = \mathbf{a} + \mathbf{b}$ and $\mathbf{q} = \mathbf{a} - \mathbf{b}$ and $|\mathbf{a}| = |\mathbf{b}| = k$ show that $\mathbf{p}$ $\mathbf{q} = 0$ and that

$$|\mathbf{p} \times \mathbf{q}| = 2\,[k^4 - (\mathbf{a} \cdot \mathbf{b})^2]^{1/2}.$$

12. The point A is (2, 1, 6) and has position vector $\mathbf{a}$ relative to the origin O; B is (1, 3, 5) and has position vector $\mathbf{b}$. The line l is given by $\mathbf{r} = \mathbf{b} + s\,(\mathbf{i} + 2\mathbf{j} + 3\mathbf{k})$ and the line m is given by $\mathbf{r} = \mathbf{a} + t\,(\mathbf{i} + 6\mathbf{j} + 5\mathbf{k})$. Show that the lines l and m intersect, and state the co-ordinates of the common point. Prove that AB is perpendicular to l, and hence write down the co-ordinates of the reflection of A in l.

13. Show that there is one and only one value of p for which the vectors

$$(p + 1)\,\mathbf{i} - 3\mathbf{j} + p\mathbf{k}, \quad p\mathbf{i} + (p + 1)\,\mathbf{j} - 3\mathbf{k}, \quad -3\mathbf{i} + p\mathbf{j} + (p + 1)\,\mathbf{k}$$

are linearly dependent and find this value of p.

14. Consider these two lines

$$\mathbf{r} = \mathbf{a} + \lambda\mathbf{l}, \qquad \mathbf{r} = \mathbf{b} + \mu\mathbf{m},$$

where

$$\mathbf{r} = 5\mathbf{i} + \mathbf{j} + 2\mathbf{k}, \qquad \mathbf{b} = -\mathbf{i} + 7\mathbf{j} + 8\mathbf{k},$$
$$\mathbf{l} = -4\mathbf{i} + \mathbf{j} - \mathbf{k}, \qquad \mathbf{m} = 2\mathbf{i} - 5\mathbf{j} - 7\mathbf{k}.$$

Show that the two lines intersect and find the position vector of the point P of their intersection. If the position vector of a point Q is $3\mathbf{i} + 7\mathbf{j} - 2\mathbf{k}$, show that PQ is perpendicular to AB where A and B are the points with position vectors $\mathbf{a}$ and $\mathbf{b}$.

15. If $\mathbf{a} = \dfrac{1}{\sqrt{3}}(\mathbf{i} + \mathbf{j} + \mathbf{k}),\ \mathbf{b} = \sqrt{\left(\dfrac{2}{3}\right)\left(\mathbf{i} - \dfrac{1}{2}\mathbf{j} - \dfrac{1}{2}\mathbf{k}\right)}$ and

$$\mathbf{c} = \frac{1}{\sqrt{2}}(\mathbf{j} - \mathbf{k}).$$

show that $\mathbf{a}$, $\mathbf{b}$ and $\mathbf{c}$ are mutually perpendicular unit vectors and express $\mathbf{i}$ in the form $p\mathbf{a} + q\mathbf{b} + r\mathbf{c}$.

16. Find the unit vectors which are perpendicular to both the vectors $\mathbf{i} + 4\mathbf{j}$ and $2\mathbf{i} + 4\mathbf{j} + 3\mathbf{k}$. Find also the angles between these unit vectors and the vector $\dfrac{1}{2}(\mathbf{i} + \mathbf{j} - \mathbf{k})$.

17. $ABCD$ is a tetrahedron; X, Y, Z are the mid-points of AB, AC, AD and P, Q, R, are the mid-points of CD, BD, BC.

Prove that

$$AB^2 + BC^2 + 2PX^2 = AC^2 + BD^2 + 2QY^2 = AD^2 + BC^2 + 2RZ^2.$$

18. Find all the sets of vectors of $\mathbf{a}$, $\mathbf{b}$ and $\mathbf{c}$ that satisfy the following conditions :

$\mathbf{a}$ is parallel to $\mathbf{i} + \mathbf{j}$ and $|\mathbf{a}| = 3$;

$\mathbf{b}$ is parallel to $\mathbf{i} - \mathbf{j}$ and $|\mathbf{b}| = 1$;

$\mathbf{c}$ satisfies the equation $\mathbf{a} = \mathbf{b} \times \mathbf{c}$ and $|\mathbf{c}| = 9$.

(Parallelism includes the case of opposite directions also.)

19. Find the general form of the vector x which satisfies the equation

$$x \times (\mathbf{i} + 2\mathbf{j} + \mathbf{k}) = \mathbf{i} - \mathbf{k}.$$

20. Points A, B, C have position vectors **a, b, c** and p, q, r are variable parameters subject to the condition $p + q + r = 1$. If the points are not collinear, prove that the plane ABC is represented by the equation $\mathbf{r} = p\mathbf{a} + q\mathbf{b} + r\mathbf{c}$.

Prove that the equation of the line of intersection of the two planes

$$\mathbf{r} = p_1\mathbf{i} + 2q_1\mathbf{j} + 3r_1\mathbf{k}, \qquad p_1 + q_1 + r_1 = 1$$

and

$$\mathbf{r} = 2p_2\mathbf{i} + q_2\mathbf{j} + 2r_2\mathbf{k}, \qquad p_2 + q_2 + r_2 = 1$$

can be written in terms of a single parametric t as

$$6\mathbf{r} = (3 + t)\,\mathbf{i} + 4t\mathbf{j} + 9\,(1 - t)\,\mathbf{k}.$$

21. The position vectors of the vertices A, B, C of a tetrahedron $OABC$ with respect to O as origin are

$$\overrightarrow{OA} = 2\mathbf{i} + \mathbf{j},\ \overrightarrow{OB} = \mathbf{j} + \mathbf{k},\ \overrightarrow{OC} = \mathbf{i} + 3\mathbf{j} - \mathbf{k}.$$

Find the angle between (*a*) the edges AB, CD (*b*) the faces OAB, OAC.

Prove that BC is perpendicular to the plane OAB and hence prove that the volume of the tetrahedron $OABC$ is 3/2.

22. Prove that the lines with vector equations

$$\mathbf{r} = (-\mathbf{i} + 2\mathbf{j}) + a\,(-\mathbf{i} + 2\mathbf{j} - \mathbf{k}).$$
$$\mathbf{r} = \mathbf{k} + b\,(\mathbf{i} + \mathbf{j} - \mathbf{k}),$$
$$\mathbf{r} = \mathbf{i} + \mathbf{j} - \mathbf{k} + c\,(\mathbf{i} + \mathbf{k})$$
$$\mathbf{r} = 2\mathbf{i} + \mathbf{j} + d\,(3\mathbf{i} - \mathbf{j}).$$

in the given order form a (skew) quadrilateral.

Prove that the ratio of the shortest distances between the two pairs of opposite sides of this quadrilateral is $\sqrt{7:1}$.

23. Referred to cartesian axes, A is the point (1, 2). If $\overrightarrow{AB} = 3\mathbf{i} - \mathbf{j}$ and $\overrightarrow{AC} = 3\mathbf{i} + 4\mathbf{j}$, write down the co-ordinates of the points B and C.

E and F are points such that $\overrightarrow{BE} = \frac{1}{2}\overrightarrow{AC}$ and $\overrightarrow{BF} = \overrightarrow{AB}$. Show that $\overrightarrow{CE} = \overrightarrow{EF}$ and state what can be deduced about the points C, E, F.

24. The line $\mathbf{r} = \mathbf{p} + t\mathbf{u}$ lies in the plane $\mathbf{a} \cdot \mathbf{r} = d$, where **a, p, u** and d are constants prove that $\mathbf{a} \cdot \mathbf{u} = 0$ and $\mathbf{a} \cdot \mathbf{p} = d$. Conversely, if $\mathbf{a} \cdot \mathbf{p} = d$ and $\mathbf{a} \cdot \mathbf{u} = 0$, show that the line lies in the plane.

25. Find the equation of the plane to which the vector $\mathbf{i} + \mathbf{j}$ is normal, and which contains the line l with equation

$$\mathbf{r} = (\mathbf{i} + \mathbf{j} + \mathbf{k}) + t\,(\mathbf{i} - \mathbf{j} - \mathbf{k}).$$

Find also the equation of the plane containing l and the point **j**. Show that $\pi/6$ is an angle between the normals of these planes.

26. Unit vectors **i** and **j** are at right angles to each other. Also

$$\mathbf{p} = 3\mathbf{i} + 4\mathbf{j},\quad \mathbf{q} = 5\mathbf{i},\quad 4\mathbf{r} = \mathbf{p} + \mathbf{q},\quad 2\mathbf{s} = \mathbf{p} - \mathbf{q}.$$

Show that

(*a*) $|\mathbf{p}| = |\mathbf{q}|$; (*b*) $|\mathbf{r}| = |\mathbf{s}|$;
(*c*) **r** is perpendicular to **s**; (*d*) $|\mathbf{r} + \mathbf{s}| = |\mathbf{r} - \mathbf{s}|$;
(*e*) $\mathbf{r} + \mathbf{a}$ is perpendicular to $\mathbf{r} - \mathbf{a}$.

Show also that, for all real k,

$$|\mathbf{r} + k\mathbf{s}| = |\mathbf{r} - k\mathbf{s}|.$$

27. Show that the lines

$$l_1 : \mathbf{r} = \mathbf{j} + s(\mathbf{i} - \mathbf{j} + \mathbf{k}), \quad l_2 : \mathbf{r} = \mathbf{i} - \mathbf{j} + t(\mathbf{i} + \mathbf{j} + \mathbf{k})$$

have no point in common. If A lies on l_1; B lies on l_2 and AB is perpendicular to both l_1 and l_2, find the length of AB and the equation of AB. Find also the equation of

(*i*) the plane π_1 perpendicular to AB and containing l_1.

(*ii*) the plane π_2 perpendicular to AB and containing l_2.

Find also the cartesian equations of the lines l_1 l_2 and the planes π_1 π_2.

28. The six faces of a rectangular cuboid have vector equations

$$\mathbf{r}\cdot\mathbf{k} = 0, \quad \mathbf{r}\cdot(\mathbf{i} + \mathbf{j}) = 2, \quad \mathbf{r}\cdot(\mathbf{j} - \mathbf{i}) = 2,$$
$$\mathbf{r}\cdot\mathbf{k} = 2, \quad \mathbf{r}\cdot(\mathbf{i} + \mathbf{j}) = 4, \quad \mathbf{r}\cdot(\mathbf{j} - \mathbf{i}) = -2.$$

where **i, j, k** represent unit vectors along perpendicular axes Ox, Oy, Oz. Prove that the acute angle between the diagonals through the points $2\mathbf{i}$ and $2\mathbf{j}$ is $\cos^{-1}(1/7)$.

Find the vector equations of the planes parallel to Oz through these diagonals.

29. From the point $\mathbf{s}_1 = \mathbf{i} + \mathbf{j} + \mathbf{k}$ the perpendicular drawn to a plane π_1 is $\mathbf{i} - \mathbf{j} + \mathbf{k}$. From the point $\mathbf{s}_2 = 3\mathbf{i} - \mathbf{j} - \mathbf{k}$ the perpendicular drawn to a plane π_2 is $\mathbf{i} + 2\mathbf{j} + \mathbf{k}$. Find the equation of the plane π_3 containing the line of intersection of the planes π_1 and π_2 and passing through the mid-point of the line joining $\mathbf{s}_1$ and $\mathbf{s}_2$. Find also the length of the perpendicular from the origin to the plane π_3.

30. Three vectors **a, b** and **c** are such that $\mathbf{a} \neq 0$ and $\mathbf{a} \times \mathbf{b} = 2\mathbf{a} \times \mathbf{c}$. Show that $\mathbf{b} - 2\mathbf{c} = \lambda\mathbf{a}$ where λ is a scalar.

Given that $|\mathbf{a}| = |\mathbf{c}| = 1$, $|\mathbf{b}| = 4$ and the angle between **b** and **c** is $\cos^{-1}\left(\frac{1}{4}\right)$, show that $\lambda = +4$ or -4. For each of these cases find the cosine of the angle between **a** and **c**.

31. Three non-collinear points A, B and C have position vectors **a, b** and **c** respectively, relative to an origin O, not necessarily in the plane ABC. Prove that the area of the triangle ABC is equal to the magnitude of the vector

$$\frac{1}{2}(\mathbf{b} \times \mathbf{c} + \mathbf{c} \times \mathbf{a} + \mathbf{a} \times \mathbf{b}).$$

When O does lie in the plane ABC interpret this result in terms of the areas of the triangles OAB, OBC and OCA, considering the cases (*i*) O is inside the triangle ABC (*ii*) O is in the region bounded by AC and BC produced.

32. Prove that the line $\mathbf{r} = \mathbf{a} + t\mathbf{b}$ and the plane $\mathbf{r}\cdot\mathbf{n} = p$ intersect in a point whose position vector is $\mathbf{a} + (p - \mathbf{a}\cdot\mathbf{n})\,\mathbf{b}/(\mathbf{b}\cdot\mathbf{n})$ given that $\mathbf{b}\cdot\mathbf{n} \neq 0$.

33. The non-zero vectors **p, q** are such that their vector product is the zero vector. State what is implied by this relation about the directions of **p** and **q.**

The three non-zero vectors **a, b** and **c** satisfy the equation $\mathbf{a} \times \mathbf{b} = \mathbf{c} \times \mathbf{a}$. Deduce that $\mathbf{b} \times \mathbf{c} = k\mathbf{a}$ where k is a scalar. If also $\mathbf{a} \times \mathbf{b} = \mathbf{b} \times \mathbf{c} \neq 0$, show that $\mathbf{a} + \mathbf{b} + \mathbf{c} = 0$.

34. Given that $\mathbf{x} + \frac{1}{\mathbf{p}^2}(\mathbf{p}\cdot\mathbf{x})\,\mathbf{p} = \mathbf{q}$, show that $\mathbf{p}\cdot\mathbf{x} = \frac{1}{2}\mathbf{p}\cdot\mathbf{q}$ and find **x** in terms of **p** and **q**.

35. In the plane of the triangle ABC, squares $ACXY$, $BCWZ$ are described, in the order given, externally to the triangle on AC and BC respectively. Taking

$$\vec{CX} = \mathbf{b},\ \vec{CA} = \mathbf{a},\ \vec{CW} = \mathbf{q},\ \vec{CB} = \mathbf{p}$$

prove that $\mathbf{a} \cdot \mathbf{p} + \mathbf{q} \cdot \mathbf{b} = 0$. Deduce that $\vec{AW} \cdot \vec{BX} = 0$.

36. In the parallelogram $OABC$, the angle AOC is acute and D is a point on BC such that $\dfrac{CD}{DB} = \dfrac{1}{2}$, $\vec{OA} = \mathbf{a}$ and $\vec{OC} = \mathbf{c}$. Prove that

$$9(AC^2 - OD^2) = 8(\mathbf{a}^2 - 3\mathbf{a} \cdot \mathbf{c}).$$

By considering the value of AD^2, prove that if $AC = OD$, then $AD = OC$.

37. Let $\mathbf{a}$ and $\mathbf{b}$ be non-zero vectors and d be a scalar.

(*i*) The equation $\mathbf{a} \cdot \mathbf{x} = d$ is to be solved for $\mathbf{x}$. Writing $\mathbf{x} = \lambda\mathbf{a} + \mathbf{e}$ where $\mathbf{e}$ is an arbitrary vector perpendicular to $\mathbf{a}$, show that the equation is satisfied if and only if λ has a particular value to be determined in terms of $\mathbf{a}$ and $\mathbf{d}$.

(*ii*) What condition must be satisfied by the vectors $\mathbf{a}$ and $\mathbf{b}$ if the equation $\mathbf{a} \times \mathbf{x} = \mathbf{b}$ has a solution ? Given that this condition is satisfied and writing

$$\mathbf{x} = \lambda\mathbf{a} + \mu\mathbf{b} + \gamma\mathbf{c}$$

where λ, μ, γ are scalars and $\mathbf{c}$ is a unit vector, perpendicular to $\mathbf{a}$ and $\mathbf{b}$ with $\mathbf{a}$, $\mathbf{b}$, $\mathbf{c}$ as right handed, show that the equation is satisfied if and only if μ, γ have particular values to be determined.

38. In the trapezium $OABC$, OA and CB are the parallel sides, and $CB = \dfrac{2}{3} OA$. M is the mid-point of the side OC, and the angle OMA is a right angle, $\vec{OA} = \mathbf{a}$ and $\vec{OC} = \mathbf{c}$.

(*i*) Express the vectors $\vec{OB}$, $\vec{AB}$ and $\vec{AM}$ in terms of $\mathbf{a}$ and $\mathbf{c}$. Prove that

(*ii*) $\mathbf{a} \cdot \mathbf{c} = \dfrac{1}{2}\mathbf{c}^2$.

(*iii*) $\vec{OB} \cdot \vec{AB} = \dfrac{1}{18}(21\,OC^2 - 4\,OA^2)$.

(*iv*) Given that the angle OAB is a right angle and that the length OC is 2 cm, calculate the length OA.

39. In the triangle OAB, $\vec{OA} = \mathbf{a}$ and $\vec{OB} = \mathbf{b}$. A point P is taken on OA such that $OP/PA = 3$, and a point Q is taken on OB such that $\dfrac{OQ}{QB} = \dfrac{1}{2}\cdot$ Express each of the vectors

$$\vec{AQ} \text{ and } \vec{BP}$$

in terms of $\mathbf{a}$ and $\mathbf{b}$.

Given that the lines AQ and BP are perpendicular, prove that

$$9\mathbf{a}^2 + 4\mathbf{b}^2 = 15\,\mathbf{a} \cdot \mathbf{b}.$$

Given also that $OA = 2$ and $OB = 3$; prove that the cosine of the angle AOB is 4/5, and calculate the length of PQ.

40. The points O, A, B, C, D are such that

$$\overrightarrow{OA} = \mathbf{a},\ \overrightarrow{OB} = \mathbf{b},\ \overrightarrow{OC} = 2\mathbf{a} + 3\mathbf{b},\ \overrightarrow{OD} = \mathbf{a} - 2\mathbf{b}.$$

Given that the length of $\overrightarrow{OA}$ is three times the length of $\overrightarrow{OB}$ show that $\overrightarrow{BD}$ and $\overrightarrow{AC}$ are perpendicular.

41. $ABCD$ is a plane quadrilateral whose sides CD, BA intersect at O. If P, Q are the mid-points of the diagonals AC and BD, prove that the area of $\Delta\, OPQ$ is one quarter of the area of the quadrilateral $ABCD$.

42. If $\mathbf{p}$ and $\mathbf{q}$ are the position vectors of two fixed points and if $\mathbf{r}$ is the position vector of a variable point, describe geometrically the loci given by the equations

$$(\mathbf{r} - \mathbf{p}) \cdot (\mathbf{r} - \mathbf{q}) = 0 \quad \text{and} \quad (\mathbf{r} - \mathbf{p}) \times (\mathbf{r} - \mathbf{q}) = 0.$$

43. AA', BB', CC' and DD' are edges of a cube perpendicular to the face $ABCD$. Find

(*i*) the perpendicular distance of D' from the diagonal AC',

(*ii*) the shortest distance between the edges DD' and the diagonal AC'.

44. In the triangle ABC, L is the mid-point of AB and M is a point in AC such that the ratio of the length of AM to the length of MC is 2 : 1. BM and LC intersect in N.

Find the position vector of N.

(*i*) Calculate the ratio of the area of triangle ANC to the area of triangle ANM.

(*ii*) Prove that the area of triangle BNC = area of triangle ANC.

(*iii*) Calculate the ratio of the length of BN to the length of NM.

(*iv*) Calculate the ratio of the area of triangle BNC to that of the triangle ABC.

45. If two opposite edges of tetrahedron are at right angles, show that every plane section of the tetrahedron parallel to this pair is a rectangle.

46. If P be any point on the circumference of the in-circle of an equilateral triangle ABC, show that $PA^2 + PB^2 + PC^2$ is constant.

47. OA, OB, OC are three mutually perpendicular straight lines through O and A, B, C also denote the angles of the triangle ABC; show that

(*i*) $\Delta\, ABC$ is acute angled.

(*ii*) $OA^2 \tan A = OB^2 \tan B = OC^2 \tan C$.

48. Show that the mid-points of the six edges of a cube which do not meet a particular diagonal are the six vertices of a regular hexagon. Also show that the plane of the hexagon bisects the diagonal normally.

49. The line AB is the common perpendicular to two skew lines AP and BQ, and C and R are the mid-points of AB and PQ respectively. Prove that CR and AB are perpendicular.

50. Let $\mathbf{a}$ be a given non-zero vector. For what vectors $\mathbf{u}$ is $\mathbf{a} \times \mathbf{u} = 0$? Let $\mathbf{a} = \mathbf{i} + 2\mathbf{j} - 3\mathbf{k}$, $\mathbf{b} = 2\mathbf{i} + \mathbf{j} - \mathbf{k}$, where $\mathbf{i}, \mathbf{j}, \mathbf{k}$ denote the usual base vectors. Determine the set of vectors $\mathbf{v}$ for which

$$\mathbf{a} \times \mathbf{v} = \mathbf{a} \times \mathbf{b}.$$

Hence, find a vector $\mathbf{v}$ satisfying

$$\mathbf{a} \times \mathbf{v} = \mathbf{a} \times \mathbf{b}, \quad \mathbf{a} \cdot \mathbf{v} = 0.$$

7

Some Miscellaneous Topics

(Products of Four Vectors)

In this chapter, we shall consider products of four vectors and the solution of some vector equations.

7.1. SCALAR PRODUCT $(\mathbf{a} \times \mathbf{b}) \cdot (\mathbf{c} \times \mathbf{d})$ OF FOUR VECTORS

We shall prove that

$$(\mathbf{a} \times \mathbf{b}) \cdot (\mathbf{c} \times \mathbf{d}) = (\mathbf{a} \cdot \mathbf{c})(\mathbf{b} \cdot \mathbf{d}) - (\mathbf{a} \cdot \mathbf{d})(\mathbf{b} \cdot \mathbf{c}) = \begin{vmatrix} \mathbf{a} \cdot \mathbf{c} & \mathbf{a} \cdot \mathbf{d} \\ \mathbf{b} \cdot \mathbf{c} & \mathbf{b} \cdot \mathbf{d} \end{vmatrix}.$$

Looking upon $(\mathbf{a} \times \mathbf{b}) \cdot (\mathbf{c} \times \mathbf{d})$ as a scalar triple product of the three vectors

$$\mathbf{a} \times \mathbf{b}, \mathbf{c}, \mathbf{d}$$

and interchanging the dot and cross, we have

$$\begin{aligned}(\mathbf{a} \times \mathbf{b}) \cdot (\mathbf{c} \times \mathbf{d}) &= [(\mathbf{a} \times \mathbf{b}) \times \mathbf{c}] \cdot \mathbf{d} \\ &= [(\mathbf{a} \cdot \mathbf{c})\, \mathbf{b} - (\mathbf{b} \cdot \mathbf{c})\, \mathbf{a}] \cdot \mathbf{d} \\ &= (\mathbf{a} \cdot \mathbf{c})(\mathbf{b} \cdot \mathbf{d}) - (\mathbf{b} \cdot \mathbf{c})(\mathbf{a} \cdot \mathbf{d}).\end{aligned}$$

7.2. VECTOR PRODUCT $(\mathbf{a} \times \mathbf{b}) \times (\mathbf{c} \times \mathbf{d})$ OF FOUR VECTORS

We shall prove that (*Patna 2003*)

$$(\mathbf{a} \times \mathbf{b}) \cdot (\mathbf{c} \times \mathbf{d}) = [\mathbf{a}\ \mathbf{b}\ \mathbf{d}]\, \mathbf{c} - [\mathbf{a}\ \mathbf{b}\ \mathbf{c}]\, \mathbf{d} = [\mathbf{a}\ \mathbf{c}\ \mathbf{d}]\, \mathbf{b} - [\mathbf{b}\ \mathbf{c}\ \mathbf{d}]\, \mathbf{a}$$

We can look upon $(\mathbf{a} \times \mathbf{b}) \times (\mathbf{c} \times \mathbf{d})$ as a vector triple product in two ways according as we put $\mathbf{a} \times \mathbf{b} = \mathbf{p}$ or $\mathbf{c} \times \mathbf{d} = \mathbf{q}$.

Thus, putting $\mathbf{a} \times \mathbf{b} = \mathbf{p}$, we have

$$\begin{aligned}(\mathbf{a} \times \mathbf{b}) \cdot (\mathbf{c} \times \mathbf{d}) &= \mathbf{p} \times (\mathbf{c} \times \mathbf{d}) \\ &= (\mathbf{p} \cdot \mathbf{d})\, \mathbf{c} - (\mathbf{p} \cdot \mathbf{c})\, \mathbf{d} = [\mathbf{a}\ \mathbf{b}\ \mathbf{d}]\, \mathbf{c} - [\mathbf{a}\ \mathbf{b}\ \mathbf{c}]\, \mathbf{d}.\end{aligned}$$

so that the vector product now appears as a linear combination of the vectors **c** and **d**.

Again, we have

$$\begin{aligned}(\mathbf{a} \times \mathbf{b}) \times (\mathbf{c} \times \mathbf{d}) &= (\mathbf{a} \times \mathbf{b}) \times \mathbf{q} \\ &= (\mathbf{a} \cdot \mathbf{q})\, \mathbf{b} - (\mathbf{b} \cdot \mathbf{q})\, \mathbf{a} = [\mathbf{a}\ \mathbf{c}\ \mathbf{d}]\, \mathbf{b} - [\mathbf{b}\ \mathbf{c}\ \mathbf{d}]\, \mathbf{a},\end{aligned}$$

so that the vector product now appears as a linear combination of the vectors **a** and **b**.

Thus, we have

$$(\mathbf{a} \times \mathbf{b}) \times (\mathbf{c} \times \mathbf{d}) = [\mathbf{a}\ \mathbf{b}\ \mathbf{d}]\ \mathbf{c} - [\mathbf{a}\ \mathbf{b}\ \mathbf{c}]\ \mathbf{d} = [\mathbf{a}\ \mathbf{c}\ \mathbf{d}]\ \mathbf{b} - [\mathbf{b}\ \mathbf{c}\ \mathbf{d}]\ \mathbf{a}.$$

7.3. LINEAR RELATION CONNECTING FOUR VECTORS

To determine the Linear Relation connecting four given vectors.

Let **a, b, c, d** be four given vectors. Considering

$$(\mathbf{a} \times \mathbf{b}) \times (\mathbf{c} \times \mathbf{d})$$

as a vector triple product in two ways as in the preceding section, we obtain

$$[\mathbf{a}\ \mathbf{b}\ \mathbf{d}]\ \mathbf{c} - [\mathbf{a}\ \mathbf{b}\ \mathbf{c}]\ \mathbf{d} = [\mathbf{a}\ \mathbf{c}\ \mathbf{d}]\ \mathbf{b} - [\mathbf{b}\ \mathbf{c}\ \mathbf{d}]\ \mathbf{a}.$$

$$\Rightarrow\ [\mathbf{a}\ \mathbf{b}\ \mathbf{d}]\ \mathbf{c} - [\mathbf{a}\ \mathbf{b}\ \mathbf{c}]\ \mathbf{d} - [\mathbf{a}\ \mathbf{c}\ \mathbf{d}]\ \mathbf{b} + [\mathbf{b}\ \mathbf{c}\ \mathbf{d}]\ \mathbf{a} = \mathbf{0}.$$

$$\Rightarrow\ [\mathbf{b}\ \mathbf{c}\ \mathbf{d}]\ \mathbf{a} - [\mathbf{a}\ \mathbf{c}\ \mathbf{d}]\ \mathbf{b} + [\mathbf{a}\ \mathbf{b}\ \mathbf{d}]\ \mathbf{c} - [\mathbf{a}\ \mathbf{b}\ \mathbf{c}]\ \mathbf{d} = \mathbf{0} \qquad ...(i)$$

which is the required relation connecting the four vectors

$$\mathbf{a}, \mathbf{b}, \mathbf{c}, \mathbf{d}$$

If [**a b, c**] ≠ 0, *i.e.*, **a, b, c** are not coplanar vectors, we can rewrite (*i*) as

$$\mathbf{d} = \frac{[\mathbf{b}\,\mathbf{c}\,\mathbf{d}]\,\mathbf{a} - [\mathbf{a}\,\mathbf{c}\,\mathbf{d}]\,\mathbf{b} + [\mathbf{a}\,\mathbf{b}\,\mathbf{d}]\,\mathbf{c}}{[\mathbf{a}\,\mathbf{b}\,\mathbf{c}]}.$$

Thus, we have expressed any given vector **d** as a linear combination of three non-coplanar vectors **a, b,** and **c**.

Another Method. If **a, b, c** are not coplanar, then we have a relation of the form

$$\mathbf{d} = \lambda\mathbf{a} + \mu\mathbf{b} + \nu\mathbf{c}. \qquad ...(i)$$

Multiplying scalarly with **b** × **c**, we obtain

$$[\mathbf{d}\ \mathbf{b}\ \mathbf{c}] = \lambda\ [\mathbf{a}\ \mathbf{b}\ \mathbf{c}]. \qquad ...(ii)$$

Similarly on multiplying (*i*) scalarly with **c** × **a** and **a** × **b**, successively, we obtain

$$[\mathbf{d}\ \mathbf{c}\ \mathbf{a}] = \mu\ [\mathbf{b}\ \mathbf{c}\ \mathbf{a}] = \mu\ [\mathbf{a}\ \mathbf{b}\ \mathbf{c}] \qquad ...(iii)$$

$$[\mathbf{d}\ \mathbf{a}\ \mathbf{b}] = \nu\ [\mathbf{c}\ \mathbf{a}\ \mathbf{b}] = \nu\ [\mathbf{a}\ \mathbf{b}\ \mathbf{c}] \qquad ...(iv)$$

Substituting in (*i*) the values of λ, μ, ν, as obtained from (*ii*), (*iii*), (*iv*), we obtain the required relation.

7.4. PRODUCT OF TWO SCALAR TRIPLE PRODUCTS

To prove that

$$[\mathbf{p}\,\mathbf{q}\,\mathbf{r}]\,[\mathbf{p}'\,\mathbf{q}'\,\mathbf{r}'] = \begin{vmatrix} \mathbf{p}\cdot\mathbf{p}' & \mathbf{p}\cdot\mathbf{q}' & \mathbf{p}\cdot\mathbf{r}' \\ \mathbf{q}\cdot\mathbf{p}' & \mathbf{q}\cdot\mathbf{q}' & \mathbf{q}\cdot\mathbf{r}' \\ \mathbf{r}\cdot\mathbf{p}' & \mathbf{r}\cdot\mathbf{q}' & \mathbf{r}\cdot\mathbf{r}' \end{vmatrix}$$

where **p, q, r; p′, q′, r′** *are any vectors.*

Writing **p′** × **q′** = **a** and considering the four vectors **p, q, r, a,** we obtain the linear relation connecting these four vectors. As in § 7.1, above we have

$$[\mathbf{q}\ \mathbf{r}\ \mathbf{a}]\ \mathbf{p} - [\mathbf{p}\ \mathbf{r}\ \mathbf{a}]\ \mathbf{q} + [\mathbf{p}\ \mathbf{q}\ \mathbf{a}]\ \mathbf{r} - [\mathbf{p}\ \mathbf{q}\ \mathbf{r}]\ \mathbf{a} = \mathbf{0} \qquad ...(i)$$

Again

$$[\mathbf{q}\ \mathbf{r}\ \mathbf{a}] = \mathbf{q} \times \mathbf{r} \,.\, \mathbf{a} = (\mathbf{q} \times \mathbf{r}) \cdot (\mathbf{p}' \times \mathbf{q}') = \begin{vmatrix} \mathbf{q} \,.\, \mathbf{p}' & \mathbf{q} \,.\, \mathbf{q}' \\ \mathbf{r} \,.\, \mathbf{p}' & \mathbf{r} \,.\, \mathbf{q}' \end{vmatrix}$$

$$[\mathbf{p}\ \mathbf{r}\ \mathbf{a}] = \mathbf{p} \times \mathbf{r} \,.\, \mathbf{a} = (\mathbf{p} \times \mathbf{r}) \cdot (\mathbf{p}' \times \mathbf{q}') = \begin{vmatrix} \mathbf{p} \,.\, \mathbf{p}' & \mathbf{p} \,.\, \mathbf{q}' \\ \mathbf{r} \,.\, \mathbf{p}' & \mathbf{r} \,.\, \mathbf{q}' \end{vmatrix}$$

$$[\mathbf{p}\ \mathbf{q}\ \mathbf{a}] = \mathbf{p} \times \mathbf{q} \,.\, \mathbf{a} = (\mathbf{p} \times \mathbf{q}) \cdot (\mathbf{p}' \times \mathbf{q}') = \begin{vmatrix} \mathbf{p} \,.\, \mathbf{p}' & \mathbf{p} \,.\, \mathbf{q}' \\ \mathbf{q} \,.\, \mathbf{p}' & \mathbf{q} \,.\, \mathbf{q}' \end{vmatrix}$$

We rewrite (i) as

$$[\mathbf{p}\ \mathbf{q}\ \mathbf{r}]\ \mathbf{p}' \times \mathbf{q}' = [\mathbf{q}\ \mathbf{r}\ \mathbf{a}]\ \mathbf{p} - [\mathbf{p}\ \mathbf{r}\ \mathbf{a}]\ \mathbf{q} + [\mathbf{p}\ \mathbf{q}\ \mathbf{a}]\ \mathbf{r}.$$

Multiplying scalarly with $\mathbf{r}'$, we obtain

$$[\mathbf{p}\ \mathbf{q}\ \mathbf{r}][\mathbf{p}'\ \mathbf{q}'\ \mathbf{r}'] = \begin{vmatrix} \mathbf{q} \,.\, \mathbf{p}' & \mathbf{q} \,.\, \mathbf{q}' \\ \mathbf{r} \,.\, \mathbf{p}' & \mathbf{r} \,.\, \mathbf{q}' \end{vmatrix} (\mathbf{r} \,.\, \mathbf{r}')$$

$$- \begin{vmatrix} \mathbf{p} \,.\, \mathbf{p}' & \mathbf{p} \,.\, \mathbf{q}' \\ \mathbf{r} \,.\, \mathbf{p}' & \mathbf{r} \,.\, \mathbf{q}' \end{vmatrix} (\mathbf{q} \,.\, \mathbf{r}') + \begin{vmatrix} \mathbf{p} \,.\, \mathbf{p}' & \mathbf{p} \,.\, \mathbf{q}' \\ \mathbf{q} \,.\, \mathbf{q}' & \mathbf{p} \,.\, \mathbf{p}' \end{vmatrix} (\mathbf{r} \,.\, \mathbf{r}')$$

$$= \begin{vmatrix} \mathbf{p} \,.\, \mathbf{p}' & \mathbf{p} \,.\, \mathbf{q}' & \mathbf{p} \,.\, \mathbf{r}' \\ \mathbf{q} \,.\, \mathbf{p}' & \mathbf{q} \,.\, \mathbf{q}' & \mathbf{q} \,.\, \mathbf{r}' \\ \mathbf{r} \,.\, \mathbf{p}' & \mathbf{r} \,.\, \mathbf{q}' & \mathbf{r} \,.\, \mathbf{r}' \end{vmatrix}.$$

Cor.

$$[\mathbf{a}\ \mathbf{b}\ \mathbf{c}]^2 = \begin{vmatrix} \mathbf{a} \,.\, \mathbf{a} & \mathbf{a} \,.\, \mathbf{b} & \mathbf{a} \,.\, \mathbf{c} \\ \mathbf{b} \,.\, \mathbf{a} & \mathbf{b} \,.\, \mathbf{b} & \mathbf{b} \,.\, \mathbf{c} \\ \mathbf{c} \,.\, \mathbf{a} & \mathbf{c} \,.\, \mathbf{b} & \mathbf{c} \,.\, \mathbf{c} \end{vmatrix}$$

7.5. TWO USEFUL DECOMPOSITIONS

7.5.1. *If* **a, b, c** *are three non-coplanar vectors, then to prove that*

$$\mathbf{b} \times \mathbf{c}, \quad \mathbf{c} \times \mathbf{a}, \quad \mathbf{a} \times \mathbf{b}$$

are also non-coplanar and to express **a, b, c** *in terms of*

$$\mathbf{b} \times \mathbf{c}, \quad \mathbf{c} \times \mathbf{a}, \quad \mathbf{a} \times \mathbf{b}.$$

We are given that $[\mathbf{a}\ \mathbf{b}\ \mathbf{c}] \neq 0$ and we shall prove that the scalar triple product

$$[\mathbf{b} \times \mathbf{c}, \quad \mathbf{c} \times \mathbf{a}, \quad \mathbf{a} \times \mathbf{b}] \neq 0$$

We have

$$(\mathbf{b} \times \mathbf{c}) \times (\mathbf{c} \times \mathbf{a}) \,.\, (\mathbf{a} \times \mathbf{b}) = [(\mathbf{b} \,.\, \mathbf{c} \times \mathbf{a})\ \mathbf{c} - (\mathbf{c} \,.\, \mathbf{c} \times \mathbf{a})\ \mathbf{b}] \,.\, (\mathbf{a} \times \mathbf{b})$$

$$= [\mathbf{a}\ \mathbf{b}\ \mathbf{c}]\ \mathbf{c} \,.\, \mathbf{a} \times \mathbf{b}, \quad \text{for} \quad \mathbf{c} \,.\, \mathbf{c} \times \mathbf{a} = \mathbf{a}$$

$$= [\mathbf{a}\ \mathbf{b}\ \mathbf{c}]\ [\mathbf{a}\ \mathbf{b}\ \mathbf{c}] \neq 0.$$

Hence, $\mathbf{b} \times \mathbf{c}$, $\mathbf{c} \times \mathbf{a}$, $\mathbf{a} \times \mathbf{b}$ are non-coplanar vectors. The vectors $\mathbf{b} \times \mathbf{c}$, $\mathbf{c} \times \mathbf{a}$, $\mathbf{a} \times \mathbf{b}$ being non-coplanar every vector and, in particular, **a, b, c** is expressible as linear combinations of the same.

Let

$$\mathbf{a} = l\mathbf{b} \times \mathbf{c} + m\mathbf{c} \times \mathbf{a} + n\mathbf{a} \times \mathbf{b}.$$

Multiplying both sides scalarly with **a, b, c,** successively, we get

$$\mathbf{a} \cdot \mathbf{a} = l\mathbf{b} \times \mathbf{c} \cdot \mathbf{a} = l\,[\mathbf{a}\ \mathbf{b}\ \mathbf{c}],\ \mathbf{a} \cdot \mathbf{b} = m\,[\mathbf{a}\ \mathbf{b}\ \mathbf{c}],\ \mathbf{a} \cdot \mathbf{c} = n\,[\mathbf{a}\ \mathbf{b}\ \mathbf{c}]$$

$$\Rightarrow \qquad l = \frac{\mathbf{a} \cdot \mathbf{a}}{[\mathbf{a}\ \mathbf{b}\ \mathbf{c}]},\ m = \frac{\mathbf{a} \cdot \mathbf{b}}{[\mathbf{a}\ \mathbf{b}\ \mathbf{c}]},\ n = \frac{\mathbf{a} \cdot \mathbf{c}}{[\mathbf{a}\ \mathbf{b}\ \mathbf{c}]},$$

$$\Rightarrow \qquad \mathbf{a} = \frac{1}{[\mathbf{a}\ \mathbf{b}\ \mathbf{c}]}[(\mathbf{a} \cdot \mathbf{a})\,\mathbf{b} \times \mathbf{c} + (\mathbf{a} \cdot \mathbf{b})\,\mathbf{c} \times \mathbf{a} + (\mathbf{a} \cdot \mathbf{c})\,\mathbf{a} \times \mathbf{b}].$$

We may similarly express **b** and **c** in terms of

$$\mathbf{b} \times \mathbf{c}, \quad \mathbf{c} \times \mathbf{a}, \quad \mathbf{a} \times \mathbf{b}.$$

7.5.2. *If* **a, b, c** *are three non-coplanar vectors, then to express* $\mathbf{b} \times \mathbf{c}$, $\mathbf{c} \times \mathbf{a}$, $\mathbf{a} \times \mathbf{b}$ *in terms of* **a, b, c.**

It has been already proved that the vectors $\mathbf{b} \times \mathbf{c}$, $\mathbf{c} \times \mathbf{a}$, and $\mathbf{a} \times \mathbf{b}$ are non-coplanar.

Let

$$\mathbf{b} \times \mathbf{c} = l\mathbf{a} + m\mathbf{b} + n\mathbf{c}.$$

Multiplying both sides scalarly with

$$\mathbf{b} \times \mathbf{c}, \quad \mathbf{c} \times \mathbf{a} \quad \text{and} \quad \mathbf{a} \times \mathbf{b}$$

successively, we get

$$(\mathbf{b} \times \mathbf{c}) \cdot (\mathbf{b} \times \mathbf{c}) = l\,[\mathbf{a}\ \mathbf{b}\ \mathbf{c}],$$
$$(\mathbf{b} \times \mathbf{c}) \cdot (\mathbf{c} \times \mathbf{a}) = m\,[\mathbf{a}\ \mathbf{b}\ \mathbf{c}],$$
$$(\mathbf{b} \times \mathbf{c}) \cdot (\mathbf{a} \times \mathbf{b}) = n\,[\mathbf{a}\ \mathbf{b}\ \mathbf{c}],$$

$$\Rightarrow \qquad l = \frac{(\mathbf{b} \times \mathbf{c}) \cdot (\mathbf{b} \times \mathbf{c})}{[\mathbf{a}\ \mathbf{b}\ \mathbf{c}]},$$

$$m = \frac{(\mathbf{b} \times \mathbf{c}) \cdot (\mathbf{c} \times \mathbf{a})}{[\mathbf{a}\ \mathbf{b}\ \mathbf{c}]},$$

$$n = \frac{(\mathbf{b} \times \mathbf{c}) \cdot (\mathbf{a} \times \mathbf{b})}{[\mathbf{a}\ \mathbf{b}\ \mathbf{c}]},$$

where the numerators can be transformed as in § 7.1 page 177.

Thus, we have expressed $\mathbf{b} \times \mathbf{c}$ in terms of **a, b, c.**

We may similarly express $\mathbf{c} \times \mathbf{a}$ and $\mathbf{a} \times \mathbf{b}$ in terms of **a, b, c.**

7.6. RECIPROCAL SYSTEM OF VECTORS (*Avadh 2005*)

If **a, b, c,** be three non-coplanar vectors so that $[\mathbf{a}\ \mathbf{b}\ \mathbf{c}] \neq 0$ and if $\mathbf{a}'$, $\mathbf{b}'$, $\mathbf{c}'$ be three other vectors such that

$$\mathbf{a}' = \frac{\mathbf{b} \times \mathbf{c}}{[\mathbf{a}\ \mathbf{b}\ \mathbf{c}]},\ \mathbf{b}' = \frac{\mathbf{c} \times \mathbf{a}}{[\mathbf{a}\ \mathbf{b}\ \mathbf{c}]},\ \mathbf{c}' = \frac{\mathbf{a} \times \mathbf{b}}{[\mathbf{a}\ \mathbf{b}\ \mathbf{c}]},$$

then $\mathbf{a}'$, $\mathbf{b}'$, $\mathbf{c}'$ are called reciprocal system to the vectors **a, b, c.**

Properties.

7.6.1. (I) *If* **a, b, c** *and* **a′, b′, c′** *be reciprocal systems of vectors, then*

$$\mathbf{a} \cdot \mathbf{a}' = \mathbf{a} \cdot \frac{\mathbf{b} \times \mathbf{c}}{[\mathbf{a}\ \mathbf{b}\ \mathbf{c}]} = \frac{[\mathbf{a}\ \mathbf{b}\ \mathbf{c}]}{[\mathbf{a}\ \mathbf{b}\ \mathbf{c}]} = 1 \qquad \textit{(Kumaon 2005)}$$

Similarly $\mathbf{b} \cdot \mathbf{b}' = \mathbf{c} \cdot \mathbf{c}' = 1$.

Note. Due to this property, the two systems of vectors are called reciprocal systems.

(II) *To show that* $\mathbf{a} \cdot \mathbf{b}' = \mathbf{a} \cdot \mathbf{c}' = \mathbf{b} \cdot \mathbf{a}' = \mathbf{b} \cdot \mathbf{c}' = \mathbf{c} \cdot \mathbf{a}' = \mathbf{c} \cdot \mathbf{b}' = 0$.

$$\mathbf{a} \cdot \mathbf{b}' = \mathbf{a} \cdot \frac{\mathbf{c} \times \mathbf{a}}{[\mathbf{a}\ \mathbf{b}\ \mathbf{c}]} = \frac{[\mathbf{a}\ \mathbf{c}\ \mathbf{a}]}{[\mathbf{a}\ \mathbf{b}\ \mathbf{c}]} = 0$$

Similarly other results follow.

(III) *To show that* $[\mathbf{a}\ \mathbf{b}\ \mathbf{c}]\,[\mathbf{a}', \mathbf{b}', \mathbf{c}'] = 1$. *(Awadh 98, 2000, Kumaon 2006)*

We have
$$[\mathbf{a}', \mathbf{b}', \mathbf{c}'] = \left[\frac{\mathbf{b} \times \mathbf{c}}{[\mathbf{a}\ \mathbf{b}\ \mathbf{c}]}, \frac{\mathbf{c} \times \mathbf{a}}{[\mathbf{a}\ \mathbf{b}\ \mathbf{c}]}, \frac{\mathbf{a} \times \mathbf{b}}{[\mathbf{a}\ \mathbf{b}\ \mathbf{c}]}\right]$$
$$= \frac{1}{[\mathbf{a}\ \mathbf{b}\ \mathbf{c}]^3}[\mathbf{b} \times \mathbf{c},\ \mathbf{c} \times \mathbf{a},\ \mathbf{a} \times \mathbf{b}]$$
$$= \frac{1}{[\mathbf{a}\ \mathbf{b}\ \mathbf{c}]^3}[\mathbf{a}\ \mathbf{b}\ \mathbf{c}]^2$$
$$= \frac{1}{[\mathbf{a}\ \mathbf{b}\ \mathbf{c}]}$$

$\therefore \quad [\mathbf{a}', \mathbf{b}', \mathbf{c}']\,[\mathbf{a}\ \mathbf{b}\ \mathbf{c}] = 1.$

Since $[\mathbf{a}\ \mathbf{b}\ \mathbf{c}] \neq 0$, hence $[\mathbf{a}', \mathbf{b}', \mathbf{c}'] \neq 0 \Rightarrow \mathbf{a}', \mathbf{b}', \mathbf{c}'$ are also non-coplanar.

(IV) *The orthogonal triad of vectors* **i, j, k** *is self-reciprocal.*

Let **i′, j′, k′** be the system of vectors reciprocal to the system **i, j, k**. Then, we have

$$\mathbf{i}' = \frac{\mathbf{j} \times \mathbf{k}}{[\mathbf{i}\ \mathbf{j}\ \mathbf{k}]} = \mathbf{i}$$

Similarly, $\mathbf{j}' = \mathbf{j}$ and $\mathbf{k}' = \mathbf{k}$.

7.6.2. Theorem

If **a, b, c** *be three non-coplanar vectors for which* $[\mathbf{a}\ \mathbf{b}\ \mathbf{c}] \neq 0$ *and* **a′, b′, c′** *constitute the reciprocal system of vectors, then any vector* **r** *can be expressed as*

$$\mathbf{r} = (\mathbf{r} \cdot \mathbf{a}')\,\mathbf{a} + (\mathbf{r} \cdot \mathbf{b}')\,\mathbf{b} + (\mathbf{r} \cdot \mathbf{c}')\,\mathbf{c}. \qquad \textit{(Avadh 2005)}$$

Since **a, b, c** are non-coplanar vectors, **r** can be expressed as a linear combination in the form

$$\mathbf{r} = x\mathbf{a} + y\mathbf{b} + z\mathbf{c} \qquad ...(1)$$

where x, y, z are some scalars.

$$(1) \Rightarrow \quad \mathbf{r} \cdot (\mathbf{b} \times \mathbf{c}) = x\mathbf{a} \cdot (\mathbf{b} \times \mathbf{c}) + y\mathbf{b} \cdot (\mathbf{b} \times \mathbf{c}) + z\mathbf{c} \cdot (\mathbf{b} \times \mathbf{c})$$

$$= x\,[\mathbf{a}\ \mathbf{b}\ \mathbf{c}]$$

$$\therefore \qquad x = \frac{\mathbf{r} \cdot (\mathbf{b} \times \mathbf{c})}{[\mathbf{a}\ \mathbf{b}\ \mathbf{c}]} = (\mathbf{r} \cdot \mathbf{a}')$$

Similarly $\quad y = \mathbf{r} \cdot \mathbf{b}', \quad z = \mathbf{r} \cdot \mathbf{c}'$

$$\therefore \qquad \mathbf{r} = (\mathbf{r} \cdot \mathbf{a}')\,\mathbf{a} + (\mathbf{r} \cdot \mathbf{b}')\,\mathbf{b} + (\mathbf{r} \cdot \mathbf{c}')\,\mathbf{c}. \qquad ...(2)$$

Cor. 1. Similarly, we can get

$$\mathbf{r} = (\mathbf{r} \cdot \mathbf{a})\,\mathbf{a}' + (\mathbf{r} \cdot \mathbf{b})\,\mathbf{b}' + (\mathbf{r} \cdot \mathbf{c})\,\mathbf{c}'$$

Cor. 2. Since the system **i, j, k** is self-reciprocal, (2) can be written as

$$\mathbf{r} = (\mathbf{r} \cdot \mathbf{i})\,\mathbf{i} + (\mathbf{r} \cdot \mathbf{j})\,\mathbf{j} + (\mathbf{r} \cdot \mathbf{k})\,\mathbf{k}.$$

EXAMPLES

Example 1. *Prove that*

$$[\mathbf{a} \times \mathbf{p},\ \mathbf{b} \times \mathbf{q},\ \mathbf{c} \times \mathbf{r}] + [\mathbf{a} \times \mathbf{q},\ \mathbf{b} \times \mathbf{r},\ \mathbf{c} \times \mathbf{p}] + [\mathbf{a} \times \mathbf{r},\ \mathbf{b} \times \mathbf{p},\ \mathbf{c} \times \mathbf{q}] = 0.$$

(Kumaon 2001)

Solution.
$$[\mathbf{a} \times \mathbf{p},\ \mathbf{b} \times \mathbf{q},\ \mathbf{c} \times \mathbf{r}] = (\mathbf{a} \times \mathbf{p}) \cdot [(\mathbf{b} \times \mathbf{q}) \times (\mathbf{c} \times \mathbf{r})]$$
$$= (\mathbf{a} \times \mathbf{p}) \cdot [\{(\mathbf{b} \times \mathbf{q}) \cdot \mathbf{r}\}\,\mathbf{c} - \{(\mathbf{b} \times \mathbf{q}) \cdot \mathbf{c}\}\,\mathbf{r}]$$
$$= (\mathbf{a} \times \mathbf{p}) \cdot \{[\mathbf{b}\ \mathbf{q}\ \mathbf{r}]\,\mathbf{c} - [\mathbf{b}\ \mathbf{q}\ \mathbf{c}]\,\mathbf{r}\}$$
$$= [\mathbf{a}\ \mathbf{p}\ \mathbf{c}]\,[\mathbf{b}\ \mathbf{q}\ \mathbf{r}] - [\mathbf{a}\ \mathbf{p}\ \mathbf{r}]\,[\mathbf{b}\ \mathbf{q}\ \mathbf{c}] \qquad ...(1)$$

Also, $\quad [\mathbf{a} \times \mathbf{q},\ \mathbf{b} \times \mathbf{r},\ \mathbf{c} \times \mathbf{p}]$
$$= (\mathbf{b} \times \mathbf{r}) \cdot [(\mathbf{c} \times \mathbf{p}) \times (\mathbf{a} \times \mathbf{q})]$$
$$= (\mathbf{b} \times \mathbf{r}) \cdot [\{(\mathbf{c} \times \mathbf{p}) \cdot \mathbf{q}\}\,\mathbf{a} - \{(\mathbf{c} \times \mathbf{p}) \cdot \mathbf{a}\}\,\mathbf{q}]$$
$$= (\mathbf{b} \times \mathbf{r}) \cdot \{[\mathbf{c}\ \mathbf{p}\ \mathbf{q}]\,\mathbf{a} - [\mathbf{c}\ \mathbf{p}\ \mathbf{a}]\,\mathbf{q}\}$$
$$= [\mathbf{b}\ \mathbf{r}\ \mathbf{a}]\,[\mathbf{c}\ \mathbf{p}\ \mathbf{q}] - [\mathbf{b}\ \mathbf{q}\ \mathbf{r}]\,[\mathbf{a}\ \mathbf{p}\ \mathbf{c}] \qquad ...(2)$$

$[\mathbf{a} \times \mathbf{r},\ \mathbf{b} \times \mathbf{p},\ \mathbf{c} \times \mathbf{q}]$
$$= (\mathbf{c} \times \mathbf{q}) \cdot [(\mathbf{a} \times \mathbf{r}) \times (\mathbf{b} \times \mathbf{p})]$$
$$= (\mathbf{c} \times \mathbf{q} \cdot [\{(\mathbf{a} \times \mathbf{r}) \cdot \mathbf{p}\}\,\mathbf{b} - \{(\mathbf{a} \times \mathbf{r}) \cdot \mathbf{b}\}\,\mathbf{p}]$$
$$= [\mathbf{c}\ \mathbf{q}\ \mathbf{b}]\,[\mathbf{a}\ \mathbf{r}\ \mathbf{p}] - [\mathbf{c}\ \mathbf{q}\ \mathbf{b}]\,[\mathbf{a}\ \mathbf{r}\ \mathbf{b}] \qquad ...(3)$$

Adding (1), (2), (3) the required result follows.

Example 2. *Prove that*

$$[\mathbf{a} \times \mathbf{b},\ \mathbf{c} \times \mathbf{d},\ \mathbf{e} \times \mathbf{f}] = [\mathbf{a}\ \mathbf{b}\ \mathbf{d}]\,[\mathbf{c}\ \mathbf{e}\ \mathbf{f}] - [\mathbf{a}\ \mathbf{b}\ \mathbf{c}]\,[\mathbf{d}\ \mathbf{e}\ \mathbf{f}]$$
$$= [\mathbf{a}\ \mathbf{b}\ \mathbf{e}]\,[\mathbf{f}\ \mathbf{c}\ \mathbf{d}] - [\mathbf{a}\ \mathbf{b}\ \mathbf{f}]\,[\mathbf{e}\ \mathbf{c}\ \mathbf{d}]$$
$$= [\mathbf{c}\ \mathbf{d}\ \mathbf{a}]\,[\ \mathbf{b}\ \mathbf{e}\ \mathbf{f}] - [\mathbf{c}\ \mathbf{d}\ \mathbf{b}]\,[\mathbf{a}\ \mathbf{e}\ \mathbf{f}]$$

(Awadh 98, Rohilkhand 2006)

Solution. $[\mathbf{a} \times \mathbf{b},\ \mathbf{c} \times \mathbf{d},\ \mathbf{e} \times \mathbf{f}]$
$$= (\mathbf{a} \times \mathbf{b}) \cdot [(\mathbf{c} \times \mathbf{d}) \times (\mathbf{e} \times \mathbf{f})]$$
$$= (\mathbf{a} \times \mathbf{b}) \cdot [\{(\mathbf{c} \times \mathbf{d}) \cdot \mathbf{f}\}\,\mathbf{e} - \{(\mathbf{c} \times \mathbf{d}) \cdot \mathbf{e}\}\,\mathbf{f}]$$
$$= (\mathbf{a} \times \mathbf{b}) \cdot \{[\mathbf{c}\ \mathbf{d}\ \mathbf{f}]\,\mathbf{e} - [\mathbf{c}\ \mathbf{d}\ \mathbf{e}]\,\mathbf{f}\}$$
$$= [\mathbf{a}\ \mathbf{b}\ \mathbf{e}]\,[\mathbf{c}\ \mathbf{d}\ \mathbf{f}] - [\mathbf{a}\ \mathbf{b}\ \mathbf{f}]\,[\mathbf{c}\ \mathbf{d}\ \mathbf{e}]$$
$$= [\mathbf{a}\ \mathbf{b}\ \mathbf{e}]\,[\mathbf{f}\ \mathbf{c}\ \mathbf{d}] - [\mathbf{a}\ \mathbf{b}\ \mathbf{f}]\,[\mathbf{e}\ \mathbf{c}\ \mathbf{d}]$$

Also,

$$\begin{aligned}[\mathbf{a}\times\mathbf{b},\ \mathbf{c}\times\mathbf{d},\ \mathbf{e}\times\mathbf{f}] &= (\mathbf{c}\times\mathbf{d})\,.\,[(\mathbf{e}\times\mathbf{f})\times(\mathbf{a}\times\mathbf{b})]\\ &= (\mathbf{c}\times\mathbf{d})\,.\,[\{(\mathbf{e}\times\mathbf{f})\,.\,\mathbf{b}\}\,\mathbf{a}-\{(\mathbf{e}\times\mathbf{f})\,.\,\mathbf{a}\}\,\mathbf{b}]\\ &= (\mathbf{c}\times\mathbf{d})\,.\,\{[\mathbf{e\ f\ b}]\,\mathbf{a}-[\mathbf{e\ f\ a}]\,\mathbf{b}\}\\ &= [\mathbf{c\ d\ a}]\,[\mathbf{e\ f\ b}]-[\mathbf{c\ d\ b}]\,[\mathbf{e\ f\ a}]\\ &= [\mathbf{c\ d\ a}]\,[\mathbf{b\ e\ f}]-[\mathbf{c\ d\ b}]\,[\mathbf{a\ e\ f}]\end{aligned}$$

Similarly,

$$\begin{aligned}[\mathbf{a}\times\mathbf{b},\ \mathbf{c}\times\mathbf{d},\ \mathbf{e}\times\mathbf{f}] &= (\mathbf{e}\times\mathbf{f})\,.\,[(\mathbf{a}\times\mathbf{b})\times(\mathbf{c}\times\mathbf{d})]\\ &= (\mathbf{e}\times\mathbf{f})\,.\,[\{\mathbf{a}\times\mathbf{b})\,.\,\mathbf{d}\}\,\mathbf{c}-\{(\mathbf{a}\times\mathbf{b})\,.\,\mathbf{c}\}\,\mathbf{d}]\\ &= (\mathbf{e}\times\mathbf{f})\,.\,\{[\mathbf{a\ b\ d}]\,\mathbf{c}-[\mathbf{a\ b\ c}]\,\mathbf{d}\}\\ &= [\mathbf{e\ f\ c}]\,[\mathbf{a\ b\ d}]-[\mathbf{e\ f\ d}]\,[\mathbf{a\ b\ c}]\\ &= [\mathbf{a\ b\ d}]\,[\mathbf{c\ e\ f}]-[\mathbf{a\ b\ c}]\,[\mathbf{d\ e\ f}]\end{aligned}$$

Example 3. *Prove that*

$$[\mathbf{a\ b\ c}]^2=\begin{vmatrix}\mathbf{a}\,.\,\mathbf{a} & \mathbf{a}\,.\,\mathbf{b} & \mathbf{a}\,.\,\mathbf{c}\\ \mathbf{b}\,.\,\mathbf{a} & \mathbf{b}\,.\,\mathbf{b} & \mathbf{b}\,.\,\mathbf{c}\\ \mathbf{c}\,.\,\mathbf{a} & \mathbf{c}\,.\,\mathbf{b} & \mathbf{c}\,.\,\mathbf{c}\end{vmatrix}$$

Solution. In § 5.2.1, Page 119 we proved that

$$(\mathbf{a}\times\mathbf{b})^2=\mathbf{a}^2\mathbf{b}^2-(\mathbf{a}\,.\,\mathbf{b})^2.$$

In the following we shall make use of this result.

We have

$$\begin{aligned}[(\mathbf{a}\times\mathbf{b})\times\mathbf{c}]^2 &= (\mathbf{a}\times\mathbf{b})^2\,\mathbf{c}^2-(\mathbf{a}\times\mathbf{b}\,.\,\mathbf{c})^2 \qquad (\S\ 5.2.1.,\ \text{P. } 119)\\ &= \{\mathbf{a}^2\mathbf{b}^2-(\mathbf{a}\,.\,\mathbf{b})^2\}\,\mathbf{c}^2-[\mathbf{a\ b\ c}]^2\\ &= \mathbf{a}^2\mathbf{b}^2\mathbf{c}^2-(\mathbf{a}\,.\,\mathbf{b})^2\,\mathbf{c}^2-[\mathbf{a\ b\ c}]^2.\end{aligned}$$

Also $[(\mathbf{a}\times\mathbf{b})\times\mathbf{c}]^2=[(\mathbf{a.c})\,\mathbf{b}-(\mathbf{b}\,.\,\mathbf{c})\,\mathbf{a}]^2$

$$=(\mathbf{a}\,.\,\mathbf{c})^2\,\mathbf{b}^2-2\,(\mathbf{a}\,.\,\mathbf{c})\,(\mathbf{b}\,.\,\mathbf{c})\,(\mathbf{a}\,.\,\mathbf{b})+(\mathbf{b}\,.\,\mathbf{c})^2\,\mathbf{a}^2$$

$$\begin{aligned}\therefore\quad &\mathbf{a}^2\mathbf{b}^2\mathbf{c}^2-(\mathbf{a}\,.\,\mathbf{b})^2\,\mathbf{c}^2-[\mathbf{a\ b\ c}]^2\\ &= (\mathbf{a}\,.\,\mathbf{c})^2\,\mathbf{b}^2-2\,(\mathbf{a}\,.\,\mathbf{c})\,(\mathbf{b}\,.\,\mathbf{c})\,(\mathbf{a}\,.\,\mathbf{b})+(\mathbf{b}\,.\,\mathbf{c})^2\,\mathbf{a}^2\end{aligned}$$

$$\begin{aligned}\Rightarrow\quad [\mathbf{a\ b\ c}]^2 &= \mathbf{a}^2\mathbf{b}^2\mathbf{c}^2-(\mathbf{a}\,.\,\mathbf{b})^2\,\mathbf{c}^2-(\mathbf{b}\,.\,\mathbf{c})^2\,\mathbf{a}^2-(\mathbf{c}\,.\,\mathbf{a})^2\,\mathbf{b}^2\\ &\qquad +2\,(\mathbf{a}\,.\,\mathbf{b})\,(\mathbf{b}\,.\,\mathbf{c})\,(\mathbf{c}\,.\,\mathbf{a})\end{aligned}$$

$$\begin{vmatrix}\mathbf{a}\,.\,\mathbf{a} & \mathbf{a}\,.\,\mathbf{b} & \mathbf{a}\,.\,\mathbf{c}\\ \mathbf{b}\,.\,\mathbf{a} & \mathbf{b}\,.\,\mathbf{b} & \mathbf{b}\,.\,\mathbf{c}\\ \mathbf{c}\,.\,\mathbf{a} & \mathbf{c}\,.\,\mathbf{b} & \mathbf{c}\,.\,\mathbf{c}\end{vmatrix}.$$

Example 4. *Decompose a vector* **r** *as a linear combination of a vector* **a** *and another vector perpendicular to* **a** *and coplanar with* **r** *and* **a.**

Solution. The vector $(\mathbf{a}\times\mathbf{r})\times\mathbf{a}$ is coplanar with **a** and **r** and perpendicular to *a*. Let

$$\mathbf{r} = x\mathbf{a} + y\,[(\mathbf{a} \times \mathbf{r}) \times \mathbf{a}]$$

Now we have

$$\mathbf{r} \,.\, \mathbf{a} = x\,\mathbf{a} \,.\, \mathbf{a} \quad \Rightarrow \quad x = (\mathbf{r} \,.\, \mathbf{a}) \,/\, (\mathbf{a} \,.\, \mathbf{a}),$$

$$\Rightarrow \quad \mathbf{r} \times \mathbf{a} = y\,[(\mathbf{a} \times \mathbf{r}) \times \mathbf{a}] \times \mathbf{a}$$

$$= y\,[(\mathbf{a} \,.\, \mathbf{a})\,\mathbf{r} - (\mathbf{a} \,.\, \mathbf{r})\,\mathbf{a}] \times \mathbf{a}$$

$$= y\,(\mathbf{a} \,.\, \mathbf{a})\,(\mathbf{r} \times \mathbf{a}).$$

$$\Rightarrow \quad y = 1/(\mathbf{a} \,.\, \mathbf{a}).$$

Thus, $$\mathbf{r} = \frac{\mathbf{r} \,.\, \mathbf{a}}{\mathbf{a} \,.\, \mathbf{a}}\,\mathbf{a} + \frac{1}{\mathbf{a} \,.\, \mathbf{a}}(\mathbf{a} \times \mathbf{r}) \times \mathbf{a}.$$

Example 5. *Prove that*

$$2\,(\mathbf{a} \times \mathbf{b}) \times (\mathbf{c} \times \mathbf{d}) = \begin{vmatrix} -\mathbf{a} & -\mathbf{b} & \mathbf{c} & \mathbf{d} \\ a_1 & b_1 & c_1 & d_1 \\ a_2 & b_2 & c_2 & d_2 \\ a_3 & b_3 & c_3 & d_3 \end{vmatrix}.$$

where $\mathbf{a} = a_1\mathbf{i} + a_2\mathbf{j} + a_3\mathbf{k}$ *etc.*

Solution. We have

$$(\mathbf{a} \times \mathbf{b}) \times (\mathbf{c} \times \mathbf{d}) = [\mathbf{a}\ \mathbf{b}\ \mathbf{d}]\,\mathbf{c} - [\mathbf{a}\ \mathbf{b}\ \mathbf{c}]\,\mathbf{d}$$

$$= \begin{vmatrix} a_1 & b_1 & d_1 \\ a_2 & b_2 & d_2 \\ a_3 & b_3 & d_3 \end{vmatrix}\mathbf{c} - \begin{vmatrix} a_1 & b_1 & c_1 \\ a_2 & b_2 & c_2 \\ a_3 & b_3 & c_3 \end{vmatrix}\mathbf{d}.$$

Also,

$$(\mathbf{a} \times \mathbf{b}) \times (\mathbf{c} \times \mathbf{d}) = [\mathbf{a}\ \mathbf{c}\ \mathbf{d}]\,\mathbf{b} - [\mathbf{b}\ \mathbf{c}\ \mathbf{d}]\,\mathbf{a}$$

$$= \begin{vmatrix} a_1 & c_1 & d_1 \\ a_2 & c_2 & d_2 \\ a_3 & c_3 & d_3 \end{vmatrix}\mathbf{b} - \begin{vmatrix} b_1 & c_1 & d_1 \\ b_2 & c_2 & d_2 \\ b_3 & c_3 & d_3 \end{vmatrix}\mathbf{a}.$$

Adding, we get the required result.

EXERCISES

1. Prove that

   (*i*) $(\mathbf{a} \times \mathbf{b}) \times (\mathbf{c} \times \mathbf{d}) + (\mathbf{a} \times \mathbf{c}) \times (\mathbf{d} \times \mathbf{b}) + (\mathbf{a} \times \mathbf{d}) \times (\mathbf{b} \times \mathbf{c}) = -2\,[\mathbf{b}\ \mathbf{c}\ \mathbf{d}]\,\mathbf{a}$ (*Garhwal 99*)

   (*ii*) $(\mathbf{b} \times \mathbf{c}) \,.\, (\mathbf{a} \times \mathbf{d}) + (\mathbf{c} \times \mathbf{a}) \,.\, (\mathbf{b} \times \mathbf{d}) + (\mathbf{a} \times \mathbf{b}) \,.\, (\mathbf{c} \times \mathbf{d}) = 0.$ (*Patna 2003*)

2. If **a** and **b** lie in a plane normal to the plane containing **c** and **d**, then show that $(\mathbf{a} \times \mathbf{b}) \,.\, (\mathbf{c} \times \mathbf{d}) = 0.$

3. If the four vectors **a, b, c, d** are coplanar, then show that
$$(\mathbf{a} \times \mathbf{b}) \times (\mathbf{c} \times \mathbf{d}) = 0.$$

4. If **a, b, c** and **a′, b′, c′** are reciprocal system of vectors, prove that

(*i*) $\mathbf{a} \cdot \mathbf{a}' + \mathbf{b} \cdot \mathbf{b}' + \mathbf{c} \cdot \mathbf{c}' = 3$.

(*ii*) $\mathbf{a}' \times \mathbf{b}' + \mathbf{b}' \times \mathbf{c}' + \mathbf{c}' \times \mathbf{a}' = \dfrac{\mathbf{a}+\mathbf{b}+\mathbf{c}}{[\mathbf{a}\ \mathbf{b}\ \mathbf{c}]}$, $[\mathbf{a}\ \mathbf{b}\ \mathbf{c}] \neq 0$.

(Kumaon 98, 2000, Avadh 99)

5. Prove that the set of vectors reciprocal to the set of vectors $2\mathbf{i} + 3\mathbf{j} - \mathbf{k}$, $\mathbf{i} - \mathbf{j} - \mathbf{k}$, $-\mathbf{i} + 2\mathbf{j} + 2\mathbf{k}$ is *(Rohilkhand 96, 2003)*

$$\frac{1}{3}(2\mathbf{i}+\mathbf{k}),\ \frac{1}{3}(-8\mathbf{i}+7\mathbf{j}-7\mathbf{k}),\ \frac{1}{3}(-7\mathbf{i}+3\mathbf{j}-5\mathbf{k}).$$

7.7. SOLUTION OF VECTOR EQUATIONS

7.7.1. *To solve for* **r**

$$\mathbf{r} \times \mathbf{b} = \mathbf{a} \times \mathbf{b}, \qquad ...(i)$$

where **a, b, c** *are two given vectors.*

Let **r** be any solution of the equation.

Rewriting the given equation as

$$(\mathbf{r} - \mathbf{a}) \times \mathbf{b} = 0.$$

we see that $(\mathbf{r} - \mathbf{a})$ is parallel to **b**, so that we must have

$$\mathbf{r} - \mathbf{a} = t\mathbf{b} \Rightarrow \mathbf{r} = \mathbf{a} + t\mathbf{b}, \qquad ...(ii)$$

where t is a scalar. Also it may be seen that $\mathbf{a} + t\mathbf{b}$ satisfies (*i*) for every value of the scalar t. Thus, (*ii*) gives the general solution of the given equation.

Geometrically, we know that the points whose position vectors are given by (*ii*) for different values of the scalar t lie on a straight line which passes through the point, with position vector **a**, and is parallel to the vector **b**. (Refer § 2.4, Ch. 2, P. 32)

7.7.2. *To solve for* **r**

$$\mathbf{r} \times \mathbf{b} = \mathbf{a},$$

where **a, b** *are two given vectors such that* **a** *is perpendicular to* **b**.

$$\mathbf{a},\quad \mathbf{b},\quad \mathbf{a} \times \mathbf{b},$$

so that every vector is expressible as a linear combination of the same.

Suppose that a solution of a given equation is

$$\mathbf{r} = x\mathbf{a} + y\mathbf{b} + z\mathbf{a} \times \mathbf{b}.$$

Substituting in a given equation, we obtain

$$(x\mathbf{a} + y\mathbf{b} + z\mathbf{a} \times \mathbf{b}) \times \mathbf{b} = \mathbf{a}$$

$$\Rightarrow \quad x\mathbf{a} \times \mathbf{b} + z\{(\mathbf{a} \cdot \mathbf{b})\mathbf{b} - (\mathbf{b} \cdot \mathbf{b})\mathbf{a}\} = \mathbf{a}$$

$$\Rightarrow \quad -\{1 + z(\mathbf{b} \cdot \mathbf{b})\}\mathbf{a} + x\mathbf{a} \times \mathbf{b} = 0, \text{ for } \mathbf{a} \cdot \mathbf{b} = 0.$$

$$\Rightarrow \quad 1 + z(\mathbf{b} \cdot \mathbf{b}) = 0, \quad x = 0.$$

$\mathbf{a}$, $\mathbf{a} \times \mathbf{b}$, being non-collinear vectors. Thus, we have

$$\mathbf{r} = y\mathbf{b} - \frac{1}{\mathbf{b} \cdot \mathbf{b}}\mathbf{a} \times \mathbf{b}. \qquad ...(i)$$

Substituting (*i*) in the given equation, we may verify that this is a solution for every value of the scalar y. It follows that

$$\mathbf{r} = -\frac{1}{\mathbf{b}\,.\,\mathbf{b}}\,\mathbf{a} \times \mathbf{b} + y\mathbf{b},$$

is the *general* solution of the equation; y being the parameter.

Geometrically speaking, the points whose position vectors satisfy the given equation lie on the straight line which passes through the point with position vector $-\frac{1}{\mathbf{b}\,.\,\mathbf{b}}(\mathbf{a} \times \mathbf{b})$ and is parallel to the vector **b**.

7.7.3. *To solve simultaneously for* **r**

$$\mathbf{r} \times \mathbf{b} = \mathbf{c} \times \mathbf{b} \qquad ...(i)$$

$$\mathbf{r}\,.\,\mathbf{a} = 0 \qquad ...(ii)$$

provided that **a** *is not perpendicular to* **b**.

Suppose that **r** is a solution of the given equations. Rewriting (i) as

$$(\mathbf{r} - \mathbf{c}) \times \mathbf{b} = 0,$$

we see that $\mathbf{r} - \mathbf{c}$ and **b** are collinear so that

$$\mathbf{r} - \mathbf{c} = t\mathbf{b} \quad \Leftrightarrow \quad \mathbf{r} = \mathbf{c} + t\mathbf{b},$$

where t is a scalar. Substituting in (ii), we obtain

$$(\mathbf{c} + t\mathbf{b})\,.\,\mathbf{a} = 0$$

$$\Leftrightarrow \qquad \mathbf{c}.\mathbf{a} + t\,\mathbf{b}\,.\,\mathbf{a} = 0 \quad \Rightarrow \quad t = -\frac{\mathbf{c}\,.\,\mathbf{a}}{\mathbf{b}\,.\,\mathbf{a}}, \text{ for } \mathbf{b}\,.\,\mathbf{a} \neq 0.$$

Thus, $$\mathbf{r} = \mathbf{c} - \left(\frac{\mathbf{c}\,.\,\mathbf{a}}{\mathbf{b}\,.\,\mathbf{a}}\right)\mathbf{b}. \qquad ...(iii)$$

We may also easily verify that, **r**, given by (iii) does satisfy the given equations.

Hence,

$$\mathbf{r} = \mathbf{c} - \frac{\mathbf{c}.\mathbf{a}}{\mathbf{b}.\mathbf{a}}\,\mathbf{b},$$

is a solution and the only solution of the given equations.

It may be seen that $\mathbf{r}\,.\,\mathbf{a} = 0$ represents the plane which passes through the origin and is normal to the vector **a**.

Thus, the solution, in question, represents the point of intersection of a line and a plane.

7.7.4. *To solve for* **r**

$$k\mathbf{r} + \mathbf{r} \times \mathbf{a} = \mathbf{b}, \qquad ...(i)$$

where, k is a given non-zero scalar and **a, b** *are two given vectors.*

Suppose that **r** is a solution. Expressing **r** as a linear combination of the non-coplanar vectors **a, b, a** × **b,** we write

$$\mathbf{r} = x\mathbf{a} + y\mathbf{b} + z\mathbf{a} \times \mathbf{b}.$$

Substituting in the given equation, we obtain

$$k\,(x\mathbf{a} + y\mathbf{b} + z\,\mathbf{a} \times \mathbf{b}) + y\,\mathbf{b} \times \mathbf{a} + z\,\{(\mathbf{a}\,.\,\mathbf{a})\,\mathbf{b} - (\mathbf{a}\,.\,\mathbf{b})\,\mathbf{a}\} = \mathbf{b}$$

$$\Rightarrow \{kx - z\,(\mathbf{a}\,.\,\mathbf{b})\}\,\mathbf{a} + (ky + z\mathbf{a}^2 - 1)\,\mathbf{b} + (kz - y)\,\mathbf{a} \times \mathbf{b} = 0$$

$$\Rightarrow kx - z\,(\mathbf{a}\,.\,\mathbf{b}) = 0, \quad ky + z\mathbf{a}^2 - 1 = 0, \quad kz - y = 0;$$

a, **b**, $\mathbf{a} \times \mathbf{b}$ being non-coplanar vectors. It follows that

$$x = \frac{\mathbf{a}\,.\,\mathbf{b}}{k\,(k^2 + \mathbf{a}^2)}, \quad y = \frac{k}{k^2 + \mathbf{a}^2}, \quad z = \frac{1}{k^2 + \mathbf{a}^2}.$$

It follows that

$$\mathbf{r} = \frac{1}{k^2 + \mathbf{a}^2}\left[\frac{\mathbf{a}\,.\,\mathbf{b}}{k}\,\mathbf{a} + k\mathbf{b} + \mathbf{a} \times \mathbf{b}\right]. \qquad ...(ii)$$

Also we may easily verify that **r** given by (*ii*) satisfies the given equation. Hence, this is a solution and the only solution.

EXERCISES

1. Solve simultaneously $\mathbf{r}\,.\,\mathbf{n}_1 = 1, \quad \mathbf{r}\,.\,\mathbf{n}_2 = 1$.
2. Find the condition for the equations

$$\mathbf{r} \times \mathbf{a} = \mathbf{b}, \quad \mathbf{r} \times \mathbf{c} = \mathbf{d}$$

to be consistent. Assuming the condition for consistency to be satisfied, solve the equations.

OBJECTIVE QUESTIONS

For each of the following questions, four alternatives are given for the answer. Only one of them is correct. Choose the correct alternative.

1. If $\mathbf{a} = 2\mathbf{i} + 3\mathbf{j} - \mathbf{k}$, $\mathbf{b} = -\mathbf{i} + 2\mathbf{j} - 4\mathbf{k}$ and $\mathbf{c} = \mathbf{i} + \mathbf{j} + \mathbf{k}$, then $(\mathbf{a} \times \mathbf{b})\,.\,(\mathbf{c} \times \mathbf{d})$ is equal to

   (*a*) 60 (*b*) 64 (*c*) 74 (*d*) – 74
2. The product $(a\mathbf{i} + b\mathbf{j}) \times (\mathbf{i} - \mathbf{j} + \mathbf{k})\,.\,(\mathbf{i} + \mathbf{j} - \mathbf{k}) \times (-\mathbf{i} + \mathbf{j} + \mathbf{k})$ is

   (*a*) 0 (*b*) $a + b$ (*c*) $b - a$ (*d*) $-2a$
3. If $\mathbf{p} = \dfrac{\mathbf{a} \times \mathbf{b}}{[\mathbf{a}\,\mathbf{b}\,\mathbf{c}]}$, $\mathbf{q} = \dfrac{\mathbf{c} \times \mathbf{a}}{[\mathbf{a}\,\mathbf{b}\,\mathbf{c}]}$, $\mathbf{r} = \dfrac{\mathbf{a} \times \mathbf{b}}{[\mathbf{a}\,\mathbf{b}\,\mathbf{c}]}$, where **a b c** are three non-coplanar vectors then the value of the expression $(\mathbf{a} + \mathbf{b} + \mathbf{c})\,.\,(\mathbf{p} + \mathbf{q} + \mathbf{r})$ is

   (*a*) 3 (*b*) 2 (*c*) 1 (*d*) 0
4. In Q. 3, value of $(\mathbf{a} + \mathbf{b})\,.\,\mathbf{p} + (\mathbf{b} + \mathbf{c})\,.\,\mathbf{q} + (\mathbf{c} + \mathbf{a})\,.\,\mathbf{r}$ is

   (*a*) 0 (*b*) 1 (*c*) 2 (*d*) 3
5. If **a, b, c** be three non-coplanar vectors for which $[\mathbf{a}\,\mathbf{b}\,\mathbf{c}] \neq$ and $\mathbf{a}'$, $\mathbf{b}'$, $\mathbf{c}'$ constitute the reciprocal system of vectors, then any vector

r can be expressed as

(*a*) $\mathbf{r} = (\mathbf{r} \times \mathbf{a}) . \mathbf{a}' + (\mathbf{r} \times \mathbf{b}) . \mathbf{b}' + (\mathbf{r} \times \mathbf{c}) . \mathbf{c}'$

(*b*) $\mathbf{r} = (\mathbf{r} \times \mathbf{a}') . \mathbf{a} + (\mathbf{r} \times \mathbf{b}') . \mathbf{b} + (\mathbf{r} \times \mathbf{c}') . \mathbf{c}$

(*c*) $\mathbf{r} = (\mathbf{r} . \mathbf{a})\, \mathbf{a}' + (\mathbf{r} . \mathbf{b})\, \mathbf{b}' + (\mathbf{r} . \mathbf{c})\, \mathbf{c}'$

(*d*) $\mathbf{r} = (\mathbf{r} . \mathbf{a}')\, \mathbf{a} + (\mathbf{r} . \mathbf{b}')\, \mathbf{b} + (\mathbf{r} . \mathbf{c}')\, \mathbf{c}$

6. Value of $\begin{vmatrix} \mathbf{a}.\mathbf{a} & \mathbf{a}.\mathbf{b} & \mathbf{a}.\mathbf{c} \\ \mathbf{b}.\mathbf{a} & \mathbf{b}.\mathbf{b} & \mathbf{b}.\mathbf{c} \\ \mathbf{c}.\mathbf{a} & \mathbf{c}.\mathbf{b} & \mathbf{c}.\mathbf{c} \end{vmatrix}$ is

(*a*) $[\mathbf{a}\ \mathbf{b}\ \mathbf{c}]$ (*b*) $[\mathbf{a}\ \mathbf{b}\ \mathbf{c}]^2$

(*c*) $[\mathbf{a}\ \mathbf{b}\ \mathbf{c}]^3$ (*d*) $3\,[\mathbf{a}\ \mathbf{b}\ \mathbf{c}]$

7. If four vectors **a, b, c, d** are coplanar, then

(*a*) $(\mathbf{a} \times \mathbf{b}) . (\mathbf{c} \times \mathbf{d}) = 0$ (*b*) $(\mathbf{a} \times \mathbf{c}) . (\mathbf{b} \times \mathbf{d}) = 0$

(*c*) $(\mathbf{a} \times \mathbf{b}) \times (\mathbf{c} \times \mathbf{d}) = 0$ (*d*) $(\mathbf{a} \times \mathbf{c}) \times (\mathbf{b} \times \mathbf{d}) = 0$

8. The value of $(\mathbf{a} \times \mathbf{b}) \times (\mathbf{b} \times \mathbf{c})$ is

(*a*) $\mathbf{a}\,[\mathbf{a}\ \mathbf{b}\ \mathbf{c}]$ (*b*) $\mathbf{b}\,[\mathbf{a}\ \mathbf{b}\ \mathbf{c}]$

(*c*) $\mathbf{c}\,[\mathbf{a}\ \mathbf{b}\ \mathbf{c}]$ (*d*) **0**

9. The value of

$(\mathbf{a} \times \mathbf{b}) . (\mathbf{c} \times \mathbf{d}) + (\mathbf{b} \times \mathbf{c}) . (\mathbf{a} \times \mathbf{d}) + (\mathbf{c} \times \mathbf{a}) . (\mathbf{b} \times \mathbf{d})$ is

(*a*) 0 (*b*) 3 (*c*) 2 (*d*) – 3

10. Value of $(\mathbf{a} \times \mathbf{b}) . (\mathbf{b} \times \mathbf{c}) + (\mathbf{c} \times \mathbf{a})$ is

(*a*) $[\mathbf{a}\ \mathbf{b}\ \mathbf{c}]$ (*b*) $[\mathbf{a}\ \mathbf{b}\ \mathbf{c}]^2$

(*c*) $[\mathbf{a}\ \mathbf{b}\ \mathbf{c}]^3$ (*d*) None of these

ANSWERS

1. (*d*)	**2.** (*d*)	**3.** (*a*)	**4.** (*d*)	**5.** (*d*)
6. (*b*)	**7.** (*c*)	**8.** (*b*)	**9.** (*a*)	**10.** (*b*)

APPENDIX II

Some Properties of Tetrahedra

1. Centre and radius of the circumsphere of a tetrahedron. *If OABC is a tetrahedron and*

$$\vec{OA} = \mathbf{a},\ \vec{OB} = \mathbf{b},\ \vec{OC} = \mathbf{c},$$

then the radius ρ *of the circumsphere of the tetrahedron is given by*

$$\begin{vmatrix} \frac{1}{2}\mathbf{a}^2 & \mathbf{a}\,.\,\mathbf{a} & \mathbf{a}\,.\,\mathbf{b} & \mathbf{a}\,.\,\mathbf{c} \\ \frac{1}{2}\mathbf{b}^2 & \mathbf{b}\,.\,\mathbf{a} & \mathbf{b}\,.\,\mathbf{b} & \mathbf{b}\,.\,\mathbf{c} \\ \frac{1}{2}\mathbf{c}^2 & \mathbf{c}\,.\,\mathbf{a} & \mathbf{c}\,.\,\mathbf{b} & \mathbf{c}\,.\,\mathbf{c} \\ \rho^2 & \frac{1}{2}\mathbf{a}^2 & \frac{1}{2}\mathbf{b}^2 & \frac{1}{2}\mathbf{c}^2 \end{vmatrix} = 0 \qquad ...(I)$$

and the position vector **r** *of the circumcentre is given by*

$$\begin{vmatrix} \mathbf{r} & \mathbf{a} & \mathbf{b} & \mathbf{c} \\ \frac{1}{2}\mathbf{a}^2 & \mathbf{a}\,.\,\mathbf{a} & \mathbf{a}\,.\,\mathbf{b} & \mathbf{a}\,.\,\mathbf{c} \\ \frac{1}{2}\mathbf{b}^2 & \mathbf{b}\,.\,\mathbf{a} & \mathbf{b}\,.\,\mathbf{b} & \mathbf{b}\,.\,\mathbf{c} \\ \frac{1}{2}\mathbf{c}^2 & \mathbf{c}\,.\,\mathbf{a} & \mathbf{c}\,.\,\mathbf{b} & \mathbf{c}\,.\,\mathbf{c} \end{vmatrix} = 0. \qquad ...(II)$$

We have

$$\rho^2 = \mathbf{r}^2 = (\mathbf{r} - \mathbf{a})^2 = (\mathbf{r} - \mathbf{b})^2 = (\mathbf{r} - \mathbf{c})^2 \qquad ...(i)$$

These give

$$\mathbf{a}^2 = 2\mathbf{r}\,.\,\mathbf{a},\ \ \mathbf{b}^2 = 2\mathbf{r}\,.\,\mathbf{b},\ \ \mathbf{c}^2 = 2\mathbf{r}\,.\,\mathbf{c}. \qquad ..(ii)$$

Let the expression for **r** in terms of **a, b, c** be

$$\mathbf{r} = x\mathbf{a} + y\mathbf{b} + z\mathbf{c}. \qquad ...(iii)$$

Multiplying scalarly with **a, b** and **c** successively and using the relations (*ii*), we get

$$\frac{1}{2}\mathbf{a}^2 = x\mathbf{a}\,.\,\mathbf{a} + y\mathbf{a}\,.\,\mathbf{b} + z\mathbf{a}\,.\,\mathbf{c}, \qquad ...(iv)$$

$$\frac{1}{2}\mathbf{b}^2 = x\mathbf{b}\,.\,\mathbf{a} + y\mathbf{b}\,.\,\mathbf{b} + z\mathbf{b}\,.\,\mathbf{c}, \qquad ...(v)$$

$$\frac{1}{2}\mathbf{c}^2 = x\mathbf{c}\,.\,\mathbf{a} + y\mathbf{c}\,.\,\mathbf{b} + z\mathbf{c}\,.\,\mathbf{c}. \qquad ...(vi)$$

Again multiplying (*iii*) scalarly with $\mathbf{r}$ and using the relations (*ii*) and the relation $\mathbf{r}^2 = \rho^2$, we get

$$\rho^2 = \frac{1}{2}x\mathbf{a}^2 + \frac{1}{2}y\mathbf{b}^2 + \frac{1}{2}z\mathbf{c}^2. \qquad ...(vii)$$

Eliminating x, y, z from (*iv*), (*v*), (*vi*) and (*vii*), we get I and elimi-nating x, y, z from (*iii*), (*iv*), (*v*) and (*vi*), we get II.

2. *If a tetrahedron OABC is such that the mid-points of its six edges lie on a sphere of radius, r, then*

(I) *the opposite edges of the tetrahedron are perpendicular.*

(II) *centre of the sphere is the centroid of the tetrahedron.*

(III) $GA^2 + GB^2 + GC^2 + GD^2 = 12r^2$; *G, being the centroid.*

(IV) $AB^2 + CD^2 = BC^2 + AD^2 = CA^2 + BD^2 = 16r^2$.

Take the centre O of the sphere through the six mid-points of the edges as the origin of reference. Let

$$\vec{OA} = \mathbf{a},\ \vec{OB} = \mathbf{b},\ \vec{OC} = \mathbf{c},\ \vec{OD} = \mathbf{d},$$

The sphere passes through the points with position vectors

$$\frac{1}{2}(\mathbf{a}+\mathbf{b}),\ \frac{1}{2}(\mathbf{c}+\mathbf{d}),\ \frac{1}{2}(\mathbf{b}+\mathbf{c}),\ \frac{1}{2}(\mathbf{a}+\mathbf{d}),\ \frac{1}{2}(\mathbf{c}+\mathbf{a}),\ \frac{1}{2}(\mathbf{b}+\mathbf{d}).$$

Thus, we have

$$4r^2 = (\mathbf{a} + \mathbf{b})^2 = (\mathbf{c} + \mathbf{d})^2 = (\mathbf{b} + \mathbf{c})^2$$
$$= (\mathbf{a} + \mathbf{d})^2 = (\mathbf{c} + \mathbf{a})^2 = (\mathbf{b} + \mathbf{d})^2 \qquad ...(i)$$

From (*i*), we have

$$(\mathbf{a} + \mathbf{b})^2 + (\mathbf{c} + \mathbf{d})^2 = (\mathbf{b} + \mathbf{c})^2 + (\mathbf{a} + \mathbf{d})^2$$

$$\Rightarrow \quad \mathbf{a}\,.\,\mathbf{b} + \mathbf{c}\,.\,\mathbf{d} = \mathbf{b}\,.\,\mathbf{c} + \mathbf{a}\,.\,\mathbf{d}$$

$$\Rightarrow \quad (\mathbf{a} - \mathbf{c})\,.\,(\mathbf{b} - \mathbf{d}) = 0 \ \Rightarrow \ AC \perp BD.$$

Similarly, $BC \perp AD$, $CA \perp AD$. Hence, the result (I).

Again, from (*i*),

$$(\mathbf{a} + \mathbf{b})^2 = (\mathbf{c} + \mathbf{d})^2$$

$$\Rightarrow \quad (\mathbf{a} + \mathbf{b} + \mathbf{c} + \mathbf{d})\,.\,[(\mathbf{a} + \mathbf{b}) - (\mathbf{c} + \mathbf{d})] = 0$$

Similarly,

$$(\mathbf{c} + \mathbf{a})^2 = (\mathbf{b} + \mathbf{d})^2 \ \Rightarrow \ (\mathbf{a} + \mathbf{b} + \mathbf{c} + \mathbf{d})\,.\,[(\mathbf{c} + \mathbf{a}) - (\mathbf{b} + \mathbf{d})] = 0.$$

$$(\mathbf{b} + \mathbf{c})^2 = (\mathbf{a} + \mathbf{d})^2 \ \Rightarrow \ (\mathbf{a} + \mathbf{b} + \mathbf{c} + \mathbf{d})\,.\,[(\mathbf{b} + \mathbf{c}) - (\mathbf{a} + \mathbf{d})] = 0.$$

The vectors

$$(\mathbf{a} + \mathbf{d}) - (\mathbf{c} + \mathbf{d}),\ (\mathbf{b} + \mathbf{c}) - (\mathbf{a} + \mathbf{d}),\ (\mathbf{c} + \mathbf{a}) - (\mathbf{b} + \mathbf{d})$$

cannot be coplanar, lying as they do along the lines joining the mid-points of the pairs of opposite edges of the tetrahedron. Therefore

$$\mathbf{a} + \mathbf{b} + \mathbf{c} + \mathbf{d} = 0$$

so that the centre of the circumsphere coincides with the centroid of the tetrahedron. Hence, the result (II).

From (*i*), by addition

$$24r^2 = 3\ \Sigma \mathbf{a}^2 + 2\ \Sigma \mathbf{a} \cdot \mathbf{b}.$$

Also

$$\Sigma \mathbf{a} = 0 \ \Rightarrow\ (\Sigma \mathbf{a})^2 = 0 \ \Rightarrow\ \Sigma \mathbf{a}^2 = -2\ \Sigma \mathbf{a} \cdot \mathbf{b}.$$

Thus, we have

$$24r^2 = 3\ \Sigma \mathbf{a}^2 - \Sigma \mathbf{a}^2 = 2\ \Sigma \mathbf{a}^2 \ \Rightarrow\ \Sigma \mathbf{a}^2 = 12r^2.$$

$$\Rightarrow \qquad \Sigma GA^2 = 12r^2.$$

Hence, the result (III).

Again

$$(\mathbf{a} + \mathbf{b})^2 + (\mathbf{c} + \mathbf{d})^2 = 8r^2$$

$$\Rightarrow \qquad 2\mathbf{a} \cdot \mathbf{b} + 2\mathbf{c} \cdot \mathbf{d} = 8r^2 - \Sigma \mathbf{a}^2 = -4r^2.$$

Thus, we have

$$(\mathbf{a} - \mathbf{b})^2 + (\mathbf{c} - \mathbf{d})^2 = \Sigma \mathbf{a}^2 - (2\mathbf{a} \cdot \mathbf{b} + 2\mathbf{c} \cdot \mathbf{d})$$

$$= 12r^2 + 4r^2 = 16r^2$$

$$\Rightarrow \qquad AB^2 + CD^2 = 16r^2.$$

Hence, the result (IV).

3. *The six mid-points of the six edges of a tetrahedron lie on a sphere, if the pairs of opposite edges of the tetrahedron are perpendicular to each other.*

Take the origin of reference at the centroid of the tetrahedron. Denoting the position vectors of the vertices A, B, C, D by **a, b, c, d,** we are given that

$$\mathbf{a} + \mathbf{b} + \mathbf{c} + \mathbf{d} = 0 \qquad ...(i)$$

$$\mathbf{a} \cdot \mathbf{b} + \mathbf{c} \cdot \mathbf{d} = \mathbf{a} \cdot \mathbf{c} + \mathbf{b} \cdot \mathbf{d} = \mathbf{b} \cdot \mathbf{c} - \mathbf{a} \cdot \mathbf{d} \qquad ...(ii)$$

From (*i*)

$$\mathbf{a} + \mathbf{b} = -(\mathbf{c} + \mathbf{d}) \ \Rightarrow\ (\mathbf{a} + \mathbf{b})^2 = (\mathbf{c} + \mathbf{d})^2.$$

Similarly

$$(\mathbf{a} + \mathbf{c})^2 = (\mathbf{b} + \mathbf{d})^2,\ (\mathbf{b} + \mathbf{c})^2 = (\mathbf{a} + \mathbf{d})^2. \qquad ...(iii)$$

Again we have

$$(\mathbf{a} + \mathbf{b})^2 + (\mathbf{c} + \mathbf{d})^2 = \Sigma \mathbf{a}^2 + 2\ (\mathbf{a} \cdot \mathbf{b} + \mathbf{c} \cdot \mathbf{d})$$

$$= \Sigma \mathbf{a}^2 + 2\ (\mathbf{a} \cdot \mathbf{c} + \mathbf{b} \cdot \mathbf{d}); \quad \text{By } (ii)$$

$$= (\mathbf{a} + \mathbf{c})^2 + (\mathbf{b} + \mathbf{d})^2.$$

Similarly

$$(\mathbf{a} + \mathbf{b})^2 = (\mathbf{c} + \mathbf{d})^2 = (\mathbf{b} + \mathbf{c})^2 + (\mathbf{a} + \mathbf{d})^2. \qquad ...(iv)$$

From (*iii*) and (*iv*), it follows that

$$(\mathbf{a} + \mathbf{b})^2 = (\mathbf{c} + \mathbf{d})^2 = (\mathbf{b} + \mathbf{c})^2 = (\mathbf{a} + \mathbf{d})^2 = (\mathbf{c} + \mathbf{a})^2 + (\mathbf{b} + \mathbf{d})^2.$$

Thus, the middle points lie on a sphere with its centre at the centroid which has been taken as the origin.

This result is the converse of the result (I) of 2.

4. Isosceles Tetrahedra. Def. *A tetrahedron, the pairs of whose opposite edges are equal in length is called an isosceles tetrahedron.*

Theorem. *The shortest distances between pairs of opposite edges of an isosceles tetrahedron lie along the joins of their middle points and the three shortest distances bisect each other at right angles.*

Take the vertex O of the isosceles tetrahedron $OABC$ as the origin of reference. Let

$$\overrightarrow{OA} = \mathbf{a}, \ \overrightarrow{OB} = \mathbf{b}, \ \overrightarrow{OC} = \mathbf{c}.$$

Now

$$\left.\begin{aligned} OA = BC &\Rightarrow \mathbf{a}^2 = \mathbf{b}^2 + \mathbf{c}^2 - 2\mathbf{b}\,.\,\mathbf{c} \Rightarrow \mathbf{b}\,.\,\mathbf{c} = \frac{1}{2}(\mathbf{b}^2 + \mathbf{c}^2 - \mathbf{a}^2) \\ OB = CA &\Rightarrow \mathbf{b}^2 = \mathbf{c}^2 + \mathbf{a}^2 - 2\mathbf{c}\,.\,\mathbf{a} \Rightarrow \mathbf{c}\,.\,\mathbf{a} = \frac{1}{2}(\mathbf{c}^2 + \mathbf{a}^2 - \mathbf{b}^2) \\ OC = AB &\Rightarrow \mathbf{c}^2 = \mathbf{a}^2 + \mathbf{b}^2 - 2\mathbf{a}\,.\,\mathbf{b} \Rightarrow \mathbf{a}\,.\,\mathbf{b} = \frac{1}{2}(\mathbf{a}^2 + \mathbf{b}^2 - \mathbf{c}^2) \end{aligned}\right\} \quad ...(i)$$

The vectors

$$\frac{1}{2}(\mathbf{b}+\mathbf{c}) - \frac{1}{2}\mathbf{a}, \ \frac{1}{2}(\mathbf{c}+\mathbf{a}) - \frac{1}{2}\mathbf{b}, \ \frac{1}{2}(\mathbf{a}+\mathbf{b}) - \frac{1}{2}\mathbf{c}$$

lie along the joins of the mid-points of the pairs of opposite edges

$$OA, BC; \quad OB, CA; \quad OC, AB,$$

respectively.

The join of the mid-points of OA and BC will be perpendicular to OA and BC, if

$$\left[\frac{1}{2}(\mathbf{b}+\mathbf{c}) - \frac{1}{2}\mathbf{a}\right]\cdot\mathbf{a} = 0 \text{ and } \left[\frac{1}{2}(\mathbf{b}+\mathbf{c}) - \frac{1}{2}\mathbf{a}\right]\cdot(\mathbf{b}-\mathbf{c}) = 0$$

$$\Rightarrow \quad \mathbf{a}^2 = \mathbf{a}\,.\,\mathbf{b} + \mathbf{a}\,.\,\mathbf{c};\ \mathbf{b}^2 - \mathbf{c}^2 = \mathbf{a}\,.\,\mathbf{b} - \mathbf{a}\,.\,\mathbf{c}. \quad ...(ii)$$

It can be at once seen that the relations (*ii*) are simple consequences of the relations (*i*). Thus, the shortest distance between OA and OB lies along the join of their middle points.

Similarly we may show that the shortest distances between the other pairs of opposite edges lie along the joins of their mid-points.

The three shortest distance lines lie along the vectors

$$\mathbf{b} + \mathbf{c} - \mathbf{a}, \ \mathbf{c} + \mathbf{a} - \mathbf{b}, \ \mathbf{a} + \mathbf{b} - \mathbf{c}$$

which join the mid-points of the opposite edges.

Now we have

$$(\mathbf{b}+\mathbf{c}-\mathbf{a}) \cdot (\mathbf{c}+\mathbf{a}-\mathbf{b}) = \mathbf{c}^2 - (\mathbf{b}-\mathbf{a})^2$$
$$= \mathbf{c}^2 - \mathbf{b}^2 - \mathbf{a}^2 + 2\mathbf{a} \cdot \mathbf{b} = 0, \quad \text{By } (i)$$

Thus, we see that the lines of shortest distance are at right angles to each other.

Surely they also all concur at the centroid.

5. In-sphere of a tetrahedron. In-centre and In-radius. It will be shown that there exists a point I inside any given tetrahedron equidistant from the four faces of the same. The sphere with centre I and radius equal to its distance from any face will touch the four faces of the tetrahedron internally. This sphere is known as the *In-sphere* and its centre I is known as the *In-centre* of the tetrahedron. The radius of the in-sphere is called the *In-radius.*

Let *OABC* be any tetrahedron and let **a, b, c** be the position vectors of *A, B, C* with reference to *O* as origin.

The equations of the three planes

OBC, OCA, OAB

through the vertex *O* are

$$\mathbf{r} \cdot \mathbf{b} \times \mathbf{c} = 0, \quad \mathbf{r} \cdot \mathbf{c} \times \mathbf{a} = 0, \quad \mathbf{r} \cdot \mathbf{a} \times \mathbf{b} = 0.$$

The bisecting planes of the angles between the planes *OBC, OCA* are

$$\frac{\mathbf{r} \cdot \mathbf{b} \times \mathbf{c}}{|\mathbf{b} \times \mathbf{c}|} \pm \frac{\mathbf{r} \cdot \mathbf{c} \times \mathbf{a}}{|\mathbf{c} \times \mathbf{a}|} = 0.$$

Of these two, the equation

$$\frac{\mathbf{r} \cdot \mathbf{b} \times \mathbf{c}}{|\mathbf{b} \times \mathbf{c}|} - \frac{\mathbf{r} \cdot \mathbf{c} \times \mathbf{a}}{|\mathbf{c} \times \mathbf{a}|} = 0 \qquad ...(i)$$

represents the internal bisecting plane, for the points *A, B* lie on opposite sides of it. By substituting **a, b** for **r** respectively in (*i*), we obtain expressions having opposite signs.

The other two internal bisecting planes are

$$\frac{\mathbf{r} \cdot \mathbf{c} \times \mathbf{a}}{|\mathbf{c} \times \mathbf{a}|} - \frac{\mathbf{r} \cdot \mathbf{a} \times \mathbf{b}}{|\mathbf{a} \times \mathbf{b}|} = 0 \qquad ...(ii)$$

$$\frac{\mathbf{r} \cdot \mathbf{a} \times \mathbf{b}}{|\mathbf{a} \times \mathbf{b}|} - \frac{\mathbf{r} \cdot \mathbf{b} \times \mathbf{c}}{|\mathbf{b} \times \mathbf{c}|} = 0. \qquad ...(iii)$$

For the sake of brevity, we write α, β, γ for

$$\frac{1}{2}|\mathbf{b} \times \mathbf{c}|, \ \frac{1}{2}|\mathbf{c} \times \mathbf{a}|, \ \frac{1}{2}|\mathbf{a} \times \mathbf{b}|,$$

respectively so that they denote the areas of the triangular faces, *OBC, OCA, OAB* respectively.

The three planes (*i*), (*ii*) and (*iii*) obviously meet in a line through *O*. The line of intersection of these planes is parallel to the vector

$$\left(\frac{\mathbf{b}\times\mathbf{c}}{\alpha}-\frac{\mathbf{c}\times\mathbf{a}}{\beta}\right)\times\left(\frac{\mathbf{c}\times\mathbf{a}}{\beta}-\frac{\mathbf{a}\times\mathbf{b}}{\gamma}\right)=\frac{[\mathbf{a}\ \mathbf{b}\ \mathbf{c}]}{\alpha\,\beta\,\gamma}\{\alpha\mathbf{a}+\beta\mathbf{b}+\gamma\mathbf{c}\}.$$

Thus, the line of intersection of these planes is

$$\mathbf{r} = t\,(\alpha\mathbf{a} + \beta\mathbf{b} + \gamma\mathbf{c}).$$

From this, we easily deduce that the point I where this line meets the plane *ABC* whose equation is

$$\mathbf{r} \cdot (\mathbf{b} \times \mathbf{c} + \mathbf{c} \times \mathbf{a} + \mathbf{a} \times \mathbf{b}) = [\mathbf{a}\ \mathbf{b}\ \mathbf{c}]$$

is

$$\frac{\alpha\mathbf{a}+\beta\mathbf{b}+\gamma\mathbf{c}}{\alpha+\beta+\gamma}$$

$$\Rightarrow \qquad \overrightarrow{OI_1}=\frac{\alpha\overrightarrow{OA}+\beta\overrightarrow{OB}+\gamma\overrightarrow{OC}}{\alpha+\beta+\gamma}.$$

We now suppose that the position vectors of the vertices *A, B, C, D* with reference to any again *O* are

$$\mathbf{a},\ \mathbf{b},\ \mathbf{c},\ \mathbf{d}$$

respectively. The position vector of I_1 being

$$(\alpha\mathbf{a} + \beta\mathbf{b} + \gamma\mathbf{c}) / (\alpha + \beta + \gamma),$$

the point dividing $O.\ I_1$ in the ratio $\alpha + \beta + \gamma : \delta$ is

$$\frac{\alpha\mathbf{a}+\beta\mathbf{b}+\gamma\mathbf{c}+\delta\mathbf{d}}{\alpha+\beta+\gamma+\delta} \qquad ...(iv)$$

δ denoting the area of the triangle *ABC*.

By symmetry, we see that the point (*iv*) lies on the line of intersection of the internal bisecting planes of the planes through every vertex.

Thus, the planes bisecting internally the angles between pairs of planes of the tetrahedron meet in the point given by (*iv*) which is the *in-centre*.

In-radius. The distance of the point (*iv*) from the face *ABC* is the in-radius. Consider the face *ABC* whose equation is

$$\mathbf{r} \cdot (\mathbf{b} \times \mathbf{c} + \mathbf{c} \times \mathbf{a} + \mathbf{a} \times \mathbf{b}) = [\mathbf{a}\ \mathbf{b}\ \mathbf{c}].$$

Thus, the in-radius is

$$\frac{\dfrac{\alpha\mathbf{a}+\beta\mathbf{b}+\gamma\mathbf{c}+\delta\mathbf{d}}{\alpha+\beta+\gamma+\delta}\cdot(\mathbf{b}\times\mathbf{c}+\mathbf{c}\times\mathbf{a}+\mathbf{a}\times\mathbf{b})-[\mathbf{a}\ \mathbf{b}\ \mathbf{c}]}{|\,\mathbf{b}\times\mathbf{c}+\mathbf{c}\times\mathbf{a}+\mathbf{a}\times\mathbf{b}\,|}$$

$$=\frac{\alpha\,|\,[\mathbf{a}\ \mathbf{b}\ \mathbf{c}]+[\mathbf{a}\ \mathbf{c}\ \mathbf{d}]+[\mathbf{a}\ \mathbf{b}\ \mathbf{d}]-[\mathbf{b}\ \mathbf{c}\ \mathbf{d}]\,|}{(\alpha+\beta+\gamma+\delta)\,2\alpha}$$

$$=\frac{6V}{2S}=\frac{3V}{S},$$

where V denotes the volume and S, the sum of the areas of the faces of the tetrahedron.

The reader may compare this formula with that for the In-radius of a triangle.

6. Regular tetrahedron. Def. *A tetrahedron whose edges are all equal in length is called a Regular tetrahedron.*

The faces of a regular tetrahedron are equilateral triangles so that the angle between any two concurrent edges of a regular tetrahedron is 60°.

The following results will be established :

I. *The angle between any two plane faces of a regular tetrahedron is* $\cos^{-1}\frac{1}{3}$.

II. *The angle between any edge and a face not containing the edge is* $\cos^{-1}\left(\sqrt{\frac{1}{3}}\right)$.

III. *Any two opposite edges are perpendicular to each other.*

IV. *The distance of any vertex from the opposite face is* $\sqrt{\frac{2}{3}}k$; *k being the length of any edge.*

I. Let $OABC$ be a regular tetrahedron. Take O as the origin of reference. Let

$$\vec{OA} = \mathbf{a}, \quad \vec{OB} = \mathbf{b}, \quad \vec{OC} = \mathbf{c}.$$

and let k be the length of each of the edges,

We have

$$|\mathbf{a}| = |\mathbf{b}| = |\mathbf{c}| = k,$$

$$|\mathbf{a} \cdot \mathbf{b}| = |\mathbf{b} \cdot \mathbf{c}| = |\mathbf{c} \cdot \mathbf{a}|$$

$$= k^2 \cos 60° = \frac{1}{2}k^2.$$

$$\mathbf{a} \cdot \mathbf{a} = \mathbf{b} \cdot \mathbf{b} = \mathbf{c} \cdot \mathbf{c} = k^2.$$

The equation of the planes OAB and OBC are

$$\mathbf{r} \cdot \mathbf{a} \times \mathbf{b} = 0, \quad \mathbf{r} \cdot \mathbf{b} \times \mathbf{c} = 0.$$

Now the angle between two planes is equal to the angle between the normals to the same. Thus, if θ denotes the angle between these two planes, we have

$$\cos\theta = \frac{(\mathbf{a} \times \mathbf{b})}{|\mathbf{a} \times \mathbf{b}|} \cdot \frac{(\mathbf{b} \times \mathbf{c})}{|\mathbf{b} \times \mathbf{c}|}.$$

Now $$(\mathbf{a} \times \mathbf{b}) \cdot (\mathbf{b} \times \mathbf{c}) = (\mathbf{a} \cdot \mathbf{b})(\mathbf{b} \cdot \mathbf{c}) - (\mathbf{b} \cdot \mathbf{b})(\mathbf{a} \cdot \mathbf{c}) = \frac{-1}{4}k^4$$

$$|\mathbf{a}\times\mathbf{b}| = |\mathbf{a}|\,|\mathbf{b}|\sin 60° = \frac{\sqrt{3}}{2}k^2 = |\mathbf{b}\times\mathbf{c}|.$$

$$\therefore \qquad \cos\theta = -\frac{1}{3}.$$

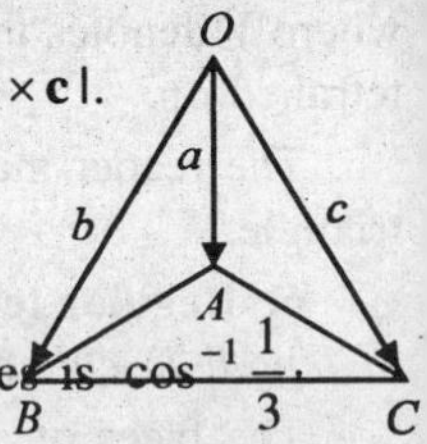

Fig. 7.1.

Thus, the acute angle between any two plane faces is $\cos^{-1}\frac{1}{3}$.

II. Consider the face ABC and the edge OA whose equations are

$$\mathbf{r}\,.\,[\mathbf{b}\times\mathbf{c}+\mathbf{c}\times\mathbf{a}+\mathbf{a}\times\mathbf{b}] = [\mathbf{a}\;\mathbf{b}\;\mathbf{c}], \quad \mathbf{r} = t\mathbf{a}$$

respectively.

The angle between a line and a plane is equal to the complement of the angle between the line and the normal to the plane. Thus, if θ denotes the angle between the face and the edge, we have

$$\sin\theta = \frac{(\mathbf{b}\times\mathbf{c}+\mathbf{c}\times\mathbf{a}+\mathbf{a}\times\mathbf{b})\,.\,\mathbf{a}}{|\mathbf{b}\times\mathbf{c}+\mathbf{c}\times\mathbf{a}+\mathbf{a}\times\mathbf{b}|\;|\mathbf{a}|}$$

$$= \frac{[\mathbf{a}\;\mathbf{b}\;\mathbf{c}]}{|\mathbf{b}\times\mathbf{c}+\mathbf{c}\times\mathbf{a}+\mathbf{a}\times\mathbf{b}|\;|\mathbf{a}|} \qquad ...(i)$$

Now $$[\mathbf{a}\;\mathbf{b}\;\mathbf{c}]^2 = \begin{vmatrix} \mathbf{a}\,.\,\mathbf{a} & \mathbf{a}\,.\,\mathbf{b} & \mathbf{a}\,.\,\mathbf{c} \\ \mathbf{b}\,.\,\mathbf{a} & \mathbf{b}\,.\,\mathbf{b} & \mathbf{b}\,.\,\mathbf{c} \\ \mathbf{c}\,.\,\mathbf{a} & \mathbf{c}\,.\,\mathbf{b} & \mathbf{c}\,.\,\mathbf{c} \end{vmatrix} = \frac{1}{2}k^4 \qquad \text{(Cor. P. 179)}$$

Also $[\mathbf{b}\times\mathbf{c}+\mathbf{c}\times\mathbf{a}+\mathbf{a}\times\mathbf{b}]$

is twice the area of the triangle ABC which is equilateral with each side k so that this is

$$= \frac{\sqrt{3}}{2}k^2.$$

Substituting in (*i*), we obtain

$$\sin\theta = \sqrt{\frac{2}{3}} \Rightarrow \cos\theta = \sqrt{\frac{1}{3}} \Rightarrow \theta = \cos^{-1}\left(\sqrt{\frac{1}{3}}\right).$$

III. We have $\vec{OA}\cdot\vec{BC} = \mathbf{a}\,.\,(\mathbf{c}-\mathbf{b}) = \mathbf{a}\,.\,\mathbf{c}-\mathbf{a}\,.\,\mathbf{b} = 0$ so that $OA \perp BC$.

Similarly,

$$OB \perp CA \quad \text{and} \quad OC \perp AB.$$

IV. The equation of the plane ABC is

$$\mathbf{r}\,.\,[\mathbf{b}\times\mathbf{c}+\mathbf{c}\times\mathbf{a}+\mathbf{a}\times\mathbf{b}] = [\mathbf{a}\;\mathbf{b}\;\mathbf{c}],$$

so that the distance of the vertex O from this plane

$$= \frac{[\mathbf{a}\;\mathbf{b}\;\mathbf{c}]}{|\mathbf{b}\times\mathbf{c}+\mathbf{c}\times\mathbf{a}+\mathbf{a}\times\mathbf{b}|} = \frac{2}{3}k.$$

7. Orthocentric tetrahedra. A *tetrahedron is called orthocentric if it* *is such that the altitude from its vertices to the opposite faces are concurrent.*

The following results will be established :

I. *If in a tetrahedron OABC, the altitudes from A and B to the opposite faces intersect then the edge AB is perpendicular to the opposite edge OC.*

II. *If the altitudes from A and B intersect and those from B and C also intersect then all the four altitudes are concurrent and the pairs of opposite edges are at right angles to each other.*

Take O as the origin of reference. Let

$$\overrightarrow{OA} = \mathbf{a},\ \overrightarrow{OB} = \mathbf{b},\ \overrightarrow{OC} = \mathbf{c}.$$

I. The equations of the altitudes from A and B to the opposite faces are

$$\mathbf{r} = \mathbf{a} + t\,(\mathbf{b} \times \mathbf{c}), \quad \mathbf{r} = \mathbf{b} + p\,(\mathbf{c} \times \mathbf{a}). \qquad ...(i)$$

Now

$$(\mathbf{a} - \mathbf{b}) \cdot (\mathbf{b} \times \mathbf{c}) \times (\mathbf{c} \times \mathbf{a}) = (\mathbf{a} + \mathbf{b}) \cdot [\mathbf{a}\ \mathbf{b}\ \mathbf{c}]$$
$$= [\mathbf{a}\ \mathbf{b}\ \mathbf{c}]\,(\mathbf{a} - \mathbf{b}) \cdot \mathbf{c}$$

so that the altitudes (i) will be coplanar if, and only if

$$(\mathbf{a} - \mathbf{b}) \cdot \mathbf{c} = 0, \quad \Rightarrow \quad AB \perp OC.$$

II. Assuming the condition as satisfied, we find the position vector of the point of intersection of the altitudes from A and B.

For this purpose, we express the vectors **a, b** as linear combination of the non-coplanar vectors **b × c, c × a, a × b.**

Let

$$\mathbf{a} = l_1\mathbf{b} \times \mathbf{c} + m_1\mathbf{c} \times \mathbf{a} + n_1\mathbf{a} \times \mathbf{b}, \quad \mathbf{b} = l_2\mathbf{b} \times \mathbf{c} + m_2\mathbf{c} \times \mathbf{a} + n_2\mathbf{a} \times \mathbf{b},$$

so that, as may be easily shown,

$$l_1 = \frac{\mathbf{a} \cdot \mathbf{a}}{[\mathbf{a}\ \mathbf{b}\ \mathbf{c}]},\ m_1 = \frac{\mathbf{a} \cdot \mathbf{b}}{[\mathbf{a}\ \mathbf{b}\ \mathbf{c}]},\ n_1 = \frac{\mathbf{a} \cdot \mathbf{c}}{[\mathbf{a}\ \mathbf{b}\ \mathbf{c}]},$$

$$l_2 = \frac{\mathbf{b} \cdot \mathbf{a}}{[\mathbf{a}\ \mathbf{b}\ \mathbf{c}]},\ m_2 = \frac{\mathbf{b} \cdot \mathbf{b}}{[\mathbf{a}\ \mathbf{b}\ \mathbf{c}]},\ n_2 = \frac{\mathbf{b} \cdot \mathbf{c}}{[\mathbf{a}\ \mathbf{b}\ \mathbf{c}]}.$$

At the point of intersection of the lines (i), we have

$$l_1\,(\mathbf{b} \times \mathbf{c}) + m_1(\mathbf{c} \times \mathbf{a}) + n_1\,(\mathbf{a} \times \mathbf{b}) + t\,(\mathbf{b} \times \mathbf{c})$$
$$= l_2\,(\mathbf{b} \times \mathbf{c}) + m_2\,(\mathbf{c} \times \mathbf{a}) + n_2\,(\mathbf{a} \times \mathbf{b}) + p\,(\mathbf{c} \times \mathbf{a})$$

$$\Rightarrow \qquad l_1 + t = l_2, \quad m_1 = m_2 + p, \quad n_1 = n_2.$$

Here $n_1 = n_2$ is equivalent o (ii), *viz.*, $(\mathbf{a} - \mathbf{b}) \cdot \mathbf{c} = 0$.

$$\therefore \qquad t = l_2 - l_1 = \frac{\mathbf{b} \cdot \mathbf{a} - \mathbf{a} \cdot \mathbf{a}}{[\mathbf{a}\ \mathbf{b}\ \mathbf{c}]},\quad p = m_1 - m_2 = \frac{\mathbf{a} \cdot \mathbf{b} - \mathbf{b} \cdot \mathbf{b}}{[\mathbf{a}\ \mathbf{b}\ \mathbf{c}]}.$$

Thus, the point of intersection of the altitudes from A and B is

$$\frac{1}{[\mathbf{a}\ \mathbf{b}\ \mathbf{c}]}\{(\mathbf{b} \cdot \mathbf{a})\,(\mathbf{b} \times \mathbf{c}) + (\mathbf{a} \cdot \mathbf{b})\,(\mathbf{c} \times \mathbf{a}) + (\mathbf{a} \cdot \mathbf{c})\,(\mathbf{a} \times \mathbf{b})\} \qquad ...(iii)$$

Again suppose that the altitudes from B and C also intersect so that $(\mathbf{b} - \mathbf{c}) \cdot \mathbf{a} = 0$. The point of intersection of these altitudes is

$$\frac{1}{[\mathbf{a}\,\mathbf{b}\,\mathbf{c}]}\{(\mathbf{a}\cdot\mathbf{b})(\mathbf{b}\times\mathbf{c})+(\mathbf{b}\cdot\mathbf{c})(\mathbf{c}\times\mathbf{a})+(\mathbf{c}\cdot\mathbf{a})(\mathbf{a}\times\mathbf{b})\} \qquad ...(iv)$$

As $\mathbf{a} \cdot \mathbf{b} = \mathbf{b} \cdot \mathbf{c} = \mathbf{c} \cdot \mathbf{a}$, the points (*iii*) and (*iv*) are the same and the common point is

$$\frac{k}{[\mathbf{a}\,\mathbf{b}\,\mathbf{c}]}\{\mathbf{b}\times\mathbf{c}+\mathbf{c}\times\mathbf{a}+\mathbf{a}\times\mathbf{b}\} \qquad ...(v)$$

where $\quad k = \mathbf{a}.\mathbf{b} = \mathbf{b} \cdot \mathbf{c} = \mathbf{c} \cdot \mathbf{a}.$

From the symmetry of (*v*), it is clear that all four altitudes concur at this point. Also the conditions

$$\mathbf{a}.\mathbf{b} = \mathbf{b}.\mathbf{c} = \mathbf{c}.\mathbf{a}$$

imply that the pairs of opposite edges of the tetrahedron are perpendicular to each other.

Also the point of concurrence is called the **Orthocentre** of the tetrahedron.

8. Equifacial Tetrahedra. Def. *A tetrahedron whose faces have equal areas is called equifacial.*

The following results will be established :

I. *If the areas of two faces OAB and OAC of a tetrahedron OABC be equal, then the line of shortest distance between OA and BC is the line through the middle point of BC perpendicular to OA.*

II. *If the areas of two faces OAB and OAC are equal and those of the faces BCA, BCO are equal then the line of shortest distance between OA and BC is the join of their middle points : also the faces equal in area are congruent.*

III. *If the areas of three triangular faces through any vertex O are equal, then the centroid and in-centre are collinear with O.*

IV. *The faces of an equifacial tetrahedron are congruent triangles. Every equifacial tetrahedron is isosceles.*

I. Take O as origin of reference. Let

$$\overrightarrow{OA} = \mathbf{a},\ \overrightarrow{OB} = \mathbf{b},\ \overrightarrow{OC} = \mathbf{c}.$$

Vector areas of the faces OAB and OAC are

$$\frac{1}{2}\overrightarrow{OA}\times\overrightarrow{OB} \text{ and } \frac{1}{2}\overrightarrow{OA}\times\overrightarrow{OC}$$

so that we are given that

$$|\,\mathbf{a}\times\mathbf{b}\,| = |\,\mathbf{a}\times\mathbf{c}\,|$$

$$\Rightarrow \quad (\mathbf{a}\times\mathbf{b})\cdot(\mathbf{a}\times\mathbf{b}) = (\mathbf{a}\times\mathbf{c})\cdot(\mathbf{a}\times\mathbf{c})$$

$$\Rightarrow \quad (\mathbf{a}\times\mathbf{b}-\mathbf{a}\times\mathbf{c})\cdot(\mathbf{a}\times\mathbf{b}+\mathbf{a}\times\mathbf{c}) = 0$$

$$\Rightarrow \quad [\mathbf{a}\times(\mathbf{b}-\mathbf{c})]\cdot[\mathbf{a}\times(\mathbf{b}+\mathbf{c})] = 0 \qquad ...(i)$$

Now $\mathbf{a} \times (\mathbf{b} - \mathbf{c})$ is parallel to the line of shortest distance between *OA* and *BC*.

Also $\frac{1}{2}(\mathbf{b}+\mathbf{c})$ is the position vector of the mid-point *D* of *BC*.

As $\vec{OA} = \mathbf{a}$ and $\vec{OD} = \frac{1}{2}(\mathbf{b}+\mathbf{c})$,

the vector $\mathbf{a} \times (\mathbf{b} + \mathbf{c})$ is normal to the plane *OAD*.

By (*i*), the line of shortest distance between *OA* and *BC* lies in the plane *OAD*. Also it meets *BC*. Hence, we have the result I.

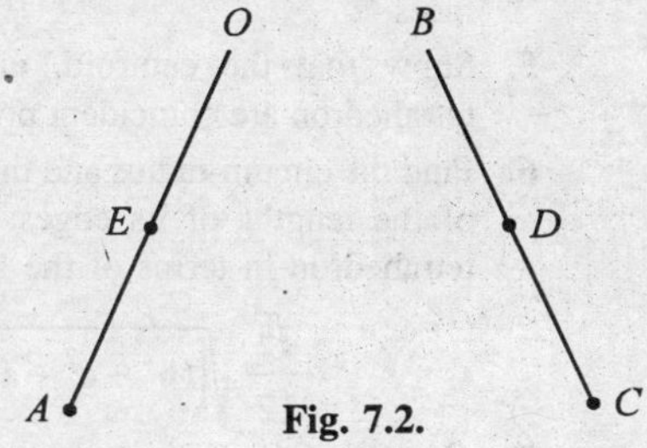

Fig. 7.2.

II. By I, the line of shortest distance passes through the mid-points of *OA* and *BC* and hence it is their join. Let *E* be the mid-point of *OA*.

The figure is symmetrical above *DE* so that

$$OB = AC \text{ and } OC = AB.$$

Hence, the triangles *OAB, OAC* are congruent and the triangles *BCO* and *BCA* are congruent.

III. If α, β, γ, δ denote the areas of the faces opposite to the vertices *A, B, C, O* the position vector of the in-centre is

$$\frac{\alpha\mathbf{a}+\beta\mathbf{b}+\gamma\mathbf{c}+\delta\mathbf{d}}{\alpha+\beta+\gamma+\delta} = \frac{3\alpha(\mathbf{a}+\mathbf{b}+\mathbf{c})}{3\alpha+\delta}, \qquad (5 \text{ Page } 193)$$

for $\alpha = \beta = \gamma$.

Also the centroid is

$$\frac{\mathbf{a}+\mathbf{b}+\mathbf{c}+\mathbf{0}}{4} = \frac{\mathbf{a}+\mathbf{b}+\mathbf{c}}{4}$$

Thus,

$$\vec{OG} = \frac{\mathbf{a}+\mathbf{b}+\mathbf{c}}{4},$$

$$\vec{OI} = \frac{3\alpha(\mathbf{a}+\mathbf{b}+\mathbf{c})}{3\alpha+\delta} = \frac{12\alpha}{3\alpha+\delta}\vec{OG}.$$

It follows that *O, I, G* are collinear.

IV is an immediate consequence of II.

EXERCISES

1. *OABC* is an orthocentric tetrahedron and α, β, γ are the angles between the edges *OA, OB, OC* taken in pairs, show that

$$\frac{OA}{\cos\alpha} = \frac{OB}{\cos\beta} = \frac{OC}{\cos\gamma}$$

2. Find the circum-radius and in-radius of a regular tetrahedron in terms of

the length of each edge.

3. Show that a tetrahedron which is isosceles and orthocentric is regular.

4. Show that the feet of the altitudes from the vertices to the opposite faces of an orthocentric tetrahedron are the orthocentres of the corresponding faces.

5. Show that the centroid, circum-centre and in-centre of an isosceles tetrahedron are coincident points.

6. Find the circum-radius and the in-radius of an isosceles tetrahedron in terms of the lengths of its edges. Also show that the volume of an isosceles tetrahedron in terms of the lengths of the edges is

$$\frac{\sqrt{2}}{12}\sqrt{\left[(\mathbf{b}^2+\mathbf{c}^2-\mathbf{a}^2)(\mathbf{c}^2+\mathbf{a}^2-\mathbf{b}^2)(\mathbf{a}^2+\mathbf{b}^2-\mathbf{c}^2)\right]}.$$

7. *OA, OB, OC* are three concurrent lines; find the axis of the right circular cone touching the planes *OBC, OCA, OAB* internally.

8. Show that in any tetrahedron, the line joining the middle points of any one pair of opposite edges is perpendicular to the shortest distance lines between either of the two other pairs of opposite edges.

9. The altitudes from the vertices *A* and *B* to the opposite faces of a tetrahedron intersect; show that *AB* is perpendicular to *CD* and the altitudes from the vertices *C, D* to the opposite faces also intersect. Prove the converse also.

10. *OABC* and *OA′B′C′* are two tetrahedra such that the plane through *OA* perpendicular to the face *B′OC′* and the two other similar planes intersect in a line; show that the plane through *OA′* perpendicular to the face *BOC* and two other similar planes also intersect in a line.

11. *ABCD* is a tetrahedron : h_1, h_2, h_3, h_4 are the perpendicular distances of the vertices from the opposite faces and s_1, s_2, s_3 are the shortest distances between the three pairs of opposite edges; prove that

$$\frac{1}{S_1^2}+\frac{1}{S_2^2}+\frac{1}{S_3^2}=\frac{1}{h_1^2}+\frac{1}{h_2^2}+\frac{1}{h_3^2}+\frac{1}{h_4^2}.$$

12. Show that the centre of the sphere through four points *A, B, C, D* with position vectors **a, b, c, d** is

$$-\frac{\mathbf{a}^2\,\overrightarrow{\Delta\,BCD}+\mathbf{b}^2\,\overrightarrow{\Delta\,CAD}+\mathbf{c}^2\,\overrightarrow{\Delta\,ABD}-\mathbf{d}^2\,\overrightarrow{\Delta\,ABC}}{\overrightarrow{AB}\times\overrightarrow{AC}\cdot\overrightarrow{AD}}.$$

13. Show that the circum-centre of the tetrahedron *OABC* is

$$\frac{\mathbf{a}^2(\mathbf{b}\times\mathbf{c})+\mathbf{b}^2(\mathbf{c}\times\mathbf{a})+\mathbf{c}^2(\mathbf{a}\times\mathbf{b})}{2\,[\mathbf{a}\;\mathbf{b}\;\mathbf{c}]},$$

where $\overrightarrow{OA}=\mathbf{a},\ \overrightarrow{OB}=\mathbf{b},\ \overrightarrow{OC}=\mathbf{c}.$

8

Statics with Vectors

Introduction. A study of Theoretical Statics with the help of Vector Algebra will be undertaken in this chapter.

In this connection it is important to remember that a force cannot be completely characterised by a vector and that we require two vectors for the purpose. As against this, a Couple is completely characterisable by one vector.

The study will be based on the **Principles of Transmissibility of Forces and the Parallelogram Law of Forces.**

8.1. FORCES REPRESENTED BY LINE VECTORS

A force acting on a rigid body is completely specified by its (*i*) *Magnitude,* (*ii*) *Line of action,* (*iii*) *Point of application, and* (*iv*) *Sense along the line of action.*

As a consequence of the principle of **Transmissibility of forces,** it can be shown that the effect of a force acting on a *rigid* body is unaltered if its point of application is changed to any other point on the line of action of the force without any change in its magnitude and sense. Thus, while specifying a force acting on a rigid body, we need only state its magnitude, line of action and sense. Unlike vectors however, the particular line of action of a force is material and the effects of two forces acting along *different* parallel lines are different even though they may have the same magnitude and sense.

Thus, a force cannot be thought of as being completely representable by a vector. For the representation of forces, we need the notion of *Line Vectors* which requires a more restricted definition of equality than that for vectors (§ 1.2.2, page 3).

Line Vectors. Def. *A directed line segment is called a line vector such that two directed line segments having the same magnitude,* **support** *and sense are equal line vectors.*

Thus, two directed line segments may be equal vectors but unequal line vectors. For the equality of line vectors, it is necessary that their supports are same and not *just* parallel as for the equality of vectors.

While two vectors are equal if the corresponding directed line segments have the same length, same or parallel supports and the same sense, the two line vectors are equal if the corresponding directed line segments have the same length, same supports and the same sense.

A line vector is also sometimes called a *Localised vector* and for the sake of distinction, the vectors, so called, are often termed **Free Vectors.**

Forces represented by line vectors. From the foregoing, it is clear that a force acting on rigid body can be represented by a line vector such that two equivalent forces are represented by equal line vectors and *vice-versa.*

Two forces may be said to be equivalent if they have the same effect.

8.2. STUDY OF STATICS WITH VECTORS

In view of the preceding, we see that an Analytical Vectorial study of Statics requires the setting up of an *Algebra of Line Vectors.* It can be shown, however, that a line vector can itself be specified by two vectors so that, on our part, we shall avoid direct reference to the notion of line vectors and show that whereas a force cannot be represented by a single vector it can be specified by two vectors, to be called

(*i*) *Vector of the Force,*

(*ii*) *Moment Vector of the Force about a given point.*

The Algebra of Vectors developed in the preceding chapters will thus form the basis of the Vectorial study of Statics.

8.2.1. Vectors of a Force

A vector whose

(*i*) length represents the magnitude of the force according to some scale,

(*ii*) support is the same or parallel to that of the line of action of the force,

(*iii*) sense is the same as that of the force,

is called the vector of the given force or *Force Vector.* The *Vector of a force* will be denoted by the same symbol as denotes the force.

It will be seen that correspondence between Forces and Vectors is *not* one-one; to each force there corresponds a single vector but to a given vector there will correspond several, not necessarily equivalent forces. Forces corresponding to the same vector will have the same magnitude, the same or *parallel* lines of action and the same sense; they may have, however, different points of application.

A force is completely specified by its Vector and its point of application.

Just as the magnitude of a force gives us only an incomplete idea of the force, so the vector of a force also gives an incomplete idea, even though this tells us more than the magnitude of the force.

8.2.2. Moment Vector about a given Point

Let **F** be any given force and let P be any point on its line of action.

Take any arbitrary point O and let

$$\overrightarrow{OP} = \mathbf{r}.$$

Then the *vector*

$$\mathbf{M} = \mathbf{r} \times \mathbf{F}$$

is called the Moment vector of the force about **O**. The symbol **F**, here denotes the vector of the force.

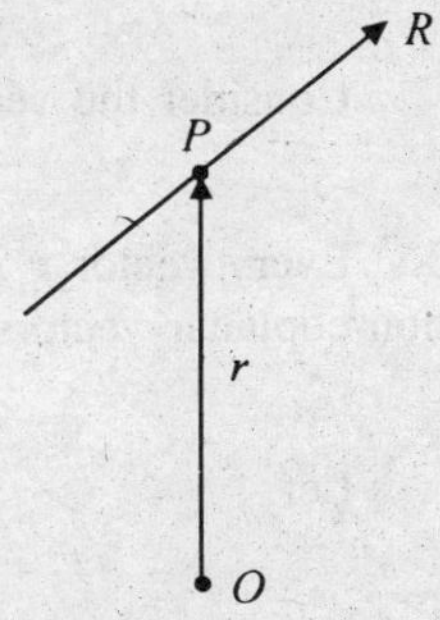

Fig. 8.1.

The moment vector about O is independent of the choice of the point P on the line of action of the force. Thus, if Q be any other point on the line of action of the force, we have

$$\overrightarrow{OQ} \times \mathbf{F} = (\overrightarrow{OP} + \overrightarrow{PQ}) \times \mathbf{F}$$

$$= \overrightarrow{OP} \times \mathbf{F} + \overrightarrow{PQ} \times \mathbf{F}$$

$$= \overrightarrow{OP} \times \mathbf{F};$$

$\overrightarrow{OP} \times \mathbf{F}$ being the zero vector, for $\overrightarrow{PQ}$ and **F** are parallel vectors.

Of course the moment vector of a force does depend upon the point, O and to lay emphasis on this point, whenever necessary, we may write, O moment vector instead of just moment vector.

The two vectors **F** and **M** are clearly perpendicular to each other so that we have

$$\mathbf{F} \cdot \mathbf{M} = 0.$$

8.2.3. Vanishing of the Moment Vector

The moment vector of a non-zero force about a point O is zero if, and only if, the point O lies on the line of action of the force.

8.3. FORCE DETERMINED BY ITS FORCE VECTOR AND MOMENT VECTOR

We shall now show that the line of action of a force is known as soon as we know **F** and **M**.

Let **F, M** be two vectors such that

$$\mathbf{F} \,.\, \mathbf{M} = 0.$$

Consider the vector equation

$$\mathbf{r} \times \mathbf{F} = \mathbf{M}. \qquad ...(i)$$

Every vector **r** is uniquely expressible as a linear combination of the non-coplanar vectors

$$\mathbf{M},\ \ \mathbf{F},\ \ \mathbf{F} \times \mathbf{M}.$$

Let

$$\mathbf{r} = \lambda_1 \mathbf{M} + \lambda_2 \mathbf{F} + \lambda_3 \mathbf{F} \times \mathbf{M}. \qquad ...(ii)$$

Substituting in (*i*) we get

$$\lambda_1 \mathbf{M} \times \mathbf{F} + \lambda_3 (\mathbf{F} \times \mathbf{M}) \times \mathbf{F} = \mathbf{M}$$

$$\Rightarrow \quad \lambda_1 [\mathbf{M} \times \mathbf{F} + \lambda_3] \,(\mathbf{F} \,.\, \mathbf{F})\, \mathbf{M} - \mathbf{M} \,.\, \mathbf{F})\, \mathbf{F}] = \mathbf{M}$$

$$\Rightarrow \quad \lambda_1 \mathbf{M} \times \mathbf{F} + [\lambda_2 (\mathbf{F} \,.\, \mathbf{F}) - 1]\, \mathbf{M} = 0, \quad \text{for} \quad \mathbf{F} \,.\, \mathbf{M} = 0,$$

$$\Leftrightarrow \quad \lambda_1 = 0, \quad \lambda_3 (\mathbf{F} \,.\, \mathbf{F}) - 1 = 0.$$

Thus, $\lambda_1 = 0$ and $\lambda_3 = 1/\mathbf{F} \,.\, \mathbf{F}$.

Substituting in (*ii*), we get

$$\mathbf{r} = \frac{1}{\mathbf{F} \cdot \mathbf{F}} (\mathbf{F} \times \mathbf{M}) + \lambda_2 \mathbf{F} \qquad ...(iii)$$

Also we may easily see that the vector **r** given by (*iii*) satisfy (*i*) for every value of the scalar λ_2.

Now the equation (*iii*) represents a straight line parallel to the force vector **F** such that for any point on this line, the equation (*i*) is true.

Hence (*iii*) is the equation of the line of action of the force which has **F** and **M** as Force Vector and Moment Vector respectively. Thus, the force is completely determined.

8.4. VECTOR SUM AND MOMENT VECTOR SUM

As in the case of a single force, we can associate two vectors to any system of forces such that, so far as the effect on a rigid body is concerned, they determined the system completely.

Consider any system of forces

$$\mathbf{F}_1,\ \mathbf{F}_2,\ \mathbf{F}_3, \ldots$$

acting on a rigid body and let

$$\mathbf{r}_1,\ \mathbf{r}_2,\ \mathbf{r}_3, \ldots$$

be the position vectors of any points on the lines of action of the forces with reference to any point O.

The force vectors of the forces are also denoted by

$$\mathbf{F}_1,\ \mathbf{F}_2,\ \mathbf{F}_3, \ldots$$

We write

$$\mathbf{R} = \mathbf{F}_1 + \mathbf{F}_2 + \mathbf{F}_3 + \ldots$$

$$\mathbf{G} = \mathbf{r}_1 \times \mathbf{F}_1 + \mathbf{r}_2 \times \mathbf{F}_2 + \mathbf{r}_3 \times \mathbf{F}_3 + \ldots$$

so that the vectors **R** and **G** are the sums of the vectors of the forces and of the moment vectors of the forces about O. For the sake of brevity, these will be referred to as

Vector sum and **Moment sum**

respectively.

8.4.1. Change in G with change in origin

Let O, O', be any two points and

$$\mathbf{r}_1, \mathbf{r}_2 \ldots, \mathbf{r}_4 \ldots;\ \ \mathbf{r}_1', \mathbf{r}_2' \ldots, \mathbf{r}_6' \ldots$$

be the position vectors of the points

$$P_1, P_2 \ldots, P_i \ldots,$$

on the lines of action of the given forces with respect to O and O' respectively.

We have

$$\overrightarrow{OP_i} = \mathbf{r}_i,\ \overrightarrow{O'P_i} = \mathbf{r}'_i$$

$$\Rightarrow \quad \mathbf{r}'_i = \overrightarrow{O'P_i}$$

$$= \overrightarrow{O'O} + \overrightarrow{OP_i} = \overrightarrow{OP_i} - \overrightarrow{OQ'} = \mathbf{r}_i - \mathbf{s},$$

where $\mathbf{s} = \overrightarrow{OO}$

If **G**, **G′** denote the moment sums about O and O', we have

$$\mathbf{G}' = \Sigma\,(\mathbf{r}_i' \times \mathbf{F}_i)$$

$$= \Sigma\,(\mathbf{r}_i - \mathbf{s}) \times \mathbf{F}_i = \Sigma\,(\mathbf{r}_i \times \mathbf{F}_i) - \Sigma \mathbf{s} \times \mathbf{F}_i$$

$$= \Sigma\,(\mathbf{r}_i \times \mathbf{F}_i) - \Sigma\,(\mathbf{s} \times \mathbf{F}_i)$$

$$= \mathbf{G} - \mathbf{s} \times \Sigma \mathbf{F}_i = \mathbf{G} - \mathbf{s} \times \mathbf{R}.$$

Thus, the *moment sum* **G′** *about any point* O' *is*

$$\mathbf{G} - \mathbf{s} \times \mathbf{R},$$

where **G** *is the moment sum about O and* **s** *is the position vector of O′ relative to O.*

Cor. 1. The moment sums **G**, **G′** are the same for two points O, O' when OO' is parallel to the vector sum **R**, for in that case

$$\mathbf{s} \times \mathbf{R} = 0.$$

Cor. 2. If the vector sum is zero and the moment sum about one point is zero then the moment sum about every point is zero.

8.5. REDUCTION OF A SYSTEM OF FORCES

An important problem of Statics consists in the determination of systems of forces which are equivalent to any given system so far as the effect on a rigid body is concerned and which are at the same time simpler to deal with than the given system. It is also of great importance to determine analytical conditions which characterise equivalent systems of forces. In this connection the two vectors **R** and **G** will be seen to play a very important part.

The two principles of reduction. The following two principles are fundamental for determining systems equivalent to a given system :

I. *Principle of transmissibility of forces* as a result of which a force can be thought of as acting at any point of its line of action.

II. *Parallelogram law of forces* as a result of which two forces $\mathbf{F}_1$ and $\mathbf{F}_2$ acting at any point O are equivalent to a single force called their esultant, whose vector is $\mathbf{F}_1 + \mathbf{F}_2$ and which also acts as O.

A system S_2 of forces which arises from another system S_1, be any series of applications of the above two principles is equivalent to S_1, in as much as the effects of the two are the same.

By successive applications of the Parallelogram law of forces, it can be shown that

(*i*) *if* $\mathbf{F}_1$, $\mathbf{F}_2$, $\mathbf{F}_3$ *be three forces acting at any point O and*

$$\overrightarrow{OA_1} = \mathbf{F}_1,\ \overrightarrow{OA_2} = \mathbf{F}_2,\ \overrightarrow{OA_3} = \mathbf{F}_3,$$

then the forces are equivalent to a single force whose line of action passes through O and whose vector is where OP is the diagonal of the parallelopiped constructed with

$$OA_1,\ OA_2,\ OA_3,$$

as adjacent sides.

(*ii*) *A system of concurrent three forces is equivalent to a single force whose point of application is the point of concurrence and whose vector is the sum of the vectors of the given forces.*

8.5.1. Zero Force

Def. *A force whose vector is the zero vector is called Zero force.*

Thus, a force whose magnitude is zero is the zero force.

Equilibrium. *A system of forces equivalent to the Zero force is said to be in equilibrium.*

Negative of a force. The force whose magnitude and line of action are the same as those of a force **F** but whose sense is opposite to that of **F** is said to be the negative of **F** and is denoted by – **F**.

The two forces **F**, – **F** form a system in equilibrium.

The moment vector of **F**, about any point is the negative of that of, – **F**, about the same point.

Theorem. *The vector sum and the moment sum about any point are the same for two equivalent systems.*

All that we need to show is that both the vector sum and the moment sum remain unaltered as a result of application of each of the two principles of reduction.

The vector sum is obviously invariant for each of the two principles.

For the moment sum, we prove the following :

The sum of the moment vectors of two forces acting at a point is equal to the moment vector of their resultant about the same point. (*Patna 2003*)

(Varignon's Theorem)

Let $\mathbf{F}_1$, $\mathbf{F}_2$, be two forces acting at any point A. The sum of the moment vectors about any point O

$$= \vec{OA} \times \mathbf{F}_1 + \vec{OA} \times \mathbf{F}_2$$

$$= \vec{OA} \times (\mathbf{F}_1 + \mathbf{F}_2)$$

= moment vector of the resultant about O.

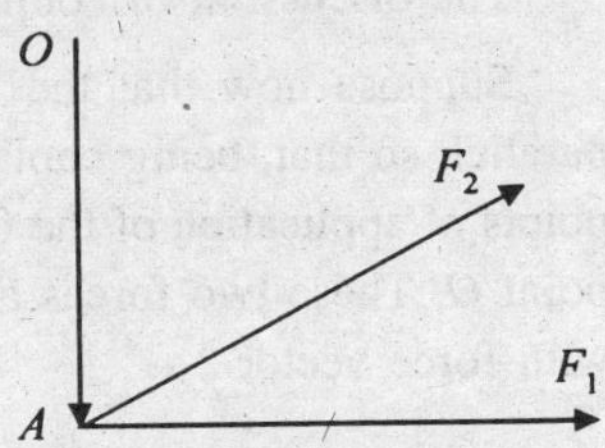

Fig. 8.2.

Also, obviously, the moment vector remains invariant as a result of the application of the first principle.

Hence the theorem.

Note. The importance of the Varignon's theorem lies in the fact that its converse is also true, but the proof of the converse can only be given after we have considered some further developments of the subject. (Refer § 8.9, Page 214)

8.6. REDUCTION OF A SYSTEM CONSISTING OF TWO PARALLEL FORCES

Couples. Let P, Q be two given parallel forces. Take two points A, B on their lines of action.

Introduce two forces **F**, – **F** at A and B acting along the join of A and B.

The forces **P** and **F** have a resultant which passes through A and whose vector is **P** + **F**.

The forces **Q** and – **F** have a resultant which passes through B and whose vector is **Q** – **F**.

Now two cases are possible : *the lines of action of these two latter forces* **P** + **F** *and* **Q** – **F** *meet or are parallel.*

Firstly, we investigate the condition for the lines of action to be parallel.

Let

$$\mathbf{P} = a\mathbf{S}, \quad \mathbf{Q} = b\mathbf{S},$$

where a, b are scalars and **S** is a unit vector.

$$\therefore \qquad \mathbf{P} + \mathbf{F} = a\mathbf{S} + \mathbf{F}, \quad \mathbf{Q} - \mathbf{F} = b\mathbf{S} - \mathbf{F}.$$

For parallelism

$$(a\mathbf{S} + \mathbf{F}) \times (b\mathbf{S} - \mathbf{F}) = \mathbf{0}$$

$$\Rightarrow \qquad (a + b)\, \mathbf{F} \times \mathbf{S} = \mathbf{0}.$$

As $\mathbf{F} \times \mathbf{S} \neq \mathbf{0}$, we have $a + b = 0$, *i.e.*, the two forces are equal in magnitude and opposite in sense.

A system of two such forces is called a *Couple.*

The discussion of couples will be taken up in § 8.7.

Suppose now that the lines of action of **P** + **F** and **Q** – **F** are not parallel, so that, being coplanar, they will meet at some point, say O. The points of application of the forces at A and B may now be transferred to the point O. These two forces at O are equivalent to a single force through O with force vector

$$\mathbf{P} + \mathbf{F} + \mathbf{Q} - \mathbf{F} = \mathbf{P} + \mathbf{Q},$$

Conclusion. *A system of two parallel forces which is not-a couple is equivalent to a single force whose vector is equal to the sum of the vectors of the given forces.*

The actual line of action of the single equivalent force will be determined in § 8.15, page 219.

8.7. REDUCTION TO THREE FORCES

Any given system of forces is equivalent to a system of three forces whose points of application are three given points.

Let L, M, N be the three given points.

Choose points

$$A_1,\ A_2,\ A_3,\ ...$$

on the lines of action of the given forces in such a manner that none of these points lies on the plane determined by the points L, M, N so that these are the points other than those where this plane meets the lines of action of the given forces.

Consider any force $\mathbf{F}_1$ acting at A_1.

Let

$$\overrightarrow{A_1B_1} = \mathbf{F}_1.$$

There exists a parallelopiped whose one diagonal is along the line A_1B_1 of application of the force $\mathbf{F}_1$ and whose three edges through A_1 lie along A_1L, A_1M and A_1N. Thus there exist three concurrent forces acting along A_1L, A_1M, A_1N equivalent to $\mathbf{F}_1$. We suppose that the points of application of these forces are transferred to L, M, N.

Let each force of the system be treated in a like manner.

Thus, we arrive at three systems of concurrent forces separately concurrent at L, M and N. Replacing these concurrent forces by their resultants, we arrive at the theorem stated.

8.8. CONDITION FOR EQUILIBRIUM

Theorem. *If for a system of forces, the vector sum and the moment sum about a point are both zero, then the system is in equilibrium.*

By Cor. 2, § 8.4.1., page 213 moment sum about every point is zero.

We replace the given system by an equivalent system of three forces having any three points L, M, N as their points of application. Let $\mathbf{F}_1$, $\mathbf{F}_2$, $\mathbf{F}_3$ be these three forces.

The vector sum and the moment sum are both zero for the new system also. (§ 8.5, page 213)

Take any three points A, B, C on the lines of action of the forces.

Equating to zero the moment sum about A, we obtain

$$\overrightarrow{AB} \times \mathbf{F}_2 + \overrightarrow{AC} \times \mathbf{F}_3 = \mathbf{0},$$

so that the vectors $\overrightarrow{AB} \times \mathbf{F}_2$ and $\overrightarrow{AC} \times \mathbf{F}_3$ are collinear. Thus, there exists a linear normal to the lines AB, AC and the lines of action of the forces $\mathbf{F}_2$ and $\mathbf{F}_3$. Accordingly the lines of action of the forces lie in a plane determined by the intersecting lines AB, AC. In other words, the lines of action of the forces $\mathbf{F}_2$ and $\mathbf{F}_3$ are coplanar and the plane containing the same passes through the point A of the line of action of $\mathbf{F}_1$. As the point A is arbitrary, we deduce that the three lines of action of the three forces are coplanar.

Now two cases arise according as some two of the three lines of action are non-parallel or all three are parallel.

In the first case, suppose that the lines of action of $\mathbf{F}_1$ and $\mathbf{F}_2$ are non-parallel so that they meet at a point say, O. Equating to zero the moment sum about O, we see that O lies on the line of action $\mathbf{F}_3$ also. Thus, the three lines of action are concurrent in this case. These three concurrent forces are equivalent to a single force whose line of action passes through O and whose vector is equal to the sum of the vectors of the forces. The latter being zero, we deduce that the system is in equilibrium.

Suppose now that the *lines of action of three forces are all parallel.*

As the vector sum of the forces is zero, the three parallel forces cannot all have the same sense.

Suppose that **P** and **Q** are having the same sense opposite to that of **R.**

The forces **P** and **Q** are equivalent to a single force parallel to the given force with vector **P + Q.**

Equating to zero the moment about any point A on the line of action of **P + Q,** we see that the point A lies on the line of action of **R.** Thus, R and **P + Q** have the same line of action. Also $\mathbf{P} + \mathbf{Q} + \mathbf{R} = 0$. Hence, the system is in equilibrium.

8.9. CRITERIA FOR EQUIVALENT SYSTEMS

Theorem. *Two systems of forces having the same vector sum and the same moment sum about a point are equivalent.*

Let there be two systems of forces

$$\mathbf{F}_1, \mathbf{F}_2, \ldots\ldots\ldots$$

$$\mathbf{F}_1', \mathbf{F}_2', \ldots\ldots\ldots$$

having the same vector sum **R** and the same moment sum **G** about a point **O**.

Consider now a new system equivalent to the first obtained by introducing the pairs of forces

$$\mathbf{F}_1', -\mathbf{F}_1'; \ \mathbf{F}_2', -\mathbf{F}_2' \ldots\ldots$$

to the same. For the part of this new system consisting of the forces

$$\mathbf{F}_1, \quad \mathbf{F}_2, \ldots\ldots\ldots$$

$$-\mathbf{F}_1', -\mathbf{F}_2', \ldots\ldots\ldots$$

the vector sum and the moment sum about O are both zero and as such this part is in equilibrium. Thus, we are left with the forces

$$\mathbf{F}_1', \mathbf{F}_2', \ldots\ldots\ldots\ldots$$

forming a system equivalent to the first.

Hence the result.

8.10. COUPLES

Def. *A system consisting of a pair of equal unlike parallel forces is called a Couple.*

The vector sum of the two forces of a couple is clearly zero.

About moment sum, we have the following important result :

The moment sum of the two forces of a couple is the same about every point.

Let the lines of action of the forces of the couple be two parallel lines. Let A, C be any two points on these parallel lines. The two forces have equal magnitudes but opposite sense.

Thus, if $\mathbf{F}$ denote the vector of one force, then $-\mathbf{F}$ denotes that of the other.

The moment sum of the forces of the couple about any point O

$$= \vec{OA} \times \mathbf{F} + \vec{OC} \times (-\mathbf{F})$$

$$= (\vec{OA} - \vec{OC}) \times \mathbf{F} = \vec{CA} \times \mathbf{F},$$

so that the moment sum is independent of the point O. Here A, C may be thought of as any two points on the lines of action of the two forces.

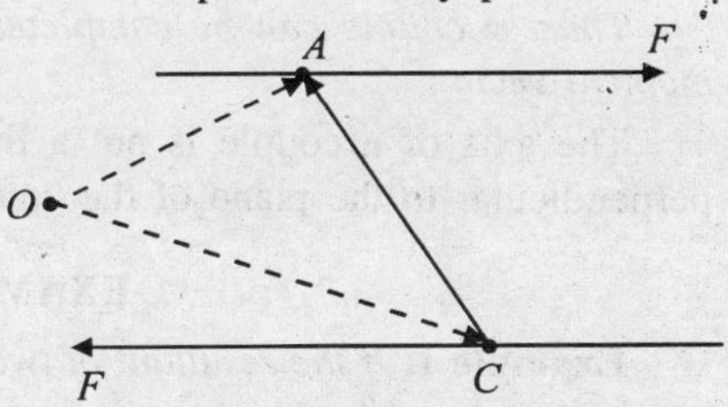

Fig. 8.3.

The result obtained above is contained in the general theorem that if the vector sum of the forces of a system is zero, then the moment sum about every point is the same.

Moment of a Couple. *The constant moment sum of the forces of a couple is called the moment of the couple.*

Axis of a Couple. *Any line perpendicular to the plane of a couple is called the axis of the couple* so that the axis of a couple is parallel to its moment vector.

Thus, the moment of a couple is a vector perpendicular to the plane of the couple and its magnitude is the product of the magnitude of either force with the perpendicular distance between the lines of action of the forces.

8.10.1. Equivalent Couples

Theorem. *Two couples with equal moments are equivalent*, for the vector and the moment sum are the same for each couple (§ 8.6, Page 209).

Two couples with equal moments necessarily lie in the same or parallel planes.

8.10.2. Composition of Couples

Theorem. *Two couples with moments* $\mathbf{M}_1$ *and* $\mathbf{M}_2$ *are equivalent to a single couple with moment* $\mathbf{M}_1 + \mathbf{M}_2$.

This result is also an immediate consequence of the Theorem of § 8.9.

8.10.3. Couples Represented by Vectors

We have seen that two couples with equal moments are equivalent and as such a couple is completely known by its moment which is a vector.

Also a system consisting of two couples is equivalent to a third couple whose moment is the vector sum of the moments of the two component couples.

Finally any vector equal to the moment vector of a couple is also the moment vector of the couple.

Thus, a couple can be completely characterised by a vector, viz., its moment vector.

The axis of a couple is not a fixed line; it can, in fact, be any line perpendicular to the plane of the couple.

EXAMPLES

Example 1. *If the resultant of two forces be equal in magnitude to one of the components and perpendicular to it in direction, find the other component.*

Solution. Let P and Q be the forces inclined at an angle θ to each other. Let $\mathbf{i}$ and $\mathbf{j}$ be the direction of P and perpendicular to it respectively. Along $\mathbf{j}$ the resultant of forces P and Q is acting.

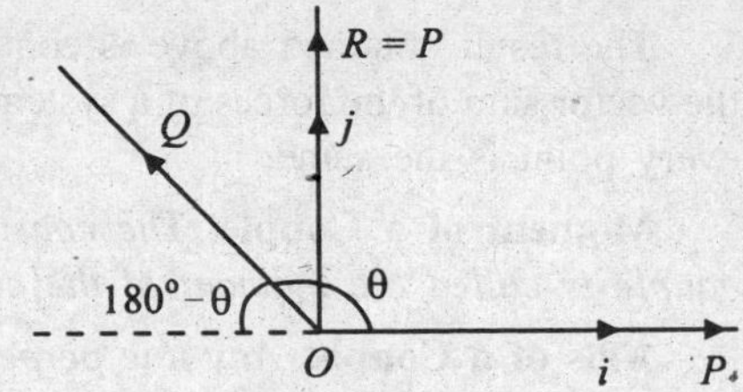

Fig. 8.4.

$$\therefore \quad P\mathbf{i} + \{-Q\cos(180^\circ - \theta)\,\mathbf{i} + Q\sin(180^\circ - \theta)\,\mathbf{j}\} = P\mathbf{j}$$

$$\Rightarrow \quad (P + Q\cos\theta)\,\mathbf{i} + (Q\sin\theta - P)\,\mathbf{j} = 0$$

Since $\mathbf{i}$ and $\mathbf{j}$ are non-collinear vectors, hence

$$P + Q\cos\theta = 0 \quad \text{and} \quad Q\sin\theta - P = 0$$

$$\therefore \quad Q\sin\theta = P \quad \text{and} \quad Q\cos\theta = -P$$

$\Rightarrow \qquad \tan \theta = -1 \;\Rightarrow\; \theta = 135°$

and $\qquad Q^2 \sin^2 \theta + Q^2 \cos^2 \theta = P^2 + P^2 \;\Rightarrow\; Q = \sqrt{2}P$

Example 2. *Forces P, Q act at O and have a resultant R. If any transversal cuts their lines of action at A, B, C respectively, show that*

$$P/OA + Q/OB = R/OC.$$

Solution. We have

$$\mathbf{P} + \mathbf{Q} = \mathbf{R} \qquad ...(1)$$

Let a transversal cuts the lines of action of P, Q and R at A, B, C. Let ON is $\perp$ to AB. Let $\hat{\mathbf{n}}$ be a unit vector along ON and $\angle BON = \alpha$, $\angle AON = \beta$ and $\angle CON = \gamma$.

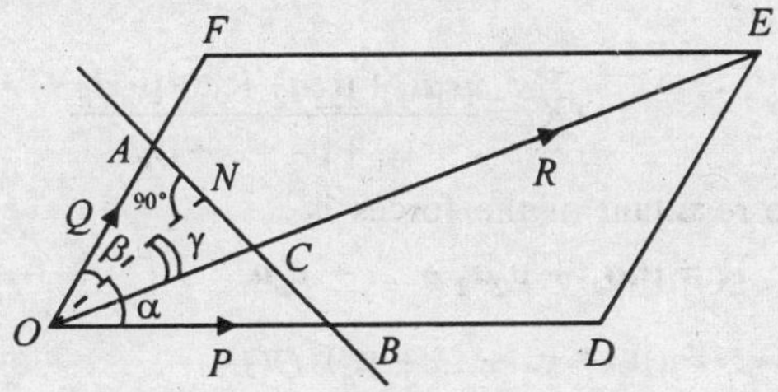

Fig. 8.5.

$(1) \Rightarrow \qquad \mathbf{P} \cdot \hat{\mathbf{n}} + \mathbf{Q} \cdot \hat{\mathbf{n}} = \mathbf{R} \cdot \hat{\mathbf{n}}$

$\Rightarrow \qquad P \cos \alpha + Q \cos \beta = R \cos \gamma$

From $\Delta\, AON$, $\Delta\, BON$ and $\Delta\, CON$,

$$\cos \alpha = \frac{ON}{OA}, \; \cos \beta = \frac{ON}{OB}, \; \cos \gamma = \frac{ON}{OC}$$

$$\therefore \qquad P\left(\frac{ON}{OA}\right) + Q\left(\frac{ON}{OB}\right) = R\left(\frac{ON}{OC}\right)$$

$$\Rightarrow \qquad \frac{P}{OA} + \frac{Q}{OB} = \frac{R}{OC}.$$

Example 3. *The forces P and Q act at a point O. Their resultant R is such that R = P. If P is doubled, show that the new resultant is at right angles to Q.*

Solution. Given $\mathbf{P} + \mathbf{Q} = \mathbf{R}$

Let $\mathbf{S}$ be the resultant of $2\mathbf{P}$ and Q.

$\therefore \qquad 2\mathbf{P} + \mathbf{Q} = \mathbf{S}$

Now $\qquad \mathbf{S} \cdot \mathbf{Q} = (2\mathbf{P} + \mathbf{Q}) \cdot \mathbf{Q}$

$\qquad\qquad = (\mathbf{R} + \mathbf{P}) \cdot (\mathbf{R} - \mathbf{P})$

$\qquad\qquad = R^2 - P^2 = 0.$

$\Rightarrow$ $\mathbf{S}$ is perpendicular to $\mathbf{Q}$.

Example 4. *A particle is acted on by a number of centres of forces some of which attract and some repel, the forces in each case varying as the distance. The intensities for different centres are different. Prove that the resultant passes through a fixed point for all positions of the particle.*

Solution. Let O be the origin. Let the given particle be situated at O. Let the position vectors of various centres of forces $A_1, A_2, \ldots, A_n$ be $\boldsymbol{a}_1, \boldsymbol{a}_2, \ldots, \boldsymbol{a}_n$. Hence, the forces acting on the particle can be taken as $\mu_1\boldsymbol{a}_1, \mu_2\boldsymbol{a}_2, \ldots, \mu_n\boldsymbol{a}_n$, where $\mu_1, \mu_2, \ldots, \mu_n$ are intensities of the centres and may be positive or negative accordingly as the centres attract or repel.

Let G be the centroid of $\boldsymbol{a}_1, \boldsymbol{a}_2, \ldots, \boldsymbol{a}_n$ with associated numbers $\mu_1, \mu_2, \ldots, \mu_n$. Then

$$\overrightarrow{OG} = \frac{\mu_1\boldsymbol{a}_1 + \mu_2\boldsymbol{a}_2 + \ldots + \mu_n\boldsymbol{a}_n}{\mu_1 + \mu_2 + \ldots + \mu_n}$$

Then **R**, the resultant of the forces is

$$\mathbf{R} = \mu_1\boldsymbol{a}_1 + \mu_2\boldsymbol{a}_2 + \ldots + \mu_n\boldsymbol{a}_n$$

$$= (\mu_1 + \mu_2 + \ldots + \mu_n)\ \overrightarrow{OG}$$

Hence, the resultant is a force passing through G, which is independent of the origin. Hence, the resultant passes through the fixed point G for all positions of the particle.

Example 5. *ABCDEF is a regular hexagon and O any point in the plane. Show that the forces represented both in magnitude and direction by OA, OC, OE, BO, DO, FO are in equilibrium.*

Solution. We have

$$\overrightarrow{OA}+\overrightarrow{OC}+\overrightarrow{OE}+\overrightarrow{BO}+\overrightarrow{DO}+\overrightarrow{FO} = (\overrightarrow{DO}+\overrightarrow{OC})+(\overrightarrow{FO}+\overrightarrow{OA})+(\overrightarrow{BO}+\overrightarrow{OE})$$

$$= \overrightarrow{DC}+\overrightarrow{FA}+\overrightarrow{BE}$$

$$= \frac{1}{2}\overrightarrow{EB}+\frac{1}{2}\overrightarrow{EB}-\overrightarrow{EB}$$

$$= \overrightarrow{EB}-\overrightarrow{EB} = 0$$

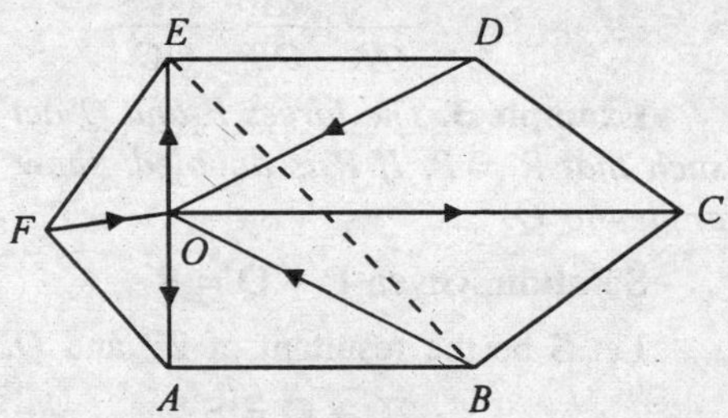

Fig. 8.6.

$\Rightarrow$ the given forces are in equilibrium.

Example 6. *Find the vector moment of the three forces* $\mathbf{i} + 2\mathbf{j} - 3\mathbf{k}$, $2\mathbf{i} + 3\mathbf{j} + 4\mathbf{k}$ *and* $-\mathbf{i} - \mathbf{j} + \mathbf{k}$ *acting on a particle at point P (0, 1, 2) about A (1, – 2, 0).*

Solution. The resultant of three forces

$$\mathbf{F} = (\mathbf{i} + 2\mathbf{j} - 3\mathbf{k}) + (2\mathbf{i} + 3\mathbf{j} + 4\mathbf{k}) + (-\mathbf{i} - \mathbf{j} + \mathbf{k})$$
$$= 2\mathbf{i} + 4\mathbf{j} + 2\mathbf{k},$$

acting at P.

$$\overrightarrow{AP} = (0, 1, 2) - (1, -2, 0) = -\mathbf{i} + 3\mathbf{j} + 2\mathbf{k}$$

$\therefore$ Moment of $\mathbf{F}$ acting at P about A

$$= \overrightarrow{AP} \times \mathbf{F} = (-\mathbf{i} + 3\mathbf{j} + 2\mathbf{k}) \times (2\mathbf{i} + 4\mathbf{j} + 2\mathbf{k})$$

$$= \begin{vmatrix} \mathbf{i} & \mathbf{j} & \mathbf{k} \\ -1 & 3 & 2 \\ 2 & 4 & 2 \end{vmatrix} = -2\mathbf{i} + 6\mathbf{j} - 10\mathbf{k}.$$

EXERCISES

1. ABC is a triangle and D, E, F are the mid-points of the sides. Forces represented by AD, $\frac{2}{3}BE$ and $\frac{1}{3}CF$ act on a particle at a point where AD and BE meet. Show that the resultant is represented both in magnitude and direction by $\frac{1}{2}AC$ and the line of action of the resultant divides BC in the ratio 2 : 1.

2. $ABCD$ is a quadrilateral and E and F are the mid-points of the sides AB and CD. Show that the resultant of forces PA, PB, PC, PD is 4 PG where G is the middle point of EF, wherever the point P may be. When will these forces be in equilibrium ?

3. Three forces P, Q, R act along three straight lines OA, OB, OC. If their resultant is parallel to the plane ABC, prove that

$$(P/OA) + (Q/OB) + (R/OC) = 0.$$

4. Forces act at the vertices of a tetrahedron along the lines joining them to the centroids of opposite faces and proportional to the lengths of the same. Show that the system is in equilibrium.

5. Forces act at the circumcentre O of a triangle ABC equal to $\overrightarrow{OA}$, $\overrightarrow{OB}$ and $\overrightarrow{OC}$. Show that their resultant is $\overrightarrow{OH}$ where H is the orthocentre of $\Delta\, ABC$.

6. Three forces $\overrightarrow{OA}, \overrightarrow{OB}, \overrightarrow{OC}$ diverge from the point O and three forces $\overrightarrow{AO'}, \overrightarrow{BO'}, \overrightarrow{CO'}$ converge to the point O'. Show that the resultant of six forces is represented in direction and magnitude by $3\overrightarrow{OO'}$.

7. H is the orthocentre of $\Delta\, ABC$. Show that the resultant of forces acting on a particle and represented by AH, HB, HC is represented by the diameter through A of the circumcentre of $\Delta\, ABC$.

8. Show that the moment about the point $\mathbf{i} + 2\mathbf{j} - \mathbf{k}$ of a force represented by $\mathbf{i} + 2\mathbf{j} + \mathbf{k}$ acting through the point $2\mathbf{i} + 3\mathbf{j} + \mathbf{k}$ is $-3\mathbf{i} + \mathbf{j} + \mathbf{k}$.

9. A force with components $(-7, 4, 5)$ acts at the point $(2, 4, -3)$. Find its moment about the origin. Find also its moment about the line $L : x = y = z$; the positive direction on the line being that in which x increases.

10. Prove that the sum of moment of two intersecting forces about a point in their plane is equal to the moment of their resultant about that point.

11. Define the moment of a vector about a point. Find the moment about a corner of a cube of three unit vectors converging on the opposite corner along three edges. Show that the sum of these moments is zero.

12. $ABCD$ is a quadrilateral. Find by the vector method the position of a point O inside the quadrilateral such that the forces represented by $\overrightarrow{OA}$, $\overrightarrow{OB}$, $\overrightarrow{OC}$ and $\overrightarrow{OD}$ may be in equilibrium.

8.11. REDUCTION TO SIMPLER SYSTEMS

8.11.1. Theorem

Every system of forces acting on a rigid body is equivalent to a single force acting at an arbitrary point together with a couple whose moment is the sum of the moments of the forces about the point.

Let O be any arbitrary point and let $\mathbf{R}$ be the vector sum and $\mathbf{G}$, the moment sum of the system about O.

Then the system consisting of

(*i*) the force whose vector is $\mathbf{R}$ and whose point of application is O, and

(*ii*) a couple whose moment is $\mathbf{G}$.

is clearly equivalent to the given system, for both the systems have the same vector sum and the same moment sum about a point O.

8.11.2. Poinsot's Central Axis. Wrench

Theorem. *Every system of forces is equivalent to a force and a couple such that the line of action of the force is the axis of the couple.*

Suppose that such a reduction is possible so that there exists a line such that the system is equivalent to a force along the line and a couple whose axis is the line. Then *the moment sum of the system about any point of this line is parallel to the vector sum of the system.*

Now the moment sum about any point O' is

$$\mathbf{G} - \mathbf{r} \times \mathbf{R} \text{ where } \overrightarrow{OO'} = \mathbf{r}.$$

This will be parallel to **R**, if

$$(\mathbf{G} - \mathbf{r} \times \mathbf{R}) \times \mathbf{R} = 0 \qquad ...(i)$$

$$\Rightarrow \quad \mathbf{G} \times \mathbf{R} - (\mathbf{r} \cdot \mathbf{R})\,\mathbf{R} + (\mathbf{R} \cdot \mathbf{R})\,\mathbf{r} = 0$$

$$\Rightarrow \quad \mathbf{r} = -\frac{\mathbf{G} \times \mathbf{R}}{\mathbf{R} \cdot \mathbf{R}} + t\mathbf{R}, \qquad ...(ii)$$

where t is a scalar. Conversely, it may be easily seen that **r** given by (*ii*) satisfies (*i*) for every value of the scalar t.

The points given by (*ii*) for different values of the scalar t, lie on a line.

Thus, we have obtained a line whose vector equation is (*ii*) such that for every point O' on the line, the moment sum about O' is parallel to the vector sum.

This establishes the existence and uniqueness of a line such that the system is equivalent to a force along the line and a couple whose axis is the line.

This line is known as the *Poinsot's Central axis.*

Its equation is

$$\mathbf{r} = -\frac{\mathbf{G} \times \mathbf{R}}{\mathbf{R} \cdot \mathbf{R}} + t\mathbf{R} = \frac{\mathbf{G} \times \mathbf{R}}{\mathbf{R} \cdot \mathbf{R}} + t\mathbf{R};$$

R being the vector sum and **G** the moment sum about the origin of reference.

Wrench. Def. *A system consisting of a force and a couple such that the line of the force is the axis of the couple is called a Wrench.*

Thus, what we have proved above is equivalent to saying that *any given system of forces is equivalent to a wrench.*

8.11.3. Moment of the Couple of the Wrench equivalent to a given System

The moment sum about any point

$$-\frac{\mathbf{G} \times \mathbf{R}}{\mathbf{R} \cdot \mathbf{R}} + t\mathbf{R}$$

of the central axis

$$= \mathbf{G} - \left(-\frac{\mathbf{G} \times \mathbf{R}}{\mathbf{R} \cdot \mathbf{R}} + t\mathbf{R} \right) \times \mathbf{R}$$

$$= \mathbf{G} + \frac{1}{\mathbf{R} \cdot \mathbf{R}} [(\mathbf{G} \cdot \mathbf{R})\,\mathbf{R} - (\mathbf{R} \cdot \mathbf{R})\,\mathbf{G}]$$

$$= \frac{\mathbf{G} \cdot \mathbf{R}}{\mathbf{R} \cdot \mathbf{R}} \mathbf{R} = p\mathbf{R}, \text{ say,}$$

where p is a scalar.

8.11.4. Intensity and Pitch of a Wrench

The magnitude $|\mathbf{R}|$ of the force **R** of a wrench is called its *Intensity.*

The moment of the couple of a wrench is necessarily the product of the force vector **R** of the wrench by a scalar. This scalar is called the *Pitch* of the wrench. It is usual to denote a wrench whose force is **R** and pitch p by the symbol

$$(\mathbf{R}, p\mathbf{R})$$

so that $p\mathbf{R}$ is the couple of the wrench.

It will be useful to remember that the wrench equivalent to a system of forces for which the vector sum is **R** and moment sum about the origin of reference O is **G** is

$$(\mathbf{R}, p\mathbf{R})$$

where

$$p = \frac{\mathbf{G}\,.\,\mathbf{R}}{\mathbf{R}\,.\,\mathbf{R}}.$$

Also the equation of the axis of the wrench is

$$\mathbf{r} = -\frac{\mathbf{G} \times \mathbf{R}}{\mathbf{R}\,.\,\mathbf{R}} + t\mathbf{R}.$$

8.11.5. Degenerate and Non-degenerate Wrenches

Def. *A wrench is said to be degenerate if either the force or the couple of the wrench is zero.*

If the force and the couple be both zero, then the system is in equilibrium.

8.12. INVARIANTS OF A SYSTEM OF FORCES

Let **R** be the vector sum and **G** the moment sum about a point O of a given system of forces.

Clearly **R** is invariant in as much as it does not depend upon the choice of the point O. The scalar $\mathbf{R}\,.\,\mathbf{R}$ is also thus an invariant.

In addition to the above, there is also another scalar invariant, *viz.*,

$$\mathbf{G}\,.\,\mathbf{R},$$

i.e., the scalar product of the moment sum and the vector sum.

The moment sum $\mathbf{G}'$ about a point O' such that

$$\overrightarrow{OO'} = \mathbf{r},$$

is given by

$$\mathbf{G}' = \mathbf{G} - \mathbf{r} \times \mathbf{R}.$$

$$\Rightarrow \qquad \mathbf{G}'\,.\,\mathbf{R} = (\mathbf{G} - \mathbf{r} \times \mathbf{R})\,.\,\mathbf{R} = \mathbf{G}\,.\,\mathbf{R}.$$

The invariance of **G . R** is also an immediate consequence of the invariance of the pitch **G . R / R . R** of the unique equivalent wrench and of the invariance of **R . R**, as seen above.

Parameter. The invariant **G . R** is called Parameter of the system and it plays an important part in investigation concerning systems of forces.

8.12.1. Geometrical Interpretation of the Parameter **G . R**

Firstly we consider a system consisting of two forces $\mathbf{F}_1, \mathbf{F}_2$. Let $\mathbf{r}_1, \mathbf{r}_2$ be the position vectors of any two points A, B on the lines of action of the forces. For these two forces, we have

$$\mathbf{R} = \mathbf{F}_1 + \mathbf{F}_2, \quad \mathbf{G} = \mathbf{r}_1 \times \mathbf{F}_1 + \mathbf{r}_2 \times \mathbf{F}_2.$$

$$\Rightarrow \quad \mathbf{R} \cdot \mathbf{G} = (\mathbf{F}_1 + \mathbf{F}_2) \cdot (\mathbf{r}_1 \times \mathbf{F}_1 + \mathbf{r}_2 \times \mathbf{F}_2)$$

$$= (\mathbf{F}_2 \times \mathbf{F}_1) \cdot (\mathbf{r}_1 - \mathbf{r}_2).$$

where

$$\mathbf{r}_2 - \mathbf{r}_1 = \overrightarrow{AB}.$$

If the two forces be denoted by the line vectors $\overrightarrow{AC}$, $\overrightarrow{BD}$ then

$$\mathbf{R} \cdot \mathbf{G} = \overrightarrow{BD} \times \overrightarrow{AC} \cdot \overrightarrow{AB},$$

so that the parameter of two forces is six times the algebraic value of the volume of the tetrahedron formed by the line vectors representing the given forces as a pair of opposite edges.

Any system of forces. Consider now a system consisting of any number of forces

$$\mathbf{F}_1, \mathbf{F}_2, \ldots\ldots.$$

Let $\overrightarrow{A_1B_1}$, $\overrightarrow{A_2B_2}, \ldots$ be the line vectors denoting the forces. For this system,

$$\mathbf{R} \cdot \mathbf{G} = (\mathbf{F}_1 + \mathbf{F}_2 + \ldots) \cdot (\overrightarrow{OA_1} \times \mathbf{F}_1 + \overrightarrow{OA_2} \times \mathbf{F}_2 + \ldots)$$

$$= \Sigma\, (\mathbf{F}_i \times \mathbf{F}_i),\ (\overrightarrow{OA_i} - \overrightarrow{OA_j}),$$

where the summation extends to every pair of forces.

$$\therefore \quad \mathbf{R} \cdot \mathbf{G} = \Sigma\, (\mathbf{F}_i \times \mathbf{F}_j) \cdot (\overrightarrow{A_i\, A_j}) = \Sigma\, (\overrightarrow{A_iB_i} \times \overrightarrow{A_jB_j} \cdot \overrightarrow{A_jA_i}),$$

so that the parameter of any system of forces is six times the algebraic sum of the volumes of the tetrahedra formed by pairs of line vectors denoting forces of the system as pairs of opposite edges.

The above geometric manner of the representation of **R . G** gives another proof of invariance of the same.

8.13. PARAMETERS FOR SOME SPECIAL SYSTEMS OF FORCES

It should be noticed that the parameter **R . G** is zero, if **R** = 0 or **G** = 0 or R is perpendicular to **G.**

I. *The parameter for a system of coplanar forces is zero.*

The moment vector of every force of the system about a point *O* of the plane is perpendicular to the plane and as such the moment sum **G** of the forces, about *O* is perpendicular to the plane. Also the vector **R** is parallel to the plane. Thus, **G . R** = 0.

II. *The parameter for a system of parallel forces is zero.*

The moment vector about any point *O* of a force of the system lies in the plane through *O* perpendicular to the force. As the forces are parallel, the plane π through *O* perpendicular to any one force is also perpendicular to the others. Thus, the moment sum **G** lies in the plane π.

Also the vector sum **R** is perpendicular to the plane π. Thus, **G . R** = 0.

III. *The parameter of a system of concurrent forces is zero.*

The proof follows from the fact that the moment vector of each force about the point of the concurrence *O* is zero and as a consequence their sum **G** is zero.

IV. *The parameter of a system consisting of any number of couples is zero.*

The proof follows from the fact that for such a system the vector sum **R** is zero.

V. *The parameter of a system consisting of two non-zero forces lying along non-coplanar lines cannot be zero.*

Let $\mathbf{F}_1$, $\mathbf{F}_2$ be two forces acting along two skew lines *AB, CD*. We have,

$$\mathbf{R} = \mathbf{F}_1 + \mathbf{F}_2$$

$$\mathbf{G} = \overrightarrow{AC} \times \mathbf{F}_2,$$

where **G** denotes moment sum about *A*.

Thus,

$$\mathbf{GR} = (\overrightarrow{AC} \times \mathbf{F}_2) \cdot (\mathbf{F}_1 + \mathbf{F}_2)$$

$$= \overrightarrow{AC} \times \mathbf{F}_2 \cdot \mathbf{F}_1 \neq 0,$$

for the lines of action of $\mathbf{F}_2$, $\mathbf{F}_1$ are not coplanar.

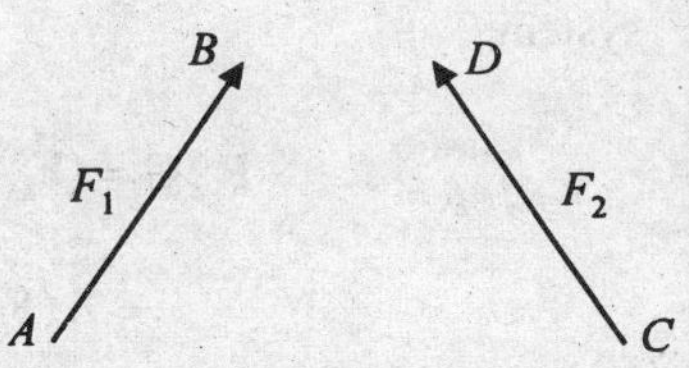

Fig. 8.7.

8.14. CRITERIA FOR THE NATURE OF A SYSTEM OF FORCES

Theorem.

I. If $\mathbf{R} \cdot \mathbf{G} \neq 0$, the system is equivalent to a non-degenerate wrench.

II. If $\mathbf{R} \cdot \mathbf{G} = 0$ then the system is

(*i*) equivalent to a couple of moment $\mathbf{G}$ if $\mathbf{R} = 0$ but $\mathbf{G} \neq 0$,

(*ii*) equivalent to a single force if $\mathbf{R} \neq 0$,

(*iii*) in equilibrium if $\mathbf{R} = 0$ as well as $\mathbf{G} = 0$.

The above criteria are immediate consequences of the result that any given systems of forces is equivalent to a wrench

$$(\mathbf{R}, p\mathbf{R}) \text{ where } p = \mathbf{G} \cdot \mathbf{R} / \mathbf{R} \cdot \mathbf{R};$$

$\mathbf{R}$ being the vector sum and $\mathbf{G}$ the moment sum of the forces of the system about a point.

Cor. A system of coplanar forces must be equivalent to a single force or a couple, unless it be in equilibrium, for the parameter of such a system is zero.

Similarly, a system of parallel forces, unless it be in equilibrium, is equivalent to a single force or a couple.

8.15. CENTRE OF A SYSTEM OF PARALLEL FORCES

Let

$$\mathbf{r}_1, \mathbf{r}_2, \ldots$$

be the position vectors of any points on the lines of action of the parallel forces

$$\mathbf{F}_1, \mathbf{F}_2, \ldots$$

Let $\mathbf{a}$ denote a unit vector parallel to the forces. We write

$$\mathbf{F}_1 = p_1\mathbf{a}, \quad \mathbf{F}_2 = p_2\mathbf{a}, \quad \text{etc.,}$$

so that $|p_1|$, $|p_2|$, etc., denote the magnitudes of the forces.

We have

$$\begin{aligned} \mathbf{R} &= \mathbf{F}_1 + \mathbf{F}_2, \ldots \\ &= (p_1 + p_2 + \ldots)\,\mathbf{a} = (\Sigma p_i)\,\mathbf{a}. \\ \mathbf{G} &= \mathbf{r}_1 \times p_1\mathbf{a} + \mathbf{r}_2 \times p_2\mathbf{a} + \ldots \\ &= p_1\mathbf{r}_1 \times \mathbf{a} + p_2\mathbf{r}_2 \times \mathbf{a} + \ldots \\ &= (\Sigma p_i\mathbf{r}_i) \times \mathbf{a}. \end{aligned}$$

If $\Sigma p_i = 0$, *i.e.*, $\mathbf{R} = 0$, the system is equivalent to a couple unless $\mathbf{G}$ is also zero, in which case the system will be in equilibrium.

Suppose now that $\Sigma p_i \neq 0$. We shall obtain points about which the moment sum is zero. Let $\mathbf{r}$ be the position vector of such a point.

The moment sum of the system about a point with position vector **r** is

$$= \mathbf{G} - \mathbf{r} \times \mathbf{R} = (\Sigma p_i \mathbf{r}_i) \times \mathbf{a} - (\Sigma p_i)\, \mathbf{r} \times \mathbf{a}$$
$$= [\Sigma p_i \mathbf{r}_i - (\Sigma p_i)\, \mathbf{r}] \times \mathbf{a}.$$

Equating it to zero, we see that **r** is given by

$$\Sigma p_i \mathbf{r}_i - (\Sigma p_i)\, \mathbf{r} = t\, \mathbf{a},$$

where t is a scalar parameter,

$$\Rightarrow \qquad \mathbf{r} = \frac{\Sigma p_i \mathbf{r}_i}{\Sigma p_i} - \frac{t}{\Sigma p_i}\, \mathbf{a}$$

which is the equation of the line of action of the resultant. Clearly this line passes through a point

$$\frac{\Sigma p_i \mathbf{r}_i}{\Sigma p_i}$$

which is independent of the actual direction of the forces and depends only on the magnitudes and the points of application of the forces.

The point is known as the *Centre of parallel forces.* Clearly this is the centroid of the weighted points

$$\mathbf{r}_1, \ \mathbf{r}_2, \ldots$$

the respective weights being

$$p_1, \ p_2, \ldots$$

8.16. MOMENT OF FORCE ABOUT A LINE

We have so far been concerned with moments of forces about points but we shall now introduce the concept of the moment of a force about a line.

Let **F** be any given force and, **1**, any directed line.

Def. *The moment of a force* **F** *about a directed line* **1** *is the projection on* **1** *of the moment vector of the force about any point on* **1**.

Thus, if O, P be two points on the line **1** such that the vector $\overrightarrow{OP}$ is of unit length, then the moment of **F** about the line **1** is the scalar triple product

$$\overrightarrow{OA} \times \mathbf{F} \,.\, \mathbf{OP}, \qquad \ldots(i)$$

where A is any point on the line of action of the force **F**.

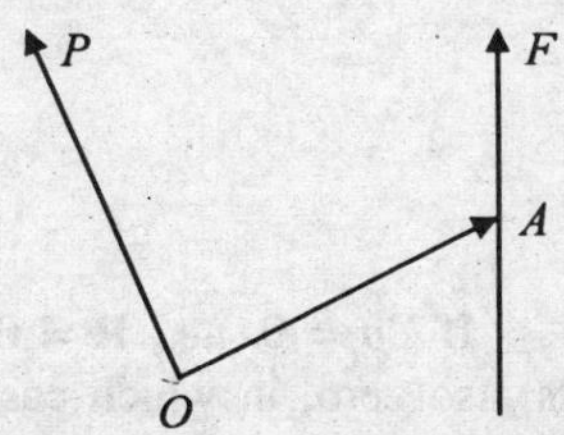

Fig. 8.8.

It may be easily seen that the moment of **F** about **1** is independent of the choice of the point O on the line **1**.

It should be carefully noted that the moments of forces about points are vectors whereas those about lines are scalars.

8.16.1. Geometrical Representation of the Moment of a Force about a Line

It may be easily seen from (*i*) above that *the moment of a force represented by a line vector* $\overrightarrow{AB}$ *about a directed line* **1** *is represented by six times the volume of the tetrahedron constructed with AB and any segment of unit length along* **I** *as a pair of opposite edges.*

It also follows from above that if α be the angle between the line of action of **F** and the line **1** and, c, be the length of shortest distance between the same, then the moment of **F** about **1** is

$$|\mathbf{F}|\, c \sin \alpha.$$

8.16.2. Vanishing of the Moment of a Force about a Line.

Theorem. *The moment of a non-zero force about a line is zero if, and only if, the line of action of the force is coplanar with the line for the moment*

$$\overrightarrow{OA} \times \mathbf{F} \cdot \overrightarrow{OP}$$

of a non-zero, **F** about the line $\overrightarrow{OP}$ will be zero if, and only if, the three vectors $\overrightarrow{OA}$, **F** and $\overrightarrow{OP}$ are all parallel to the same plane; *i.e.*, the line of action of **F** and the line OP are coplanar.

8.16.3. Moment of a System of Forces about a Line is, by definition, the sum of the moments of the forces of the system about the line. Thus, if $\overrightarrow{OP}$ be a unit vector along the given line and $\mathbf{F}_1$, $\mathbf{F}_2$, ... be the forces of the system and $\mathbf{r}_1$, $\mathbf{r}_2$, ... the position vectors of any points on their lines of action with respect to O then the moment of the system about the line

$$= \mathbf{r}_1 \times \mathbf{F}_1 \cdot \overrightarrow{OP} + \mathbf{r}_2 \times \mathbf{F}_2 \cdot \overrightarrow{OP} + \ldots$$

$$= (\mathbf{r}_1 \times \mathbf{F}_1 + \mathbf{r}_2 \times \mathbf{F}_2 + \ldots) \cdot \overrightarrow{OP} = \mathbf{G} \cdot \overrightarrow{OP},$$

so that the moment of a system about any line is equal to the projection on the line of the moment sum of the system about any point of the line.

From this it follows that the moment about a line of a system remains unchanged when we pass from a given system to another equivalent system.

8.17. NULL LINES. NULL PLANES AND NULL POINTS

8.17.1. Null Lines

Def. *Any line about which the moment of a system is zero is called a Null line.*

Clearly any given line will be a null line if the moment sum of the system about a point of the line is perpendicular to the line.

Null lines through a given point. Any line perpendicular to the moment sum about the point is a null line and as such we see that the *null lines through a point lie in a plane which passes through the point and is perpendicular to the moment sum about the point.*

This plane is called the *Null plane of the point.*

If $\mathbf{a}$ be the position vector of the given point, then the moment sum of the system about the point is $\mathbf{G} - \mathbf{a} \times \mathbf{R}$ and accordingly the equation of the null plane of the point is

$$(\mathbf{r} - \mathbf{a}) \,.\, (\mathbf{G} - \mathbf{a} \times \mathbf{R}) = 0$$

$$\Rightarrow \qquad \mathbf{r} \,.\, (\mathbf{G} - \mathbf{a} \times \mathbf{R}) = \mathbf{a} \,.\, \mathbf{G}.$$

Also a point is called the *null point of its null plane.*

It is clear that a null line lies in the null plane of every point of the line.

8.17.2. Theorem

If the null plane of a point A passes through a point B then the null plane of B passes through A.

As the null plane of *A* passes through *B*, the line *AB, i.e., BA* is a null line and as such it lies in the null plane of the point *B*. Hence, the null plane of *B* passes through *A*.

Analytically, we may see that the condition for the null plane

$$\mathbf{r} \,.\, (\mathbf{G} - \mathbf{a} \times \mathbf{R}) = \mathbf{a} \,.\, \mathbf{G}$$

of a point with position vector $\mathbf{a}$ to pass through the point with position vector $\mathbf{b}$ is

$$\mathbf{b} \,.\, (\mathbf{G} - \mathbf{a} \times \mathbf{R}) = \mathbf{a} \,.\, \mathbf{G}$$

$$\Rightarrow \qquad \mathbf{b} \,.\, \mathbf{G} - \mathbf{a} \,.\, \mathbf{G} = [\mathbf{b}\ \mathbf{a}\ \mathbf{R}] \qquad ...(i)$$

which is also the condition for the null plane of $\mathbf{b}$, to pass through $\mathbf{a}$. In fact, interchanging $\mathbf{a}$ and $\mathbf{b}$ in (*i*), we arrive against at (*i*).

Cor. *Null planes of collinear points have a common line of intersection.*

8.17.3. Theorem

If a system of forces is equivalent to two forces then every line intersecting the lines of action of the two forces is a null line of the system.

This follows from the fact that the moment of each of the two forces about every line intersecting their lines of action is zero.

8.18. CONJUGATE FORCES

Theorem. *Any system of forces is, in general, equivalent to two forces one of which acts along a given line.*

Let OA be the given line.

The system is equivalent to a force $\mathbf{R}$ whose line of action OB passes through O and a couple of moment $\mathbf{G}$.

We suppose that

(*i*) OA is not a null line, (*ii*) $\mathbf{R} \neq \mathbf{0}$, and (*iii*) OB does not coincide with OA.

The null line of O is the plane of the couple $\mathbf{G}$.

Let the plane through OA and OB meet the null plane of O in a line OC.

The force $\mathbf{R}$ along OB is equivalent to two forces along OA, OC. Let these forces be $\mathbf{F}_1$, $\mathbf{F}_2$ respectively.

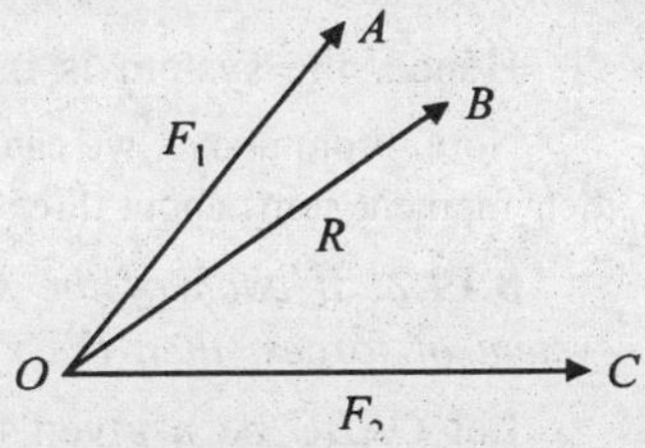

Fig. 8.9.

The couple $\mathbf{G}$ can be replaced by two parallel forces one of which is $-\mathbf{F}_2$ acting along CO and the other acts along another line in the plane of $\mathbf{G}$, *i.e.*, in the null plane of O. The two forces along OC and CO are in equilibrium and as such may be removed.

Thus, we are left with a force along the given line OA and another force whose line of action LM lies in the null plane of O.

As the system is equivalent to two forces along the lines OA and LM, every line intersecting these lines, is a null line of the system. Thus, the null plane of every point on one of these lines passes through the other.

8.18.1. Conjugate Forces and Conjugate Lines

Def. Two forces equivalent to a system of forces are called conjugate forces and their lines of action are called conjugate lines.

The line conjugate to any given line can be obtained as the line of intersection of the null planes of any two points of the given line.

8.19. OTHER FORMS OF CRITERIA FOR EQUILIBRIUM

8.19.1. *If the moment sum of a system about each of three non-collinear points is zero, then the system is in equilibrium.*

Let O, A, B be three non-collinear points. Let

$$\overrightarrow{OA} = \mathbf{a}, \quad \overrightarrow{OB} = \mathbf{b}.$$

Let **R** be the vector sum and **G**, the moment sum about O. Then the moment sums about A and B are

$$\mathbf{G} - \mathbf{a} \times \mathbf{R}, \quad \mathbf{G} - \mathbf{b} \times \mathbf{R}.$$

Equating to zero the moment sums about O, A, B, we have

$$\mathbf{0} = \mathbf{G} = \mathbf{G} - \mathbf{a} \times \mathbf{R} = \mathbf{G} - \mathbf{b} \times \mathbf{R},$$

$$\Rightarrow \quad \mathbf{G} = \mathbf{0}, \quad \mathbf{a} \times \mathbf{R} = \mathbf{0}, \quad \mathbf{b} \times \mathbf{R} = \mathbf{0}.$$

Now **R** cannot be parallel to both the vectors **a** and **b.** Thus,

$$\mathbf{R} = \mathbf{0}.$$

Also

$$\mathbf{G} = \mathbf{0}.$$

Hence, the system is in equilibrium.

Note. From above, we can deduce that two systems of forces are equivalent if their moment sums about three non-collinear points are equal.

8.19.2. *If each of the six edges of a tetrahedron is a null line of a system of forces, then the system is in equilibrium.*

Let $OABC$ be a given tetrahedron and let

$$\overrightarrow{OA} = \mathbf{a}, \quad \overrightarrow{OB} = \mathbf{b}, \quad \overrightarrow{OC} = \mathbf{c}.$$

Let **R** be the vector sum and **G**, the moment sum about O of the system of forces.

Equating to zero the moments about the lines OA, OB, OC, we obtain

$$\mathbf{G} \cdot \frac{\mathbf{a}}{|\mathbf{a}|} = 0, \quad \mathbf{G} \cdot \frac{\mathbf{b}}{|\mathbf{b}|} = 0, \quad \mathbf{G} \cdot \frac{\mathbf{c}}{|\mathbf{c}|} = 0,$$

$$\Rightarrow \quad \mathbf{G} \cdot \mathbf{a} = 0, \quad \mathbf{G} \cdot \mathbf{b} = 0, \quad \mathbf{G} \cdot \mathbf{c} = 0. \qquad \ldots(i)$$

Again, as the moment sums about A, B, C are

$$\mathbf{G} - \mathbf{a} \times \mathbf{R}, \quad \mathbf{G} - \mathbf{b} \times \mathbf{R}, \quad \mathbf{G} - \mathbf{c} \times \mathbf{R},$$

we have, equating to zero the moments about AB, BC, CA,

$$(\mathbf{G} - \mathbf{a} \times \mathbf{R}) \cdot \frac{\mathbf{b} - \mathbf{a}}{|\mathbf{b} - \mathbf{a}|} = 0, \quad (\mathbf{G} - \mathbf{b} \times \mathbf{R}) \cdot \frac{\mathbf{c} - \mathbf{b}}{|\mathbf{c} - \mathbf{b}|} = 0,$$

$$(\mathbf{G} - \mathbf{c} \times \mathbf{R}) \cdot \frac{\mathbf{a} - \mathbf{c}}{|\mathbf{a} - \mathbf{c}|} = 0 \qquad \ldots(ii)$$

The vector **a, b, c** being non-coplanar, we have, from (i)

$$\mathbf{G} = \mathbf{0}. \qquad (\S\ 2.7, \text{Page } 45)$$

Now the equations (*ii*) give

$$\mathbf{a} \times \mathbf{R} \,.\, (\mathbf{b} - \mathbf{a}) = 0, \;\; \mathbf{b} \times \mathbf{R} \,.\, (\mathbf{c} - \mathbf{b}) = 0, \;\; \mathbf{c} \times \mathbf{R} \,.\, (\mathbf{a} - \mathbf{c}) = 0.$$

which are equivalent to

$$\mathbf{R} \,.\, \mathbf{a} \times \mathbf{b} = 0, \;\; \mathbf{R} \,.\, \mathbf{b} \times \mathbf{c} = 0, \;\; \mathbf{R} \,.\, \mathbf{c} \times \mathbf{a} = 0.$$

The vectors $\mathbf{a} \times \mathbf{b}, \mathbf{b} \times \mathbf{c}, \mathbf{c} \times \mathbf{a}$ being non-coplanar, we obtain

$$\mathbf{R} = \mathbf{0}. \qquad \text{(§ 2.7, Page 45)}$$

Hence, the system is in equilibrium.

Note. The result obtained above enables us to state the criterion for the equivalence of two systems of forces in terms of moments about lines as follows :

Two systems of forces are equivalent if the moments of the two systems about each of the six edges of a tetrahedron are the same.

8.20. FORMULATION IN TERMS OF RECTANGULAR CARTESIAN AXES

Let

$$\mathbf{F}_1, \mathbf{F}_2, ..., \mathbf{F}_p, ...$$

be forces acting on a rigid body and let

$$P_1, P_2, ..., P_p, ...$$

be any points on their lines of action.

Let X_p, Y_p, Z_p, be the resolved parts of $\mathbf{F}_p$ along the three co-ordinate axes and (x_p, y_p, z_p) the co-ordinates of the point P_p. We have

$$\mathbf{F}_p = X_p\mathbf{i} + Y_p\mathbf{j} + Z_p\mathbf{k},$$
$$\mathbf{r}_p = x_p\mathbf{i} + y_p\mathbf{j} + z_p\mathbf{k},$$

where $\mathbf{r}_p$ is the position vector of the point $\mathbf{P}_p$ with reference to the origin **0**.

Moment of $\mathbf{F}_p$ *about the origin*

$$= \mathbf{r}_p \times \mathbf{F}_p$$
$$= (x_p\mathbf{i} + y_p\mathbf{j} + z_p\mathbf{k}) \times (Z_p\mathbf{i} + Y_p\mathbf{j} + Z_p\mathbf{k})$$
$$= (y_pZ_p - z_pY_p)\,\mathbf{i} + (z_pX_p - x_pZ_p)\,\mathbf{j} + (x_pY_p - x_pX_p)\,\mathbf{k}$$

Moment of $\mathbf{F}_p$ *about the coordinate axes.* We have :

$$\mathbf{r}_p \times \mathbf{F}_p \,.\, \mathbf{i} = y_pZ_p - z_pY_p, \;\; \mathbf{r}_p \times \mathbf{F}_p \,.\, \mathbf{j} = z_pX_p - x_pZ_p,$$
$$\mathbf{r}_p \times \mathbf{F}_p \,.\, \mathbf{k} = x_pY_p - y_pX_p$$

$$\Rightarrow \qquad y_pZ_p - z_pY_p, \;\; z_pX_p - x_pZ_p, \;\; x_pY_p - y_pX_p$$

are the moments of $\mathbf{F}_p$ about the co-ordinate axes.

The vector sum and the moment sum about O,

$$\mathbf{R} = \mathbf{i}\,\Sigma X_p + \mathbf{j}\,\Sigma Y_p + \mathbf{k}\,\Sigma Z_p$$

$$= \mathbf{i}X + \mathbf{j}Y + \mathbf{k}Z, \text{ say.}$$

$$\mathbf{G} = \mathbf{i}\,\Sigma\,(y_pZ_p - z_pY_p) + \mathbf{j}\,\Sigma\,(z_pX_p - x_pZ_p) + \mathbf{k}\,\Sigma\,(x_pY_p - y_pX_p)$$

$$= \mathbf{i}L + \mathbf{j}M + \mathbf{k}N.$$

Here L, M, N are the moments of the system about the co-ordinate axes.

Moment sum about a point, A (a_1, a_2, a_3).

The position vector of the point A being

$$\mathbf{a} = a_1\mathbf{i} + a_2\mathbf{j} + a_3\mathbf{k},$$

the moment sum about A

$$= \mathbf{G} - \mathbf{a} \times \mathbf{R}$$

$$= (L\mathbf{i} + M\mathbf{j} + N\mathbf{k}) - (a_1\mathbf{i} + a_2\mathbf{j} + a_3\mathbf{k}) \times (X\mathbf{i} + Y\mathbf{j} + Z\mathbf{k})$$

$$= [L - (a_2Z - a_3Y)]\,\mathbf{i} + [M - (a_3X - a_1Z)]\,\mathbf{j} + [N - (a_1Y - a_2X)]\,\mathbf{k}$$

Thus, the moments about the lines through A parallel to the co-ordinate axes are

$$L - (a_2Z - a_3Y),\quad M - (a_3X - a_1Z),\quad N - (a_1Y - a_2X).$$

Parameter of the system :

$$\mathbf{R}\,.\,\mathbf{G} = LX + MY + NZ.$$

Equations of the central axis. The vector equation of the axis is

$$\mathbf{r} = -\frac{\mathbf{G} \times \mathbf{R}}{\mathbf{R}\,.\,\mathbf{R}} + t\mathbf{R}$$

$$\Rightarrow \quad x\mathbf{i} + y\mathbf{j} + z\mathbf{k} = -\frac{(L\mathbf{i} + M\mathbf{j} + N\mathbf{k}) \times (X\mathbf{i} + Y\mathbf{j} + Z\mathbf{k})}{X^2 + Y^2 + Z^2} + t\,(X\mathbf{i} + Y\mathbf{j} + Z\mathbf{k})$$

$$= -\frac{1}{X^2 + Y^2 + Z^2}\{(MZ - NY)\,\mathbf{i} + (NX - LZ)\,\mathbf{j} + (LY - MX)\,\mathbf{k}\} + t\,(X\mathbf{i} + Y\mathbf{j} + Z\mathbf{k})$$

Thus,

$$\frac{x + \dfrac{MZ - NY}{\Sigma X^2}}{X} = \frac{y + \dfrac{NX - LZ}{\Sigma X^2}}{Y} = \frac{\dfrac{LY - MX}{\Sigma X^2}}{Z}$$

are the cartesian equations of the central axis.

The pitch $= \dfrac{\mathbf{R}\,.\,\mathbf{G}}{\mathbf{R}\,.\,\mathbf{R}} = \dfrac{LX + MY + NZ}{X^2 + Y^2 + Z^2}.$

The equations of the null plane of a point A (a_1, a_2, a_3).

If **a** denote the position vector of the point **A**, the vector equation of the null plane of A is

$$\mathbf{r} \cdot (\mathbf{G} - \mathbf{a} \times \mathbf{R}) = \mathbf{a} \cdot \mathbf{G}.$$

Making substitutions, we obtain

$$[L - (a_2 Z - a_3 Y)]\, x + [M - (a_3 X - a_1 Z)]\, y + [N - (a_1 Y - a_2 X)]\, z = a_1 L + a_2 M + a_3 N$$

as the required cartesian equation of the null plane.

Conditions for equilibrium :

$$X = 0,\quad Y = 0,\quad Z = 0;\quad L = 0,\quad M = 0,\quad N = 0.$$

Conditions for the system to be equivalent to a single force

$$LX + MY + NZ = 0,\; X^2 + Y^2 + Z^2 \neq 0.$$

Conditions for the system to be equivalent to a single couple

$$LX + MY + NZ = 0,\quad X^2 + Y^2 + Z^2 = 0,\quad L^2 + M^2 + N^2 \neq 0.$$

EXAMPLES

Example 1. *One system of force reduces to a force* **R** *acting at a point O and a couple* **G** *and a second system to a force* **R′** *at O and a couple* **G′**. *Prove that if each system is reduced to a wrench and if*

$$\mathbf{R} \cdot \mathbf{G}' + \mathbf{R}' \cdot \mathbf{G} = (p + p')\, \mathbf{R} \cdot \mathbf{R}'.$$

where p and p′ are the pitches of the two wrenches then the axes of the two wrenches are coplanar.

Solution. With O as the origin of reference, the equations of the axes of the two wrenches are

$$\mathbf{r} = \frac{\mathbf{R} \times \mathbf{G}}{\mathbf{R} \cdot \mathbf{R}} + t\mathbf{R},\quad \mathbf{r} = \frac{\mathbf{R}' \times \mathbf{G}'}{\mathbf{R}' \cdot \mathbf{R}'} + k\mathbf{R}'.$$

They will be coplanar if the scalar triple product

$$\mathbf{R} \times \mathbf{R}' \cdot \left(\frac{\mathbf{R} \times \mathbf{G}}{\mathbf{R} \cdot \mathbf{R}} - \frac{\mathbf{R}' \times \mathbf{G}'}{\mathbf{R}' \cdot \mathbf{R}'} \right) = 0$$

$$\Rightarrow \qquad \frac{(\mathbf{R} \times \mathbf{R}') \cdot (\mathbf{R} \times \mathbf{G})}{\mathbf{R} \cdot \mathbf{R}} - \frac{(\mathbf{R} \times \mathbf{R}') \cdot (\mathbf{R}' \times \mathbf{G}')}{\mathbf{R}' \cdot \mathbf{R}'} = 0$$

$$\Rightarrow \qquad \frac{(\mathbf{R}' \cdot \mathbf{R}')(\mathbf{R}' \cdot \mathbf{G}) - (\mathbf{R} \cdot \mathbf{G})(\mathbf{R}' \cdot \mathbf{R})}{\mathbf{R} \cdot \mathbf{R}}$$

$$- \frac{(\mathbf{R}' \cdot \mathbf{R})(\mathbf{R}' \cdot \mathbf{G}') - (\mathbf{R} \cdot \mathbf{G}')(\mathbf{R}' \cdot \mathbf{R}')}{\mathbf{R}' \cdot \mathbf{R}'} = 0.$$

Putting

$$p = \frac{\mathbf{R}\,.\,\mathbf{G}}{\mathbf{R}\,.\,\mathbf{R}'},\; p' = \frac{\mathbf{R}'\,.\,\mathbf{G}'}{\mathbf{R}\,.\,\mathbf{R}'}$$

we obtain

$$\mathbf{R}'\,.\,\mathbf{G} - p\,(\mathbf{R}'\,.\,\mathbf{R}) - (\mathbf{R}\,.\,\mathbf{R}')\,p' + \mathbf{R}\,.\,\mathbf{G}' = 0,$$

$$\Rightarrow \quad \mathbf{R}\,.\,\mathbf{G}' + \mathbf{R}'\,.\,\mathbf{G} = (p + p')\,(\mathbf{R}\,.\,\mathbf{R}'),$$

which is the required condition.

Example 2. *Forces act at the centroids of the faces of a tetrahedron along the outward drawn normals to the faces and proportional to their areas; show that these forces are in equilibrium.*

Solution. Let *OABC* be the tetrahedron. Take O as the origin of reference.

Let

$\vec{OA} = \mathbf{a}$, $\vec{OB} = \mathbf{b}$, $\vec{OC} = \mathbf{c}$.

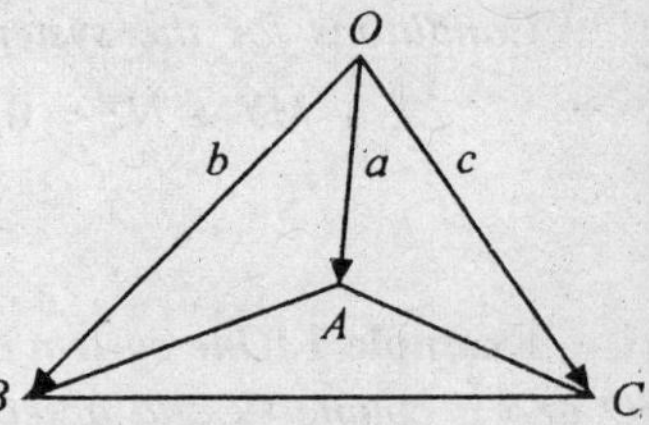

Fig. 8.10.

The centroids of the faces are

$\frac{1}{3}(\mathbf{a}+\mathbf{b}+\mathbf{c})$,

$\frac{1}{3}(\mathbf{b}+\mathbf{c})$, $\frac{1}{3}(\mathbf{a}+\mathbf{c})$, $\frac{1}{3}(\mathbf{a}+\mathbf{b})$,

Outward drawn normal to the plane *ABC* is parallel to the vector

$$\vec{AB} \times \vec{AC} = (\mathbf{b}-\mathbf{a}) \times (\mathbf{c}-\mathbf{a})$$

$$= \mathbf{b} \times \mathbf{c} + \mathbf{c} \times \mathbf{a} + \mathbf{a} \times \mathbf{b}$$

Also the outward drawn normal vectors to the faces

OBA, OAC, OCB

are parallel to the vectors

$$\mathbf{b} \times \mathbf{a}, \quad \mathbf{a} \times \mathbf{c}, \quad \mathbf{c} \times \mathbf{b}$$

respectively. The magnitudes of each of these outward drawn normal vectors is equal to twice the area of the corresponding face.

Thus, omitting the constant of proportionality, we have the forces

$$\mathbf{b} \times \mathbf{a}, \quad \mathbf{a} \times \mathbf{c}, \quad \mathbf{c} \times \mathbf{b}, \quad \mathbf{b} \times \mathbf{c} + \mathbf{c} \times \mathbf{a} + \mathbf{a} \times \mathbf{b}$$

acting at the points

$$\frac{1}{3}(\mathbf{b}+\mathbf{a}),\ \frac{1}{3}(\mathbf{a}+\mathbf{c}),\ \frac{1}{3}(\mathbf{c}+\mathbf{b}),\ \frac{1}{3}(\mathbf{a}+\mathbf{b}+\mathbf{c})$$

respectively.

Now we have

$$\mathbf{R} = 0$$

and $$\mathbf{G} = \frac{1}{3}(\mathbf{b}+\mathbf{a})\times(\mathbf{b}\times\mathbf{a}) + \frac{1}{3}(\mathbf{a}+\mathbf{c})\times(\mathbf{a}\times\mathbf{c}) + \frac{1}{3}(\mathbf{c}+\mathbf{b})\times(\mathbf{c}\times\mathbf{b})$$

$$+\frac{1}{3}(\mathbf{a}+\mathbf{b}+\mathbf{c})\times(\mathbf{b}\times\mathbf{c}+\mathbf{c}\times\mathbf{a}+\mathbf{a}\times\mathbf{b})$$

$$= \frac{1}{3}[\mathbf{a}\times(\mathbf{b}\times\mathbf{c}) + \mathbf{b}\times(\mathbf{c}\times\mathbf{a}) + \mathbf{c}\times(\mathbf{a}\times\mathbf{b})] = 0.$$

Hence, the system is in equilibrium.

Example 3. *Equal forces act along the edges BC, CA, AB, DA, DB, DC of a regular tetrahedron; show that the central axis of the system is perpendicular from D to the plane ABC and the pitch of the equivalent wrench is* $\frac{1}{2\sqrt{2}}k$*, where k is an edge of the tetrahedron.*

Solution. Take D as the origin of reference.

Let

$$\overrightarrow{DA} = \mathbf{a},\ \overrightarrow{DB} = \mathbf{b},\ \overrightarrow{DC} = \mathbf{c},$$

$$\Rightarrow \quad \overrightarrow{BC} = \mathbf{c}-\mathbf{b},\ \overrightarrow{CA} = \mathbf{a}-\mathbf{c},\ \overrightarrow{AB} = \mathbf{b}-\mathbf{a}.$$

If we neglect the factor of proportionality, we may represent the forces themselves by

$$\mathbf{a},\ \mathbf{b},\ \mathbf{c},\ \mathbf{c}-\mathbf{b},\ \mathbf{a}-\mathbf{c},\ \mathbf{b}-\mathbf{a}.$$

The points of application of these forces may be taken to be

$$D,\ D,\ D,\ B,\ C,\ A$$

whose position vectors are **0, 0, 0, b, c, a.**

We have

$$\mathbf{R} = \mathbf{a}+\mathbf{b}+\mathbf{c}+\mathbf{c}-\mathbf{b}+\mathbf{a}-\mathbf{c}+\mathbf{b}-\mathbf{a} = \mathbf{a}+\mathbf{b}+\mathbf{c},$$

$$\mathbf{G} = \mathbf{0}\times\mathbf{a}+\mathbf{0}\times\mathbf{b}+\mathbf{0}\times\mathbf{c}+\mathbf{b}\times(\mathbf{c}-\mathbf{b})+\mathbf{c}\times(\mathbf{a}-\mathbf{c})+\mathbf{a}\times(\mathbf{b}-\mathbf{a})$$

$$= \mathbf{a}\times\mathbf{b}+\mathbf{b}\times\mathbf{c}+\mathbf{c}\times\mathbf{a}.$$

We represent $\mathbf{a}+\mathbf{b}+\mathbf{c}$ in terms of the vectors

$$\mathbf{b}\times\mathbf{c},\quad \mathbf{c}\times\mathbf{a},\quad \mathbf{a}\times\mathbf{b}.$$

Let

$$\mathbf{a}+\mathbf{b}+\mathbf{c} = l\,\mathbf{b}\times\mathbf{c}+m\,\mathbf{c}\times\mathbf{a}+n\,\mathbf{a}\times\mathbf{b}.$$

Multiplying scalarly by **a, b, c** respectively, we obtain

$$l = \frac{\mathbf{a}.\mathbf{a}+\mathbf{a}.\mathbf{b}+\mathbf{a}.\mathbf{c}}{[\mathbf{a}\,\mathbf{b}\,\mathbf{c}]},\ m = \frac{\mathbf{b}.\mathbf{a}+\mathbf{b}.\mathbf{b}+\mathbf{b}.\mathbf{c}}{[\mathbf{a}\,\mathbf{b}\,\mathbf{c}]},\ n = \frac{\mathbf{c}.\mathbf{a}+\mathbf{c}.\mathbf{b}+\mathbf{c}.\mathbf{c}}{[\mathbf{a}\,\mathbf{b}\,\mathbf{c}]}.$$

The tetrahedron being regular of edge k, we have

$$\mathbf{a}\cdot\mathbf{a}=\mathbf{b}\cdot\mathbf{b}=\mathbf{c}\cdot\mathbf{c}=k^2,\ \mathbf{a}\cdot\mathbf{b}=\mathbf{b}\cdot\mathbf{c}=\mathbf{c}\cdot\mathbf{a}=\frac{1}{2}k^2$$

$$[\mathbf{a}\,\mathbf{b}\,\mathbf{c}]^2=\begin{vmatrix}\mathbf{a}\cdot\mathbf{a} & \mathbf{a}\cdot\mathbf{b} & \mathbf{a}\cdot\mathbf{c}\\ \mathbf{b}\cdot\mathbf{a} & \mathbf{b}\cdot\mathbf{b} & \mathbf{b}\cdot\mathbf{c}\\ \mathbf{c}\cdot\mathbf{a} & \mathbf{c}\cdot\mathbf{b} & \mathbf{c}\cdot\mathbf{c}\end{vmatrix}=\frac{1}{2}k^6$$

so that $l = m = n = \dfrac{2\sqrt{2}}{k}\cdot$

$$\therefore \qquad \mathbf{R}=\frac{2\sqrt{2}}{k}=(\mathbf{b}\times\mathbf{c}+\mathbf{c}\times\mathbf{a}+\mathbf{a}\times\mathbf{b})=\frac{2\sqrt{2}}{k}\mathbf{G}$$

$$\Rightarrow \qquad \mathbf{G}=\frac{k}{2\sqrt{2}}\mathbf{R}.$$

Now it is known that the vector $\mathbf{b}\times\mathbf{c}+\mathbf{c}\times\mathbf{a}+\mathbf{a}\times\mathbf{b}$ is normal to the plane ABC.

Thus, **R** and **G** about the point D have the same direction and are in fact both perpendicular to the plane ABC.

Thus, the line through D normal to the plane ABC is the central axis of the system.

Also $\mathbf{G}=\dfrac{k}{2\sqrt{2}}\mathbf{R}$ so that the pitch $=\dfrac{k}{2\sqrt{2}}\cdot$

Example 4. *Forces act along the edges BC, CA, AB, AO, BO, CO of a regular tetrahedron OABC such that the forces acting along any pair of opposite edges are of the same magnitude; show that the pitch of the resultant wrench is* $\dfrac{k}{2\sqrt{2}}$ *where k is the length of each edge.*

Solution. Take O as the origin of reference and let

$$\overrightarrow{OA}=\mathbf{a},\ \overrightarrow{OB}=\mathbf{b},\ \overrightarrow{OC}=\mathbf{c}.$$

The forces are

$$-\lambda_1\mathbf{a},\ \lambda_1(\mathbf{c}-\mathbf{b});\ -\lambda_2\mathbf{b},\ \lambda_2(\mathbf{a}-\mathbf{c});\ -\lambda_3\mathbf{c},\ \lambda_3(\mathbf{b}-\mathbf{a})$$

acting at the points

$$O,\ B;\ \ O,\ C;\ \ O,\ A$$

with position vectors

$$\mathbf{0},\ \mathbf{b};\ \ \mathbf{0},\ \mathbf{c};\ \ \mathbf{0},\ \mathbf{a};$$

where λ_1/k, λ_2/k, λ_3/k, are the magnitudes of the forces along the opposite pairs of edges. We have

$$\begin{cases} \mathbf{R} = (\lambda_2 - \lambda_3 - \lambda_1)\,\mathbf{a} + (\lambda_3 - \lambda_1 - \lambda_2)\,\mathbf{b} + (\lambda_1 - \lambda_2 - \lambda_3)\,\mathbf{c}; \\ \mathbf{G} = \lambda_1 \mathbf{b} \times \mathbf{c} + \lambda_2 \mathbf{c} \times \mathbf{a} + \lambda_3 \mathbf{a} \times \mathbf{b}. \end{cases}$$

$$\mathbf{G} \,.\, \mathbf{R} = [\lambda_1(\lambda_2 - \lambda_3 - \lambda_1) + \lambda_3(\lambda_3 - \lambda_1 - \lambda_2) + \lambda_3(\lambda_1 - \lambda_2 - \lambda_3)]\,(\mathbf{abc})$$

$$= -(\lambda_1^2 + \lambda_2^2 + \lambda_3^2)\,[\mathbf{a\,b\,c}] = -(\lambda_1^2 + \lambda_2^2 + \lambda_3^2)k^2 / \sqrt{2}.$$

Also

$$\mathbf{R} \,.\, \mathbf{R} = [(\lambda_2 - \lambda_3 - \lambda_1)^2 + (\lambda_3 - \lambda_1 - \lambda_2)^2 + (\lambda_1 - \lambda_2 - \lambda_3)^2]\,k^2$$

$$+2\,(\lambda_2 - \lambda_3 - \lambda_1)(\lambda_3 - \lambda_1 - \lambda_2) + (\lambda_3 - \lambda_1 - \lambda_2)(\lambda_1 - \lambda_2 - \lambda_3)$$

$$+ (\lambda_1 - \lambda_2 - \lambda_3)(\lambda_2 - \lambda_3 - \lambda_1)]\frac{1}{2}k^2$$

$$= 2\,(\lambda_1^2 + \lambda_2^2 + \lambda_3^2)\,k^2.$$

Thus, the required pitch $= \dfrac{\mathbf{G} \,.\, \mathbf{R}}{\mathbf{R} \,.\, \mathbf{R}} = \dfrac{k}{2\sqrt{2}}$.

Example 5. *Show that any system of forces can be reduced to a set of forces along the edge of a given tetrahedron OABC.*

If these forces are

$$\lambda \overrightarrow{BC},\ \mu \overrightarrow{CA},\ \nu \overrightarrow{AB} \text{ and } \lambda' \overrightarrow{OA},\ \mu' \overrightarrow{OB},\ \nu' \overrightarrow{OC},$$

prove that the forces reduce to a single force or a couple, if

$$\lambda\lambda' + \mu\mu' + \nu\nu' = 0,$$

Show that the forces reduce to a couple if

$$\lambda' + \mu - \nu = 0,\ \ \mu' + \nu - \lambda = 0,\ \ \nu' + \lambda - \mu = 0.$$

and prove that the axis of this couple makes angles α, β, γ *with OA, OB, OC respectively where*

$$\cos\alpha : \cos\beta : \cos\gamma = \frac{\lambda}{OA} : \frac{\mu}{OB} : \frac{\nu}{OC}.$$

Solution. We know that any system of forces can be replaced by three forces whose lines of action pass through three given points.

In the present case, we replace the given system by three forces whose lines of action pass through *A, B, C* and regard *A, B, C* as their points of application. Now the force at *A* can be replaced by three forces acting along the three edges *AB, AC, AO* through *A* and so also in relation to the other two forces. Hence, any system of forces can be reduced to a set of forces acting along the edges of a given tetrahedron.

Let

$$\overrightarrow{OA} = \mathbf{a},\ \overrightarrow{OB} = \mathbf{b},\ \overrightarrow{OC} = \mathbf{c}.$$

We have forces

$$\lambda\ (\mathbf{c} - \mathbf{b}),\ \mu\ (\mathbf{a} - \mathbf{c}),\ \ \nu\ (\mathbf{b} - \mathbf{c}),\ \ \lambda'\mathbf{a},\ \mu'\mathbf{b},\ \nu'\mathbf{c}$$

acting the points with position vectors

$$\mathbf{b},\ \mathbf{c},\ \mathbf{a},\qquad \mathbf{0},\ \mathbf{0},\ \mathbf{0}.$$

We have

$$\mathbf{R} = \lambda\ (\mathbf{c} - \mathbf{b}) + \mu\ (\mathbf{a} - \mathbf{c}) + \nu\ (\mathbf{b} - \mathbf{a}) + \lambda'\mathbf{a} + \mu'\mathbf{b} + \nu'\mathbf{c}$$

$$= (\lambda' + \mu - \nu)\ \mathbf{a} + (\mu' + \nu - \lambda)\ \mathbf{b} + (\nu' + \lambda - \mu)\ \mathbf{c},$$

and

$$\mathbf{G} = \mathbf{b} \times \lambda\ (\mathbf{c} - \mathbf{b}) + \mathbf{c} \times \mu\ (\mathbf{a} - \mathbf{c}) + \mathbf{a} \times \nu\ (\mathbf{b} - \mathbf{a})$$

$$= \lambda\ \mathbf{b} \times \mathbf{c} + \mu\ \mathbf{c} \times \mathbf{a} + \nu\ \mathbf{a} \times \mathbf{b}.$$

Thus,

$$\mathbf{R} \cdot \mathbf{G} = [\ \lambda\ (\lambda' + \mu - \nu) + \mu\ (\mu' + \nu - \lambda) + \nu\ (\nu' + \lambda - \mu)]\ [\mathbf{a}\ \mathbf{b}\ \mathbf{c}]$$

$$= (\lambda\lambda' + \mu\mu' + \nu\nu')\ [\mathbf{a}\ \mathbf{b}\ \mathbf{c}].$$

Thus, the system will reduce to a single force or a couple if

$$\lambda\lambda' + \mu\mu' + \nu\nu' = 0. \qquad ...(i)$$

Also the system will reduce to a couple if $\mathbf{R} = 0$ which will be so if, and only if,

$$\lambda' + \mu - \nu = 0,\ \ \mu' + \nu - \lambda = 0,\ \ \nu' + \lambda - \mu = 0. \qquad ...(ii)$$

Axis of the couple is parallel to **G.** Now,

$$\mathbf{G} \cdot \overrightarrow{OA} = |\mathbf{G}|\ OA \cos \alpha. \qquad \text{(By def.)}$$

Also

$$\mathbf{G} \cdot \overrightarrow{OA} = \lambda\ [\mathbf{a}\ \mathbf{b}\ \mathbf{c}].$$

$$\Rightarrow \qquad \lambda [\mathbf{a}\ \mathbf{b}\ \mathbf{c}] = |\mathbf{G}|\ OA \cos \alpha$$

$$\Rightarrow \qquad \frac{\lambda}{OA \cos \alpha} = \frac{|\mathbf{G}|}{[\mathbf{a}\ \mathbf{b}\ \mathbf{c}]}.$$

Thus,

$$\frac{\lambda}{OA \cos \alpha} = \frac{\mu}{OB \cos \beta} = \frac{\nu}{OC \cos \gamma}.$$

Hence the result.

It may be easily seen that the condition (*i*) is a consequence of the conditions (*ii*).

Example 6. *Prove that the axis of the wrench equivalent to two forces* $\mathbf{R}_1$, $\mathbf{R}_2$ *acting along two non-coplanar lines intersects the line of shortest* *distance between the lines perpendicularly and divides the same in the ratio*

$$\mathbf{R}_1 \cdot (\mathbf{R}_1 + \mathbf{R}_2) \,/\, \mathbf{R}_2 \cdot (\mathbf{R}_1 + \mathbf{R}_2).$$

Solution. Take the mid-point O of the line of shortest distance AC between the lines of action of the forces as the origin of reference.

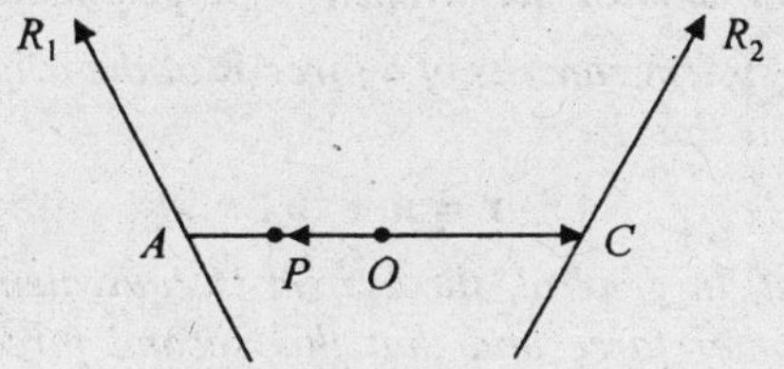

Fig. 8.11.

Let $\overrightarrow{OA} = \mathbf{a}$, so that $\overrightarrow{OC} = -\mathbf{a}$. We have

$$\mathbf{a} \cdot \mathbf{R}_1 = 0, \qquad \mathbf{a} \cdot \mathbf{R}_2 = 0.$$

The vector sum $\mathbf{R}$ and the moment sum $\mathbf{G}$ about the origin O are given by

$$\mathbf{R} = \mathbf{R}_1 + \mathbf{R}_2$$

$$\mathbf{G} = \mathbf{a} \times \mathbf{R}_1 - \mathbf{a} \times \mathbf{R}_2 = \mathbf{a} \times (\mathbf{R}_1 - \mathbf{R}_2).$$

Now $\quad \mathbf{R} \times \mathbf{G} = (\mathbf{R}_1 + \mathbf{R}_2) \times [\mathbf{a} \times (\mathbf{R}_1 - \mathbf{R}_2)]$

$$= [(\mathbf{R}_1 + \mathbf{R}_2) \cdot (\mathbf{R}_1 - \mathbf{R}_2)]\, \mathbf{a} - [(\mathbf{R}_1 + \mathbf{R}_2) \cdot \mathbf{a}]\, (\mathbf{R}_1 - \mathbf{R}_2)$$

$$= (\mathbf{R}_1{}^2 - \mathbf{R}_2{}^2)\, \mathbf{a}.$$

Thus, the central axis is

$$\mathbf{r} = \frac{\mathbf{R} \times \mathbf{G}}{\mathbf{R} \cdot \mathbf{R}} + t\mathbf{R} = \frac{(\mathbf{R}_1^2 - \mathbf{R}_2^2)\, \mathbf{a}}{(\mathbf{R}_1 + \mathbf{R}_2)^2} + t(\mathbf{R}_1 + \mathbf{R}_2)$$

so that it passes through a point, P, whose position vector is

$$\frac{(\mathbf{R}_1^2 - \mathbf{R}_2^2)\, \mathbf{a}}{(\mathbf{R}_1 + \mathbf{R}_2)^2}$$

which clearly lies on the line of shortest distance and is such that

$$\overrightarrow{OP} = \frac{\mathbf{R}_1^2 - \mathbf{R}_2^2}{(\mathbf{R}_1 + \mathbf{R}_2)^2}\, \overrightarrow{OA}$$

$$PA = OA - OP$$

$$= OA\left[1 - \frac{\mathbf{R}_1^2 - \mathbf{R}_2^2}{(\mathbf{R}_1 + \mathbf{R}_2)^2}\right] = \frac{2\mathbf{R}_2\, (\mathbf{R}_1 + \mathbf{R}_2)}{(\mathbf{R}_1 + \mathbf{R}_2)^2}\, OA.$$

$$CP = CO + OP$$

$$= OA + OP = OA\left[1 + \frac{\mathbf{R}_1^2 - \mathbf{R}_2^2}{(\mathbf{R}_1 + \mathbf{R}_2)^2}\right]$$

$$= \frac{2\mathbf{R}_1 \cdot (\mathbf{R}_1 + \mathbf{R}_2)}{(\mathbf{R}_1 + \mathbf{R}_2)^2} \; OA.$$

$$\therefore \quad \frac{\mathbf{CP}}{\mathbf{PA}} = \frac{\mathbf{R}_1 \cdot (\mathbf{R}_1 + \mathbf{R}_2)}{\mathbf{R}_2 \cdot (\mathbf{R}_1 + \mathbf{R}_2)^2}.$$

Also clearly the axis of the wrench is perpendicular to AC.

Example 7. *A system consists of a force* **R** *at the origin O and a couple of moment* **G** *and a line l*

$$\mathbf{r} = \mathbf{a} + t\mathbf{b}$$

is given. Show that, in general, the system is equivalent to a force along the line l and another force and that this second force is given by

$$\frac{\mathbf{G} \times (\mathbf{R} \times \mathbf{b}) - [\mathbf{a}\ \mathbf{R}\ \mathbf{b}]\,\mathbf{R}}{(\mathbf{G} - \mathbf{a} \times \mathbf{R}) \cdot \mathbf{b}}$$

where **b** *is a unit vector.*

Solution. If $p\mathbf{b}$ be the force along the given line, then the vector of the second force is $\mathbf{R} - p\mathbf{b}$. Suppose that **c** is the position vector of any point on the line of action of second force. We have, taking moments about the origin O.

$$\mathbf{r} \times p\mathbf{b} + \mathbf{c} \times (\mathbf{R} - p\mathbf{b}) = \mathbf{G}.$$

Multiplying scalarly with $\mathbf{R} - p\mathbf{b}$, we obtain

$$(\mathbf{a} \times p\mathbf{b}) \cdot (\mathbf{R} - p\mathbf{b}) = \mathbf{G} \cdot (\mathbf{R} - p\mathbf{b})$$

$$\Rightarrow \quad p\,[\mathbf{a}\ \mathbf{b}\ \mathbf{R}] = \mathbf{G} \cdot \mathbf{R} - p\,\mathbf{G} \cdot \mathbf{b}$$

$$\Rightarrow \quad p = \frac{\mathbf{G} \cdot \mathbf{R}}{[\mathbf{a}\ \mathbf{b}\ \mathbf{R}] + \mathbf{G} \cdot \mathbf{b}}$$

Thus, the second force is

$$= \mathbf{R} - \frac{\mathbf{G} \cdot \mathbf{R}}{[\mathbf{a}\ \mathbf{b}\ \mathbf{R}] + \mathbf{G} \cdot \mathbf{b}}\,\mathbf{b}$$

$$= \frac{\{[\mathbf{a}\ \mathbf{b}\ \mathbf{R}] + \mathbf{G} \cdot \mathbf{b}\}\,\mathbf{R} - (\mathbf{G} \cdot \mathbf{R})\,\mathbf{b}}{[\mathbf{a}\ \mathbf{b}\ \mathbf{R}] + \mathbf{G} \cdot \mathbf{b}} = \frac{[\mathbf{a}\ \mathbf{b}\ \mathbf{R}]\,\mathbf{R} + \mathbf{G} \times (\mathbf{R} \times \mathbf{b})}{(\mathbf{G} - \mathbf{a} \times \mathbf{R}) \cdot \mathbf{b}}$$

Example 8. *Show that the null point of the plane* $\mathbf{r} \cdot \mathbf{n} = p$, *is* $(\mathbf{n} \times \mathbf{G} + p\mathbf{R}) / (\mathbf{G} \cdot \mathbf{R})$, *for the system of forces equivalent to a couple* **G** *and force* **R**, *whose line of action passes through the origin of reference O.*

Solution. Given plane is $\mathbf{r} \cdot \mathbf{n} = p$ ...(1)

Let the null point be **a**. Then, the equation of the null plane of the point **a** is

$$\mathbf{r} \cdot (\mathbf{G} - \mathbf{a} \times \mathbf{R}) = \mathbf{a} \cdot \mathbf{G} \qquad ...(2)$$

Comparing (1) and (2), we get

$$\mathbf{G} - \mathbf{a} \times \mathbf{R} = \mathbf{n} \quad ...(3)$$

and

$$\mathbf{a} \cdot \mathbf{G} = p \quad ...(4)$$

$$(3) \Rightarrow \quad \mathbf{G} \times \mathbf{G} - (\mathbf{a} \times \mathbf{R}) \times \mathbf{G} = \mathbf{n} \times \mathbf{G}$$

$$\Rightarrow \quad -[(\mathbf{a} \cdot \mathbf{G})\,\mathbf{R} - (\mathbf{R} \cdot \mathbf{G})\,\mathbf{a}] = \mathbf{n} \times \mathbf{G}$$

$$\Rightarrow \quad -p\mathbf{R} + (\mathbf{R} \cdot \mathbf{G})\,\mathbf{a} = \mathbf{n} \times \mathbf{G}$$

$$\Rightarrow \quad \mathbf{a} = [\mathbf{n} \times \mathbf{G} + p\mathbf{R}] / (\mathbf{G} \cdot \mathbf{R}).$$

EXERCISES

1. Force proportional to the lengths of the edges of a tetrahedron act along the edges *OA, OB, OC, BC, CA, AB* of a tetrahedron *OABC;* show that the sum of the moment vectors about *O* is perpendicular to the plane *ABC.* Also show that the central axis is parallel to the line *OG; G* being the centroid of the tetrahedron.

2. Forces act at the vertices of a tetrahedron along the outward drawn normals to the opposite faces and proportional to their areas; show that the forces are in equilibrium.

3. Forces are acting at the vertices of a tetrahedron along the lines joining them to the centroids of the opposite faces and proportional to the lengths of the same; show that the system is in equilibrium.

4. The lines of action of two non-intersecting forces are at right angles to one another; show that their distances from the central axis are in the ratio of the squares of the magnitudes of the forces.

5. Every line whatsoever of the null plane of a point for a given system is a null line; show that the parameter of the system is zero.

6. Show that the parameter of any system of forces is the sum of the parameters of the pairs of forces of the system.

7. *P* is the null point of a plane and *A, B* are two points where any two conjugate lines for a system of forces meet the plane; show that the points *A, B, P* are collinear.

8. Show that for a system of forces, the conjugate line of the line of intersection of two planes is the line joining their null points.

9. If the null point of a plane π_1 is a point of a plane π_2, then show that the null point of π_2 is a point of π_1.

10. Show that the conjugate lines of a system of coplanar lines is a system of concurrent lines.

11. Prove that the line of shortest distance between two conjugate lines of a system of forces intersects perpendicularly the central axis of the system.

12. Prove that a system of forces may, in general, be reduced to two forces one of which passes through a given point and the other lies in a given plane.

    (If *Q* be the null point of a given plane π, then the conjugate of the line joining the given point to *Q* lies in the plane π).

13. A system of forces is equivalent to two forces one of which acts at a fixed point and lies in a fixed plane through the point; prove that the other passes through another fixed point and lies in a fixed plane.

14. A system of four forces is in equilibrium; show that the parameter of any two of them equals the parameter of the other two and the parameter of any three is zero.

15. The sum of the moments of a system of forces about each of three non-coplanar lines is zero; show that the system cannot be equivalent to a couple.

16. Equal forces are acting along three non-intersecting edges of a rectangular parallelopiped; find the condition for the system of forces to be equivalent to a single force.

17. Forces act along the edges *BA, BC, OA, OC* of a tetrahedron *OABC* in the directions indicated by the order of the letters and each force is inversely proportional to the length of the edge in which it acts. Prove that, if the forces have a single resultant,

$$AO \, . \, BC = AB \, . \, CO.$$

18. Wrenches of the same pitch p act along the edges of a regular tetrahedron *ABCD* of side a. If the intensities of the wrenches along *AB, DC* are the same, and also those along *BC, DA* and *DB, CA,* are the same show that the pitch of the equivalent wrench is $p + a/2\sqrt{2}$.

19. Wrenches of equal intensities act along the edges *BC, CA, AB, AD, BD, CD* of a regular tetrahedron *ABCD*. Prove that the pitch of the resultant wrench is $\frac{1}{2}(p+q+r)+\frac{1}{2}k$, where p, q, r are the pitches of the wrenches which act along *AD, BC; BD, CA; AB, CD;* and k is the length of the shortest distance between any pair of opposite edges of the tetrahedron.

20. *OX, OY, OZ* is a system of right-handed rectangular coordinate axes and the axes of two wrenches are along *OX* and *OY*. I_1, I_2 are the intensities and p_1, p_2 are the pitches of these wrenches. Show that the intensity and the pitch of the resultant wrench are

$$\sqrt{(I_1^2 + I_2^2)} \text{ and } (p_1 I_1^2 + p_2 I_2^2)/(I_1^2 + I_2^2),$$

and show that the axis of the resultant wrench meets *OZ* at a point distant

$$(p_2 - p_1)\,[I_1 I_2 / \sqrt{(I_1^2 + I_2^2)}]$$

from *O* and makes an angle

$$\tan^{-1}(I_2/I_1)$$

with *OX*.

21. A system of forces is equivalent to a couple **G** and a force **R** through the origin *O*. Show that the line which is conjugate to the line

$$\mathbf{r} = \mathbf{a} + t\mathbf{b};$$

b being a unit vector, is given by

$$\mathbf{G} - \mathbf{r} \times \mathbf{R} = \lambda\,(\mathbf{a} - \mathbf{r}) \times \mathbf{b},$$

where $$\lambda = \frac{\mathbf{G} \cdot \mathbf{R}}{\mathbf{G} \cdot \mathbf{b} + [\mathbf{a}\ \mathbf{b}\ \mathbf{R}]}.$$

22. The line of shortest distance between the axes of the two wrenches ($\mathbf{R}_1$, $p_1\ \mathbf{R}_1$) and ($\mathbf{R}_2$, $p_2\ \mathbf{R}_2$) meets the axes in the points A, C; show that the axis of the resultant equivalent wrench divides AC in the ratio

$$\frac{\mathbf{d}^2\mathbf{R}_2 \cdot (\mathbf{R}_1 + \mathbf{R}_2) - (p_2 - p_1)\,V}{\mathbf{d}^2\mathbf{R}_2 \cdot (\mathbf{R}_1 + \mathbf{R}_2) + (p_2 - p_1)\,V}$$

and its pitch is

$$\frac{V + p_1\mathbf{R}_1 \cdot (\mathbf{R}_1 + \mathbf{R}_2) + p_2\mathbf{R}_2 \cdot (\mathbf{R}_1 + \mathbf{R}_2)}{(\mathbf{R}_1 + \mathbf{R}_2)^2}$$

where

$$V = [\mathbf{R}_1\mathbf{R}_2\ \overrightarrow{AC}] \text{ and } d = AC.$$

23. Coplanar forces all drawn inwards (or outwards) are acting at the middle points of the sides of a plane convex polygon perpendicular and proportional to the corresponding sides; show that the system is in equilibrium.

24. Prove that three forces represented in magnitude, direction and line of action by three sides of a triangle, taken in order, are equivalent to a couple, whose magnitude is twice the area of the triangle. Hence, or otherwise, show that the system of forces represented in magnitude, direction and line of action by the four sides taken in order, of a plane quadrilateral of area A is equivalent to a couple of moment $2A$.

25. Forces represented by $\overrightarrow{AC}$ and $\overrightarrow{DB}$ act along the diagonals AC and DB of a parallelogram $ABCD$. Show that their resultant is represented in magnitude, and direction by $2\overrightarrow{AB}$. Determine the line of action of the resultant of the forces represented in magnitude, direction and line of action by

$$\overrightarrow{AB},\ \overrightarrow{BC},\ \overrightarrow{CD},\ \overrightarrow{DA},\ \overrightarrow{AC} \text{ and } \overrightarrow{DB}.$$

26. Calculate the moment about the point $C(1, 1, 1)$ of a force of 5 lb wt. acting along the line $\overrightarrow{AB}$, where A, B are the points (2, 3, 4), (3, 5, 6) respectively, the distances being measured in feet.

27. A force of magnitude 6 in the direction of the vector $\mathbf{i} - 2\mathbf{j} + 2\mathbf{k}$ acts at the point whose position vector is $\mathbf{i} - \mathbf{j}$. A second force acting at the point with position vector $\mathbf{j} - \mathbf{k}$ forms a couple with the first force. Find the vector moment of the couple.

28. A force of unit magnitude acts along a line which passes through the origin O of a rectangular system of co-ordinate axes Ox, Oy and Oz and has equal moments about each of the points (1, 1, 0) and (2, 0, 1). Find the possible values of the components of the force in the directions Ox, Oy and Oz.

SUMMARY

1. The moment **M** of a force **F** about a point O is, by def., $\mathbf{r} \times \mathbf{F}$, where $\mathbf{r} = \overrightarrow{OP}$ and P is a point on the line of action of the force **F**.

2. The equation of the line of action of the force **F** whose moment about a point O is **M** is

$$\mathbf{r} = \frac{-1}{\mathbf{F}.\mathbf{F}}(\mathbf{F} \times \mathbf{M}) + t\mathbf{F};$$

O being the origin of reference.

3. The force vector sum **R** and the moment vector sum **G** about point O of a system are given as follows :

$$\mathbf{R} = \mathbf{F}_1 + \mathbf{F}_2 + \mathbf{F}_3 + \dots$$

and
$$\mathbf{G} = \mathbf{r}_1 \times \mathbf{F}_1 + \mathbf{r}_2 \times \mathbf{F}_2 + \mathbf{r}_3 \times \mathbf{F}_3 + \dots$$

$\mathbf{r}_1, \mathbf{r}_2, \mathbf{r}_3, \dots$ being the position vectors of the points $P_1, P_2, P_3 + \dots$ on the lines of action of the forces $\mathbf{F}_1, \mathbf{F}_2, \mathbf{F}_3, \dots$.

4. If **G** denotes the moment about a point O, then the moment about O' is

$$\mathbf{G} - \mathbf{s} \times \mathbf{R} \text{ where } \mathbf{s} = \overrightarrow{OO'}.$$

5. The moment of a force about a line. The moment of a force **F** about a directed line $\overrightarrow{OI}$ is

$$\mathbf{r} \times \mathbf{F} \cdot \overrightarrow{OI}$$

where **r** is the position vector $\overrightarrow{OP}$ of a point P on the line of action of the force **F** and $\overrightarrow{OI}$ is a unit vector.

The moment of a force about a line is a scalar.

6. The moment of a system of forces about a line is the sum of the moments of the forces of the system about the line.

7. Null lines. Any line about which the moment of a system is zero, is called a null line of the system.

8. Null planes. Null lines through a point lie on a plane through the point called a null plane.

9. A pair of equal, unlike parallel forces is called a *couple.* The moment sum of the forces of a couple about every point is the same and is perpendicular to the plane of the couple. The vector so determined is called the *Axis of the couple.*

10. Poinsot's Central Axis. Every system of forces is equivalent to a force and a couple, such that the line of action of the force is the axis of the couple. This axis is called the *Poinsot's Central Axis* and its equation is

$$\mathbf{r} = \frac{\mathbf{R} \times \mathbf{G}}{\mathbf{R} \cdot \mathbf{R}} + t\mathbf{R}.$$

11. Wrench. A system consisting of a force and a couple such that the line of the force is the axis of the couple is called a *wrench.*

12. Conditions of Equilibrium. A system is in equilibrium when any one of the following conditions hold :

(*i*) $\mathbf{R} = \mathbf{O}$ and $\mathbf{G} = \mathbf{O}$.

(*ii*) The moment sum of the system about each of three non-collinear points is zero.

(*iii*) Each of the six edges of a tetrahedron is a null line for the system of forces.

OBJECTIVE QUESTIONS

For each of the following questions, four alternatives are given for the answer. Only one of them is correct. Choose the correct alternative.

1. If the resultant of two forces P and Q be equal in magnitude to one of the components P and perpendicular to it in direction, then the value of Q is

   (*a*) P (*b*) $\frac{1}{\sqrt{2}} P$ (*c*) $\sqrt{2}\, P$ (*d*) $\sqrt{3}\, P$

2. A, B, C are fixed points and P a variable point such that the resultant of forces at P represented by $\overrightarrow{PA}$ and $\overrightarrow{PB}$ always passes through C and D is the mid-point of AB, then locus of P is

   (*a*) circle passing through A, B, C and D

   (*b*) straight line passing through C and D

   (*c*) straight line passing through A and B

   (*d*) None of these

3. Moment about the point $\mathbf{i} + 2\mathbf{j} - \mathbf{k}$ of a force represented by $\mathbf{i} + 2\mathbf{j} + \mathbf{k}$ acting through the point $2\mathbf{i} + 3\mathbf{j} + \mathbf{k}$ is

   (*a*) $3\mathbf{i} + \mathbf{j} - \mathbf{k}$ (*b*) $3\mathbf{i} - \mathbf{j} + \mathbf{k}$

   (*c*) $-3\mathbf{i} + \mathbf{j} + \mathbf{k}$ (*d*) $3\mathbf{i} + \mathbf{j} + \mathbf{k}$

4. Sum of moments of two intersecting forces about a point in their plane is equal to

(*a*) Moment of their resultant about that point
(*b*) Negative of moment of their resultant about that point
(*c*) zero
(*d*) None of these

5. A single force and coplanar couple together acting on a rigid body are equivalent to
(*a*) a couple
(*b*) a single force equal to the given force and in any direction
(*c*) a single force equal and parallel to the given force
(*d*) None of these

6. Any number of coplanar forces acting on a rigid body can be reduced to
(*a*) a single force
(*b*) a single couple
(*c*) either to a single force or a single couple
(*d*) a single force together with a couple

7. If in a wrench either the force or the couple of the wrench is zero, then wrench is called
(*a*) null wrench (*b*) degenerate wrench
(*c*) non-degenerate wrench (*d*) central wrench

8. The quantities which are invariants for any given system of forces acting on a rigid body are
(*a*) **R** and **G** (*b*) **K** and **G**
(*c*) **R** and $\mathbf{R} \times \mathbf{G}$ (*d*) $\mathbf{R} \cdot \mathbf{R}$ and $\mathbf{G} \cdot \mathbf{R}$

9. If in a system of forces acting on a rigid body $\mathbf{R} \cdot \mathbf{G} \neq 0$, then the given system of forces is equivalent to
(*a*) degenerate wrench (*b*) non-degenerate wrench
(*c*) a couple (*d*) a single force

10. A line about which the moment of a given system of forces is zero is known as
(*a*) central axis (*b*) null line
(*c*) line of force (*d*) arm of couple

ANSWERS

1. (*c*)	**2.** (*b*)	**3.** (*c*)	**4.** (*a*)	**5.** (*c*)
6. (*d*)	**7.** (*b*)	**8.** (*d*)	**9.** (*b*)	**10.** (*b*)

9

Vector Valued Functions of Scalar Variables

Introduction. In this chapter, we shall introduce the concepts of Vector Valued Function of a single and of several scalar variables and those of their limits, continuity and derivability. Basic properties of derivatives will also be obtained. In addition to these, the notion of *'Indefinite integral'* of a vector valued function of a single scalar variable will also be introduced.

The notion of *'Derivatives'* is a development of the concept of *'Instantaneous Rate of Change'*.

It will be seen that the treatment of fundamentals in this chapter is a good deal analogous to that of Scalar Calculus and in fact a number of theorems of this latter subject will be used to prove the corresponding theorems in Vector Calculus.

It should be noted that the symbols **i, j, k** denote a fixed triad of mutually orthogonal unit vectors.

9.1. VECTOR VALUED FUNCTIONS OF A SINGLE SCALAR VARIABLE

If to each value of a scalar variable t in some interval $[a, b]$, there corresponds, by any law whatsoever, a value of a variable vector $\mathbf{r}$ then we say that $\mathbf{r}$ is a vector function of the scalar variable 't' defined in the interval $[a, b]$ and write

$$\mathbf{r} = \mathbf{f}\,(t),$$

where $\mathbf{f}$ denotes the law of correspondence, in question. Also as usual, $\mathbf{f}\,(c)$ denotes the particular vector which the law associates to the value 'c' of t.

In the following, we shall not, in general, mention the intervals of definitions of vector functions and it will always be understood that, for any given function, the values of the scalar variable which we consider are included in the interval of definition of the function.

Illustrations.

(*i*) If a particle moves in any manner so that at any given time t, it is at a point P whose position vector relative to an origin O is $\mathbf{r}$ then the vector $\mathbf{r}$ is a function of the scalar variable t. The velocity and acceleration of a moving particle are also vector functions of the time t which is scalar.

(*ii*) Analytically a curve can be given by expressing the position vector $\mathbf{r}$ of any point on the curve as a function of some scalar variable t known as parameter. In particular, we may take the arc length c of the point from some fixed point on the curve as the parameter.

Thus,

$$\mathbf{r} = a \cos t\, \mathbf{i} + b \sin t\, \mathbf{j} + 0\mathbf{k},$$

$$\mathbf{r} = at^2\, \mathbf{i} + 2at\, \mathbf{j} + 0\mathbf{k},$$

are the vector equations of an ellipse and a parabola respectively; a, b being constant scalars.

9.1.2. Decomposition of a Vector Functions

Decomposing the vector $\mathbf{f}(t)$ as a linear combination of **i, j, k,** we may write

$$\mathbf{f}(t) = f_1(t)\, \mathbf{i} + f_2(t)\, \mathbf{j} + f_3(t)\, \mathbf{k},$$

where $f_1(t)$, $f_2(t)$, $f_3(t)$ are three scalar functions of t.

9.2. LIMIT OF A VECTOR FUNCTION

$$\lim_{t \to c} \mathbf{f}(t) = \mathbf{1}.$$

Def. *A vector function* $\mathbf{f}(t)$ *is said to tend, as its limit, to a vector* $\mathbf{1}$ *when t tends to c, if, to any preassigned positive number* ε *however small, there corresponds a positive number* δ *such that*

$$|\mathbf{f}(t) - \mathbf{1}| < \varepsilon \text{ when } 0 < |t - c| \le \delta.$$

If $\mathbf{f}(t)$ tends to a limit when t tends to c, we say that

$$\lim_{t \to c} \mathbf{f}(t)$$

exists.

Note. In the investigation concerning limits, we shall need the following results :

(*i*) If **a, b, c** be three mutually orthogonal vectors, then

$$|\mathbf{a}| < |\mathbf{a} + \mathbf{b} + \mathbf{c}|, \quad |\mathbf{b}| < |\mathbf{a} + \mathbf{b} + \mathbf{c}|, \quad |\mathbf{c}| < |\mathbf{a} + \mathbf{b} + \mathbf{c}|.$$

This follows from the fact that the length of a diagonal of a *rect-angular* parallelopiped exceeds that of any edge of the same.

(*ii*) For *any* three vectors **a, b, c,**

$$|\mathbf{a} + \mathbf{b} + \mathbf{c}| \le |\mathbf{a}| + |\mathbf{b}| + |\mathbf{c}|.$$

This follows from the fact that the length of any diagonal of a parallelopiped is less than or equal to the sum of the lengths of its three coterminous edges.

(*iii*) $$|\mathbf{a} \cdot \mathbf{b}| = |\mathbf{a}|\ |\mathbf{b}|\cos\theta| \le |\mathbf{a}|\ |\mathbf{b}|.$$

$$|\mathbf{a} \times \mathbf{b}| = |\mathbf{a}|\ |\mathbf{b}|\sin\theta| \le |\mathbf{a}|\ |\mathbf{b}|.$$

9.2.1. Theorem

If

(*i*) *lim* $\mathbf{f}(t) = \mathbf{l}$ *when* $t \to c$,

(*ii*) $\mathbf{f}(t) = f_1(t)\,\mathbf{i} + f_2(t)\,\mathbf{j} + f_3(t)\,\mathbf{k}$,

and (*iii*) $\mathbf{l} = l_1\mathbf{i} + l_2\mathbf{j} + l_3\mathbf{k}$,

then, when $t \to c$,

lim $f_1(t) = l_1$, *lim* $f_2(t) = l_2$, *lim* $f_3(t) = l_3$,

and conversely.

Let ε be any preassigned positive number.

Since

$$\lim_{t\to c} \mathbf{f}(t) = \mathbf{l},$$

there exists a positive number δ such that

$$|\mathbf{f}(t) - \mathbf{l}| < \varepsilon \text{ when } 0 < |t - c| \le \delta.$$

Thus, for

$0 < |t - c| \le \delta$, we have

$$|\overline{(f_1(t)\,\mathbf{i} + f_2(t)\,\mathbf{j} + f_3(t)\,\mathbf{k})} - \overline{(l_1\mathbf{i} + l_2\mathbf{j} + l_3\mathbf{k})}| < \varepsilon$$

or

$$|[f_1(t) - l_1]\,\mathbf{i} + [f_2(t) - l_2]\,\mathbf{j} + [f_3(t) - l_3]\,\mathbf{k} < \varepsilon.$$

Therefore, by (*i*) of the note after § 9.2, we have for $0 < |t - c| \le \delta$,

$$|f_1(t) - l_1| < \varepsilon,\ |f_2(t) - l_2| < \varepsilon,\ |f_3(t) - l_3| < \varepsilon.$$

Hence, when $t \to c$,

$$\lim f_1(t) = l_1,\ \lim f_2(t) = l_2,\ \lim f_3(t) = l_3.$$

To prove the converse, we suppose that, when $t \to c$,

$$\lim f_1(t) = l_1,\ \lim f_2(t) = l_2,\ \lim f_3(t) = l_3,$$

so that if ε be any preassigned positive number, there exists positive number δ_1, δ_2, δ_3 such that

$$|f_1(t) - l_1| < \frac{1}{3}\varepsilon \text{ when } 0 < |t - c| \le \delta_1,$$

$$|f_2(t) - l_2| < \frac{1}{3}\varepsilon \text{ when } 0 < |t - c| \le \delta_2,$$

$$|f_3(t) - l_3| < \frac{1}{3}\varepsilon \text{ when } 0 < |t - c| \le \delta_3.$$

Let

$$\delta = \min(\delta_1, \delta_2, \delta_3).$$

Then each of the three inequalities above holds for $0 < | t - c | \leq \delta$. Thus, $0 < | t - c | \leq \delta$,

$$| \mathbf{f}(t) - \mathbf{l} | = | [f_1(t) - l_1] \mathbf{i} + [f_2(t) - l_2] \mathbf{j} + [f_3(t) - l_3] \mathbf{k} |$$
$$\leq | f_1(t) - l_1 | + | f_2(t) - l_2 | + | f_3(t) - l_3 |,$$

[Refer (*ii*) of note after § 9.1.2]

since

$$| \mathbf{i} | = | \mathbf{j} | = | \mathbf{k} | = 1.$$

Hence,

$$| \mathbf{f}(t) - \mathbf{l} | < \varepsilon \text{ when } 0 < | t - c | \leq \delta.$$

Thus,

$$\lim \mathbf{f}(t) = \mathbf{l} \text{ when } t \to c.$$

9.2.2. The Limiting Operation in Relation to Algebraic Operations

We shall now obtain the fundamental theorems on limits which correlate the limiting operation with algebraic operations.

Theorems. *If* **f** (*t*) *and* **F** (*t*) *are two vector functions of a scalar 't' such that, when $t \to c$,*

$$\lim \mathbf{f}(t) = \mathbf{l}, \text{ limit } \mathbf{F}(t) = \mathbf{m};$$

then

(*i*) $\lim [\mathbf{f}(t) \pm \mathbf{F}(t)] = \mathbf{l} \pm \mathbf{m}$,

(*ii*) $\lim [\mathbf{f}(t) \,.\, \mathbf{F}(t)] = \mathbf{l} \,.\, \mathbf{m}$,

(*iii*) $\lim [\mathbf{f}(t) \times \mathbf{F}(t)] = \mathbf{l} \times \mathbf{m}$.

Also if $\phi(t)$ is any scalar function of 't' such that, when $t \to c$,

$$\lim \phi(t) = p,$$

then

(*iv*) $\lim [\phi(t) \mathbf{f}(t)] = p\mathbf{l}$.

Finally,

(*v*) $\lim | \mathbf{f}(t) | = | \mathbf{l} |$.

The proofs as based on the corresponding theorems in Scalar Calculus and the theorem of § 9.2.1 above will be given here and independent proofs are given in the appendix to this chapter. The theorem in the Scalar Calculus which will be used is that the limit of the sum, difference and product of two functions is equal to the sum, difference and product of the corresponding limits of the functions.

Let

$\mathbf{f}(t) = f_1(t)\,\mathbf{i} + f_2(t)\,\mathbf{j} + f_3(t)\,\mathbf{k}, \quad \mathbf{F}(t) = F_1(t)\,\mathbf{i} + F_2(t)\,\mathbf{j} + F_3(t)\,\mathbf{k};$

$\mathbf{l} = l_1\mathbf{i} + l_2\,\mathbf{j} + l_3\mathbf{k}, \quad \mathbf{m} = m_1\mathbf{i} + m_2\,\mathbf{j} + m_3\mathbf{k}.$

As $\lim \mathbf{f}(t) = \mathbf{l}, \quad \therefore \quad \lim f_1(t) = l_1, \lim f_2(t) = l_2, \lim f_3(t) = l_3.$

As $\lim \mathbf{F}(t) = \mathbf{m}, \quad \therefore \quad \lim F_1(t) = m_1, \lim F_2(t) = m_2, \lim F_3(t) = m_3.$

$\therefore \quad \lim (f_1 \pm F_1) = l_1 \pm m_1, \lim (f_2 \pm F_2) = l_2 \pm m_2,$

$$\lim (f_3 \pm F_3) = l_3 \pm m_3,$$

omitting t, here and subsequently.

$$\lim (f_1F_1) = l_1m_1, \quad \lim (f_2F_2) = l_2m_2, \quad \lim (f_3F_3) = l_3m_3$$

$\therefore$ by § 9.2.1,

(*i*) $$\lim [\mathbf{f}(t) \pm \mathbf{F}(t)] = (l_1 \pm m_1)\,\mathbf{i} + (l_2 \pm m_2)\,\mathbf{j} + (l_3 \pm m_3)\,\mathbf{k} = \mathbf{l} \pm \mathbf{m}.$$

(*ii*) $$\begin{aligned}\lim [\mathbf{f}(t) \,.\, \mathbf{F}(t)] &= \lim (f_1F_1 + f_2F_2 + f_3F_3)\\ &= l_1m_1 + l_2m_2 + l_3m_3\\ &= (l_1\mathbf{i} + l_2\,\mathbf{j} + l_3\mathbf{k}) \,.\, (m_1\mathbf{i} + m_2\,\mathbf{j} + m_3\mathbf{k})\\ &= \mathbf{l} \,.\, \mathbf{m}.\end{aligned}$$

(*iii*) $$\begin{aligned}\lim [\mathbf{f}(t) \times \mathbf{F}(t)] &= \lim [(f_2F_3 - f_3F_2)\,\mathbf{i} + (f_3F_1 - f_1F_3)\,\mathbf{j} + (f_1F_2 - f_2F_1)\,\mathbf{k}]\\ &= (l_2m_3 - l_3m_2)\,\mathbf{i} + (l_3m_1 - l_1m_3)\,\mathbf{j} + (l_1m_2 - l_2m_1)\,\mathbf{k}\\ &= \mathbf{l} \times \mathbf{m}.\end{aligned}$$

(*iv*) $$\begin{aligned}\lim [\phi(t)\,\mathbf{f}(t)] &= \lim (\phi f_1\mathbf{i} + \phi f_2\,\mathbf{j} + \phi f_3\mathbf{k})\\ &= pl_1\mathbf{i} + pl_2\,\mathbf{j} + pl_3\mathbf{k}\\ &= p\mathbf{l}.\end{aligned}$$

(*v*) $$\begin{aligned}\lim |\,\mathbf{f}(t)\,| &= \lim \sqrt{f_1^2 + f_2^2 + f_3^2)}\\ &= \sqrt{(l_1^2 + l_2^2 + l_3^2)}\\ &= |\,\mathbf{l}\,|.\end{aligned}$$

9.3. CONTINUITY

Def. *A vector function* $\mathbf{f}(t)$ *is said to be continuous for a value c, if*

$$\lim_{t \to c} \mathbf{f}(t) = \mathbf{f}(c),$$

Also $\mathbf{f}(t)$ *is said to be a continuous function, if it is continuous for every value of the interval of definition of the function.*

With the help of the theorem of § 9.2.2, we can easily show that *the sum, difference, scalar product and vector product of two continuous functions are also continuous and the product of a continuous vector function with a continuous scalar function is also continuous.*

Thus, for example, let us take the case of vector product.

The functions $\mathbf{f}(t)$ and $\mathbf{F}(t)$ being continuous for $t = c$, we have

$$\lim_{t\to c} [\mathbf{f}(t)\times \mathbf{F}(t)] = \lim_{t\to c} \mathbf{f}(t)\times \lim_{t\to c} \mathbf{F}(t)$$

$$= \mathbf{f}(c) \times \mathbf{F}(c)$$

$$= \text{value of } \mathbf{f}(t) \times \mathbf{F}(t) \text{ for } t = c.$$

Hence, $\mathbf{f}(t) \times \mathbf{F}(t)$ is continuous for $t = c$.

With the help of the theorem of § 9.2.1, we at once see that if

$$\mathbf{f}(t) = f_1(t)\,\mathbf{i} + f_2(t)\,\mathbf{j} + f_3(t)\,\mathbf{k}$$

is continuous, then $f_1(t)$, $f_2(t)$ and $f_3(t)$ are also continuous scalar functions and conversely.

9.4. DERIVATIVE

Def. *A vector function* $\mathbf{f}(t)$ *of a scalar variable 't' is said to be derivable for a value c if*

$$\lim_{h\to 0} \frac{\mathbf{f}(c+h)-\mathbf{f}(c)}{h}$$

exists and the limit, called the derivative or *differential coefficient* of $\mathbf{f}(t)$ at c, is denoted by $\mathbf{f}'(t)$. The process is called derivation or differentiation.

We can also write

$$\mathbf{f}'(c) = \lim_{t\to c} \frac{\mathbf{f}(t)-\mathbf{f}(c)}{t-c},$$

A function is said to be derivable if it is derivable for every value of the interval of definition of the function.

The derivative of a derivable function $\mathbf{f}(t)$ which is itself a vector function is denoted by the symbol

$$\mathbf{f}'(t) \text{ or } \frac{d\mathbf{f}}{dt},$$

For a derivable function $\mathbf{f}(t)$, we have

$$\frac{d\mathbf{f}}{dt} = \mathbf{f}'(t) = \lim_{h\to 0} \frac{\mathbf{f}(t+h)-\mathbf{f}(t)}{h},$$

the equality holding for every value of 't' belonging to the interval of definition of the function.

If as is usual in Scalar Calculus, we denote the changes in t and $\mathbf{f}(t)$ by δt and $\delta\mathbf{f}$ respectively, we have

$$\mathbf{f}'(t) = \frac{d\mathbf{f}}{dt} = \lim \frac{\delta \mathbf{f}}{\delta t}.$$

9.4.1. Differentials

The relation

$$d\mathbf{f} = \mathbf{f}'(t)\, dt.$$

We can write

$$\frac{\delta \mathbf{f}}{\delta t} = \frac{\mathbf{f}\,(t+\delta t) - \mathbf{f}(t)}{\delta t} = \mathbf{f}'(t) + \mathbf{h},$$

where $\mathbf{h}$ is a vector which $\to 0$ as $\delta t \to 0$.

Thus, we have

$$\delta \mathbf{f} = \delta t \mathbf{f}'(t) + \mathbf{h}\delta t.$$

As in the scalar calculus the change δt in the independent variable is called the differential of t denoted by dt and principal part of the change in $\delta \mathbf{f}$. *viz.*, $\delta t \mathbf{f}'(t)$ is called the differential of $\mathbf{f}$ denoted by $d\mathbf{f}$. Thus, in terms of our definitions of differentials,

$$d\mathbf{f} = \mathbf{f}'(t)\, dt.$$

9.4.2. Derivability in Relation to Continuity

Theorem. *Every derivable function is continuous,*

For any value 'c' of t, we have

$$\mathbf{f}(t) - \mathbf{f}(c) = (t-c)\frac{\mathbf{f}(t) - \mathbf{f}(c)}{t-c},$$

so that

$$\lim_{t \to c} [\mathbf{f}(t) - \mathbf{f}(c)] = \lim (t-c)\frac{\mathbf{f}(t) - \mathbf{f}(c)}{t-c}$$
$$= 0\mathbf{f}'(c) = 0.$$

Hence,

$$\lim_{t \to c} \mathbf{f}(t) = \mathbf{f}(c),$$

so that $\mathbf{f}\,(t)$ is continuous at c. At c may be any value, we see that *every derivable function is continuous.*

Note. The converse of the above theorem is not true. Thus

$$\mathbf{f}\,(t) = |\,t\,|\,\mathbf{i},$$

is continuous for every value of t but not derivable for $t = 0$. In fact, we have

$$|\,\mathbf{f}\,(t) - \mathbf{f}\,(0)\,| = |\,|\,t\,|\,\mathbf{i} - 0\,| = |\,t\,|$$

so that

$$\lim_{t \to 0} \mathbf{f}(t) = 0 = \mathbf{f}(0).$$

Also

$$\frac{\mathbf{f}(t) - \mathbf{f}(0)}{t - 0} = \frac{|t|\,\mathbf{i}}{t}$$

so that the limit is **i** or – **i** according as t tends to zero through positive or through negative values. Hence $\mathbf{f}'(0)$ does not exist.

9.4.3. Theorem

If

$$\mathbf{f}(t) = f_1(t)\,\mathbf{i} + f_2(t)\,\mathbf{j} + f_3(t)\,\mathbf{k},$$

is a derivable function, then $f_1(t)$, $f_2(t)$, $f_3(t)$ *are also derivable functions and* *(Nagpur 2005)*

$$\frac{d\mathbf{f}}{dt} = \frac{df_1}{dt}\mathbf{i} + \frac{df_2}{dt}\mathbf{j} + \frac{df_3}{dt}\mathbf{k}.$$

Also conversely.

We have

$$\frac{\mathbf{f}(t+h) - \mathbf{f}(t)}{h} = \frac{f_1(t+h) - f_1(t)}{h}\mathbf{i} + \frac{f_2(t+h) - f_2(t)}{h}\mathbf{j} + \frac{f_3(t+h) - f_3(t)}{h}\mathbf{k}. \qquad ...(1)$$

so that by the theorem of § 9.2.1,

$$\lim_{h \to 0} \frac{\mathbf{f}(t+h) - \mathbf{f}(t)}{h},$$

exists, if and only if, the three limits

$$\lim_{h \to 0} \frac{f_1(t+h) - f_1(t)}{h}, \quad \lim_{h \to 0} \frac{f_2(t+h) - f_2(t)}{h}, \quad \lim_{h \to 0} \frac{f_3(t+h) - f_3(t)}{h},$$

all exist. Also if these limits exist, we have, from (1),

$$\frac{d\mathbf{f}}{dt} = \frac{df_1}{dt}\mathbf{i} + \frac{df_2}{dt}\mathbf{j} + \frac{df_3}{dt}\mathbf{k}. \qquad (\S\ 9.2.1)$$

9.5. DERIVABILITY IN RELATION TO ALGEBRAIC OPERATIONS

Theorem. *If* $\mathbf{f}(t)$ *and* $\mathbf{F}(t)$ *are two derivable functions of t then*

$$\mathbf{f}(t) \pm \mathbf{F}(t), \ \mathbf{f}(t) \cdot \mathbf{F}(t), \ \mathbf{f}(t) \times \mathbf{F}(t).$$

are also derivable and

(*i*) $$\frac{d(\mathbf{f} \pm \mathbf{F})}{dt} = \frac{d\mathbf{f}}{dt} \pm \frac{d\mathbf{F}}{dt}.$$

(*ii*) $$\frac{d(\mathbf{f} \cdot \mathbf{F})}{dt} = \mathbf{f} \cdot \frac{d\mathbf{F}}{dt} + \mathbf{F} \cdot \frac{d\mathbf{f}}{dt}.$$ *(Kolkata, 97)* (*iii*) $\dfrac{d(\mathbf{f}\times\mathbf{F})}{dt}=\mathbf{f}\times\dfrac{d\mathbf{F}}{dt}+\dfrac{d\mathbf{f}}{dt}\times\mathbf{F}.$ (*Patna 2003, Kolkata, 97*)

Also if $\phi(t)$ *is a derivable scalar function, then* $\phi(t)\,\mathbf{f}(t)$ *is also derivable and*

(*iv*) $\dfrac{d(\phi\mathbf{f})}{dt}=\mathbf{f}\,\dfrac{d\mathbf{F}}{dt}+\dfrac{d\phi}{dt}\,\mathbf{f}.$

The order of factors in the case of the derivative of vector product is *not* to be changed.

(*i*) Let

$$\mathbf{G}(t)=\mathbf{f}(t)\pm\mathbf{F}(t).$$

$$\therefore\quad \frac{\mathbf{G}(t+h)-\mathbf{G}(t)}{h}=\frac{\mathbf{f}(t+h)-\mathbf{f}(t)}{h}\pm\frac{\mathbf{F}(t+h)-\mathbf{F}(t)}{h}.$$

Taking limits, when $h\to 0$, we have, by § 9.2.1

$$\frac{d(\mathbf{f}\pm\mathbf{F})}{dt}=\frac{d\mathbf{G}}{dt}=\frac{d\mathbf{f}}{dt}\pm\frac{d\mathbf{F}}{dt}.$$

(*ii*) Let

$$g(t)=\mathbf{f}(t)\,.\,\mathbf{F}(t).$$

$$\therefore\quad \frac{g(t+h)-g(t)}{h}=\frac{\mathbf{f}(t+h)\cdot\mathbf{F}(t+h)-\mathbf{f}(t)\cdot\mathbf{F}(t)}{h}$$

$$=\mathbf{f}(t+h)\cdot\frac{\mathbf{F}(t+h)-\mathbf{F}(t)}{h}$$

$$+\mathbf{F}(t)\cdot\frac{\mathbf{f}(t+h)-\mathbf{f}(t)}{h}$$

Taking limits, (§ 9.2.1), we obtain

$$\frac{d(\mathbf{f}\cdot\mathbf{F})}{dt}=\frac{dg}{dt}=\mathbf{f}\cdot\frac{d\mathbf{F}}{dt}+\mathbf{F}\cdot\frac{d\mathbf{f}}{dt}.$$

Here

$$\lim_{h\to 0}\mathbf{f}(t+h)=\mathbf{f}(t),$$

for, being derivable, the function $\mathbf{f}(t)$ is necessarily continuous.

(*iii*) Let

$$\mathbf{G}(t)=\mathbf{f}(t)\times\mathbf{F}(t).$$

$$\therefore\quad \frac{\mathbf{G}(t+h)-\mathbf{G}(t)}{h}=\frac{\mathbf{f}(t+h)\times\mathbf{F}(t+h)-\mathbf{f}(t)\times\mathbf{F}(t)}{h}$$

$$=\mathbf{f}(t+h)\times\frac{\mathbf{F}(t+h)-\mathbf{F}(t)}{h}+\frac{\mathbf{f}(t+h)-\mathbf{f}(t)}{h}\times\mathbf{F}(t)$$

Taking limits, (§ 9.2.1), we obtain

$$\frac{d(\mathbf{f}\times\mathbf{F})}{dt}=\frac{d\mathbf{G}}{dt}=\mathbf{f}\times\frac{d\mathbf{F}}{dt}+\frac{d\mathbf{f}}{dt}\times\mathbf{F}.$$

(*iv*) Let

$$\mathbf{G}(t)=\phi(t)\,\mathbf{f}(t).$$

$$\therefore\quad \frac{\mathbf{G}(t+h)-\mathbf{G}(t)}{t}=\frac{\phi(t+h)\,\mathbf{f}(t+h)-\phi(t)\,\mathbf{f}(t)}{h}$$

$$=\phi(t+h)\,\frac{\mathbf{f}(t+h)-\mathbf{f}(t)}{h}+\mathbf{f}(t)\,\frac{\phi(t+h)-\phi(t)}{h}$$

Taking limits, (§ 9.2.1), we obtain

$$\frac{d(\phi\mathbf{f})}{dt}=\phi\frac{d\mathbf{f}}{dt}+\mathbf{f}\frac{d\phi}{dt}.$$

Cor.

$$\mathbf{f}\cdot\frac{d\mathbf{f}}{dt}=|\,\mathbf{f}\,|\frac{d|\,\mathbf{f}\,|}{dt}$$

We have

$$\mathbf{f}\,.\,\mathbf{f}=|\,\mathbf{f}\,|^2,$$

so that by (*ii*) above

$$\mathbf{f}\cdot\frac{d\mathbf{f}}{dt}+\frac{d\mathbf{f}}{dt}\cdot\mathbf{f}=2\,|\,\mathbf{f}\,|\,\frac{d|\,\mathbf{f}\,|}{dt}.$$

or

$$\mathbf{f}\cdot\frac{d\mathbf{f}}{dt}=|\,\mathbf{f}\,|\,\frac{d|\,\mathbf{f}\,|}{dt}.$$

9.5.1. Derivation of Triple Products

If **f**, **g**, **h** *be three derivable functions of t, then*

$$(i)\quad \frac{d[\mathbf{f}\;\mathbf{g}\;\mathbf{h}]}{dt}=\left[\frac{d\mathbf{f}}{dt}\mathbf{g}\;\mathbf{h}\right]+\left[\mathbf{f}\frac{d\mathbf{g}}{dt}\mathbf{h}\right]+\left[\mathbf{f}\;\mathbf{g}\frac{d\mathbf{h}}{dt}\right].$$

(*Awadh 99, Kolkata 97*)

$$(ii)\quad \frac{d[\mathbf{f}\times(\mathbf{g}\times\mathbf{h})]}{dt}=\frac{d\mathbf{f}}{dt}\times(\mathbf{g}\times\mathbf{h})+\mathbf{f}\times\left(\frac{d\mathbf{g}}{dt}\times h\right)+\mathbf{f}\times\left(\mathbf{g}\times\frac{d\mathbf{h}}{dt}\right).$$

(*i*) We have

$$\frac{d[\mathbf{f}\;\mathbf{g}\;\mathbf{h}]}{dt}=\frac{d(\mathbf{f}\times\mathbf{g}\,.\,\mathbf{h})}{dt}$$

$$=\frac{d(\mathbf{f}\times\mathbf{g})}{dt}\cdot\mathbf{h}+(\mathbf{f}\times\mathbf{g})\cdot\frac{d\mathbf{h}}{dt}\qquad\text{[By (ii) of § 9.5]}$$

$$= \left(\frac{d\mathbf{f}}{dt} \times \mathbf{g} + \mathbf{f} \times \frac{d\mathbf{g}}{dt}\right) \cdot \mathbf{h} + (\mathbf{f} \times \mathbf{g}) \cdot \frac{d\mathbf{h}}{dt} \quad \text{[By (iii) of § 9.5]}$$

$$= \frac{d\mathbf{f}}{dt} \times \mathbf{g} \cdot \mathbf{h} + \mathbf{f} \times \frac{d\mathbf{g}}{dt} \cdot \mathbf{h} + (\mathbf{f} \times \mathbf{g}) \cdot \frac{d\mathbf{h}}{dt}$$

$$= \left[\frac{d\mathbf{f}}{dt}\,\mathbf{g}\,\mathbf{h}\right] + \left[\mathbf{f}\,\frac{d\mathbf{g}}{dt}\,\mathbf{h}\right] + \left[\mathbf{f}\,\mathbf{g}\,\frac{d\mathbf{h}}{dt}\right].$$

(*ii*) It can be established as above by repeated application of the result (*iii*) of § 9.1 above for the derivative of vector product.

9.5.2. Derivability of Function of a Function

Theorem. *If $f(t)$ is a derivable vector function of t and t is a derivable scalar function $\phi(u)$ of u then* **f** *is a derivable function of u and*

$$\frac{d\mathbf{f}}{du} = \frac{d\mathbf{f}}{dt}\frac{dt}{du} = \frac{d\mathbf{f}}{dt}\frac{d\phi}{du}.$$

Consider corresponding values of u, t and let δu, δ', $\delta\mathbf{f}$ be the corresponding changes in u, t and **f**.

As $t = \phi(u)$ is a derivable function of u, we have

$$\delta t = \delta u\, \phi'(u) + k\, \delta u,$$

where $k \to 0$ as $\delta u \to 0$.

Again, as $\mathbf{f}(t)$ is a derivable vector function of t, we have

$$\delta\mathbf{f} = \delta t\mathbf{f}'(t) + \mathbf{h}\delta t, \text{ [§ 9.4.1]} \qquad ...(2)$$

where $\mathbf{h} \to 0$ as $\delta t \to 0$.

From (1) and (2),

$$\delta\mathbf{f} = \delta u\mathbf{f}'(t)\,\phi'(u) + \mathbf{h}\delta u\phi'(u) + k\delta u\mathbf{f}'(t) + k\mathbf{h}\delta u$$

or
$$\frac{d\mathbf{f}}{du} = \mathbf{f}'(t)\phi'(u) + \mathbf{h}\phi'(u) + k\mathbf{f}'(t) + k\mathbf{h}.$$

Thus,
$$\lim \frac{\delta\mathbf{f}}{\delta u} = \mathbf{f}'(t)\phi'(u),$$

i.e.,
$$\frac{d\mathbf{f}}{du} = \frac{d\mathbf{f}}{dt}\frac{dt}{du}.$$

9.6. CONSTANT VECTORS

Vectors with constant magnitude or with constant direction. A vector will change if either its magnitude changes or direction changes or both direction and magnitude change. We will now find criteria for a vector function to remain constant in magnitude or in direction or in both.

9.6.1. Constant Vectors

Theorem. *The necessary and sufficient condition for* $\mathbf{f}(t)$ *to be constant is* (*Avadh 2005*)

$$\frac{d\mathbf{f}}{dt} = \mathbf{0}.$$

If $\mathbf{f}$ is constant then

$$\mathbf{f}(t+h) - \mathbf{f}(t) = \mathbf{0},$$

for every h and accordingly

$$\frac{d\mathbf{f}}{dt} = \lim_{h \to 0} \frac{\mathbf{f}(t+h) - \mathbf{f}(t)}{h} = \mathbf{0}.$$

Suppose now that the derivative is zero. Let

$$\mathbf{f}(t) = f_1(t)\,\mathbf{i} + f_2(t)\,\mathbf{j} + f_3(t)\,\mathbf{k},$$

so that

$$\mathbf{0} = \frac{d\mathbf{f}}{dt} = \frac{df_1}{dt}\mathbf{i} + \frac{df_2}{dt}\mathbf{j} + \frac{df_3}{dt}\mathbf{k},$$

$$\therefore \qquad \frac{df_1}{dt} = 0, \quad \frac{df_2}{dt} = 0, \quad \frac{df_3}{dt} = 0,$$

so that from Scalar Calculus, $f_1(t)$, $f_2(t)$ and $f_3(t)$ are constant scalars.

Hence, $\mathbf{f}(t)$ is constant.

9.6.2. Vectors with Constant Magnitude.

Theorem. *The necessary and sufficient condition for* $\mathbf{f}(t)$ *to have constant magnitude is*

$$\mathbf{f} \cdot \frac{d\mathbf{f}}{dt} = 0. \qquad (\textit{Kumaon, 99})$$

We have

$$\mathbf{f}(t) \,.\, \mathbf{f}(t) = |\,\mathbf{f}(t)\,|^2,$$

so that

$$\mathbf{f} \cdot \frac{d\mathbf{f}}{dt} = |\,\mathbf{f}\,| \frac{d\,|\,\mathbf{f}\,|}{dt}.$$

Thus

$$\mathbf{f} \cdot \frac{d\mathbf{f}}{dt} = 0,$$

if and only if, $\dfrac{d\,|\,\mathbf{f}\,|}{dt}$ is zero, *i.e.*, $|\,\mathbf{f}\,|$ is constant for, without any loss of generality we may suppose that $\mathbf{f}$ is not constantly zero.

9.6.3. Vectors with Direction Constant

Theorem. *The necessary and sufficient condition for* $\mathbf{f}(t)$ *to have constant direction is*

$$\mathbf{f} \times \frac{d\mathbf{f}}{dt} = \mathbf{0}. \qquad (\textit{Lucknow, 99})$$

Let

$$\mathbf{f}\,(t) = \phi\,(t)\,\mathbf{F}\,(t),$$

where $\phi\,(t)$ denotes the magnitude of $\mathbf{f}\,(t)$ and accordingly $\mathbf{F}\,(t)$ is a vector function with modulus unity for every value of t.

$$\therefore \qquad \frac{d\mathbf{f}}{dt} = \phi\frac{d\mathbf{F}}{dt} + \frac{d\phi}{dt}\mathbf{F},$$

$$\mathbf{f} \times \frac{d\mathbf{f}}{dt} = \phi\mathbf{F} \times \left(\phi\frac{d\mathbf{F}}{dt} + \frac{d\phi}{dt}\mathbf{F}\right) = \phi^2\mathbf{F} \times \frac{d\mathbf{F}}{dt}. \qquad \text{...(1)}$$

Firstly suppose that the direction of **f** remains constant. Then **F** is constant and accordingly by § 9.6.1.

$$\frac{d\mathbf{F}}{dt} = \mathbf{0}.$$

Thus, in this case

$$\mathbf{f} \times \frac{d\mathbf{f}}{dt} = \mathbf{0}.$$

Suppose now that $\quad \mathbf{f} \times \dfrac{d\mathbf{f}}{dt} = \mathbf{0}.$

$\therefore$ From (1),

$$\phi^2\mathbf{F} \times \frac{d\mathbf{F}}{dt} = \mathbf{0}.$$

Without any loss of generality we may suppose that ϕ is not constantly zero. Thus,

$$\mathbf{F} \times \frac{d\mathbf{F}}{dt} = \mathbf{0}. \qquad \text{...(2)}$$

Also **F** being of constant length we have, by § 9.6.2,

$$\mathbf{F} \cdot \frac{d\mathbf{F}}{dt} = \mathbf{0}.$$

From (2) and (3), we have $\dfrac{d\mathbf{F}}{dt} = 0. \qquad \text{...(3)}$

$\therefore$ by § 9.6.1,

F is constant,

i.e., the direction of **f** remains the same.

Note. The student would do well to compare the results obtained in the preceding three sub-sections.

9.7. DERIVATIVES OF SECOND AND HIGHER ORDERS

We suppose that the derivative $d\mathbf{f}/dt$ of the derivable function $\mathbf{f}\,(t)$ is itself a derivable function. Then the derivative

$$\frac{d}{dt}\left(\frac{d\mathbf{f}}{dt}\right),$$

denoted by

$$\frac{d^2\mathbf{f}}{dt^2},$$

is called the second order derivative of **f**.

Higher order derivatives

$$\frac{d^3\mathbf{f}}{dt^3}, \frac{d^4\mathbf{f}}{dt^4}, \text{ etc.}$$

may also be defined as in the Scalar Calculus.

9.8. INDEFINITE INTEGRATION

If $\mathbf{F}(t)$ be the derivative of a function $\mathbf{f}(t)$, we say that $\mathbf{f}(t)$ is an indefinite integral or simply integral of $\mathbf{F}(t)$ with respect to t and write

$$\int \mathbf{F}(t)\, dt = \mathbf{f}(t).$$

As in Scalar Calculus, indefinite integral of a function is not unique. In fact, *if* $\mathbf{f}(t)$ *be any indefinite integral of* $\mathbf{F}(t)$, *then* $\mathbf{f}(t) + \mathbf{c}$, *where* $\mathbf{c}$ *is any constant vector, is the general indefinite integral of* $\mathbf{F}(t)$.

Firstly we see that

$$\frac{d[\mathbf{f}(t)+\mathbf{c}]}{dt} = \frac{d\mathbf{f}(t)}{dt},$$

for the derivative of the constant $\mathbf{c}$ is zero.

Secondly, if $\mathbf{g}(t)$ be any other indefinite integral of $\mathbf{F}(t)$, besides $\mathbf{f}(t)$ so that

$$\frac{d[\mathbf{f}(t)+\mathbf{f}(t)]}{dt} = \frac{d\mathbf{g}(t)}{dt} - \frac{d\mathbf{f}(t)}{dt} = \mathbf{F}(t) - \mathbf{F}(t) = 0,$$

then, by § 9.6.1,

$$\mathbf{g}(t) - \mathbf{f}(t),$$

is a constant say, $\mathbf{c}$. Thus,

$$\mathbf{g}(t) - \mathbf{f}(t) = \mathbf{c} \quad \text{or} \quad \mathbf{g}(t) = \mathbf{f}(t) + \mathbf{c}.$$

Hence, $\mathbf{g}(t)$ is of the form $\mathbf{f}(t) + \mathbf{c}$.

Note. The reader would be advised to bear in mind the following two useful results concerning indefinite integrals :

$$2\int \mathbf{f}\cdot\frac{d\mathbf{f}}{dt}\,dt = \mathbf{f}\cdot\mathbf{f} + c = \mathbf{f}^2 + c.$$

$$\int \mathbf{f}\times\frac{d^2\mathbf{f}}{dt^2}\,dt = \mathbf{f}\times\frac{d\mathbf{f}}{dt} + c.$$

9.8.1. Definite Integral

If $\int \mathbf{f}(t)\,dt = \mathbf{F}(t)$, then $\mathbf{F}(b) - \mathbf{F}(a)$ is called the definite integral of $\mathbf{f}(t)$ between the limits $t = a$ and $t = b$ and is denoted as

$$\int_a^b \mathbf{f}(t)\,dt.$$

$$\therefore \qquad \int_a^b \mathbf{f}(t)\,dt = [\mathbf{F}(t)]_a^b = \mathbf{F}(b) - \mathbf{F}(a).$$

EXAMPLES

Example 1. *A particle moves along the curve $x = 2t^2$, $y = t^2 - 4t$, $z = 3t - 5$, where t is the time. Find the component of its velocity and acceleration at time $t = 1$ in the direction* $\mathbf{i} - 3\mathbf{j} + 2\mathbf{k}$.

Solution. $\mathbf{r} = 2t^2\mathbf{i} + (t^2 - 4t)\,\mathbf{j} + (3t - 5)\,\mathbf{k}$

$$\therefore \quad \text{Velocity} = \frac{d\mathbf{r}}{dt} = 4t\,\mathbf{i} + (2t - 4)\,\mathbf{j} + 3\mathbf{k}$$

$$= 4\mathbf{i} - 2\mathbf{j} + 3\mathbf{k} \text{ at } t = 1.$$

we know that $\frac{\mathbf{A}\cdot\mathbf{B}}{|\mathbf{A}|}$ is the projection of **B** in the direction of **A.**

Hence, component of velocity vector in the direction of $\mathbf{i} - 3\mathbf{j} + 2\mathbf{k}$

$$= \frac{(\mathbf{i} - 3\mathbf{j} + 2\mathbf{k})}{\sqrt{(1+9+4)}} \cdot (4\mathbf{i} - 2\mathbf{j} + 3\mathbf{k}) = \frac{16}{\sqrt{14}}.$$

Again acceleration $\frac{d^2\mathbf{r}}{dt^2} = 4\mathbf{i} + 2\mathbf{j}$

Hence, component of acceleration in the given direction is

$$\frac{\mathbf{i} - 3\mathbf{j} + 2\mathbf{k}}{\sqrt{14}} \cdot (4\mathbf{i} + 2\mathbf{j}) = -\frac{2}{\sqrt{14}}$$

Example 2. *If* **e** *is the unit vector making an angle θ with x-axis, show that $\frac{d\mathbf{e}}{d\theta}$ is a unit vector obtained by rotating* **e** *through a right angle in θ increasing direction.*

Solution. Since **e** is a unit vector making an angle θ with *x*-axis.

$$\mathbf{e} = \cos\theta\,\mathbf{i} + \sin\theta\,\mathbf{j} \qquad \text{...(1)}$$

Replacing θ by 90° + θ,

$$\mathbf{f} = -\sin\theta\,\mathbf{i} + \cos\theta\,\mathbf{j} \qquad \text{...(2)}$$

Clearly **f** is also a unit vector. Again

$$\frac{d\mathbf{e}}{d\theta} = -\sin\theta\,\mathbf{i} + \cos\theta\,\mathbf{j} = \mathbf{f}.$$

Example 3. *Evaluate* $\frac{d^2}{dt^2}\left[\mathbf{r}\ \frac{d\mathbf{r}}{dt}\ \frac{d^2\mathbf{r}}{dt^2}\right]$

(*Kolkata 97, Kumaon 98, Garhwal 2004, Avadh, 2003*)

Solution. $$\frac{d}{dt}\left[\mathbf{r}\ \frac{d\mathbf{r}}{dt}\ \frac{d^2\mathbf{r}}{dt^2}\right] = \left[\frac{d\mathbf{r}}{dt}\ \frac{d\mathbf{r}}{dt}\ \frac{d^2\mathbf{r}}{dt^2}\right] + \left[\mathbf{r}\ \frac{d^2\mathbf{r}}{dt^2}\ \frac{d^2\mathbf{r}}{dt^2}\right] + \left[\mathbf{r}\ \frac{d\mathbf{r}}{dt}\ \frac{d^3\mathbf{r}}{dt^3}\right]$$

$$= \left[\mathbf{r}\ \frac{d\mathbf{r}}{dt}\ \frac{d^3\mathbf{r}}{dt^3}\right] \quad \text{as first two scalar triple products vanish}$$

$$\therefore \quad \frac{d^2}{dt}\left[\mathbf{r}\ \frac{d\mathbf{r}}{dt}\ \frac{d^2\mathbf{r}}{dt^2}\right] = \frac{d}{dt}\left[\mathbf{r}\ \frac{d\mathbf{r}}{dt}\ \frac{d^3\mathbf{r}}{dt^3}\right]$$

$$= \left[\frac{d\mathbf{r}}{dt}\ \frac{d\mathbf{r}}{dt}\ \frac{d^3\mathbf{r}}{dt^3}\right] + \left[\mathbf{r}\ \frac{d^2\mathbf{r}}{dt^2}\ \frac{d^3\mathbf{r}}{dt^3}\right] + \left[\mathbf{r}\ \frac{d\mathbf{r}}{dt}\ \frac{d^4\mathbf{r}}{dt^4}\right]$$

$$= \left[\mathbf{r}\ \frac{d^2\mathbf{r}}{dt^2}\ \frac{d^3\mathbf{r}}{dt^3}\right] + \left[\mathbf{r}\ \frac{d\mathbf{r}}{dt}\ \frac{d^4\mathbf{r}}{dt^4}\right].$$

Example 4. *If* **r** = **a** *cos* ω*t* + **b** *sin* ω*t, show that*

$$\mathbf{r} \times \frac{d\mathbf{r}}{dt} = \omega\,\mathbf{a} \times \mathbf{b} \quad \textit{and} \quad \frac{d^2\mathbf{r}}{dt^2} = -\omega^2\mathbf{r} \qquad (\textit{Awadh, 98})$$

where **a, b** *and* ω *are constants.*

Solution. $$\frac{d\mathbf{r}}{dt} = -\mathbf{a}\,\omega\sin\omega t + \mathbf{b}\,\omega\cos\omega t$$

$$\frac{d^2\mathbf{r}}{dt^2} = -\mathbf{a}\,\omega^2\cos\omega t - \mathbf{b}\,\omega^2\sin\omega t$$

$$= -\omega^2(\mathbf{a}\cos\omega t + \mathbf{b}\sin\omega t) = -\omega^2\mathbf{r}$$

$$\mathbf{r} \times \frac{d\mathbf{r}}{dt} = (\mathbf{a}\cos\omega t + \mathbf{b}\sin\omega t) \times (-\mathbf{a}\,\omega\sin\omega t + \mathbf{b}\,\omega\cos\omega t)$$

$$= \omega\,[\mathbf{a} \times \mathbf{b}\cos^2\omega t - \mathbf{b} \times \mathbf{a}\sin^2\omega t]$$

$$= \omega\,(\mathbf{a} \times \mathbf{b})(\cos^2\omega t + \sin^2\omega t) = \omega\,(\mathbf{a} \times \mathbf{b})$$

Example 5. *If* **r** *is a vector function of a scalar t and module of* **r** *be r, and* **a** *is a constant vector, then differentiate the following w.r.t t.*

$$(i)\ \frac{\mathbf{r}\times\mathbf{a}}{\mathbf{r}\cdot\mathbf{a}}, \qquad (ii)\ \frac{\mathbf{r}\,\mathbf{a}}{\mathbf{a}\times\mathbf{r}},$$

Solution. (*i*) $\frac{d}{dt}\left(\frac{\mathbf{r}\times\mathbf{a}}{\mathbf{r}\cdot\mathbf{a}}\right)=\frac{d}{dt}(\mathbf{r}\times\mathbf{a})\frac{1}{\mathbf{r}\cdot\mathbf{a}}+(\mathbf{r}\times\mathbf{a})\frac{d}{dt}\left(\frac{1}{\mathbf{r}\cdot\mathbf{a}}\right)$

$$=(\mathbf{r}\times\mathbf{a})\cdot\frac{1}{\mathbf{r}\cdot\mathbf{a}}+(\mathbf{r}\times\mathbf{a})\left\{-\frac{1}{(\mathbf{r}\cdot\mathbf{a})^2}(\mathbf{r}\cdot\mathbf{a})\right\}$$

$$=\frac{\mathbf{r}\times\mathbf{a}}{\mathbf{r}\cdot\mathbf{a}}-\frac{(\mathbf{r}\times\mathbf{a})(\mathbf{r}\cdot\mathbf{a})}{(\mathbf{r}\cdot\mathbf{a})^2}$$

$$=\frac{(\mathbf{r}\times\mathbf{a})(\mathbf{r}\cdot\mathbf{a})-(\mathbf{r}\times\mathbf{a})(\mathbf{r}\cdot\mathbf{a})}{(\mathbf{r}\cdot\mathbf{a})^2}$$

(*ii*) $\frac{d}{dt}\left(\frac{\mathbf{r}\,\mathbf{a}}{\mathbf{a}\times\mathbf{r}}\right)=\frac{d}{dt}(\mathbf{r}\,\mathbf{a})\cdot\frac{1}{\mathbf{a}\times\mathbf{r}}+(\mathbf{r}\,\mathbf{a})\frac{d}{dt}\left(\frac{1}{\mathbf{a}\times\mathbf{r}}\right)$

$$=(\mathbf{r}\,\mathbf{a}+0)\cdot\frac{1}{\mathbf{a}\times\mathbf{r}}+(\mathbf{r}\,\mathbf{a})\left[-\frac{1}{(\mathbf{a}\times\mathbf{r})^2}\{(\mathbf{a}\times\mathbf{r})+0\}\right].$$

$$=\frac{\mathbf{r}\,\mathbf{a}}{\mathbf{a}\times\mathbf{r}}-\frac{(\mathbf{r}\,\mathbf{a})\cdot(\mathbf{a}\times\mathbf{r})}{(\mathbf{a}\times\mathbf{r})^2}$$

Example 6. *If* $\hat{\mathbf{r}}$ *denotes a unit vector, prove that*

$\hat{\mathbf{r}}\times d\hat{\mathbf{r}}=(\mathbf{r}\times d\mathbf{r})/r^2$ *where* $r=r\hat{\mathbf{r}}$. *(Rohilkhand 97)*

Solution. $\hat{\mathbf{r}}=\frac{\mathbf{r}}{r}\quad\therefore\quad d\hat{\mathbf{r}}=\frac{1}{r}d\mathbf{r}+\mathbf{r}\left(-\frac{1}{r^2}\right)dr$

$$\therefore\qquad \hat{\mathbf{r}}\times d\hat{\mathbf{r}}=\hat{\mathbf{r}}\times\left[\frac{1}{r}d\mathbf{r}-\frac{\mathbf{r}}{r^2}dr\right]$$

$$=\frac{\mathbf{r}}{r}\times\left[\frac{1}{r}d\mathbf{r}-\frac{\mathbf{r}}{r^2}dr\right]=\frac{\mathbf{r}\times d\mathbf{r}}{r^2}.$$

Example 7. *Find* **r** *from the equation* $\frac{d^2\mathbf{r}}{dt^2}=\mathbf{a}t+\mathbf{b}$ *given that both* **r** *and* $\frac{d\mathbf{r}}{dt}$ *vanish when t = 0.* *(Rohilkhand 2000)*

Solution. Integrating the given relation w.r.t. *t*

$$\frac{d\mathbf{r}}{dt}=\frac{1}{2}\mathbf{a}t^2+\mathbf{b}t+\mathbf{c},$$

where **c** is a constant vector.

when $t = 0$, $\frac{d\mathbf{r}}{dt} = 0$; $\therefore$ $\mathbf{c} = 0$

$$\therefore \quad \frac{d\mathbf{r}}{dt} = \frac{1}{2}\mathbf{a}t^2 + \mathbf{b}t$$

Integrating, $\mathbf{r} = \frac{1}{6}\mathbf{a}t^3 + \frac{1}{2}\mathbf{b}t^2 + \mathbf{d}$

when $t = 0$, $\mathbf{r} = 0$; $\therefore$ $\mathbf{d} = 0$

$$\therefore \quad \mathbf{r} = \frac{1}{6}\mathbf{a}t^3 + \frac{1}{2}\mathbf{b}t^2.$$

Example 8. *Given that* $\mathbf{r}(t) = 2\mathbf{i} - \mathbf{j} + 2\mathbf{k}$, *when* $t = 2$

$= 4\mathbf{i} - 2\mathbf{j} + 3\mathbf{k}$, *when* $t = 3$

show that $\int_2^3 \mathbf{r} \cdot \frac{d\mathbf{r}}{dt} dt = 10.$ *(Rohilkhand 98, 2003 Kumaon 2000)*

Solution. $\int_2^3 \mathbf{r} \cdot \frac{d\mathbf{r}}{dt} dt = \frac{1}{2}\left[r^2\right]_2^3$

$$= \frac{1}{2}\left[\left(r^2 \text{ at } t = 3\right) - \left(r^2 \text{ at } t = 2\right)\right]$$

$\mathbf{r}^2$ at $t = 2$ is $(2\mathbf{i} - \mathbf{j} + 2\mathbf{k})^2 = 4 + 1 + 4 = 9$

$\mathbf{r}^2$ at $t = 3$ is $(4\mathbf{i} - 2\mathbf{j} + 3\mathbf{k})^2 = 16 + 4 + 9 = 29$

$\therefore$ given integral $\frac{1}{2}[29 - 9] = 10.$

Example 9. *Evaluate* $\int \mathbf{a} \cdot \left(\mathbf{r} \times \frac{d^2\mathbf{r}}{dt^2}\right) dt$

Solution. We have $\frac{d}{dt}\left(\mathbf{r} \times \frac{d\mathbf{r}}{dt}\right) = \frac{d\mathbf{r}}{dt} \times \frac{d\mathbf{r}}{dt} + \mathbf{r} \times \frac{d^2\mathbf{r}}{dt^2} = \mathbf{r} \times \frac{d^2\mathbf{r}}{dt^2}$

$$\therefore \quad \int \left(\mathbf{r} \times \frac{d^2\mathbf{r}}{dt^2}\right) dt = \mathbf{r} \times \frac{d\mathbf{r}}{dt} + \mathbf{r} \times \frac{d^2\mathbf{r}}{dt^2} + \mathbf{c}.$$

where $\mathbf{c}$ is an arbitrary constant vector.

Now, $\int \mathbf{a} \cdot \left(\mathbf{r} \cdot \frac{d^2\mathbf{r}}{dt^2}\right) dt = \mathbf{a} \cdot \int \mathbf{r} \times \frac{d^2\mathbf{r}}{dt^2} dt$

$$= \mathbf{a} \cdot \left[\mathbf{r} \times \frac{d\mathbf{r}}{dt} + \mathbf{c}\right] = \mathbf{a} \cdot \mathbf{r} \times \frac{d\mathbf{r}}{dt} + d$$

where $d = \mathbf{a} \cdot \mathbf{c}$ is an arbitrary constant scalar.

Example 10. *The equation of motion of a particle P of mass m is given by* $m\left(d^2\mathbf{r}/dt^2\right)=f(r)\,\hat{\mathbf{r}}$, *where* **r** *is the position vector of P measured from an origin O,* $\hat{\mathbf{r}}$ *is a unit vector in the direction of* **r** *and f (r) is a function of the distance of P from O, show that* $\mathbf{r}\times\left(\dfrac{d\mathbf{r}}{dt}\right)=\mathbf{c}$, *where* **c** *is a constant vector.*

Solution. We have

$$m\left(\frac{d^2\mathbf{r}}{dt^2}\right)=f(r)\,\hat{\mathbf{r}}$$

$$\Rightarrow\qquad m\mathbf{r}\times\left(\frac{d^2\mathbf{r}}{dt^2}\right)=f(r)\,\mathbf{r}\times\hat{\mathbf{r}}$$

$$\Rightarrow\qquad m\mathbf{r}\times\left(\frac{d^2\mathbf{r}}{dt^2}\right)=0 \;\Rightarrow\; \mathbf{r}\times\left(\frac{d^2\mathbf{r}}{dt^2}\right)=0 \qquad ...(1)$$

$$\left[\because\; \mathbf{r}\times\hat{\mathbf{r}}=\mathbf{r}\times\frac{\mathbf{r}}{r}=0\right]$$

But, $$\frac{d}{dt}\left(\mathbf{r}\frac{d\mathbf{r}}{dt}\right)=\frac{d\mathbf{r}}{dt}\times\frac{d\mathbf{r}}{dt}+\mathbf{r}\times\frac{d^2\mathbf{r}}{dt^2}=\mathbf{r}\times\frac{d^2\mathbf{r}}{dt^2}.$$

Hence, (1) becomes

$$\frac{d}{dt}\left(\mathbf{r}\times\frac{d\mathbf{r}}{dt}\right)=0$$

Integrating, $$\mathbf{r}\times\frac{d\mathbf{r}}{dt}=\mathbf{c}.$$

EXERCISES

1. If

$$\mathbf{r}=\mathbf{a}\cos\omega t+\mathbf{b}\sin\omega t,$$

show that

$$\mathbf{r}\times\frac{d\mathbf{r}}{dt}=\omega\mathbf{a}\times\mathbf{b}.$$

and

$$\frac{d^2\mathbf{r}}{dt^2}=-\omega^2\mathbf{r};$$

a, b, ω being constants.

2. If

$$\mathbf{r}=a\cos t\,\mathbf{i}+a\sin t\,\mathbf{j}+at\tan\alpha\,\mathbf{k},$$

find

$$\left|\frac{d\mathbf{r}}{dt}\times\frac{d^2\mathbf{r}}{dt^2}\right| \text{ and } \left[\frac{d\mathbf{r}}{dt}\;\frac{d^2\mathbf{r}}{dt^2}\;\frac{d^2\mathbf{r}}{dt^2}\right].$$ *(Rohilkhand, 2000, Delhi 2005, Garhwal 2000)*

3. Given $r = 4a \sin^3 \theta \mathbf{i} + 4a \cos^3 \theta \mathbf{j} + 3b \cos 2\theta \mathbf{k}$,

find

$$\left| \frac{d\mathbf{r}}{d\theta} \times \frac{d^2\mathbf{r}}{d\theta^2} \right| \text{ and } \left[\frac{d\mathbf{r}}{d\theta} \, \frac{d^2\mathbf{r}}{d\theta^2} \, \frac{d^2\mathbf{r}}{d\theta^2} \right].$$

4. Evaluate

(*i*) $\dfrac{d}{dt}\left[\mathbf{r} \, \dfrac{d\mathbf{r}}{dt} \times \dfrac{d^2\mathbf{r}}{dt^2}\right]$ (*ii*) $\dfrac{d}{dt}\left\{\left(\mathbf{r} \times \dfrac{d\mathbf{r}}{dt}\right) \times \dfrac{d^2\mathbf{r}}{dt^2}\right\}$

(*iii*) $\dfrac{d^2}{dt^2}\left\{\left(\mathbf{r} \times \dfrac{d\mathbf{r}}{dt}\right) \times \dfrac{d^2\mathbf{r}}{dt^2}\right\}$. (*Avadh 2003*)

$\mathbf{r}$ being a vector function of t.

5. Find the derivatives of the following :

(*i*) $\dfrac{\mathbf{r}}{r}$. (*ii*) $\mathbf{r}^n\mathbf{r}$.

(*iii*) $r^2 + \dfrac{1}{r^2}$. (*iv*) $\dfrac{\mathbf{r}}{\mathbf{a} \cdot \mathbf{r}} - \dfrac{\mathbf{r} \times \mathbf{a}}{r}$.

[Here $\mathbf{r}$ is the magnitude $|\mathbf{r}|$ of $\mathbf{r}$.]

6. A particle moves along the curve $x = t^3 + 1$, $y = t^2$, $z = 2t + 5$, where t is time. Find the components of its velocity and acceleration at $t = 1$ in the direction $\mathbf{i} + \mathbf{j} + 3\mathbf{k}$.

7. If $\mathbf{a} = \sin \theta \mathbf{i} + \cos \theta \mathbf{j} + \theta \mathbf{k}$, $\mathbf{b} = \cos \theta \mathbf{i} - \sin \theta \mathbf{j} - 3\mathbf{k}$ and $\mathbf{c} = \mathbf{i} + 2\mathbf{j} - 3\mathbf{k}$ then find $\dfrac{d}{d\theta}\{\mathbf{a} \times (\mathbf{b} \times \mathbf{c})\}$ at $\theta = \pi/2$.

8. If $\mathbf{r} = \mathbf{a}\, e^{nt} + \mathbf{b}\, e^{-nt}$ where $\mathbf{a}$ and $\mathbf{b}$ are constant vectors then prove that $\dfrac{d^2\mathbf{r}}{dt^2} = n^2\mathbf{r}$. (*Kumaon, 2000, 2005, 2006*)

9. If $\dfrac{d\mathbf{a}}{dt} = \mathbf{c} \times \mathbf{a}$ and $\dfrac{d\mathbf{b}}{dt} = \mathbf{c} \times \mathbf{b}$, then prove that

$$\frac{d}{dt} = (\mathbf{a} \times \mathbf{b}) = \mathbf{c} \times (\mathbf{a} \times \mathbf{b}).$$

(*Ajmer 98, Rohilkhand 98, Kumaon 2001, 2003*)

10. A particle P is moving on a circle of radius $\mathbf{r}$ with constant angular velocity $\omega = d\theta/dt$. Show that the acceleration is $-\omega^2\mathbf{r}$. (*Purvanachal 2004*)

11. If $\mathbf{r}(t) = 5t^2\mathbf{i} + t\mathbf{j} - t^3\mathbf{k}$, prove that (*Rohilkhand 99, Kumaon 2000, 2001*)

$$\int_1^2 \mathbf{r} \times \frac{d^2\mathbf{r}}{dt^2}\, dt = -14\mathbf{i} + 75\mathbf{j} - 15\mathbf{k}.$$

12. The acceleration of a particle at any time t is

$$e^t\mathbf{i} + e^{2t}\mathbf{j} + \mathbf{k},$$

find $\mathbf{v}$ given that $\mathbf{v} = \mathbf{i} + \mathbf{j}$ at $t = 0$.

13. Integrate $\mathbf{a}\times\dfrac{d^2\mathbf{r}}{dt^2}=\mathbf{b}$, where **a** and **b** are constant vectors.

14. Evaluate $\int_1^2 \mathbf{a}\cdot(\mathbf{b}\times\mathbf{c})\,dt$, whose

$\mathbf{a} = t\mathbf{i} - 3\mathbf{j} + 2t\mathbf{k}$, $\mathbf{b} = \mathbf{i} - 2\mathbf{j} + 2\mathbf{k}$, $\mathbf{c} = 3\mathbf{i} + t\mathbf{j} - \mathbf{k}$.

15. A particle moves with constant velocity **v**, starting from a point with position vector $\mathbf{r}_0$ at time $t = 0$. Given the position vector **r** of the particle at time t. Show that the particle will be closer to the co-ordinate origin at time $t = -(\mathbf{r}_0 \cdot \mathbf{v}) / (\mathbf{v} \cdot \mathbf{v})$.

16. Solve

$$\frac{d^2\mathbf{r}}{dt^2}=\mathbf{a},$$

where **a** is a constant vector; given that $\mathbf{r} = 0$ and $d\mathbf{r}/dt = 0$ when $t = 0$ Is the solution unique ? *(Avadh 2005, Rohilkhand 2000)*

17. Solve

$$\frac{d^2\mathbf{r}}{dt^2}=\mathbf{a}t+\mathbf{b},$$

where **a** and **b** are constant vectors, given that $\mathbf{r} = 0$ and $d\mathbf{r}/dt = \mathbf{u}$ when $t = 0$. Show that the solution is unique.

18. Show that

$$\mathbf{r} = \mathbf{a}e^{kt} + \mathbf{b}e^{l}t,$$

is the solution of the linear differential equation

$$\frac{d^2\mathbf{r}}{dt^2}+p\frac{d\mathbf{r}}{dt}+q\mathbf{r}=0, \qquad (p,\ q \text{ constant scalars})$$

where k, l are the distinct roots of the equation

$$m^2 + pm + q = 0,$$

and **a, b** are arbitrary constant vectors.

Also show that if the two roots of $m^2 + pm + a = 0$ are both equal then

$$(\mathbf{a} + l\mathbf{b})\, e^{kt}$$

is the solution of the given differential equation; k being the root.

19. The vector **r** satisfies the equation

$$m\frac{d^2\mathbf{r}}{dt^2}=e\mathbf{E}+\frac{e}{c}\frac{d\mathbf{r}}{dt}\times\mathbf{H},$$

where $\mathbf{E} = E\mathbf{j}$, $\mathbf{H} = H\mathbf{k}$ and e, m, c, E, H are constants.

Find the solution that satisfies the conditions

$$\mathbf{r}=0,\ \frac{d\mathbf{r}}{dt}=0 \text{ when } t=0.$$

20. If **r** is a unit vector, show that

$$\left|\mathbf{r}\times\frac{d\mathbf{r}}{dt}\right|=\left|\frac{d\mathbf{r}}{dt}\right|.$$

(Awadh 2000, Gorakhpur 2000, Lucknow 98)

9.9. VECTOR VALUED FUNCTIONS OF SEVERAL SCALAR VARIABLES

Like vector valued functions of a single scalar variable, we can also consider vector functions **f** $(u, v, w, ...)$ of several scalar variables $u, v, w,$ ... . We shall, however, generally attend to the developments in relation to functions of two variables only and the extension to the case of more than two variables can be made without any further difficulty.

Thus, we shall consider functions

$$\mathbf{f}(u, v),$$

so that to each pair of values of the scalar variables, u, v belonging to a given aggregate of pairs of the same, there corresponds a vector $\mathbf{f}(u, v)$.

Illustrations.

(*i*) If

$$\mathbf{r} = \mathbf{f}(u, v),$$

be thought of as the position vector of a point P then the locus of P for different pairs of values of u, v, is a surface and we say that $\mathbf{r} = \mathbf{f}(u, v)$ is the parametric vectorial equation or simply the equation of the surface.

(*ii*) If we consider any fluid in motion, then the velocity **q** of a fluid particle at any point and at any time is a vector function of four scalar variables x, y, z, t, where x, y, z are the co-ordinates of the point and t is the time.

9.10. DECOMPOSITION

Decomposition the vector $\mathbf{f}(u, v)$, as a linear combination of **i, j, k,** we have

$$\mathbf{f}(u, v) = f_1(u, v)\,\mathbf{i} + f_2(u, v)\,\mathbf{j} + f_3(u, v)\,\mathbf{k},$$

where f_1, f_2, f_3, are three scalar functions of the two variables u, v.

9.11. LIMIT OF A VECTOR FUNCTION

$$\lim_{(u,v)\to(c,d)} \mathbf{f}(u, v) = \mathbf{l}.$$

Def. *A vector function* **f** (u, v) *is said to tend to a limit* **l** *when u tends to c and v tends to d, if to every preassigned positive number* ε, *there corresponds a positive number* δ *such that*

$$|\mathbf{f}(u, v) - \mathbf{l}| < \varepsilon \text{ when } 0 < |u - c| \le \delta,\ \ 0 < |v - d| \le \delta.$$

As in the case of vector functions of a single variable (§ 9.2.1), we can show that *if*

(*i*) $\lim \mathbf{f}(u, v) = \mathbf{l}$ *when* $(u, v) \to (c, d)$, *(ii)* $\mathbf{f}(u, v) = f_1(u, v)\,\mathbf{i} + f_2(u, v)\,\mathbf{j} + f_3(u, v)\,\mathbf{k}$,

and *(iii)* $\mathbf{l} = l_1\mathbf{i} + l_2\,\mathbf{j} + l_3\mathbf{k}$,

when $(u, v) \to (c, d)$,

$$\lim f_1(u, v) = l_1, \quad \lim f_2(u, v) = l_2, \quad \lim f_3(u, v) = l_3$$

and conversely.

Again we can also easily establish theorems for vector valued functions of several variables corresponding to those for functions of a single variable given in § 9.2.2.

9.12. CONTINUITY

Def. A function $\mathbf{f}(u, v)$ is said to be continuous for $u = c$, $v = d$, if

$$\lim_{(u,v)\to(c,d)} \mathbf{f} = (u, v) = \mathbf{f}(c, d).$$

Again $\mathbf{f}(u, v)$ is said to be continuous function, if it is continuous for every pair of values u, v belonging to the region of definition of the function.

It can be easily shown (Refer § 9.11) that

$$\mathbf{f}(u, v) = f_1(u, v)\,\mathbf{i} + f_2(u, v)\,\mathbf{j} + f_3(u, v)\,\mathbf{k},$$

is continuous if, and only if,

$$f_1(u, v), \quad f_2(u, v), \quad f_3(u, v)$$

are continuous.

Again we may show that if $\mathbf{f}(u, v)$, $\mathbf{F}(u, v)$ are two continuous functions, then the functions

$$\mathbf{f} \pm \mathbf{F}, \quad \mathbf{f} \cdot \mathbf{F}, \quad \mathbf{f} \times \mathbf{F},$$

are also continuous.

9.13. PARTIAL DERIVATIVES

First order partial derivatives. If $\mathbf{f}(u, v)$ is any function, then

$$\lim_{h\to 0} \frac{\mathbf{f}(c+h, d) - \mathbf{f}(c, d)}{h}$$

if it exists, is called the partial derivatives of $\mathbf{f}(u, v)$ with respect to u for $u = c$, $v = d$.

Similarly

$$\lim_{h\to 0} \frac{\mathbf{f}(c, d+k) - \mathbf{f}(c, d)}{k}$$

if it exists, is the partial derivative of $\mathbf{f}(u, v)$ with respect to v for $u = c$, $v = d$.

The two partial derivatives are denoted by the symbols

$$\mathbf{f}_u(c, d), \quad \mathbf{f}_v(c, d)$$

respectively.

If we suppose that $\mathbf{f}(u, v)$ possesses partial derivatives with respect to u and v for every pair of values u, v then we obtain two vector functions denoted by

$$\frac{\partial \mathbf{f}}{\partial u}, \frac{\partial \mathbf{f}}{\partial v}$$

of (u, v) such that the value of $\partial \mathbf{f}/\partial u$ for $u = c$, $v = d$ is the partial derivative $\mathbf{f}$ with respect to u for $u = c$, $v = d$ and similarly for $\partial \mathbf{f}/\partial v$.

We may show as in § 9.4.3, that if

$$\mathbf{f}(u, v) = f_1(u, v)\,\mathbf{i} + f_2(u, v)\,\mathbf{j} + f_3(u, v)\,\mathbf{k},$$

then,

$$\frac{\partial \mathbf{f}}{\partial u} = \frac{\partial f_1}{\partial u}\mathbf{i} + \frac{\partial f_2}{\partial u}\mathbf{j} + \frac{\partial f_3}{\partial u}\mathbf{k},$$

and

$$\frac{\partial \mathbf{f}}{\partial v} = \frac{\partial f_1}{\partial v}\mathbf{i} + \frac{\partial f_2}{\partial v}\mathbf{j} + \frac{\partial f_3}{\partial v}\mathbf{k}.$$

9.13.1. Partial Derivatives of Second and Higher Orders

We suppose that each of the two vector functions

$$\frac{\partial \mathbf{f}}{\partial u}, \frac{\partial \mathbf{f}}{\partial v}$$

possesses partial derivatives w.r. to u and v. Then these partial derivatives, *viz.*,

$$\frac{\partial}{\partial u}\left(\frac{\partial \mathbf{f}}{\partial u}\right), \frac{\partial}{\partial v}\left(\frac{\partial \mathbf{f}}{\partial u}\right); \frac{\partial}{\partial u}\left(\frac{\partial \mathbf{f}}{\partial v}\right), \frac{\partial}{\partial u}\left(\frac{\partial \mathbf{f}}{\partial v}\right)$$

are denoted by the symbols

$$\frac{\partial^2 \mathbf{f}}{\partial u^2}, \frac{\partial^2 \mathbf{f}}{\partial v \partial u}, \frac{\partial^2 \mathbf{f}}{\partial v \partial u}, \frac{\partial^2 \mathbf{f}}{\partial v^2}$$

respectively and are called second order partial derivatives of $\mathbf{f}$. The two partial derivatives

$$\frac{\partial^2 \mathbf{f}}{\partial v \partial u}, \frac{\partial^2 \mathbf{f}}{\partial u \partial v},$$

involving inverted orders of derivation, are generally equal. In this connection, we shall have a theorem in the following subsection.

The first order partial derivatives of the second order partial derivatives of $\mathbf{f}(u, v)$ and so on for higher order derivatives of $\mathbf{f}$.

9.13.2. Inversion of the Order of Derivation

Theorem. *The two second order partial derivatives*

$$\frac{\partial^2 \mathbf{f}}{\partial u \partial v}, \frac{\partial^2 \mathbf{f}}{\partial v \partial u}$$

are equal, if each of them is a continuous function.

The proof will be based on the corresponding theorem in the Scalar Calculus which states that if the two derivatives

$$\frac{\partial^2 \phi}{\partial u \partial v}, \frac{\partial^2 \phi}{\partial v \partial u}.$$

of scalar function $\phi\,(u, v)$ are both continuous, then they are equal.

Let

$$\mathbf{f}\,(u, v) = f_1(u, v)\,\mathbf{i} + f_2(u, v)\,\mathbf{j} + f_3(u, v)\,\mathbf{k}.$$

$$\therefore \qquad \frac{\partial \mathbf{f}}{\partial u} = \frac{\partial f_1}{\partial u}\mathbf{i} + \frac{\partial f_2}{\partial u}\mathbf{j} + \frac{\partial f_3}{\partial u}\mathbf{k}, \qquad ...(1)$$

$$\frac{\partial \mathbf{f}}{\partial v} = \frac{\partial f_1}{\partial v}\mathbf{i} + \frac{\partial f_2}{\partial v}\mathbf{j} + \frac{\partial f_3}{\partial v}\mathbf{k}. \qquad ...(2)$$

Also

$$\frac{\partial^2 \mathbf{f}}{\partial v \partial u} = \frac{\partial}{\partial v}\left(\frac{\partial \mathbf{f}}{\partial u}\right) = \frac{\partial^2 f_1}{\partial v \partial u}\mathbf{i} + \frac{\partial^2 f_2}{\partial v \partial u}\mathbf{j} + \frac{\partial^2 f_3}{\partial v \partial u}\mathbf{k}, \qquad ...(3)$$

and

$$\frac{\partial^2 \mathbf{f}}{\partial u \partial v} = \frac{\partial}{\partial u}\left(\frac{\partial \mathbf{f}}{\partial v}\right) = \frac{\partial^2 f_1}{\partial u \partial v}\mathbf{i} + \frac{\partial^2 f_2}{\partial u \partial v}\mathbf{j} + \frac{\partial^2 f_3}{\partial u \partial v}\mathbf{k} \qquad ...(4)$$

As $\partial^2\mathbf{f}/\partial u \partial v$ is a continuous function, if follows that

$$\frac{\partial^2 f_1}{\partial u \partial v}, \frac{\partial^2 f_2}{\partial u \partial v}, \frac{\partial^2 f_3}{\partial u \partial v},$$

are also continuous. Similarly

$$\frac{\partial^2 f_1}{\partial v \partial u}, \frac{\partial^2 f_2}{\partial v \partial u}, \frac{\partial^2 f_3}{\partial v \partial u},$$

are continuous functions.

Thus, by the theorem quoted,

$$\frac{\partial^2 f_1}{\partial u \partial v} = \frac{\partial^2 f_1}{\partial v \partial u}, \frac{\partial^2 f_2}{\partial u \partial v} = \frac{\partial^2 f_2}{\partial v \partial u}, \frac{\partial^2 f_3}{\partial u \partial v} = \frac{\partial^2 f_3}{\partial v \partial u}. \qquad ...(5)$$

Hence, from (3), (4) and (5),

$$\frac{\partial^2 \mathbf{f}}{\partial u \partial v} = \frac{\partial^2 \mathbf{f}}{\partial v \partial u}.$$

9.13.3. Derivation of Composite Functions

Theorem. *If*

$$\mathbf{r} = \mathbf{f}\,(u, v), \quad u = \phi\,(t, s), \quad v = \psi\,(t, s)$$

are three functions possessing continuous first order partial derivatives with respect to the respective variables then r, is a function of t, s possessing continuous first order partial derivatives with respect to the same and

$$\frac{\partial \mathbf{r}}{\partial t} = \frac{\partial \mathbf{f}}{\partial u}\frac{\partial u}{\partial t} + \frac{\partial \mathbf{f}}{\partial v}\frac{\partial v}{\partial t},$$

$$\frac{\partial \mathbf{r}}{\partial s} = \frac{\partial \mathbf{f}}{\partial u}\frac{\partial u}{\partial s} + \frac{\partial \mathbf{f}}{\partial v}\frac{\partial v}{\partial s}.$$

Let

$$\mathbf{r} = \mathbf{f}\,(u, v) = f_1(u, v)\,\mathbf{i} + f_2(u, v)\,\mathbf{j} + f_3(u, v)\,\mathbf{k}.$$

As f_1, f_2, f_3 possess continuous first order partial derivatives with respect to u, v we see, by the corresponding theorem in Scalar Calculus, that these functions possess continuous first order partial derivatives with respect to t and s and

$$\frac{\partial f_1}{\partial t} = \frac{\partial f_1}{\partial u}\frac{\partial u}{\partial t} + \frac{\partial f_1}{\partial v}\frac{\partial v}{\partial t}, \quad \frac{\partial f_1}{\partial s} = \frac{\partial f_1}{\partial u}\frac{\partial u}{\partial s} + \frac{\partial f_1}{\partial v}\frac{\partial v}{\partial s},$$

with similar expressions in relation to the functions f_2 and f_3. Thus, $\partial \mathbf{r}/\partial t$ and $\partial \mathbf{r}/\partial s$ are continuous and

$$\frac{\partial \mathbf{r}}{\partial t} = \frac{\partial \mathbf{f}}{\partial t} = \frac{\partial f_1}{\partial t}\mathbf{i} + \frac{\partial f_2}{\partial t}\mathbf{j} + \frac{\partial f_3}{\partial t}\mathbf{k}$$

$$= \left(\frac{\partial f_1}{\partial u}\frac{\partial u}{\partial t} + \frac{\partial f_1}{\partial v}\frac{\partial v}{\partial t}\right)\mathbf{i} + \left(\frac{\partial f_2}{\partial u}\frac{\partial u}{\partial t} + \frac{\partial f_2}{\partial v}\frac{\partial v}{\partial t}\right)\mathbf{j}$$

$$+ \left(\frac{\partial f_3}{\partial u}\frac{\partial u}{\partial t} + \frac{\partial f_3}{\partial v}\frac{\partial u}{\partial t}\right)\mathbf{k}.$$

$$= \left(\frac{\partial f_1}{\partial u}\mathbf{i} + \frac{\partial f_2}{\partial u}\mathbf{j} + \frac{\partial f_3}{\partial u}\mathbf{k}\right)\frac{\partial u}{\partial t} + \left(\frac{\partial f_1}{\partial v}\mathbf{i} + \frac{\partial f_2}{\partial v}\mathbf{j} + \frac{\partial f_3}{\partial v}\mathbf{k}\right)\frac{\partial v}{\partial t},$$

$$= \frac{\partial \mathbf{f}}{\partial u}\frac{\partial u}{\partial t} + \frac{\partial \mathbf{f}}{\partial v}\frac{\partial v}{\partial t}.$$

Similarly for $\partial \mathbf{r}/\partial s$.

9.13.4. Total Differential

The relation

$$d\mathbf{f} = \frac{\partial \mathbf{f}}{\partial u}du + \frac{\partial \mathbf{f}}{\partial v}dv.$$

Let

$$\mathbf{f}\,(u, u) = f_1(u, v)\,\mathbf{i} + f_2(u, v)\,\mathbf{j} + f_3(u, v)\,\mathbf{k}.$$

We suppose that **f** possesses continuous first order partial derivatives so that f_1, f_2, f_3 also do the same.

Now, by a theorem in Scalar Calculus, there exists scalars

$$\varepsilon_1, \rho_1;\ \varepsilon_2, \rho_2;\ \varepsilon_3, \rho_3$$

which $\to 0$ as $\delta u \to 0$ and $\delta v \to 0$ and such that

$$\delta f_1 = \frac{\partial f_1}{\partial u}\delta u + \frac{\partial f_1}{\partial b}\delta v + \varepsilon_1 \delta u + \rho_1 \delta v,$$

$$\delta f_2 = \frac{\partial f_2}{\partial u}\delta u + \frac{\partial f_2}{\partial u}\delta v + \varepsilon_2 \delta u + \rho_2 \delta v,$$

$$\delta f_3 = \frac{\partial f_3}{\partial u}\delta u + \frac{\partial f_3}{\partial u}\delta v + \varepsilon_3 \delta u + \rho_3 \delta v.$$

$$\therefore \quad \delta \mathbf{f} = \mathbf{f}\,(u + \delta u, v + \delta v) - \mathbf{f}\,(u, v)$$

$$= \delta f_1 \mathbf{i} + \delta f_2\,\mathbf{j} + \delta f_3 \mathbf{k}$$

$$= \frac{\partial \mathbf{f}}{\partial u}\delta u + \frac{\partial \mathbf{f}}{\partial v}\delta v + (\varepsilon_1 \mathbf{i} + \varepsilon_2 \mathbf{j} + \varepsilon_3 \mathbf{k})\,\delta u\,(\rho_1 \mathbf{i} + \rho_2 \mathbf{j} + \rho_3 \mathbf{k})\,\delta v.$$

Here $\varepsilon_1 \mathbf{i} + \varepsilon_2\,\mathbf{j} + \varepsilon_3 \mathbf{k}$ and $\rho_1 \mathbf{i} + \rho_2\,\mathbf{j} + \rho_3 \mathbf{k}$ both tend to zero when δu and δv tend to 0.

The increments δu, δv of the independent variables are also called the differentials of u, v denoted by du, dv respectively. Also the principal part

$$\frac{\partial \mathbf{f}}{\partial u}\partial u + \frac{\partial \mathbf{f}}{\partial v}\partial v$$

of the increment $\delta\mathbf{f}$ in **f** is called the differential of **f**, denoted by $d\mathbf{f}$. Thus, we have

$$d\mathbf{f} = \frac{\partial \mathbf{f}}{\partial u}du + \frac{\partial \mathbf{f}}{\partial v}dv.$$

EXAMPLES

Example 1. *If* $\mathbf{A} = (2x^2y - x^4)\,\mathbf{i} + (e^{xy} - y \sin x)\,\mathbf{j} + x^2 \cos y\,\mathbf{k}$, verify that $\dfrac{\partial^2 \mathbf{A}}{\partial y \partial x} = \dfrac{\partial^2 \mathbf{A}}{\partial x \partial y}$.

Solution. $\dfrac{\partial \mathbf{A}}{\partial x} = \dfrac{\partial}{\partial x}(2x^2y - x^4)\,\mathbf{i} + \dfrac{\partial}{\partial x}(e^{xy} - y \sin x)\,\mathbf{j} + \dfrac{\partial}{\partial x}(x^2 \cos y)\,\mathbf{k}$

$$= (4xy - 4x^3)\,\mathbf{i} + (ye^{xy} - y \cos x)\,\mathbf{j} + 2x \cos y\,\mathbf{k}$$

$$\frac{\partial^2 \mathbf{A}}{\partial y \partial x} = \frac{\partial}{\partial y}\left(\frac{\partial^2 \mathbf{A}}{\partial x}\right) = 4x\mathbf{i} + (xye^{xy} - \cos x)\,\mathbf{j} - 2x \sin y\,\mathbf{k} \quad ...(1)$$

Again,

$$\frac{\partial \mathbf{A}}{\partial y} = 2x^2\mathbf{i} + (xe^{xy} - \sin x)\,\mathbf{j} - x^2 \sin y\,\mathbf{k}$$

$$\frac{\partial^2 \mathbf{A}}{\partial x \partial y} = \frac{\partial}{\partial x}\left(\frac{\partial \mathbf{A}}{\partial y}\right) = 4x\mathbf{i} + (xye^{xy} - \cos x)\,\mathbf{j} - 2x \sin y\,\mathbf{k}$$

From (1) and (2), the result follows.

Example 2. *If* $\phi(x, y, z) = xy^2z$ *and* $\mathbf{A} = xz\,\mathbf{i} - xy\,\mathbf{j} + yz^2\,\mathbf{k}$ *find* $\dfrac{\partial^3(\phi\mathbf{A})}{\partial x^2 \partial z}$ *at* $(2, -1, 1)$.

Solution. $\phi\mathbf{A} = xy^2z\,(xz\mathbf{i} - xy\mathbf{j} + yz^2\mathbf{k})$

$= x^2y^2z^2\,\mathbf{i} - x^2y^2z\,\mathbf{j} + xy^3z^3\,\mathbf{k}$

Now

$$\frac{\partial}{\partial z}(\phi\mathbf{A}) = 2x^2y^2z\,\mathbf{i} - x^2y^3\mathbf{j} + 3xy^3z^2\mathbf{k}$$

$$\frac{\partial^2}{\partial x \partial z}(\phi\mathbf{A}) = 4xy^2z\,\mathbf{i} - 2xy^3\mathbf{j} + 3y^3z^2\mathbf{k}$$

and

$$\frac{\partial^3}{\partial x^2 \partial z}(\phi\mathbf{A}) = 4y^2z\,\mathbf{i} - 2y^3\mathbf{j}$$

$$= 4(-1)^2(1)\,\mathbf{i} - 2(-1)^3\,\mathbf{j}$$

$$= 4\mathbf{i} + 2\mathbf{j} \text{ at the point } (2, -1, 1).$$

EXERCISES

1. Find

$$\frac{\dfrac{\partial \mathbf{r}}{\partial u} \times \dfrac{\partial \mathbf{r}}{\partial v}}{\left|\dfrac{\partial \mathbf{r}}{\partial u} \times \dfrac{\partial \mathbf{r}}{\partial v}\right|},$$

when

(*i*) $\mathbf{r} = u \cos v\mathbf{i} + u \sin v\,\mathbf{j} + ae^{mv}\mathbf{k}$,

(*ii*) $\mathbf{r} = \dfrac{1}{2}a(u+v)\,\mathbf{i} + \dfrac{1}{2}b(u-v)\,\mathbf{j} + \dfrac{1}{2}uv\,\mathbf{k}$.

(*iii*) $\mathbf{r} = u \cos v\,\mathbf{i} + u \sin v\,\mathbf{j} + c \log\left[u + \sqrt{(u^2 - c^2)}\right]\mathbf{k}$.

2. Find

$$\left[\frac{\partial \mathbf{r}}{\partial u}\,\frac{\partial \mathbf{r}}{\partial v}\,\frac{\partial^2 \mathbf{r}}{\partial u^2}\right], \left[\frac{\partial \mathbf{r}}{\partial u}\,\frac{\partial \mathbf{r}}{\partial v}\,\frac{\partial^2 \mathbf{r}}{\partial u \partial v}\right], \left[\frac{\partial \mathbf{r}}{\partial u}\,\frac{\partial \mathbf{r}}{\partial v}\,\frac{\partial^2 \mathbf{r}}{\partial v^2}\right]$$

are each of the functions given in the Ex. 1, above.

APPENDIX III

The proofs of the theorems on limits given in § 9.2.2, which are independent of the corresponding theorems in scalar calculus will now be given.

(*i*) Let ε be any positive number. As

$$\lim_{t \to c} \mathbf{f}(t) = \mathbf{l}, \quad \lim_{t \to c} \mathbf{F}(t) = \mathbf{m},$$

there exists a positive number δ such that

$$\left| \mathbf{f}(t) - \mathbf{l} \right| < \frac{1}{2}\varepsilon, \left| \mathbf{F}(t) - \mathbf{m} \right| < \frac{1}{2}\varepsilon$$

when

$$0 < | t - c | \le \delta.$$

Thus, for $0 < | t - c | \le \delta$,

$$\begin{aligned} | [\mathbf{f}(t) + \mathbf{F}(t)] - [\mathbf{l} + \mathbf{m}] | &= | [\mathbf{f}(t) - \mathbf{l}] + [\mathbf{F}(t) - \mathbf{m}] | \\ &\le | \mathbf{f}(t) - \mathbf{l} | + [\mathbf{F}(t) - \mathbf{m} | \\ &< \frac{1}{2}\varepsilon + \frac{1}{2}\varepsilon = \varepsilon. \end{aligned}$$

and

$$\begin{aligned} | [\mathbf{f}(t) - \mathbf{F}(t)] - [\mathbf{l} - \mathbf{m}] | &= | [\mathbf{f}(t) - \mathbf{l}] - [\mathbf{F}(t) - \mathbf{m}] | \\ &\le | \mathbf{f}(t) - \mathbf{l} | + [\mathbf{F}(t) - \mathbf{m} | \\ &< \frac{1}{2}\varepsilon + \frac{1}{2}\varepsilon = \varepsilon. \end{aligned}$$

Hence,

$$\lim [\mathbf{f}(t) \pm \mathbf{F}(t)] = \mathbf{l} \pm \mathbf{m} \text{ when } t \to c.$$

(*ii*) We write

$$\mathbf{f}(t) \,.\, \mathbf{F}(t) - \mathbf{l} \,.\, \mathbf{m} = \mathbf{f}(t) \,.\, [\mathbf{F}(t) - \mathbf{m}] + \mathbf{m} \,.\, [\mathbf{f}(t) - \mathbf{l}],$$

so that for every value of t,

$$\begin{aligned} | \mathbf{f}(t) \,.\, \mathbf{F}(t) - \mathbf{l} \,.\, \mathbf{m} | &\le | \mathbf{f}(t) \,.\, [\mathbf{F}(t) - \mathbf{m}] | + | \mathbf{m} \,.\, [\mathbf{f}(t) - \mathbf{l}] | \\ &\le | \mathbf{f}(t) | \; | \mathbf{F}(t) - \mathbf{m} | + | \mathbf{m} | \; | \mathbf{f}(t) - \mathbf{l}] | . \end{aligned}$$

As $\lim_{t \to c} \mathbf{f}(t) = \mathbf{l}$, there exists a positive number δ_1 such that

$$| \mathbf{f}(t) - \mathbf{l} | < 1 \text{ when } 0 < | t - c | \le \delta_1.$$

But

$$| \mathbf{f}(t) - \mathbf{l} | \ge | \mathbf{f}(t) | - | \mathbf{l} |, \text{ for every } t.$$

Thus, $0 < | t - c | \le \delta_1$, we have

$$| \mathbf{f}(t) | - | \mathbf{l} | \le | \mathbf{f}(t) - \mathbf{l} | < 1.$$

or

$$| \mathbf{f}(t) | < | \mathbf{l} | + 1.$$

$\therefore$ for $0 < |t - c| \le \delta_1$, we have

$|\mathbf{f}(t) \,.\, \mathbf{F}(t) - \mathbf{l} \,.\, \mathbf{m}| < (|\mathbf{l}| + 1)\, |\mathbf{F}(t) - \mathbf{m}| + |\mathbf{m}|\; |\mathbf{f}(t) - \mathbf{l}|$.

Let now ε be any positive number. Then there exist positive number δ_2 and δ_3 such that

$$|\mathbf{F}(t) - \mathbf{m}| < \frac{\varepsilon}{2(|\mathbf{l}|+1)} \text{ when } 0 < |t-c| \le \delta_2,$$

and $$|\mathbf{f}(t) - \mathbf{l}| < \frac{\varepsilon}{2(|\mathbf{m}|+1)} \text{ when } 0 < |t-c| \le \delta_3.$$

If $\delta = \min.\,(\delta_1, \delta_2, \delta_3)$, we have, for $0 < |t - c| \le \delta$,

$$|\mathbf{f}(t) \,.\, \mathbf{F}(t) - \mathbf{l} \,.\, \mathbf{m}| < (|\mathbf{l}|+1) \frac{\varepsilon}{2(|\mathbf{l}|+1)} + |\mathbf{m}| \frac{2}{2(|\mathbf{m}|+1)} < \varepsilon,$$

i.e., $$|\mathbf{f}(t) \,.\, \mathbf{F}(t) - \mathbf{l} \,.\, \mathbf{m}| < \varepsilon.$$

Hence, $$\lim_{t \to c} [\mathbf{f}(t) \,.\, \mathbf{F}(t)] = \mathbf{l} \,.\, \mathbf{m}$$

(*iii*) We write

$$\mathbf{f}(t) \times \mathbf{F}(t) - \mathbf{l} \times \mathbf{m} = \mathbf{f}(t) \times [\mathbf{F}(t) - \mathbf{m}] + [\mathbf{f}(t) - \mathbf{l}] \times \mathbf{m},$$

so that for every value of t,

$$|\mathbf{f}(t) \times \mathbf{F}(t) - \mathbf{l} \times \mathbf{m}| \le |\mathbf{f}(t)|\; |\mathbf{F}(t) - \mathbf{m}| + |\mathbf{f}(t) - \mathbf{l}|\; |\mathbf{m}|.$$

The further process is now the same as in the case of scalar product above.

(*iv*) We write

$$\phi(t)\, \mathbf{f}(\mathbf{l}) - p\mathbf{l} = \mathbf{f}(t)\, [\phi(t) - p] + p\, [\mathbf{f}(t) - \mathbf{l}],$$

so that for every value of t,

$$|\phi(t)\, \mathbf{f}(t) - p\mathbf{l}| \le |\mathbf{f}(t)|\; |\phi(t) - p| + |p|\; |\mathbf{f}(t) - \mathbf{l}|.$$

Again the further processes is the same as in the two cases above.

(*v*) We have

$$|\mathbf{f}(t)| - |\mathbf{l}| \le |\mathbf{f}(t) - \mathbf{l}|, \qquad ...(1)$$

for every value of t.

As $\lim \mathbf{f}(t) = \mathbf{l}$, when $t \to c$,

there exists a positive number δ corresponding to any preassigned positive ε such that

$$|\mathbf{f}(t) - \mathbf{l}| < \varepsilon \text{ when } 0 < |t - c| \le \delta. \qquad ...(2)$$

From (1) and (2), we have

$$\big|\, |\mathbf{f}(t)| - |\mathbf{l}| \,\big| < \varepsilon \text{ when } 0 < |t - c| \le \delta.$$

Hence,

$$\lim_{t \to c} |\mathbf{f}(t)| = |\mathbf{l}|.$$

OBJECTIVE QUESTIONS

For each of the following questions, four alternatives are given for the answer. Only one of them is correct. Choose the correct alternative.

1. Let $\mathbf{f}(t)$ and $\mathbf{g}(t)$ be two vector functions such that $\lim_{t \to a} \mathbf{f}(t) = \mathbf{L}$ and $\lim_{t \to a} \mathbf{g}(t) = \mathbf{M}$, then $\lim_{t \to 0} [\mathbf{f}(t) \times \mathbf{g}(t)]$

(*a*) $\mathbf{L}$ (*b*) $\mathbf{M}$ (*c*) $\mathbf{L} \times \mathbf{M}$ (*d*) $\mathbf{L} \,.\, \mathbf{M}$

2. If $\mathbf{r} = \mathbf{f}(s)$ be a continuous and single-valued function of the scalar variable s, then $d\mathbf{r}/ds$ is a unit vector along

(*a*) Normal (*b*) Radius vector

(*c*) Tangent (*d*) None of these

3. Derivative of a vector of constant magnitude of (Rohilkhand 2005)

(*a*) zero (*b*) parallel to the vector

(*c*) perpendicular to the vector (*d*) None of these

4. Necessary and sufficient condition for the vector $\mathbf{v}(t)$ to have a constant direction is (Garhwal 2006, Rohilkhand 2003)

(*a*) $\mathbf{v} \cdot \dfrac{d\mathbf{v}}{dt} = 0$ (*b*) $\mathbf{v} \times \dfrac{d\mathbf{v}}{dt} = 0$

(*c*) $\mathbf{v} = 0$ (*d*) $\dfrac{d\mathbf{v}}{dt} = 0$

5. Necessary and sufficient condition for the vector $\mathbf{v}(t)$ to have a constant magnitude is

(*a*) $\dfrac{d\mathbf{v}}{dt} = 0$ (*b*) $\mathbf{v} \times \dfrac{d\mathbf{v}}{dt} = 0$

(*c*) $\mathbf{v} \cdot \dfrac{d\mathbf{v}}{dt} = 0$ (*d*) None of these

6. If $\mathbf{r} = \mathbf{a} \cos \omega t + \mathbf{b} \sin \omega t$, then value of $\dfrac{d^2\mathbf{r}}{dt^2}$ is

(*a*) $\omega (\mathbf{a} \times \mathbf{b})$ (*b*) $-\omega^2 \mathbf{r}$ (*c*) $\omega^2 \mathbf{r}$ (*d*) 0

7. If $\mathbf{r} \times d\mathbf{r} = 0$, then

(*a*) $\hat{\mathbf{r}}$ = constant (*b*) $\hat{\mathbf{r}} = \mathbf{0}$

(*c*) $\hat{\mathbf{r}}$ is indeterminate (*d*) None of these

8. A particle P is moving on a circle of radius r with constant angular velocity $\omega = d\theta/dt$. The acceleration is

(*a*) 0 (*b*) $\omega^2 \mathbf{r}$ (*c*) $-\omega^2 \mathbf{r}$ (*d*) $\omega \mathbf{r}$

9. If $\frac{d\mathbf{a}}{dt} = \lambda\mathbf{b}$ and $\frac{d\mathbf{b}}{dt} = \mu\mathbf{a}$, then $\frac{d}{dt} = (\mathbf{a} \times \mathbf{b})$ is

(*a*) $\lambda\mu\ (\mathbf{a} \times \mathbf{b})$ (*b*) $\lambda\mu\ (\mathbf{b} \times \mathbf{a})$

(*c*) $\frac{\lambda}{\mu}(\mathbf{a} \times \mathbf{b})$ (*d*) zero

10. If $\mathbf{r} \,.\, d\mathbf{r} = 0$, then

(*a*) $\hat{\mathbf{r}} = \mathbf{0}$ (*b*) $\hat{\mathbf{r}}$ is constant

(*c*) $|\,\mathbf{r}\,| =$ constant (*d*) $|\,\mathbf{r}\,| = 0$

ANSWERS

1. (*c*) **2.** (*c*) **3.** (*c*) **4.** (*b*) **5.** (*c*)

6. (*b*) **7.** (*a*) **8.** (*c*) **9.** (*d*) **10.** (*c*)

10

Differential Operators

Introduction. This chapter is devoted to a study of the Differential Operators — *Gradient, Divergence* and *Curl* — which, as a result of operation on point functions, leads to new point functions. These operators play a very important part in the study of Hydrodynamics and Electrodynamics and other branches of Applied Mathematics. The operators will be defined and the rules of combination of the same will be obtained in terms of rectangular cartesian axes and a section will also be devoted to the determination of the expressions of these operators in terms of general orthogonal curvilinear co-ordinates.

10.1. POINT FUNCTIONS

10.1.1. Scalar Valued Point Functions

Consider any region R of space and suppose that to each point P of the same, there corresponds, by any law whatsoever, a scalar denoted by $\phi(P)$. We then say that ϕ is a scalar point function defined for the region R.

Illustrations. Consider some material body occupying a region R and let $\phi(P)$ denote the density of the body at any point P thereof. Then ϕ is a scalar function defined for the region R.

The temperature at any point of a body occupying a certain region is a scalar point function defined for the region.

10.1.2. Vector Valued Point Functions

If to any point P of a region there corresponds a vector $\mathbf{f}(P)$, then we say that $\mathbf{f}$ is a vector function defined for the region.

Illustration. Consider any body in motion. At any given time t if $\mathbf{f}(P)$ denote the velocity of the particle at a point P, then $\mathbf{f}$ is a vector point function defined for the region occupied by the body at time t.

Notes. 1. If **r** denotes the position vector of a point P relative to any origin, then we may write

$$f(\mathbf{r}) \text{ and } \mathbf{f}(\mathbf{r}) \text{ instead of } f(P) \text{ and } \mathbf{f}(P)$$

respectively.

2. In the following, the regions of definition of point functions will not, in general, be mentioned.

10.2. LIMITS

10.2.1. A Scalar Point Function $\phi(P)$ *is said to tend to a limit l when P tends to a given point A, if to every pre-assigned positive number ε, there corresponds a positive number δ such that*

$$|\phi(P) - l| < \varepsilon,$$

for all points P within the sphere with centre A and radius δ, except possibly for the point A.

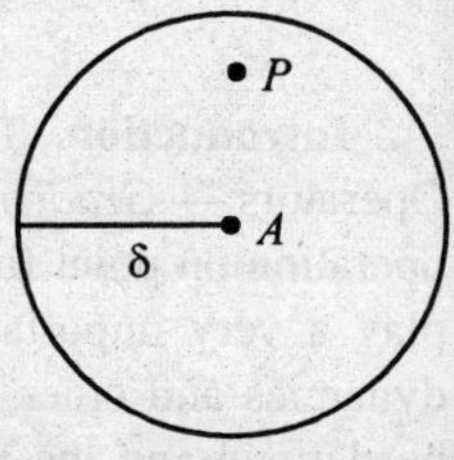

Fig. 10.1.

Also, we then write

$$\lim_{P \to A} \phi(P) = l.$$

This means that to every pre-assigned positive number ε, there corresponds a sphere with its centre at A such that for every point P within the sphere (except possibly A), $\phi(P)$ differs from l numerically by a number less than ε.

10.2.2. A Vector Point Function f (P), *is said to tend to a limit* **l** *when P tends to a given point A, if to every pre-assigned positive number ε, there corresponds a positive number δ such that*

$$|\mathbf{f}(P) - \mathbf{l}| < \varepsilon,$$

for all points P within the sphere with centre A and radius δ, except possibly for the point A.

We write

$$\lim_{P \to A} \mathbf{f}(P) = \mathbf{l}.$$

10.2.3. Continuity. A scalar point function $\phi(P)$ is said to be continuous at a point A, if

$$\lim_{P \to A} \phi(P) = \phi(A).$$

Also a scalar point function is said to be continuous, if it is continuous at every point of its region of definition.

The continuity of vector point functions may be similarly defined.

10.3. DIRECTIONAL DERIVATIVES

10.3.1. Scalar Point Functions

Let $\phi(P)$ be any scalar point function and A any given point of its region of direction. Also, let l be any given semi-line through A, *i.e.*, drawn on one side of A only and let P be any point on the semi-line.

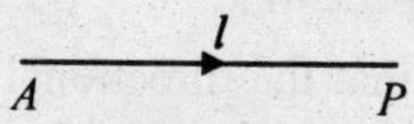

Fig. 10.2.

Then

$$\lim \frac{\phi(P)-\phi(A)}{AP}, \text{ when } P \to A,$$

if it exists, is called the *directional derivative* of the scalar function at A along the direction of the line l.

The length AP is to be regarded positive.

10.3.2. Vector Point Functions

In terms of the above notation

$$\lim \frac{\mathbf{f}(P)-\mathbf{f}(A)}{AP}, \text{ when } P \to A,$$

if it exists, is called the *directional derivative* of the vector function at A along the line l.

The directional derivative of a vector (scalar) point function at a point along any direction is a vector (scalar).

10.4. CARTESIAN REPRESENTATION OF POINT FUNCTIONS AND THEIR DIRECTIONAL DERIVATIVES

We introduce a system of orthogonal cartesian axes OX, OY, OZ.

Let (x, y, z) be the co-ordinates of a point P. We may now write

$$\phi(P) = \phi(x, y, z),$$

so that a scalar point function appears as a scalar function of three scalar variables x, y, z.

We similarly have

$$\mathbf{f}(P) = \mathbf{f}(x, y, z).$$

10.4.1. Directional Derivatives of Scalar Point Functions along Co-ordinate Axes

Take a point $P(x + \delta x, y, z)$ on a line through $A(x, y, z)$ drawn parallel to the positive direction of x-axis. We have AP = δx and δx is positive.

We have

$$\frac{\phi(P)-\phi(A)}{AP}=\frac{\phi(x+\delta x, y, z)-\phi(x, y, z)}{\delta x}.$$

This shows that the directional derivative at any point A (x, y, z) along the positive direction of x-axis is the partial derivative $\partial\phi/\partial x$ of the function with respect to x.

Similarly, we may see that the directional derivatives at A (x, y, z) along the positive directions of y-axis and z-axis are the partial derivatives $\partial\phi/\partial y$ and $\partial\phi/\partial z$ with respect to y and z respectively.

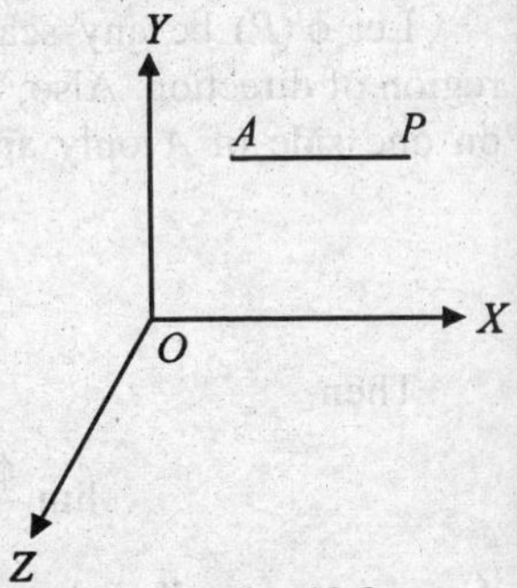

Fig. 10.3.

10.4.2. Directional Derivative along any Line

Consider now *any* semi-line through A (x, y, z) with direction cosines l, m, n.

The co-ordinates of the point P on this line at a distance r from A are

$$(x + lr,\ \ y + mr,\ \ z + nr).$$

We have

$$\frac{\phi(P)-\phi(A)}{AP}$$

$$=\frac{\phi(x+lr,\ y+mr,\ z+nr)-\phi(x, y, z)}{r}.$$

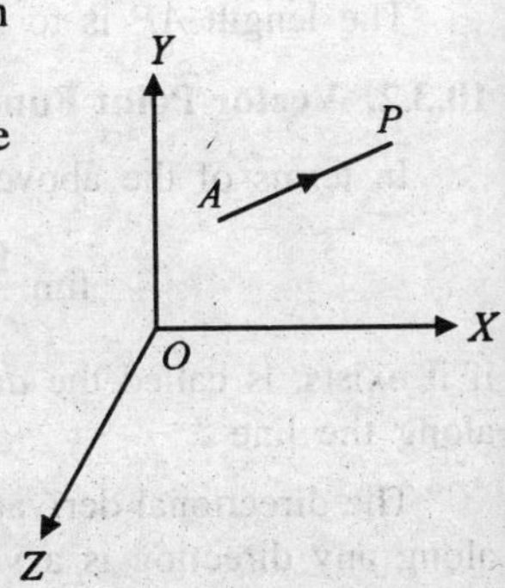

Fig. 10.4.

We now suppose that ϕ (x, y, z) *possesses continuous first order partial derivatives.* Then by Taylor's theorem for functions of three scalar variables, we have

$$\phi(x + lr, y + mr, z + nr) = \phi(x, y, z) + lr\phi_x(\xi, \eta, \zeta)$$
$$+ mr\phi_y(\xi, \eta, \zeta) + nr\phi_z(\xi, \eta, \zeta),$$

where (ξ, η, ζ) is a point of the line AP, joining A (x, y, z) to P $(x + lr, y + mr, z + nr)$. Thus,

$$\frac{\phi(P)-\phi(A)}{AP}=l\phi_x(\xi, \eta, \zeta)+m\phi_y(\xi, \eta, \zeta)+n\phi_z(\xi, \eta, \zeta)$$

so that when $P \to A$, we obtain

$$\frac{\phi(P)-\phi(A)}{AP}=l\phi_x(x, y, z)+m\phi_y(x, y, z)+n\phi_z(x, y, z)$$
$$= l\phi_x + m\phi_y + n\phi_z.$$

Thus, *the directional derivative at A (x, y, z) along a line with direction cosines l, m, n is*

$$l\phi_x + m\phi_y + n\phi_z,$$

so that we can write the directional derivative along any line in terms of those along the three co-ordinate axes.

We have seen that if $\phi(x, y, z)$ possesses continuous first order partial derivatives, then it possesses directional derivative along every direction and we say that $\phi(x, y, z)$ is *continuously differentiable.*

Note. The direction cosines of any semi-line through a point are equal and opposite to those of the semi-line produced on to the other side. Thus, if l, m, n are the direction cosines of AP, then $-l, -m, -n$ are those of AQ. Hence, the directional derivatives along the semi-lines AP, AQ are

$$l\phi_x + m\phi_y + n\phi_z, \quad -l\phi_x - m\phi_y - n\phi_z$$

respectively so that each is the negative of the other.

From above or otherwise directly, we may see that the directional derivatives along the negative directions of the coordinate axes are

$$-\phi_x, -\phi_y, -\phi_z.$$

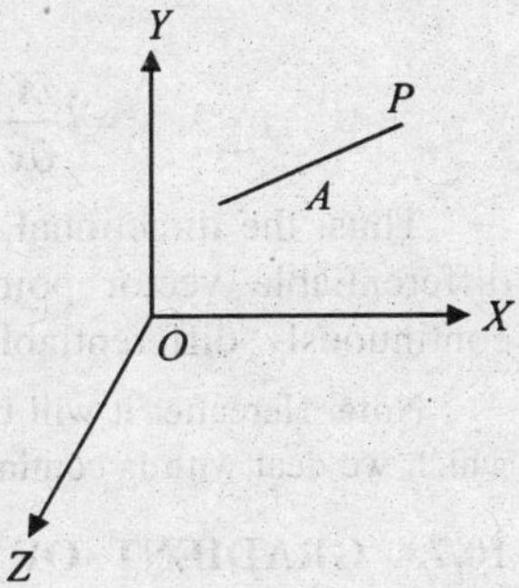

Fig. 10.5.

10.5. Directional Derivatives of Vector Point Functions along Co-ordinate Axes

As in the case of scalar point functions, we may show that the directional derivatives of a vector point function $\mathbf{f}(\mathbf{r}) = \mathbf{f}(x, y, z)$ at any point (x, y, z) along the positive directions of co-ordinate axes are

$$\frac{\partial \mathbf{f}}{\partial x}, \frac{\partial \mathbf{f}}{\partial y}, \frac{\partial \mathbf{f}}{\partial z}.$$

10.6. Directional Derivatives along any Line

Let l, m, n be the direction cosines of any semi-line through $A(x, y, z)$. We write

$$\mathbf{f}(x, y, z) = \mathbf{i} f_1(x, y, z) + \mathbf{j} f_2(x, y, z) + \mathbf{k} f_3(x, y, z)$$

and suppose that $\mathbf{f}(x, y, z)$ possesses continuous first order partial derivatives. Then, by § 9.2, 9.3 the scalar functions f_1, f_2 and f_3 also possesses continuous first order partial derivatives. In this case we say that the vector functions $\mathbf{f}$ is continuously differentiable.

We have

$$\frac{\mathbf{f}(P)-\mathbf{f}(A)}{AP} = \mathbf{i}\frac{f_1(P)-f_1(A)}{AP} + \mathbf{j}\frac{f_2(P)-f_2(A)}{AP} + \mathbf{k}\frac{f_3(P)-f_3(A)}{AP}$$

Let $P \to A$. Then by § 10.5.2,

$$\lim \frac{\mathbf{f}(P)-\mathbf{f}(A)}{AP} = \mathbf{i}\left[l\frac{\partial f_1}{\partial x}+m\frac{\partial f_1}{\partial y}+n\frac{\partial f_1}{\partial z}\right]+\mathbf{j}\left[l\frac{\partial f_2}{\partial x}+m\frac{\partial f_2}{\partial y}+n\frac{\partial f_2}{\partial z}\right]+$$
$$\mathbf{k}\left[l\frac{\partial f_3}{\partial x}+m\frac{\partial f_3}{\partial y}+n\frac{\partial f_3}{\partial z}\right]$$
$$= l\left[\mathbf{i}\frac{\partial f_1}{\partial x}+\mathbf{j}\frac{\partial f_2}{\partial x}+\mathbf{k}\frac{\partial f_3}{\partial x}\right]+m\left[\mathbf{i}\frac{\partial f_1}{\partial y}+\mathbf{j}\frac{\partial f_2}{\partial y}+\mathbf{k}\frac{\partial f_3}{\partial y}\right]+$$
$$n\left[\mathbf{i}\frac{\partial f_1}{\partial z}+\mathbf{j}\frac{\partial f_2}{\partial z}+\mathbf{k}\frac{\partial f_3}{\partial z}\right]$$
$$= l\frac{\partial \mathbf{f}}{\partial x}+m\frac{\partial \mathbf{f}}{\partial y}+n\frac{\partial \mathbf{f}}{\partial z}.$$

Thus, the directional derivative along any direction of a continuously differentiable vector point function is of the same form as that of a continuously differentiable scalar point function.

Note. Hereafter it will be assumed that every point function, scalar or vector, which we deal with is continuously differentiable.

10.7. GRADIENT OF A SCALAR POINT FUNCTION

Let, ϕ, be any continuously differentiable scalar point function.

We have

$$l\frac{\partial \phi}{\partial x}+m\frac{\partial \phi}{\partial y}+n\frac{\partial \phi}{\partial z} = (\mathbf{i}l+\mathbf{j}m+\mathbf{k}n)\cdot\left(\mathbf{i}\frac{\partial \phi}{\partial x}+\mathbf{j}\frac{\partial \phi}{\partial y}+\mathbf{k}\frac{\partial \phi}{\partial z}\right)$$

so that the directional derivative on the left appears as a scalar product of the two vectors

$$\mathbf{i}l+\mathbf{j}m+\mathbf{k}n \text{ and } \mathbf{i}\frac{\partial \phi}{\partial x}+\mathbf{j}\frac{\partial \phi}{\partial y}+\mathbf{k}\frac{\partial \phi}{\partial z}$$

of which the former is the unit vector along the line with direction cosines l, m, n and the latter is such that it depends only on the point (x, y, z) and not on any direction. Thus, we see that *the directional derivative along any direction can be obtained on multiplying the vector*

$$\mathbf{i}\frac{\partial \phi}{\partial x}+\mathbf{j}\frac{\partial \phi}{\partial y}+\mathbf{k}\frac{\partial \phi}{\partial z},$$

scalarly with the unit vector along the direction in question.

Thus, we are lead to introduce the notion of Gradient as follows :

Gradient. Def. *The vector function*

$$\mathbf{i}\frac{\partial \phi}{\partial x}+\mathbf{j}\frac{\partial \phi}{\partial y}+\mathbf{k}\frac{\partial \phi}{\partial z},$$

is called the gradient of the scalar point function ϕ *and is written as grad* ϕ. Thus,

$$\text{grad}\ \phi = \mathbf{i}\frac{\partial \phi}{\partial x} + \mathbf{j}\frac{\partial \phi}{\partial y} + \mathbf{k}\frac{\partial \phi}{\partial z}\cdot$$

Here, we have assumed that ϕ is a continuously differentiable function.

Thus, the directional derivative of the scalar point function ϕ along the direction of the unit vector **a** can be written as

$$\mathbf{a} \cdot \text{grad}\ \phi.$$

10.7.1. Gradient of a Constant

If ϕ is constant then $\partial\phi/\partial x$, $\partial\phi/\partial y$, $\partial\phi/\partial z$ are all zero so that grad $\phi = 0$. Conversely if, grad ϕ is zero, the partial derivatives are all zero and accordingly the function is constant. *Thus,*

$$\text{grad}\ \phi = 0,$$

if, and only if, the function ϕ *is constant.*

10.7.2. Derivative along a Curve

Let

$$\mathbf{r} = \mathbf{F}\,(s) = \mathbf{i}\phi_1(s) + \mathbf{j}\phi_2(s) + \mathbf{k}\phi_3(s)$$

be any curve. Along the curve $\phi\ (x, y, z)$ is a function of s. Then if A, P be two points on the curve, we have

$$\lim_{P \to A} \frac{\phi\,(P) - \phi\,(A)}{AP} = \lim_{\delta s \to 0} \frac{\phi\,(s + \delta s) - \phi\,(s)}{\delta s}$$

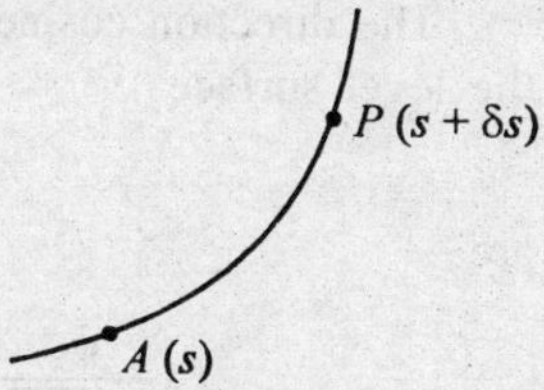

Fig. 10.6.

$$= \frac{d\phi}{ds}$$

$$= \frac{\partial \phi}{\partial x}\frac{dx}{ds} + \frac{\partial \phi}{\partial y}\frac{dy}{ds} + \frac{\partial \phi}{\partial z}\frac{dz}{ds}$$

$$= \left(\mathbf{i}\frac{dx}{ds} + \mathbf{j}\frac{dy}{ds} + \mathbf{k}\frac{dz}{ds}\right)\cdot\left(\mathbf{i}\frac{\partial \phi}{\partial x} + \mathbf{j}\frac{\partial \phi}{\partial y} + \mathbf{k}\frac{\partial \phi}{\partial z}\right)$$

$$= \mathbf{t} \cdot \text{grad}\ \phi,$$

where **t** denotes unit tangent to the curve.

It can be shown *that the derivative*

$$\mathbf{t} \cdot \text{grad}\ \phi$$

along a curve is zero if, and only if, the function ϕ *is constant along the curve.*

10.7.3. Differential dφ

We have

$$d\phi = \frac{\partial \phi}{\partial x}dx + \frac{\partial \phi}{\partial y}dy + \frac{\partial \phi}{\partial z}dz$$

$$= (\mathbf{i}dx + \mathbf{j}dy + \mathbf{k}dz) \cdot \left(\mathbf{i}\frac{\partial \phi}{\partial x} + \mathbf{j}\frac{\partial \phi}{\partial y} + \mathbf{k}\frac{\partial \phi}{\partial z}\right)$$

$$= d\mathbf{r} \,.\, \text{grad } \phi.$$

Thus,

$$d\phi = d\mathbf{r} \,.\, \text{grad } \phi.$$

10.8. GEOMETRICAL INTERPRETATION OF Grad φ

Level surfaces. The locus of point P such that

$$\phi\,(P) = \phi\,(A),$$

where A is any given point, is called the level surface of the scalar point function ϕ through the point A. This surface is such that the value of the function at any point thereof is the same as that at A. The level surface for any point A is also the level surface for any point B of the surface.

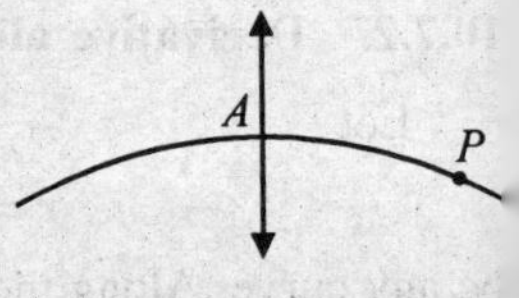

Fig. 10.7.

The direction cosines of the two sides of the normal at $A\ (x, y, z)$ to the level surface

$$\phi\,(X, Y, Z) = \phi\,(x, y, z)$$

are

$$\frac{\phi_x}{\sqrt{\Sigma\phi_x^2}}, \frac{\phi_y}{\sqrt{\Sigma\phi_x^2}}, \frac{\phi_z}{\sqrt{\Sigma\phi_x^2}}; \qquad \ldots(1)$$

$$-\frac{\phi_x}{\sqrt{\Sigma\phi_x^2}}, -\frac{\phi_y}{\sqrt{\Sigma\phi_x^2}}, -\frac{\phi_z}{\sqrt{\Sigma\phi_x^2}} \qquad \ldots(2)$$

so that the direction derivatives of the function ϕ along these two sides are

$$\frac{\phi_x}{\sqrt{\Sigma\phi_x^2}}\phi_x + \frac{\phi_y}{\sqrt{\Sigma\phi_x^2}}\phi_y + \frac{\phi_z}{\sqrt{\Sigma\phi_x^2}}\phi_z = \sqrt{\Sigma\phi_x^2} = |\,\text{grad } \phi\,| \qquad \ldots(3)$$

$$-\frac{\phi_x}{\sqrt{\Sigma\phi_x^2}}\phi_x - \frac{\phi_y}{\sqrt{\Sigma\phi_x^2}}\phi_y - \frac{\phi_z}{\sqrt{\Sigma\phi_x^2}}\phi_z = -\sqrt{\Sigma\phi_z^2} = -\,|\,\text{grad } \phi\,| \qquad \ldots(4)$$

The directional derivative | grad ϕ | along the side of the normal given by (1) being positive, we see that this side of the normal is one along which ϕ is increasing.

Also we see that the direction cosines of the vector grade ϕ are the same as those given in (1).

Hence, we have the following :

10.8.1. *The vector, grade ϕ, at any point is parallel (including sense) to the side along which ϕ increases to the normal to the level surface through the point and the magnitude of grade ϕ is the directional derivative of ϕ along the same side.*

10.8.2. Again, if **a** be any unit vector so that

$$\mathbf{a} \text{ grad } \phi$$

is the directional derivative of ϕ along the same, we have

$$\mathbf{a} \text{ grad } \phi \leq | \text{ grad } \phi |,$$

for all unit vectors **a.** Thus, we see that the *direction of the vector, grad ϕ, at any point coincides with the direction along which the directional derivative of ϕ is greatest.*

This fact may also be seen otherwise as follows. Consider two level surfaces through two points A, C and let B be the point on the level surface through C where the normal at A meets the same. We have

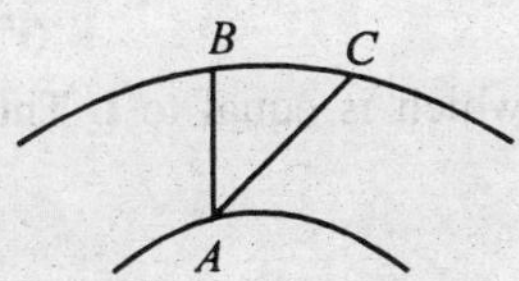

Fig. 10.8.

$$\phi(B) - \phi(A) = \phi(C) - \phi(A)$$

for $\phi(B) = \phi(C)$,

and $$AB < AC.$$

$$\therefore \quad \frac{\phi(B) - \phi(A)}{AB} > \frac{\phi(C) - \phi(A)}{AC}.$$

The result now follows on taking limits.

10.9. CHARACTER OF GRADIENT AS A POINT FUNCTION

Theorem. *The gradient of a scalar point function is a vector point function.*

The grad ϕ at any point (x, y, z) is the vector

$$\mathbf{i}\frac{\partial \phi}{\partial x} + \mathbf{j}\frac{\partial \phi}{\partial y} + \mathbf{k}\frac{\partial \phi}{\partial z},$$

where

$$\mathbf{i}, \mathbf{j}, \mathbf{k}$$

are three mutually perpendicular vectors and

$$\partial\phi/\partial x,\ \partial\phi/\partial y,\ \partial\phi/\partial z$$

are the directional derivatives of ϕ at (x, y, z) along the lines parallel to **i, j, k.** It may thus appear that the vector grad ϕ at any point depends upon the choice of the mutually perpendicular directions through the point but, as will now be seen, this is certainly not the case, *i.e.*, grad ϕ at any point is independent of the choice of perpendicular directions and is as such a vector point function.

Let $\mathbf{i}', \mathbf{j}', \mathbf{k}'$ be any other set of mutually orthogonal unit vectors. The directional derivatives of ϕ along these vectors being

$$\mathbf{i}'.\text{ grad } \phi, \quad \mathbf{j}'.\text{ grad } \phi, \quad \mathbf{k}'.\text{ grad } \phi,$$

we see that the expression for the gradient in terms of $\mathbf{i}', \mathbf{j}', \mathbf{k}'$ is

$$\mathbf{i}'(\mathbf{i}'.\text{ grad } \phi) + \mathbf{j}'(\mathbf{j}'.\text{ grad } \phi) + \mathbf{k}'(\mathbf{k}'.\text{ grad } \phi). \quad ...(1)$$

Replacing grade ϕ by

$$\mathbf{i}\frac{\partial\phi}{\partial x} + \mathbf{j}\frac{\partial\phi}{\partial y} + \mathbf{k}\frac{\partial\phi}{\partial z},$$

We see that the coefficient of $\partial\phi/\partial x$ in (1) is

$$\mathbf{i}'(\mathbf{i}'.\mathbf{i}') + \mathbf{j}'(\mathbf{j}'.\mathbf{i}') + \mathbf{k}'(\mathbf{k}'.\mathbf{i}')$$

which is equal to **i**. Thus, the expression (1) is the same as

$$\mathbf{i}\frac{\partial\phi}{\partial x} + \mathbf{j}\frac{\partial\phi}{\partial y} + \mathbf{k}\frac{\partial\phi}{\partial z}.$$

Hence, the theorem.

The theorem also follows from the fact that, as shown in § 10.8.1, *the vector, grad ϕ, at any point is completely describable in terms of the level surface through the point* and this description is independent of the choice of co-ordinate axes.

10.10. THE OPERATOR ∇

It is found very convenient to introduce a symbol called *Del* or *Nabla* and written as ∇ in connection with our dealings with differential operators. We write

$$\nabla = \mathbf{i}\frac{\partial}{\partial x} + \mathbf{j}\frac{\partial}{\partial y} + \mathbf{k}\frac{\partial}{\partial z}, \quad ...(1)$$

and state that it operates distributively. In terms of ∇, we write

$$\text{grad } \phi = \mathbf{i}\frac{\partial\phi}{\partial x} + \mathbf{j}\frac{\partial\phi}{\partial y} + \mathbf{k}\frac{\partial\phi}{\partial z}$$

$$= \left(\mathbf{i}\frac{\partial}{\partial x} + \mathbf{j}\frac{\partial}{\partial y} + \mathbf{k}\frac{\partial}{\partial z}\right)\phi = \nabla\phi.$$

The interest and usefulness of the introduction of this symbol lies specially in the fact that it can be *formally assumed* to have the character of a vector and as such it facilitates the manipulations with expressions involving differential operators. Thus, *formally* $\nabla\phi$ being the product of a vector ∇ by a scalar ϕ is a vector. (Refer § 10.12, Page 291).

The operator ∇ is also known as *Differential Operator*. It should be clearly understood that the expression (1) for ∇ holds only when we are dealing with cartesian rectangular co-ordinates.

10.11. OPERATOR

$$\mathbf{a} \cdot \nabla,$$

a, *being any vector.*

The operator $\mathbf{a} \cdot \nabla$ is defined by the equality

$$\mathbf{a} \cdot \nabla = \mathbf{a} \cdot \mathbf{i}\frac{\partial}{\partial x} + \mathbf{a} \cdot \mathbf{j}\frac{\partial}{\partial y} + \mathbf{a} \cdot \mathbf{k}\frac{\partial}{\partial z},$$

so that we have

$$(\mathbf{a} \cdot \nabla)\,\phi = \mathbf{a} \cdot \mathbf{i}\frac{\partial \phi}{\partial x} + \mathbf{a} \cdot \mathbf{j}\frac{\partial \phi}{\partial y} + \mathbf{a} \cdot \mathbf{k}\frac{\partial \phi}{\partial z},$$

$$(\mathbf{a} \cdot \nabla)\,\mathbf{f} = \mathbf{a} \cdot \mathbf{i}\frac{\partial \mathbf{f}}{\partial x} + \mathbf{a} \cdot \mathbf{j}\frac{\partial \mathbf{f}}{\partial y} + \mathbf{a} \cdot \mathbf{k}\frac{\partial \mathbf{f}}{\partial z} \cdot$$

For a unit vector **a**,

$$\mathbf{a} \cdot \mathbf{i}, \ \mathbf{a} \cdot \mathbf{j}, \ \mathbf{a} \cdot \mathbf{k}$$

are the direction cosines of **a** and hence by § 10.5 and § 10.6.

$$(\mathbf{a} \cdot \nabla)\,\phi, \ (\mathbf{a} \cdot \nabla)\,\mathbf{f}$$

stand for the directional derivatives of the respective functions along the direction of the unit vector,

We have, as may be easily seen,

$$(\mathbf{a} \cdot \nabla)\,\phi = \mathbf{a} \cdot \nabla\,\phi = \mathbf{a} \cdot \text{grad}\,\phi.$$

As, however, we have not defined $\nabla\,\mathbf{f}$, where **f** is a vector, we cannot write

$$\mathbf{a} \cdot \nabla\,\mathbf{f} \text{ in place of } (\mathbf{a} \cdot \nabla)\,\mathbf{f},$$

unless, of course, we say that, by def.,

$$\mathbf{a} \cdot \nabla\,\mathbf{f} \text{ means } (\mathbf{a} \cdot \nabla)\,\mathbf{f}.$$

Corresponding to the relation $d\phi = d\mathbf{r} \cdot \nabla\phi$ obtain § 10.7.3. for scalar functions ϕ, we have the relation

$$d\mathbf{f} = (d\mathbf{r} \cdot \nabla)\,\mathbf{f} = d\mathbf{r} \cdot \nabla\,\mathbf{f}$$

for vector functions **f**.

We have,

$$d\mathbf{f} = \frac{\partial \mathbf{f}}{\partial x}dx + \frac{\partial \mathbf{f}}{\partial y}dy + \frac{\partial \mathbf{f}}{\partial z}dz$$

$$= d\mathbf{r}\cdot\mathbf{i}\frac{\partial \mathbf{f}}{\partial x} + d\mathbf{r}\cdot\mathbf{j} + \frac{\partial \mathbf{f}}{\partial y} + d\mathbf{r}\cdot\mathbf{k}\frac{\partial \mathbf{f}}{\partial z}.$$

$$= d\mathbf{r} \,.\, \nabla\, \mathbf{f}.$$

EXAMPLES

Example 1. *Find grad log | **r** |.* (*Rohilkhand, 2000, Garhwal 2000*)

Solution. We have $r = \sqrt{(x^2 + y^2 + z^2)}$

$$\log |\,\mathbf{r}\,| = \frac{1}{2}\log(x^2 + y^2 + z^2)$$

Now, $\dfrac{\partial}{\partial x}\log|\mathbf{r}| = \dfrac{1}{2}\cdot\dfrac{2x}{(x^2 + y^2 + z^2)} = \dfrac{x}{r^2}$

Similarly, $\dfrac{\partial}{\partial y}\log|\mathbf{r}| = \dfrac{y}{r^2},\ \dfrac{\partial}{\partial z}\log|\mathbf{r}| = \dfrac{z}{r^2}$

$\therefore$ $\quad$ grad log $|\,\mathbf{r}\,| = \dfrac{1}{r^2}(\mathbf{i}x + \mathbf{j}y + \mathbf{k}z) = \dfrac{r}{r^2}$

Example 2. *Prove that grad f (r) × **r** = 0.*

(*Rohilkhand 2004, Avadh 2004*)

Solution. Let $\phi = f(r) = f(x^2 + y^2 + z^2)^{1/2}$

$$\frac{\partial \phi}{\partial x} = f'(r)\frac{\partial r}{\partial x} = f'(r)\frac{x}{r}$$

Similarly, $\dfrac{\partial \phi}{\partial y} = f'(r)\dfrac{y}{r},\ \dfrac{\partial \phi}{\partial z} = f'(r)\dfrac{z}{r}$

$\therefore$ $\quad$ grad $f(r) = \Sigma\ \mathbf{i}\dfrac{\partial \phi}{\partial x}$

$$= \frac{1}{r}f'(r)\cdot(\mathbf{i}x + \mathbf{j}y + \mathbf{k}z) = \frac{f'(r)}{r}\mathbf{r}$$

$\therefore$ grad $f(r) \times \mathbf{r} = \dfrac{f'(r)}{r}\mathbf{r}\times\mathbf{r} = 0.$

Example 3. *Find grad r^m where r is the distance of any point from the origin.* (*Rohilkhand 99, Kumaon 2000, Garhwal 2005*)

Solution. We have

$$\phi(x, y, z) = r^m = (x^2 + y^2 + z^2)^{\frac{m}{2}}$$

$$\frac{\partial \phi}{\partial x} = mx\,(x^2 + y^2 + z^2)^{\frac{m}{2}-1} = mxr^{m-2}$$

$$\frac{\partial \phi}{\partial y} = myr^{m-2},$$

$$\frac{\partial \phi}{\partial z} = mzr^{m-2}$$

Thus,

$$\text{grad } r^m = \Sigma\, \mathbf{i}\, \frac{\partial \phi}{\partial x}$$

$$= mr^{m-2}\, \Sigma\, \mathbf{i}\, x$$

$$= mr^{m-2}\mathbf{r},$$

where, $\mathbf{r}$, is the position vector of any point. We can also write

$$\text{grad } r^m = mr^{m-1}\left(\frac{\mathbf{r}}{r}\right),$$

where $\mathbf{r}/r$ is the unit vector joining the origin to any point.

Another method. The level surfaces of the given scalar point function are spheres with centre at the origin so that if P be any point, the normal to the level surface at P is the line OP. Also as r stands for the distance measured along the normal OP, the directional derivative of r^m along OP is mr^{m-1}. Hence, by § 10.1,

$$\text{grad } r^m = mr^{m-1}\, \frac{\mathbf{r}}{r}.$$

Example 4. *If* $\phi\,(x, y) = \log \sqrt{(x^2 + y^2)}$, *show that*

$$\text{grad } \phi = \frac{\mathbf{r} - (\mathbf{k}\cdot\mathbf{r})\mathbf{k}}{\{\mathbf{r} - (\mathbf{k}\cdot\mathbf{r})\,\mathbf{k}\}\cdot\{\mathbf{r} - (\mathbf{k}\cdot\mathbf{r})\,\mathbf{k}\}}$$

Solution. We have

$$\mathbf{r} = x\mathbf{i} + y\mathbf{j} + z\mathbf{k} \quad \therefore \quad \mathbf{r}\,.\,\mathbf{k} = z = \mathbf{k}\,.\,\mathbf{r} \qquad ...(1)$$

$$\phi = \frac{1}{2}\log\,(x^2 + y^2) \quad \therefore \quad \frac{\partial \phi}{\partial x} = \frac{1}{2\,(x^2 + y^2)}\cdot 2x = \frac{x}{x^2 + y^2}$$

$$\therefore \quad \text{grad } \phi = \Sigma\, \mathbf{i}\, \frac{\partial \phi}{\partial x} = \frac{x\mathbf{i} + y\mathbf{j} + z\mathbf{k}}{x^2 + y^2}$$

$$= \frac{\mathbf{r} - z\mathbf{k}}{(x\mathbf{i} + y\mathbf{j})\cdot(x\mathbf{i} + y\mathbf{j})} \qquad \text{[from (1)]}$$

$$= \frac{\mathbf{r} - z\mathbf{k}}{(r - z\mathbf{k})\cdot(r - z\mathbf{k})}$$

Now replacing z by $\mathbf{r}\,.\,\mathbf{k} = \mathbf{k}\,.\,\mathbf{r}$ in the above we get the required result.

Example 5. *If* ϕ *and* ψ *are two scalar point functions and* ϕ *is a function of* ψ*, then prove that* $\nabla\phi = \nabla\psi \dfrac{d\phi}{d\psi}$.

Solution. $$\nabla\phi = \Sigma \mathbf{i} \frac{\partial\phi}{\partial x} = \Sigma \mathbf{i} \frac{d\phi}{d\psi} \cdot \frac{\partial\psi}{\partial x},$$

$\because$ ϕ is a function of ψ

$$= \frac{d\phi}{d\psi} \Sigma \mathbf{i} \frac{\partial\psi}{\partial x} = \frac{d\phi}{d\psi} \nabla\psi$$

$$= \nabla\psi \frac{d\phi}{d\psi}.$$

Example 6. *Prove that (i) grad* $(\mathbf{r} \cdot \mathbf{a}) = \mathbf{a}$ *(ii) grad* $[\mathbf{r}\ \mathbf{a}\ \mathbf{b}] = \mathbf{a} \times \mathbf{b}$, *where* **a** *and* **b** *are constant vectors.*

Solution. (*i*) Let $\mathbf{r} = x\mathbf{i} + y\mathbf{j} + z\mathbf{k}$ and $\mathbf{a} = a_1\mathbf{i} + a_2\mathbf{j} + a_3\mathbf{k}$, where, a_1, a_2, a_3 are constants.

$\therefore$ $$\mathbf{r} \cdot \mathbf{a} = xa_1 + ya_2 + za_3$$

$\therefore$ $$\text{grad}\ (\mathbf{r} \cdot \mathbf{a}) = \text{grad}\ (xa_1 + ya_2 + za_3)$$

$$= \Sigma \mathbf{i} \frac{\partial}{\partial x} (\mathbf{i}a_1 + \mathbf{j}a_2 + \mathbf{k}a_3)$$

$$\Sigma\, a_1\mathbf{i} = a_1\mathbf{i} + a_2\mathbf{j} + a_3\mathbf{k} = \mathbf{a}$$

(*ii*) $[\mathbf{r}\ \mathbf{a}\ \mathbf{b}] = \mathbf{r} \cdot (\mathbf{a} \times \mathbf{b})$

Proceed as in part (*i*) taking $\mathbf{a} \times \mathbf{b}$ in placed of **a**.

$\therefore$ grad $[\mathbf{r}\ \mathbf{a}\ \mathbf{b}] = \mathbf{r} \cdot (\mathbf{a} \times \mathbf{b})$

Example 7. *Find the directional derivative of* $\phi = x^2yz + 4xz^2$ *at* $(1, -2, -1)$ *in the direction* $2\mathbf{i} - \mathbf{j} - 2\mathbf{k}$. *In which direction the directional derivative will be maximum and what is its magnitude. Also find a unit normal to the surface* $x^2yz + 4xz^2 = 6$ *at the point* $(1, -2, -1)$, *find the equation of tangent plane and normal at the point* $(1, -2, -1)$.

Solution. We have

$$\frac{\partial\phi}{\partial x} = 2xyz + 4z^2,\ \frac{\partial\phi}{\partial y} = x^2z,\ \frac{\partial\phi}{\partial z} = x^2y + 8xz.$$

$$\text{grad}\ \phi = \Sigma\, \mathbf{i} \frac{\partial\phi}{\partial x} = (2xyz + 4z^2)\, \mathbf{i} + x^2z\ \mathbf{j} + (x^2y + 8xz)\, \mathbf{k}.$$

or, $$\text{grad}\ \phi = 8\mathbf{i} - \mathbf{j} - 10\mathbf{k} \text{ at the point } (1, -2, -1).$$

Now, if $\hat{\mathbf{a}}$ be a unit vector then directional derivative of ϕ along the direction of $\hat{\mathbf{a}}$ is $\hat{\mathbf{a}} \cdot \text{grad}\ \phi$.

A unit vector in the direction $(2\mathbf{i} - \mathbf{j} - 2\mathbf{k})$ is

$$\frac{2\mathbf{i}-\mathbf{j}-2\mathbf{k}}{\sqrt{(4+1+4)}} = \frac{1}{3}(2\mathbf{i}-\mathbf{j}-2\mathbf{k})$$

$\therefore$ Directional derivative

$$= \frac{1}{3}(2\mathbf{i}-\mathbf{j}-2\mathbf{k}) \cdot (8\mathbf{i}-\mathbf{j}-10\mathbf{k})$$

$$= \frac{1}{3}(16+1+20) = \frac{37}{3}.$$

We know that the directional derivative is maximum in the direction of normal which is the direction of grad ϕ. Hence, directional derivative is maximum along grad $\phi = 8\mathbf{i} - \mathbf{j} - 10\mathbf{k}$. Maximum value of this directional derivative is $|\text{ grad } \phi \,|$

$$= \sqrt{(64+1+100)} = \sqrt{(165)}.$$

A unit normal to the surface $= \dfrac{\text{grad } \phi}{|\text{grad } \phi|}$

$$= \frac{8\mathbf{i}-\mathbf{j}-10\mathbf{k}}{\sqrt{(165)}}.$$

Tangent plane to the surface at (1, – 2, – 1).

Let P be any point on the tangent plane at A $(1, -2, -1)$ and the position vectors of P and A be $\mathbf{r}$ and $\mathbf{r}_0$.

$\therefore$ $\mathbf{r} = x\mathbf{i} + y\mathbf{j} + z\mathbf{k}$ and $\mathbf{r}_0 = \mathbf{i} - 2\,\mathbf{j} - \mathbf{k}$

$$\overrightarrow{AP} = \mathbf{r} - \mathbf{r}_0 = (x-1)\,\mathbf{i} + (y+2)\,\mathbf{j} + (z+1)\,\mathbf{k}.$$

grad ϕ is normal to the surface and as such it is perpendicular to AP.

$\therefore$ $\text{grad } \phi \,.\, \overrightarrow{AP} = 0.$

$\Rightarrow$ $(8\mathbf{i} - \mathbf{j} - 10\mathbf{k}) \,.\, [(x-1)\,\mathbf{i} + (y+2)\,\mathbf{j} + (z+1)\,\mathbf{k}] = 0$

$\Rightarrow$ $8\,(x-1) - (y+2) - 10\,(z+1) = 0$

$\Rightarrow$ $8x - y - 10z = 20.$

Normal to the surface.

If ρ be any point on the normal line at A, then $(\mathbf{r} - \mathbf{r}_0)$ lies along the normal, *i.e.*, along grad ϕ, hence

$$(\mathbf{r} - \mathbf{r}_0) \times \text{grad } \phi = 0$$

$$\Rightarrow \quad \begin{vmatrix} \mathbf{i} & \mathbf{j} & \mathbf{k} \\ x-1 & y+2 & z+1 \\ 8 & -1 & -10 \end{vmatrix} = 0$$

$$\Rightarrow \mathbf{i}\{(z+1)-10(y+2)\}+\mathbf{j}\{10(x-1)+8(z+1)\}$$
$$-\mathrm{k}\{(x-1)+8(y+2)\}=0$$

Equating to zero the coefficients of **i, j** and **k**, we get

$$\frac{z+1}{10}=\frac{y+2}{1};\ \frac{z+1}{10}=\frac{x-1}{-8};\ \frac{x-1}{-8}=\frac{y+2}{1}$$

$$\therefore \qquad \frac{x-1}{-8}=\frac{y+2}{1}=\frac{z+1}{10}.$$

Note. It could be written directly as the equation of a line through (1, – 2, – 1) and with direction ratios as 8, – 1, – 10.

Example 8. *Show that the directional derivative of a scalar function of any point along any tangent line to the level surface at the point is zero.*

Solution. We know that directional derivative along any direction $\hat{\mathbf{a}}$ is $\hat{\mathbf{a}}$. grad ϕ.

Unit vector $\hat{\mathbf{a}}$ is along the direction of tangent line as given. Also grad ϕ is along the normal, hence

$$\text{directional derivative} = \hat{\mathbf{a}}\,.\,\text{grad}\,\phi = 0.$$

Example 9. *Find the values of constants a, b, c so that the directional derivative of* $\phi = axy^2 + byz + cz^2x^3$ *at (1, 2, – 1) has a maximum magnitude 64 in a direction parallel to z-axis.* (*Purvanchal 2004*)

Solution. The directional derivative is maximum along the normal, *i.e.*, along grad ϕ.

$$\text{grad}\,\phi = (ay^2 + 3cz^2x^2)\,\mathbf{i} + (2axy + bz)\,\mathbf{j} + (by + 2czx^3)\,\mathbf{k}$$
$$= (4a + 3c)\,\mathbf{i} + (4a - b)\,\mathbf{j} + (2b - 2c)\,\mathbf{k} \text{ at } (1, 2, -1)$$

But directional derivative is maximum along z-axis. Hence, the coefficient of **i** and **j** should be zero.

$$\therefore \qquad 4a + 3c = 0 \quad \text{and } 4a - b = 0$$

$$\therefore \qquad \text{grad}\,\phi = (2b - 2c)\,\mathbf{k}$$

Also, maximum value of directional derivative

$$= |\text{ grad}\,\phi\,|$$

$$\therefore \quad 2(b - c) = 64 \quad \Rightarrow \quad b - c = 32.$$

Solving the three equations we get

$$a = 6, \quad b = 24, \quad c = -8.$$

Example 10. *If*

$$\mathbf{F}\left(y\frac{\partial f}{\partial z}-z\frac{\partial f}{\partial y}\right)\mathbf{i}+\left(z\frac{\partial f}{\partial x}-x\frac{\partial f}{\partial z}\right)\mathbf{j}+\left(x\frac{\partial f}{\partial y}-y\frac{\partial f}{\partial x}\right)\mathbf{k}$$

prove that (*i*) $\mathbf{F} = \mathbf{r} \times \nabla f$, (*ii*) $\mathbf{F}\,.\,\mathbf{r} = 0$, (*iii*) $\mathbf{F}\,.\,\nabla f = 0$.

Solution. Given

$$\mathbf{F} = \left(y\frac{\partial f}{\partial z} - z\frac{\partial f}{\partial y} \right)\mathbf{i} + \left(z\frac{\partial f}{\partial x} - x\frac{\partial f}{\partial z} \right)\mathbf{j} + \left(x\frac{\partial f}{\partial y} - y\frac{\partial f}{\partial x} \right)\mathbf{k} \qquad \text{...(1)}$$

Also, $$\nabla \mathbf{f} = \frac{\partial f}{\partial x}\mathbf{i} + \frac{\partial f}{\partial y}\mathbf{j} + \frac{\partial f}{\partial z}\mathbf{k}$$

and $$\mathbf{r} = x\mathbf{i} + y\mathbf{j} + z\mathbf{k}.$$

(*i*) $$\mathbf{r} \times \nabla \mathbf{f} = \begin{vmatrix} \mathbf{i} & \mathbf{j} & \mathbf{k} \\ x & y & z \\ \frac{\partial f}{\partial x} & \frac{\partial f}{\partial y} & \frac{\partial f}{\partial z} \end{vmatrix}$$

$$= \left(y\frac{\partial f}{\partial z} - z\frac{\partial f}{\partial y} \right)\mathbf{i} + \left(z\frac{\partial f}{\partial x} - x\frac{\partial f}{\partial z} \right)\mathbf{j} + \left(x\frac{\partial f}{\partial y} - y\frac{\partial f}{\partial x} \right)\mathbf{k}$$

$$= \mathbf{F}.$$

(*ii*) $\mathbf{F} \cdot \mathbf{r} = (\mathbf{r} \times \nabla \mathbf{f}) \cdot \mathbf{r}$ [by part (*i*)]

$= 0$ [scalar triple product with two equal elements **r**]

(*iii*) $\mathbf{F} \cdot \nabla \mathbf{f} = (\mathbf{r} \times \nabla \mathbf{f}) \cdot \nabla \mathbf{f}$

$= 0$, by the same cause.

EXERCISES

1. Find grad ϕ if (*i*) $\phi = \frac{1}{r}$; (*ii*) $\mathbf{r} \cdot \mathbf{r}$; (*iii*) $\phi = 3x^2 y - y^3 z^2$ at point $(1, -2, -1)$; (*iv*) $\phi = 2x^2 y^3 - 3y^2 z^3$ at the point $(1, -1, 1)$. (*Rohilkhand, 98*)

2. Prove that $\nabla \int \phi (u)\, du = \phi (u) \nabla u$.

   (*i*) Find the directional derivative of $\phi = x^3 + y^3 + z^3$ at the point $(1, -1, 2)$ in the direction of the vector $\mathbf{i} + 2\mathbf{i} + \mathbf{k}$.

   (*ii*) Find the directional derivative of $\phi = xy + yz + zx$ in the direction of the vector $\mathbf{i} + 2\mathbf{j} + 2\mathbf{k}$ at $(1, 2, 0)$. (*Kumaon* 99)

   (*iii*) Find the directional derivative of $\phi = x^2 - 2y^2 + 4z^2$ at $(1, 1, -1)$ in the direction $2\mathbf{i} + \mathbf{j} - \mathbf{k}$. In what direction is the directional derivative from the point $(1, 1, -1)$, is maximum and what is its value.

3. (*i*) Find the unit vector normal to $\phi = x^2 - y^2 + z$ at the point $(1, -1, 2)$.

   (*ii*) Find the unit vector normal to the surface $z = x^2 + y^2$ at the point $(-1, -2, 5)$. (*Kumaon* 2005

4. Find the angle between the surfaces $x^2 + y^2 + z^2 = 9$ and $z = x^2 + y^2 - 3$ at the point (2, – 1, 2).

5. Find the values of the constants *a, b, c* so that the directional derivative of $\phi = ax^2 - by^2 + cz^2$ at (1, 1, 2) has a maximum magnitude 4 in the direction parallel to *y*-axis.

6. Find the direction along which the directional derivative of the function $xy + 2yz + 3xz$ is greatest at the point (1, 1, 1). Also find the greatest directional derivatives.

   Do the same for the function $x(y + z) - y(x + y)$ and the point (0, – 1, 2).

7. Find grad [sin (*xy*) + sin (*yz*) + sin (zx)] and $\nabla e^{(x^2+y^2+z^2)}$.

8. Find the directional derivative of the function $xy^2 + yz^2 + zx^2$ along the tangent to the curve $x = t, y = t^2, z = t^2$ at (1, 1, 1) and at (– 1, 1, – 1).

10.12. DIVERGENCE AND CURL

We have seen how, starting with a scalar point function ϕ, we can construct a vector point function grad ϕ. We shall now start with a vector point function and construct therewith two more point functions, one a scalar and the other a vector, respectively called **Divergence** and **Curl** of the given function.

Def. *If* **f** *be any given continuously differentiable vector point function, then the functions*

$$\mathbf{i}\cdot\frac{\partial \mathbf{f}}{\partial x}+\mathbf{j}\cdot\frac{\partial \mathbf{f}}{\partial y}+\mathbf{k}\cdot\frac{\partial \mathbf{f}}{\partial z},$$

$$\mathbf{i}\times\frac{\partial \mathbf{f}}{\partial x}+\mathbf{j}\times\frac{\partial \mathbf{f}}{\partial y}+\mathbf{k}\times\frac{\partial \mathbf{f}}{\partial z},$$

are respectively called Divergence and Curl of **f**, *and are denoted by*

div **f**, *curl* **f**.

Clearly *div* **f** *is scalar and curl* **f** *a vector.*

In terms of the operator ∇, we write

$$\text{div}\,\mathbf{f}=\mathbf{i}\cdot\frac{\partial \mathbf{f}}{\partial x}+\mathbf{j}\cdot\frac{\partial \mathbf{f}}{\partial y}+\mathbf{k}\cdot\frac{\partial \mathbf{f}}{\partial z}$$

$$=\left(\mathbf{i}\frac{\partial}{\partial x}+\mathbf{j}\frac{\partial}{\partial y}+\mathbf{k}\frac{\partial}{\partial z}\right)\cdot\mathbf{f}$$

$$=\nabla\,.\,\mathbf{f}$$

$$\text{Curl}\,\mathbf{f}=\mathbf{i}\times\frac{\partial \mathbf{f}}{\partial x}+\mathbf{j}\times\frac{\partial \mathbf{f}}{\partial y}+\mathbf{k}\times\frac{\partial \mathbf{f}}{\partial z}$$

$$= \left(\mathbf{i}\frac{\partial}{\partial x} + \mathbf{j}\frac{\partial}{\partial y} + \mathbf{k}\frac{\partial}{\partial z}\right) \times \mathbf{f}$$

$$= \nabla \times \mathbf{f}$$

If terms of the *formal* character of ∇, as a vector, we see that the scalar $\nabla \cdot \mathbf{f}$ is the scalar product of the vector ∇ and $\mathbf{f}$ and the vector $\nabla \times \mathbf{f}$ is the vector product of the vectors ∇ and $\mathbf{f}$.

10.12.1. Expressions for div f and curl f in terms of the components of f

Let

$$\mathbf{f} = \mathbf{i}f_1 + \mathbf{i}f_2 + \mathbf{i}f_3$$

$$\therefore \quad \text{div } \mathbf{f} = \Sigma \mathbf{i} \cdot \frac{\partial \mathbf{f}}{\partial x}$$

$$= \Sigma \mathbf{i} \cdot \left[\mathbf{i}\frac{\partial f_1}{\partial x} + \mathbf{j}\frac{\partial f_2}{\partial x} + \mathbf{k}\frac{\partial f_3}{\partial x}\right]$$

$$= \frac{\partial f_1}{\partial x} + \frac{\partial f_2}{\partial y} + \frac{\partial f_3}{\partial z}.$$

and

$$\text{curl } \mathbf{f} = \Sigma \mathbf{i} \times \frac{\partial \mathbf{f}}{\partial x}$$

$$= \Sigma \mathbf{i} \times \left[\mathbf{i}\frac{\partial f_1}{\partial x} + \mathbf{j}\frac{\partial f_2}{\partial x} + \mathbf{k}\frac{\partial f_3}{\partial x}\right]$$

$$= \mathbf{i}\left(\frac{\partial f_3}{\partial y} - \frac{\partial f_2}{\partial z}\right) + \mathbf{j}\left(\frac{\partial f_1}{\partial z} - \frac{\partial f_3}{\partial x}\right) + \mathbf{k}\left(\frac{\partial f_2}{\partial x} - \frac{\partial f_1}{\partial y}\right)$$

$$= \begin{vmatrix} \mathbf{i} & \mathbf{j} & \mathbf{k} \\ \dfrac{\partial}{\partial x} & \dfrac{\partial}{\partial y} & \dfrac{\partial}{\partial z} \\ f_1 & f_2 & f_3 \end{vmatrix}$$

10.12.2. Characters of div f and curl f as Point Functions

Theorem. *The functions div* **f** *and curl* **f** *are point functions.*

We have to show that div **f** and curl **f** are independent of the choice of the orthogonal unit vectors in terms of which we may formulate the same.

Let $\mathbf{i}', \mathbf{j}', \mathbf{k}'$ be any other set of mutually orthogonal unit vectors. By § 10.11, Page 284, the directional derivative of **f** along the directions of the vectors $\mathbf{i}', \mathbf{j}', \mathbf{k}'$ are

$$(\mathbf{i}' \cdot \nabla)\, \mathbf{f}, \quad (\mathbf{j}' \cdot \nabla)\, \mathbf{f}, \quad (\mathbf{k}' \cdot \nabla)\, \mathbf{f}$$

i.e.,

$$\mathbf{i}' \cdot \mathbf{i}\frac{\partial \mathbf{f}}{\partial x} + \mathbf{i}' \cdot \mathbf{j}\frac{\partial \mathbf{f}}{\partial y} + \mathbf{i}' \cdot \mathbf{k}\frac{\partial \mathbf{f}}{\partial z}, \text{ etc.}$$

Thus, the expression for div **f** in terms of **i′**, **j′**, **k′** is

$$= \Sigma \mathbf{i}' \cdot \left[\mathbf{i}' \cdot \mathbf{i}\frac{\partial \mathbf{f}}{\partial x} + \mathbf{i}' \cdot \mathbf{j}\frac{\partial \mathbf{f}}{\partial y} + \mathbf{i}' \cdot \mathbf{k}\frac{\partial \mathbf{f}}{\partial z} \right]$$

$$= \Sigma \mathbf{i}' \cdot [(\mathbf{i}' \cdot \mathbf{i})\, \mathbf{i}' + (\mathbf{j}' \cdot \mathbf{i})\, \mathbf{j}' + (\mathbf{k}' \cdot \mathbf{i})\, \mathbf{k}'] \cdot \frac{\partial \mathbf{f}}{\partial x}$$

$$= \Sigma \mathbf{i} \cdot \frac{\partial \mathbf{f}}{\partial x}.$$

Thus, we see that div **f** is essentially a point function. The case of curl **f** can be treated in a similar manner.

10.13. ILLUSTRATIONS OF Curl f AND Div f.

10.13.1. Let ω be the angular velocity of a rigid body moving about a fixed point. Then the velocity **v** of any particle of the body with position vector **r** relative to the fixed point is given by

$$\mathbf{v} = \omega \times \mathbf{r}.$$

Here, **v** is a vector point function and the vector ω is the same for all the points.

We have

$$\text{curl } \mathbf{v} = \text{curl } (\omega \times \mathbf{r})$$

$$= \Sigma \mathbf{i} \times \frac{\partial \mathbf{f}}{\partial x}(\omega \times \mathbf{r})$$

$$= \Sigma \mathbf{i} \times \left(\omega \times \frac{\partial \mathbf{r}}{\partial x} \right) = \Sigma\, \mathbf{i} \times (\omega \times \mathbf{i})$$

$$= \Sigma\, (\mathbf{i} \cdot \mathbf{i})\, \omega - \Sigma\, (\mathbf{i} \cdot \omega)\, \mathbf{i} = 3\omega - \omega = 2\omega.$$

Thus, the *curl of the velocity of any particle of a rigid body equals twice the angular velocity of the body.*

This result justifies the term Rotation which is sometimes used for curl.

10.13.2. Let a vector point function **q** denote the velocity of any homogeneous and incompressible fluid at any time *t*.

Consider any point *P* (*x*, *y*, *z*) and enclose the same in a rectangular parallelopiped with edges parallel to the co-ordinate axes and of lengths δ*x*, δ*y*, δ*z*.

Let

$$\mathbf{q} = q_1\mathbf{i} + q_2\mathbf{j} + q_3\mathbf{k}.$$

The velocity component parallel to x-axis at any point of the face $ABCD$

$$= q_1\left(x+\frac{1}{2}\delta x, y, z\right)$$

$$= q_1(x, y, z)+\frac{1}{2}\delta x\frac{\partial q_1}{\partial x}$$

omitting powers of δx higher than one.

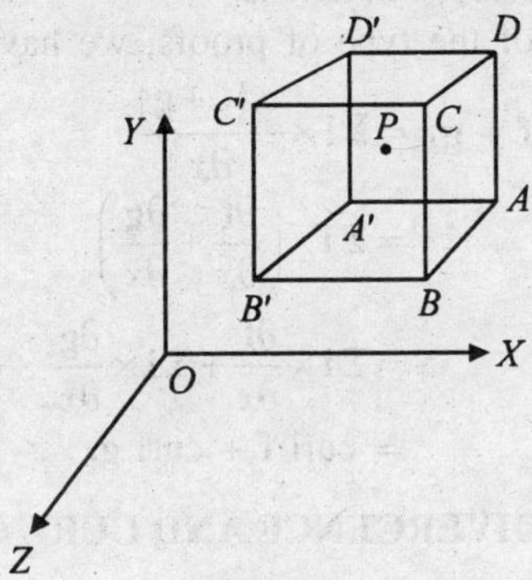

Fig. 10.9.

Thus, the mass of fluid that moves out of the face $ABCD$ in time δt

$$= \rho\left[q_1+\frac{1}{2}\delta x\frac{\partial q_1}{\partial x}\right]\delta y\,\delta z\,\delta t.$$

Similarly the mass of fluid that enters through the face $A'B'C'D'$ in time δt

$$= \rho\left[q_1-\frac{1}{2}\delta x\frac{\partial q_1}{\partial x}\right]\delta y\,\delta z\,\delta t.$$

Thus, the total mass of fluid that moves out through the face $ABCD$, $A'B'C'D'$ perpendicular to x-axis.

$$= \rho\frac{\partial q_1}{\partial x}\delta x\,\delta y\,\delta z\,\delta t.$$

Taking into consideration the other faces also, we see that the total mass of fluid that moves out of the parallelopiped in time δt

$$= \rho\left(\frac{\partial q_1}{\partial x}+\frac{\partial q_2}{\partial y}+\frac{\partial q_3}{\partial z}\right)\delta x\,\delta y\,\delta z\,\delta t$$

$$= \rho\,\delta x\,\delta y\,\delta z\,\delta t \text{ div } \mathbf{q}.$$

The volume of the parallelopiped being δx, δy, δz, we see, on taking limit, when δx, δy, δz, δt all tend to zero, that the amount of fluid per unit volume per unit time that passes through a point (x, y, z)

$$= \rho \text{ div } \mathbf{q}.$$

10.14. GRADIENT, DIVERGENCE AND CURL OF SUMS

The following results are immediately deducible from the definitions :

$$\text{grad}\ (\phi \pm \psi) = \text{grad}\ \phi \pm \text{grad}\ \psi,$$

$$\text{div}\ (\mathbf{f} \pm \mathbf{g}) = \text{div}\ \mathbf{f} \pm \text{div}\ \mathbf{g}, \qquad (\textit{Kumaon 2004})$$

$$\text{curl}\ (\mathbf{f} \pm \mathbf{g}) = \text{curl}\ \mathbf{f} \pm \text{curl}\ \mathbf{g}.$$

Thus, each of three differential operations is distributive with respect to the sum and difference of functions.

As an illustration of the type of proofs, we have

$$\begin{aligned} \text{curl}\ (\mathbf{f} + \mathbf{g}) &= \Sigma\, \mathbf{i} \times \frac{\partial (\mathbf{f} + \mathbf{g})}{\partial x} \\ &= \Sigma\, \mathbf{i} \times \left(\frac{\partial \mathbf{f}}{\partial x} + \frac{\partial \mathbf{g}}{\partial x} \right) \\ &= \Sigma\, \mathbf{i} \times \frac{\partial \mathbf{f}}{\partial x} + \Sigma\, \mathbf{i} \times \frac{\partial \mathbf{g}}{\partial x} \\ &= \text{curl}\ \mathbf{f} + \text{curl}\ \mathbf{g}. \end{aligned}$$

10.15. GRADIENT, DIVERGENCE AND CURL OF PRODUCTS

We shall now see how the gradient, divergence and curl of products of point functions can be expressed in terms of those of the factors. Now remembering that we have defined gradient of a scalar point function and divergence and curl of a vector point function, we see that we have to consider the following cases only.

$$\text{grad}\ (\phi\psi),\ \text{grad}\ (\mathbf{f} \cdot \mathbf{g});$$

$$\text{div}\ (\phi\mathbf{f}),\ \text{div}\ (\mathbf{f} \times \mathbf{g});$$

$$\text{curl}\ (\phi\mathbf{f}),\ \text{curl}\ (\mathbf{f} \times \mathbf{g}).$$

The following results will be obtained :

I. $\text{grad}\ (\phi\psi) = \phi\ \text{grad}\ \psi + \psi\ \text{grad}\ \phi.$

II. $\text{grad}\ (\mathbf{f} \cdot \mathbf{g}) = \mathbf{f} \times \text{curl}\ \mathbf{g} + \mathbf{g} \times \text{curl}\ \mathbf{f} + (\mathbf{f} \cdot \nabla)\ \mathbf{g} + (\mathbf{g} \cdot \nabla)\ \mathbf{f}.$

(*Kumaon 2000, U.P.P.C.S. 2004*)

III. $\text{div}\ (\phi\mathbf{f}) = \phi\ \text{div}\ \mathbf{f} + \mathbf{f} \cdot \text{grad}\ \phi.$

(*Gorakhpur 98, Lucknow 98, Garhwal 2001, 2005 Kumaon 2004, Rohilkhand 2003, U.P.P.C.S. 2004*)

IV. $\text{div.}\ (\mathbf{f} \times \mathbf{g}) = \mathbf{g} \cdot \text{curl}\ \mathbf{f} - \mathbf{f} \cdot \text{curl}\ \mathbf{g}.$

(*Ajmer 97, 99, Gorakhpur 99, Rohilkhand 2000, 2001 Kumaon 99, 2005, Lucknow 97*)

V. $\text{curl}\ (\phi\mathbf{f}) = \text{grad}\ \phi \times \mathbf{f} + \phi\ \text{curl}\ \mathbf{f}.$ (*Gorakhpur 2000, Avadh 2001, Rohilkhand 2002, Garhwal 2002, 2006, Utkal 2003, Patna 2003*)

VI. $\text{curl}\ (\mathbf{f} \times \mathbf{g}) = \mathbf{f}\ \text{div}\ \mathbf{g} - \mathbf{g}\ \text{div}\ \mathbf{f} + (\mathbf{g} \cdot \nabla)\ \mathbf{f} - (\mathbf{f} \cdot \nabla)\ \mathbf{g}.$

(*Kolkata 96, Lucknow 97, Patna 2003, Rohilkhand 2002, Avadh 2004, Garhwal 2001, 2005*)

We shall now prove these one by one.

I. $$\text{grad}\,(\phi\psi) = \Sigma\,\mathbf{i}\,\frac{\partial(\phi\psi)}{\partial x}$$

$$= \Sigma\,\mathbf{i}\left(\phi\frac{\partial\psi}{\partial x} + \phi\frac{\partial\phi}{\partial x}\right)$$

$$= \Sigma\,\mathbf{i}\,\phi\frac{\partial\psi}{\partial x} + \Sigma\,\mathbf{i}\psi\frac{\partial\phi}{\partial x}$$

$$= \phi\,\Sigma\,\mathbf{i}\frac{\partial\psi}{\partial x} + \psi\,\Sigma\,\mathbf{i}\frac{\partial\phi}{\partial x}$$

$$= \phi\,\text{grad}\,\psi + \psi\,\text{grad}\,\phi.$$

II. $$\text{grad}\,(\mathbf{f}\,.\,\mathbf{g}) = \Sigma\,\mathbf{i}\,\frac{\partial(\mathbf{f}\,.\,\mathbf{g})}{\partial x}$$

$$= \Sigma\,\mathbf{i}\left(\mathbf{f}\cdot\frac{\partial\mathbf{g}}{\partial x} + \frac{\partial\mathbf{f}}{\partial x}\cdot\mathbf{g}\right)$$

$$= \Sigma\,\mathbf{i}\left(\mathbf{f}\cdot\frac{\partial\mathbf{g}}{\partial x}\right) + \Sigma\,\mathbf{i}\left(\frac{\partial\mathbf{f}}{\partial x}\cdot\mathbf{g}\right). \qquad ...(1)$$

Now

$$\mathbf{f}\times\left(\mathbf{i}\times\frac{\partial\mathbf{g}}{\partial x}\right) = \left(\mathbf{f}\cdot\frac{\partial\mathbf{g}}{\partial x}\right)\mathbf{i} - (\mathbf{f}\,.\,\mathbf{i})\frac{\partial\mathbf{g}}{\partial x},$$

or, $$\left(\mathbf{f}\cdot\frac{\partial\mathbf{g}}{\partial x}\right)\mathbf{i} = \mathbf{f}\left(\mathbf{i}\times\frac{\partial\mathbf{g}}{\partial x}\right) + (\mathbf{f}\,.\,\mathbf{i})\frac{\partial\mathbf{g}}{\partial x}.$$

$\therefore$ $$\Sigma\left(\mathbf{f}\cdot\frac{\partial\mathbf{g}}{\partial x}\right)\mathbf{i} = \mathbf{f}\times\left(\Sigma\,\mathbf{i}\times\frac{\partial\mathbf{g}}{\partial x}\right) + \Sigma\,(\mathbf{f}\,.\,\mathbf{i})\frac{\partial\mathbf{g}}{\partial x}$$

$$= \mathbf{f}\times\text{curl}\,\mathbf{g} + (\mathbf{f}\,.\,\nabla)\,\mathbf{g}. \qquad ...(2)$$

Interchanging **f** and **g**, we have

$$\Sigma\left(\mathbf{g}\cdot\frac{\partial\mathbf{f}}{\partial x}\right)\mathbf{i} = \mathbf{g}\times\text{curl}\,\mathbf{f} + (\mathbf{g}\,.\,\nabla)\,\mathbf{f}. \qquad ...(3)$$

From (1), (2) and (3), we have

$$\text{grad}\,(\mathbf{f}\,.\,\mathbf{g}) = \mathbf{f}\times\text{curl}\,\mathbf{g} + \mathbf{g}\times\text{curl}\,\mathbf{f} + (\mathbf{f}\,.\,\nabla)\,\mathbf{g} + (\mathbf{g}\,.\,\nabla)\,\mathbf{f}.$$

III. $$\text{div}\,(\phi\mathbf{f}) = \Sigma\,\mathbf{i}\cdot\frac{\partial(\phi\mathbf{f})}{\partial x}$$

$$= \Sigma\,\mathbf{i}\cdot\left(\phi\frac{\partial\mathbf{f}}{\partial x} + \frac{\partial\phi}{\partial x}\mathbf{f}\right)$$

$$= \phi \, \Sigma \mathbf{i} \cdot \frac{\partial \mathbf{f}}{\partial x} + \left(\Sigma \mathbf{i} \, \frac{\partial \phi}{\partial x} \right) \cdot \mathbf{f}$$

$$= \phi \text{ div } \mathbf{f} + \mathbf{f} \text{ . grad } \phi.$$

IV. $$\text{div } (\mathbf{f} \times \mathbf{g}) = \Sigma \mathbf{i} \cdot \frac{\partial (\mathbf{f} \times \mathbf{g})}{\partial x}$$

$$= \Sigma \mathbf{i} \cdot \left(\frac{\partial \mathbf{f}}{\partial x} \times \mathbf{g} + \mathbf{f} \times \frac{\partial \mathbf{g}}{\partial x} \right)$$

$$= \Sigma \mathbf{i} \cdot \frac{\partial \mathbf{f}}{\partial x} \times \mathbf{g} + \Sigma \mathbf{i} \cdot \mathbf{f} \times \frac{\partial \mathbf{g}}{\partial x}$$

$$= \Sigma \mathbf{i} \times \frac{\partial \mathbf{f}}{\partial x} \cdot \mathbf{g} - \Sigma \mathbf{i} \times \frac{\partial \mathbf{g}}{\partial x} \cdot \mathbf{f}$$

$$= \left(\Sigma \mathbf{i} \times \frac{\partial \mathbf{f}}{\partial x} \right) \cdot \mathbf{g} - \left(\Sigma \mathbf{i} \times \frac{\partial \mathbf{g}}{\partial x} \right) \cdot \mathbf{f}$$

$$= \text{curl } \mathbf{f} \text{ . } \mathbf{g} - \text{curl } \mathbf{g} \text{ . } \mathbf{f}.$$

V. $$\text{curl } (\phi \, \mathbf{f}) = \Sigma \mathbf{i} \times \frac{\partial (\phi \mathbf{f})}{\partial x}$$

$$= \Sigma \mathbf{i} \times \left(\frac{\partial \phi}{\partial x} \mathbf{f} + \phi \frac{\partial \mathbf{f}}{\partial x} \right)$$

$$= \Sigma \mathbf{i} \times \frac{\partial \phi}{\partial x} \times \mathbf{f} + \Sigma \mathbf{i} \times \phi \frac{\partial \mathbf{f}}{\partial x}$$

$$= \Sigma \mathbf{i} \times \frac{\partial \phi}{\partial x} \times \mathbf{f} + \left(\Sigma \mathbf{i} \times \frac{\partial \mathbf{f}}{\partial x} \right) \phi$$

$$= \text{grad } \phi \times \mathbf{f} + \phi \text{ curl } \mathbf{f}.$$

VI. $$\text{curl } (\mathbf{f} \times \mathbf{g}) = \Sigma \mathbf{i} \times \frac{\partial (\mathbf{f} \times \mathbf{g})}{\partial x}$$

$$= \Sigma \mathbf{i} \times \left(\mathbf{f} \times \frac{\partial \mathbf{g}}{\partial x} + \frac{\partial \mathbf{f}}{\partial x} \times \mathbf{g} \right)$$

$$= \Sigma \mathbf{i} \times \left(\mathbf{f} \times \frac{\partial \mathbf{g}}{\partial x} \right) + \Sigma \mathbf{i} \times \left(\frac{\partial \mathbf{f}}{\partial x} \times \mathbf{g} \right)$$

$$= \Sigma \left(\mathbf{i} \cdot \frac{\partial \mathbf{g}}{\partial x} \right) \mathbf{f} - \Sigma \, (\mathbf{i} \cdot \mathbf{f}) \frac{\partial \mathbf{g}}{\partial x} + \Sigma \, (\mathbf{i} \cdot \mathbf{g}) \frac{\partial \mathbf{f}}{\partial x} - \left(\Sigma \mathbf{i} \cdot \frac{\partial \mathbf{f}}{\partial x} \right) \mathbf{g}.$$

$$=\left(\Sigma \mathbf{i}\cdot\frac{\partial \mathbf{g}}{\partial x}\right)\mathbf{f}-\left(\Sigma \mathbf{i}\cdot\frac{\partial \mathbf{f}}{\partial x}\right)\mathbf{g}+\Sigma\left(\mathbf{i}\cdot\mathbf{g}\right)\frac{\partial \mathbf{f}}{\partial x}-\Sigma\left(\mathbf{i}\cdot\mathbf{f}\right)\frac{\partial \mathbf{g}}{\partial x}$$

$$= \mathbf{f}\text{ div }\mathbf{g} - \mathbf{g}\text{ div }\mathbf{f} + (\mathbf{g}\,.\,\nabla)\,\mathbf{f} - (\mathbf{f}\,.\,\nabla)\,\mathbf{g}.$$

10.16. SECOND ORDER DIFFERENTIAL OPERATORS

The gradient of a scalar point function being a vector point function, we may again form its divergence and curl,

i.e.,
$$\text{div grad } \phi = \nabla\,.\,\nabla\,\phi,$$
$$\text{curl grad } \phi = \nabla \times \nabla\,\phi.$$

Also curl of a vector point function being a vector point function, we may again form its divergence and curl, *i.e.*,

$$\text{div curl } \mathbf{f} = \nabla\,.\,\nabla \times \mathbf{f},$$
$$\text{curl curl } \mathbf{f} = \nabla \times \nabla \times \mathbf{f}.$$

Finally divergence of a vector point function being a scalar, we may have

$$\text{grad div } \mathbf{f} = \nabla\,\nabla\,.\,\mathbf{f}.$$

We shall now prove the following :

I. $\text{div grad } \phi = \nabla\cdot\nabla\phi = \dfrac{\partial^2\phi}{\partial x^2}+\dfrac{\partial^2\phi}{\partial y^2}+\dfrac{\partial^2\phi}{\partial z^2}\cdot$ (*U.P.P.C.S. 2003*)

II. $\text{curl grad } \phi = \nabla \times \nabla \times \phi = 0.$ (*Gorakhpur, 99*)

III. $\text{div curl } \mathbf{f} = \nabla\,.\,\nabla \times \mathbf{f} = 0.$ (*Gorakhpur, 2000, 2006, Utkal 2003, U.P.P.C.S. 99, Rohilkhand 2005*)

IV, V. $\text{grad div } \mathbf{f} = \text{curl curl } \mathbf{f} + \Sigma\,\dfrac{\partial^2 \mathbf{f}}{\partial x^2},$ (*U.P.P.C.S. 2003, Garhwal 2004, Kumaon 2001*)

i.e.,
$$\nabla\,\nabla\cdot\mathbf{f} = \nabla\times\nabla\times\mathbf{f}+\Sigma\,\frac{\partial^2\mathbf{f}}{\partial x^2}\cdot$$

These will now be proved one by one.

I.
$$\nabla\phi = \Sigma\mathbf{i}\,\frac{\partial\phi}{\partial x} = \mathbf{i}\frac{\partial\phi}{\partial x}+\mathbf{j}\frac{\partial\phi}{\partial y}+\mathbf{k}\frac{\partial\phi}{\partial z}\cdot$$

$\therefore$
$$\nabla\cdot\nabla\phi = \Sigma\mathbf{i}\cdot\frac{\partial}{\partial x}\left(\mathbf{i}\frac{\partial\phi}{\partial x}+\mathbf{j}\frac{\partial\phi}{\partial y}+\mathbf{k}\frac{\partial\phi}{\partial z}\right)$$
$$=\frac{\partial^2\phi}{\partial x^2}+\frac{\partial^2\phi}{\partial y^2}+\frac{\partial^2\phi}{\partial z^2}\cdot$$

II.
$$\text{curl grad } \phi = \Sigma\mathbf{i}\times\frac{\partial}{\partial x}\left(\mathbf{i}\frac{\partial\phi}{\partial x}+\mathbf{j}\frac{\partial\phi}{\partial y}+\mathbf{k}\frac{\partial\phi}{\partial z}\right)$$

$$= \Sigma \mathbf{i} \times \left(\mathbf{i} \frac{\partial^2 \phi}{\partial x^2} + \mathbf{j} \frac{\partial^2 \phi}{\partial x \partial y} + \mathbf{k} \frac{\partial^2 \phi}{\partial x \partial z} \right)$$

$$= \Sigma \left(\mathbf{k} \frac{\partial^2 \phi}{\partial x \partial y} - \mathbf{j} \frac{\partial^2 \phi}{\partial x \partial z} \right)$$

$$= \mathbf{0},$$

terms cancelling in pairs.

III. $\text{div . curl } \mathbf{f} = \Sigma \mathbf{i} \cdot \frac{\partial}{\partial x} \left(\mathbf{i} \times \frac{\partial \mathbf{f}}{\partial x} + \mathbf{j} \times \frac{\partial \mathbf{f}}{\partial y} + \mathbf{k} \times \frac{\partial \mathbf{f}}{\partial z} \right)$

$$= \Sigma \mathbf{i} \cdot \left(\mathbf{i} \times \frac{\partial^2 \mathbf{f}}{\partial x^2} + \mathbf{j} \times \frac{\partial^2 \mathbf{f}}{\partial x \partial y} + \mathbf{k} \times \frac{\partial^2 \mathbf{f}}{\partial x \partial z} \right)$$

$$= \Sigma \left(\mathbf{i} \times \mathbf{i} \cdot \frac{\partial^2 \mathbf{f}}{\partial x^2} + \mathbf{i} \times \mathbf{j} \cdot \frac{\partial^2 \mathbf{f}}{\partial x \partial y} + \mathbf{i} \times \mathbf{k} \cdot \frac{\partial^2 \mathbf{f}}{\partial x \partial z} \right)$$

$$= \Sigma \left(\mathbf{k} \cdot \frac{\partial^2 \mathbf{f}}{\partial x \partial y} - \mathbf{j} \cdot \frac{\partial^2 \mathbf{f}}{\partial x \partial z} \right)$$

$$= 0.$$

IV, V. $\nabla \times (\nabla \times \mathbf{f}) = \Sigma \mathbf{i} \times \frac{\partial}{\partial x} \left(\mathbf{i} \times \frac{\partial \mathbf{f}}{\partial x} + \mathbf{j} \times \frac{\partial \mathbf{f}}{\partial y} + \mathbf{k} \times \frac{\partial \mathbf{f}}{\partial z} \right)$

$$= \Sigma \mathbf{i} \times \left(\mathbf{i} \times \frac{\partial^2 \mathbf{f}}{\partial x^2} + \mathbf{j} \times \frac{\partial^2 \mathbf{f}}{\partial x \partial y} + \mathbf{k} \times \frac{\partial^2 \mathbf{f}}{\partial x \partial z} \right)$$

$$= \Sigma \left[\left(\mathbf{i} \cdot \frac{\partial^2 \mathbf{f}}{\partial x^2} \mathbf{i} - \mathbf{i} \cdot \mathbf{i} \frac{\partial^2 \mathbf{f}}{\partial x^2} \right) + \left(\mathbf{i} \cdot \frac{\partial^2 \mathbf{f}}{\partial x \partial y} \mathbf{j} - \mathbf{i} \cdot \mathbf{j} \frac{\partial^2 \mathbf{f}}{\partial x \partial y} \right) + \left(\mathbf{i} \cdot \frac{\partial^2 \mathbf{f}}{\partial x \partial z} \mathbf{k} - \mathbf{i} \cdot \mathbf{k} \frac{\partial^2 \mathbf{f}}{\partial x \partial z} \right) \right]$$

$$= \Sigma \left(\mathbf{i} \cdot \frac{\partial^2 \mathbf{f}}{\partial x^2} \mathbf{i} + \mathbf{i} \cdot \frac{\partial^2 \mathbf{f}}{\partial x \partial y} \mathbf{j} + \mathbf{i} \cdot \frac{\partial^2 \mathbf{f}}{\partial x \partial z} \mathbf{k} \right) - \Sigma \frac{\partial^2 \mathbf{f}}{\partial x^2} \cdot$$

Also $\nabla \nabla \cdot \mathbf{f} = \Sigma \mathbf{i} \frac{\partial}{\partial x} \left(\mathbf{i} \cdot \frac{\partial \mathbf{f}}{\partial x} + \mathbf{j} \cdot \frac{\partial \mathbf{f}}{\partial y} + \mathbf{k} \cdot \frac{\partial \mathbf{f}}{\partial z} \right)$

$$= \Sigma \mathbf{i}\left(\mathbf{i}\cdot\frac{\partial^2 \mathbf{f}}{\partial x^2}+\mathbf{j}\cdot\frac{\partial^2 \mathbf{f}}{\partial x \partial y}+\mathbf{k}\cdot\frac{\partial^2 \mathbf{f}}{\partial x \partial z}\right)$$

$$= \Sigma\left(\mathbf{i}\cdot\frac{\partial^2 \mathbf{f}}{\partial x^2}\mathbf{i}+\mathbf{i}\cdot\frac{\partial^2 \mathbf{f}}{\partial x \partial y}\mathbf{j}+\mathbf{i}\cdot\frac{\partial^2 \mathbf{f}}{\partial x \partial z}\mathbf{k}\right)$$

$$= \nabla\times(\nabla\times\mathbf{f})+\Sigma\frac{\partial^2 \mathbf{f}}{\partial x^2}.$$

$$\therefore \qquad \text{grad div } \mathbf{f} = \text{curl curl } \mathbf{f} + \Sigma\frac{\partial^2 \mathbf{f}}{\partial x^2}$$

10.17. THE LAPLACIAN OPERATOR, ∇^2

Looking upon ∇ formally as a vector, we have

$$\nabla^2 = \nabla\cdot\nabla = \left(\Sigma \mathbf{i}\frac{\partial}{\partial x}\right)\cdot\left(\Sigma \mathbf{i}\frac{\partial}{\partial x}\right)$$

$$= \frac{\partial^2}{\partial x^2}+\frac{\partial^2}{\partial y^2}+\frac{\partial^2}{\partial z^2}.$$

Now, by def., the operator ∇^2, called Laplacian, is given by

$$\nabla^2\phi = \frac{\partial^2\phi}{\partial x^2}+\frac{\partial^2\phi}{\partial y^2}+\frac{\partial^2\phi}{\partial z^2}.$$

$$\nabla^2\mathbf{f} = \frac{\partial^2\mathbf{f}}{\partial x^2}+\frac{\partial^2\mathbf{f}}{\partial y^2}+\frac{\partial^2\mathbf{f}}{\partial z^2}.$$

Note on the Role of ∇ as a symbolic vector. The proper role of the symbol ∇ is that of a differential operator and the symbol has actually been introduced as such. However we should notice that it helps greatly in reproducing the various formulae concerning differential operations, if ∇ is *thought* of as a vector and this role of ∇ is constantly borne in mind.

Thus, in the first instance we see that the vectorial character of ∇ is consistent with the following statements :

(*i*) $\nabla\, \phi$ is a vector.

(*ii*) $\nabla\, .\, \mathbf{f}$ is a scalar.

(*iii*) $\nabla \times \mathbf{f}$ is a vector.

(*iv*) $\nabla\, \mathbf{f}$ is meaningless.

(*v*) curl grad $\phi = \nabla \times \nabla\phi = 0$.

(*vi*) div curl $\mathbf{f} = \nabla\, .\, \nabla \times \mathbf{f} = 0$.

(*vii*) div grad $\phi = \nabla\, .\, \nabla\phi = \nabla^2\phi$.

(*viii*) curl curl $\mathbf{f} = \nabla \times (\nabla \times \mathbf{f})$

$= \nabla\, \nabla\, .\, \nabla\, \mathbf{f} - \nabla\, .\, \nabla\, \mathbf{f}$

$=$ grad div . $\mathbf{f} - \nabla^2\mathbf{f}$.

The following suggested procedure involving the role of ∇ as a vector proves very helpful in reproducing the various formulae concerning the operation of ∇ on a product of two functions (§ 10.15).

I. Regarding the two factors separately as constants, express the result of the opeartion of ∇ as a sum of two terms.

II. Transform each of the two terms, taking care that the operator always appears before (and not after) a function and bearing in mind whether the result of the operation is a vector or a scalar. The properties of triple products may also be sometimes needed to carry out the transformation.

III. Restore the change so far as regarding the functions as constants is concerned.

We shall now see how the three-step procedure suggested above can actually be employed to reproduce the various formulae of § 10.15.

I. $$\begin{aligned} \text{grad}\,(\phi\psi) &= \nabla(\phi\psi) \\ &= \nabla(\phi_c\psi + \phi\psi_c) \\ &= \phi_c\nabla\psi + \psi_c\nabla\phi \\ &= \phi\nabla\psi + \psi\nabla\phi. \end{aligned}$$

II. $$\begin{aligned} \text{grad}\,(\mathbf{f}\,.\,\mathbf{g}) &= \nabla(\mathbf{f}\,.\,\mathbf{g}) \\ &= \nabla(\mathbf{f}_c\,.\,\mathbf{g}) + \nabla(\mathbf{f}\,.\,\mathbf{g}_c) \\ &= \mathbf{f}_c \times (\nabla \times \mathbf{g}) + (\mathbf{f}_c\,.\,\nabla)\,\mathbf{g} + \mathbf{g}_c \times (\nabla \times \mathbf{f}) + (\mathbf{g}_c\,.\,\nabla)\,\mathbf{f}\ ? \\ &= \mathbf{f} \times (\nabla \times \mathbf{g}) + \mathbf{g} \times (\nabla \times \mathbf{f}) + (\mathbf{f}\,.\,\nabla)\,\mathbf{g} + (\mathbf{g}\,.\,\nabla)\,\mathbf{f}\ ? \end{aligned}$$

III. $$\begin{aligned} \text{div}\,(\phi\mathbf{f}) &= \nabla\,.\,(\phi\mathbf{f}) \\ &= \nabla\,.\,(\phi_c\mathbf{f}) + \nabla\,.\,(\phi\mathbf{f}_c) \\ &= \phi_c\nabla\,.\,\mathbf{f} + \nabla\phi\,.\,\mathbf{f}_c \\ &= \phi\nabla\,.\,\mathbf{f} + \nabla\phi\,.\,\mathbf{f}. \end{aligned}$$ *(Ajmer 98)*

IV. $$\begin{aligned} \text{div}\,(\mathbf{f} \times \mathbf{g}) &= \nabla\,.\,(\mathbf{f} \times \mathbf{g}) \\ &= \nabla\,.\,(\mathbf{f}_c \times \mathbf{g}) + \nabla\,.\,(\mathbf{f} \times \mathbf{g}_c) \\ &= -\nabla \times \mathbf{g}\,.\,\mathbf{f}_c + \nabla \times \mathbf{f}\,.\,\mathbf{g}_c \\ &= -\nabla \times \mathbf{g}\,.\,\mathbf{f} + \nabla \times \mathbf{f}\,.\,\mathbf{g}. \end{aligned}$$

V. $$\begin{aligned} \text{curl}\,(\phi\mathbf{f}) &= \nabla \times (\phi\mathbf{f}) \\ &= \nabla \times (\phi_c\mathbf{f}) + \nabla \times (\phi\mathbf{f}_c) \\ &= \phi_c\nabla \times \mathbf{f} + \nabla\phi \times \mathbf{f}_c \\ &= \phi\nabla \times \mathbf{f} + \nabla\phi \times \mathbf{f}. \end{aligned}$$

VI. $$\begin{aligned} \text{curl}\,(\mathbf{f} \times \mathbf{g}) &= \nabla \times (\mathbf{f} \times \mathbf{g}) \\ &= \nabla \times (\mathbf{f}_c \times \mathbf{g}) + \nabla \times (\mathbf{f} \times \mathbf{g}_c) \\ &= \nabla\,.\,\mathbf{g}\,\mathbf{f}_c - (\mathbf{f}_c\,.\,\nabla)\,\mathbf{g} + (\mathbf{g}_c\,.\,\nabla)\,\mathbf{f} - \nabla\mathbf{f}\,\mathbf{g}_c \\ &= \mathbf{f}\,\nabla\,.\,\mathbf{g} - \mathbf{g}\nabla\,.\,\mathbf{f} + (\mathbf{g}\,.\,\nabla)\,\mathbf{f} - (\mathbf{f}\,.\,\nabla)\,\mathbf{g}. \end{aligned}$$

EXAMPLES

Example 1. *If* $\mathbf{f} = x^2z\mathbf{i} - 2y^3z^2\mathbf{j} + xy^2z\mathbf{k}$. *Find div* **f**, *curl* **f**, *at (1, – 1, 1).*

Solution. $\text{div } \mathbf{f} = \Sigma \mathbf{i} \cdot \frac{\partial \mathbf{f}}{\partial x} = \nabla \cdot \mathbf{f}$

$$= \left(\mathbf{i}\frac{\partial}{\partial x} + \mathbf{j}\frac{\partial}{\partial y} + \mathbf{k}\frac{\partial}{\partial z}\right) \cdot \left(x^2z\mathbf{i} - 2y^3z^2\mathbf{j} + xy^2z\mathbf{k}\right)$$

$$= \frac{\partial}{\partial x}\left(x^2z\right) - \frac{\partial}{\partial y}\left(2y^3z^2\right) + \frac{\partial}{\partial x}\left(xy^2z\right)$$

$$= 2xz - 6y^2z^2 + xy^2$$

$$= -3 \text{ at } (1, -1, 1)$$

$$\text{curl } \mathbf{f} = \nabla \times \mathbf{f} = \begin{vmatrix} \mathbf{i} & \mathbf{j} & \mathbf{k} \\ \frac{\partial}{\partial x} & \frac{\partial}{\partial y} & \frac{\partial}{\partial z} \\ x^2z & -2y^3z^2 & xy^2z \end{vmatrix}$$

$$= \mathbf{i}\,(2xyz + 4y^3z) + \mathbf{j}\,(x^2 - y^2z) + \mathbf{k}\,(0 - 0)$$

$$= -6\mathbf{i} \quad \text{at} \quad (1, -1, 1).$$

Example 2. *Prove that* $r^n\mathbf{r}$ *is irrotational. Find n when it is solenoidal.*

(Ajmer 97, Gorakhpur 98, Rohilkhand 98)

Solution. *A vector is said to be irrotational if curl* **f** = *0 and solenoidal if div* **f** = *0.*

$$r^n\mathbf{r} = r^nx\mathbf{i} + r^ny\mathbf{j} + r^nz\mathbf{k} = f_1\mathbf{i} + f_2\mathbf{j} + f_3\mathbf{k}$$

$$\therefore \qquad \text{div } r^n\mathbf{r} = \frac{\partial f_1}{\partial x} + \frac{\partial f_2}{\partial y} + \frac{\partial f_3}{\partial z}$$

Now, $\qquad f_1 = r^nx \Rightarrow \frac{\partial f_1}{\partial x} = r^n + xnr^{n-1}\frac{\partial r}{\partial x}$

$$= r^n + nxr^{n-1}\frac{x}{r} = r^n + nr^{n-2}x^2 \text{ etc.}$$

$$\therefore \quad \text{div } r^n\mathbf{r} = \Sigma\frac{\partial f_1}{\partial x} = 3r^n + nr^{n-2}\,\Sigma x^2$$

$$= 3r^n + nr^{n-2}r^2 = (n+3)\,r^n$$

For $r^n\mathbf{r}$ to be solenoidal, $(n + 3)\,r^n = 0$.

$\Rightarrow \quad n + 3 = 0 \Rightarrow n = -3.$

$$\text{curl } \mathbf{f} = \begin{vmatrix} \mathbf{i} & \mathbf{j} & \mathbf{k} \\ \dfrac{\partial}{\partial x} & \dfrac{\partial}{\partial y} & \dfrac{\partial}{\partial z} \\ r^n x & r^n y & r^n z \end{vmatrix}$$

coeff. of **i** is $\dfrac{\partial}{\partial y}\left(r^n z\right)-\dfrac{\partial}{\partial z}\left(r^n y\right)$

$$= z.nr^{n-1}\frac{\partial r}{\partial y} - y.nr^{n-1}\frac{\partial r}{\partial z}$$

$$= nr^{n-1}\left[z\cdot\frac{y}{r} - y\cdot\frac{z}{r}\right] = 0$$

Similarly, coeff. of **j** = coeff. of **k** = 0

$\therefore$ curl $r^n\mathbf{r} = 0$.

$\Rightarrow$ $r^n\mathbf{r}$ is a irrotational vector function.

Example 3. *Prove that* $\nabla^2 f(r) = f''(r) + \dfrac{2}{r} f'(r).$ *(Ajmer 97)*

Solution. Let $\phi = f(r)$; $\therefore$ $\nabla^2 f(r) = \nabla^2 \phi = \Sigma \dfrac{\partial^2 \phi}{\partial x^2}.$

$$\frac{\partial \phi}{\partial x} = f'(r)\,\frac{\partial r}{\partial x} = \frac{x}{r} f'(r)$$

$$\therefore \quad \frac{\partial^2 \phi}{\partial x^2} = \frac{r\left\{1\cdot f'(r) + x f''(r)\cdot\dfrac{x}{r}\right\} - x f'(r)\cdot\dfrac{x}{r}}{r^2}$$

$$= \frac{1}{r} f'(r) + \frac{x^2}{r^2} f''(r) - \frac{x^2}{r^3} f'(r)$$

$$\therefore \quad \Sigma\frac{\partial^2 \phi}{\partial x^2} = \frac{3}{r} f'(r) + \frac{\Sigma x^2}{r^2} f''(r) - \frac{\Sigma x^2}{r^3} f'(r)$$

$$= \frac{3}{r} f'(r) + f''(r) - \frac{1}{r} f'(r)$$

$$= f''(r) + \frac{2}{r} f'(r).$$

Example 4. *Prove that* $\nabla^2\left(\dfrac{x}{r^2}\right) = \dfrac{-2x}{x^4}.$ *(Avadh 2003, Kumaon 2006)*

Solution. $$\frac{\partial^2}{\partial x^2}\left(\frac{x}{r^2}\right)=\frac{\partial}{\partial x}\left[\frac{\partial}{\partial x}\left(\frac{x}{r^2}\right)\right]$$

$$=\frac{\partial}{\partial x}\left[1\cdot\frac{1}{r^2}-\frac{2x}{r^3}\cdot\frac{x}{r}\right] \qquad \left[\because\ \frac{\partial r}{\partial x}=\frac{x}{r}\right]$$

$$=\frac{\partial}{\partial x}\left[\frac{1}{r^2}-\frac{2x^2}{r^4}\right]$$

$$=\left[-\frac{2}{r^3}\cdot\frac{x}{r}-\frac{4x}{r^4}+\frac{8x^2}{r^5}\cdot\frac{x}{r}\right]$$

$$=\left[-\frac{2}{r^4}x-\frac{4x}{r^4}+\frac{8x^3}{r^6}\right]$$

$$=-\frac{6x}{r^4}+\frac{8x^3}{r^6}=-2x\left[\frac{3}{r^4}-\frac{4x^2}{r^6}\right] \qquad ...(1)$$

$$\frac{\partial^2}{\partial y^2}\left(\frac{x}{r^2}\right)=\frac{\partial}{\partial y}\left[\frac{\partial}{\partial y}\left(\frac{x}{r^2}\right)\right]$$

$$=\frac{\partial}{\partial y}\left[-\frac{2x}{r^3}\cdot\frac{\partial r}{\partial y}\right]$$

$$=\frac{\partial}{\partial y}\left[-\frac{2x}{r^3}\cdot\frac{y}{r}\right]=\frac{\partial}{\partial y}\left[-\frac{2xy}{r^4}\right]$$

$$=-2x\left[\frac{1}{r^4}-\frac{4y}{r^5}\cdot\frac{y}{r}\right]=-2x\left[\frac{1}{r^4}-\frac{4y^2}{r^6}\right] \qquad ...(2)$$

Similarly, $$\frac{\partial^2}{\partial z^2}\left(\frac{x}{r^2}\right)=-2x\left[\frac{1}{r^4}-\frac{4z^2}{r^6}\right] \qquad ...(3)$$

$$\therefore\quad \nabla^2\left(\frac{x}{r^2}\right)=\Sigma\,\frac{\partial^2}{\partial x^2}\left(\frac{x}{r^2}\right)$$

$$=-2x\left[\frac{3}{r^4}+\frac{1}{r^4}+\frac{1}{r^4}-\frac{4}{r^6}\left(x^2+y^2+z^2\right)\right]$$

by (1), (2) and (3)

$$=-2x\left[\frac{5}{r^4}-\frac{4}{r^4}\right]=-\frac{2x}{r^4}.$$

Example 5. *If* **f** *and* **g** *are irrotational then show that* **f** × **g** *is a solenoidal vector.*

Solution. Given that curl **f** = 0, curl **g** = 0.

$$\text{div}\,(\mathbf{f}\times\mathbf{g}) = (\text{curl}\,\mathbf{f})\,.\,\mathbf{g} + (\text{curl}\,\mathbf{g})\,.\,\mathbf{f}$$
$$= 0 + 0 = 0.$$

Hence, **f** × **g** is solenoidal.

Example 6. *Prove that div* [(**r** × **a**) × **b**) = – 2**b** . **a**.

(Gorakhpur 99, Lucknow 97)

Solution.
$$\mathbf{r}\times\mathbf{a} = \begin{vmatrix} \mathbf{i} & \mathbf{j} & \mathbf{k} \\ x & y & z \\ a_1 & a_2 & a_3 \end{vmatrix}$$
$$= \mathbf{i}\,(ya_3 - za_2) + \mathbf{j}\,(za_1 - xa_3) + \mathbf{k}\,(xa_2 - ya_1)$$

$$\therefore \quad (\mathbf{r}\times\mathbf{a})\times\mathbf{b} = \begin{vmatrix} \mathbf{i} & \mathbf{j} & \mathbf{k} \\ ya_3 - za_2 & za_1 - xa_3 & xa_2 - ya_1 \\ b_1 & b_2 & b_3 \end{vmatrix}$$
$$= \Sigma\,\mathbf{i}\left[b_3(za_1 - xa_3) - b_2(xa_3 - ya_1)\right]$$
$$= f_1\mathbf{i} + f_2\mathbf{j} + f_3\mathbf{k}$$

$$\therefore \quad \frac{\partial f_1}{\partial x} = (-b_3a_3 - b_2a_2) \text{ etc.}$$

Now,
$$\text{div}\,.\,\mathbf{f} = \frac{\partial f_1}{\partial x} + \frac{\partial f_2}{\partial y} + \frac{\partial f_3}{\partial z}$$
$$= (-b_3a_3 - b_2a_2) + (-b_2a_2 - b_1a_1) + (-b_1a_1 - b_3a_3)$$
$$= -2\,(a_1b_1 + a_2b_2 + a_3b_3)$$
$$= -2\,(\mathbf{a}\,.\,\mathbf{b}) = -2\,(\mathbf{b}\,.\,\mathbf{a}).$$

Example 7. *Prove that div* $\left(\frac{f(r)}{r}\mathbf{r}\right) = \frac{1}{r^2}\frac{d}{dr}\left[r^2 f(r)\right]$.

Solution.
$$\frac{f(r)}{r}\mathbf{r} = \frac{f(r)}{r}(\mathbf{i}x + \mathbf{j}y + \mathbf{k}z)$$
$$= \frac{f(r)}{r}x\mathbf{i} + \frac{f(r)}{r}y\mathbf{j} + \frac{f(r)}{r}z\mathbf{k}$$
$$= f_1\mathbf{i} + f_2\mathbf{j} + f_3\mathbf{k}$$

$$\therefore \quad \text{div } \frac{f(r)}{r}\mathbf{r} = \frac{\partial f_1}{\partial x} + \frac{\partial f_2}{\partial y} + \frac{\partial f_3}{\partial z} = \Sigma \frac{\partial f_1}{\partial x}$$

$$f_1 = \frac{f(r)}{r}x, \therefore \frac{\partial f_1}{\partial x} = \frac{f(r)}{r}\cdot 1 + x\cdot\frac{rf'(r) - f(r)}{r^2}\cdot\frac{\partial r}{\partial x}$$

$$= \frac{f(r)}{r} + x\cdot\frac{rf'(r) - f(r)}{r^2}\cdot\frac{x}{r}$$

$$= \frac{f(r)}{r} + \frac{x^2}{r^2}f'(r) - \frac{x^2}{r^3}f(r)$$

$$\therefore \quad \Sigma\frac{\partial f_1}{\partial x} = \frac{3f(r)}{r} + \frac{\Sigma x^2}{r^2}f'(r) - \frac{\Sigma x^2}{r^3}f(r)$$

$$= \frac{3f(r)}{r} + f'(r) - \frac{1}{r}f(r)$$

$$= \frac{2f(r)}{r} + f'(r) = \frac{2rf(r) + r^2 f'(r)}{r^2}$$

$$= \frac{1}{r^2}\frac{d}{dr}\left[r^2 f(r)\right].$$

Example 8. *Find div grad r^m and verify that*

$$\nabla \times \nabla r^m = 0. \quad \textit{(Rohilkhand 2001,2002, 2004)}$$

Solution. We have

$$\text{grad } r^m = mr^{m-2}\,\mathbf{r}$$

$$= mr^{m-2}(\mathbf{i}x + \mathbf{j}y + \mathbf{k}z)$$

$$\therefore \quad \text{div grad } r^m = \frac{\partial}{\partial x}\left(mr^{m-2}x\right) + \frac{\partial}{\partial y}\left(mr^{m-2}y\right) + \frac{\partial}{\partial z}\left(mr^{m-2}z\right).$$

Also

$$\frac{\partial}{\partial x}\left(mr^{m-2}x\right) = m(m-2)\,r^{m-3} + \frac{\partial r}{\partial x}x + mr^{m-2}$$

$$= m(m-2)\,r^{m-4}x^2 + mr^{m-2}, \text{ for } \partial r/\partial x = x/r.$$

Similarly,

$$\frac{\partial}{\partial y}\left(mr^{m-2}y\right) = m(m-2)\,r^{m-1}y^2 + mr^{m-2},$$

$$\frac{\partial}{\partial z}\left(mr^{m-2}z\right) = m(m-2)\,r^{m-4}z^2 + mr^{m-2}.$$

$$\therefore \quad \text{div grad } r^m = m(m-2)\,r^{m-4}r^2 + 3mr^{m-2}.$$

$$= m(m+1)\,r^{m-2}.$$

Again curl grad $r^m = \begin{vmatrix} \mathbf{i} & \mathbf{j} & \mathbf{k} \\ \dfrac{\partial}{\partial x} & \dfrac{\partial}{\partial y} & \dfrac{\partial}{\partial z} \\ mr^{m-2}x & mr^{m-2}y & mr^{m-2}z \end{vmatrix}$

Thus, the coefficient of **i**, in curl grad r^m

$$= \frac{\partial}{\partial y}\left(mr^{m-2}z\right) - \frac{\partial}{\partial z}\left(mr^{m-2}y\right)$$

$$= m(m-2)\,r^{m-3}\frac{\partial r}{\partial y}z - m(m-2)\,r^{m-3}\frac{\partial r}{\partial z}y$$

$$= m(m-2)\,r^{m-3}\frac{yz}{r} - m(m-2)\,r^{m-3}\frac{zy}{r} = 0.$$

Hence, curl grad $r^m = \nabla \times \nabla r^m = 0.$

Example 9. *Show that*

$$curl\,\frac{\mathbf{a}\times\mathbf{r}}{r^3} = -\frac{\mathbf{a}}{r^3} + \frac{3\mathbf{r}}{r^3}(\mathbf{a}\,.\,\mathbf{r}),$$

where **a** *is a constant vector.*

(Kolkata 99, Gorakhpur 98, 2000, Lucknow 97)

Solution. Now,

$$\text{curl}\,\frac{\mathbf{a}\times\mathbf{r}}{r^3} = \Sigma\mathbf{i}\times\frac{\partial}{\partial x}\left(\frac{\mathbf{a}\times\mathbf{r}}{r^3}\right)$$

Also

$$\frac{\partial}{\partial x}\left(\frac{\mathbf{a}\times\mathbf{r}}{r^3}\right) = -\frac{3}{r^4}\frac{\partial r}{\partial x}(\mathbf{a}\times\mathbf{r}) + \frac{1}{r^3}\left(\mathbf{a}\times\frac{\partial \mathbf{r}}{\partial x}\right)$$

$$= -\frac{3x}{r^5}(\mathbf{a}\times\mathbf{r}) + \frac{1}{r^3}\mathbf{a}\times\mathbf{i}.$$

$$\therefore \quad \mathbf{i}\times\frac{\partial}{\partial x}\left(\frac{\mathbf{a}\times\mathbf{r}}{r^3}\right) = -\frac{3x}{r^5}\left[(\mathbf{i}\,.\,\mathbf{r})\,\mathbf{a} - (\mathbf{i}\,.\,\mathbf{a})\,\mathbf{r}\right] + \frac{1}{r^3}\left[(\mathbf{i}\,.\,\mathbf{i})\mathbf{a} - (\mathbf{i}\,.\,\mathbf{a})\,\mathbf{i}\right]$$

$$= -\frac{3x^2}{r^5}\mathbf{a} + \frac{3\mathbf{r}}{r^5}(\mathbf{i}x\cdot\mathbf{a}) + \frac{1}{r^3}\mathbf{a} - \frac{1}{r^3}(\mathbf{i}\cdot\mathbf{a})\,\mathbf{i}.$$

$$\therefore \quad \Sigma \mathbf{i} \times \frac{\partial}{\partial x}\left(\frac{\mathbf{a} \times \mathbf{r}}{r^3}\right) = -\frac{3\mathbf{a}}{r^5} r^2 + \frac{3\mathbf{r}}{r^5} \mathbf{r} \cdot \mathbf{a} + \frac{3}{r^3}\mathbf{a} - \frac{1}{r^3}\mathbf{a}$$

$$= -\frac{\mathbf{a}}{r^3} + \frac{3\mathbf{r}}{r^5}\mathbf{a} \,.\, \mathbf{r}.$$

The question could also be done by employing the results of § 10.15 V, VI.

Example 10. *If* **F** *and* ϕ *are point functions, show that the components of the former normal and tangential to the surface* $\phi = 0$ *at any point thereof are*

$$\frac{(\mathbf{F} \,.\, \nabla\phi)\,\nabla\phi}{(\nabla\phi)^2} \quad \textit{and} \quad \frac{\nabla\phi \times (\mathbf{F} \times \nabla\phi)}{(\nabla\phi)^2}.$$

Solution. The unit vector normal to the surface $\phi = 0$ being $\nabla\phi \,/\, |\nabla\phi|$, the magnitude of the component of **F** along this normal

$$= \mathbf{F} \cdot \frac{\nabla\phi}{|\nabla\phi|} = \frac{(\mathbf{F} \cdot \nabla\phi)}{|\nabla\phi|}$$

Thus, the vector **F** is equal to the sum of vector

$$\frac{(\mathbf{F} \cdot \nabla\phi)}{|\nabla\phi|}\frac{\nabla\phi}{|\nabla\phi|} = \frac{(\mathbf{F} \cdot \nabla\phi)}{(\nabla\phi)^2}\nabla\phi \qquad \text{...(1)}$$

along the normal and the vector

$$\mathbf{F} - \frac{\mathbf{F} \cdot \nabla\phi}{(\nabla\phi)^2}\nabla\phi \qquad \text{...(2)}$$

along the tangent. This latter

$$= \frac{\mathbf{F}\,(\nabla\phi \cdot \nabla\phi) - (\mathbf{F} \cdot \nabla\phi)\,\nabla\phi}{(\nabla\phi)^2}$$

$$= \frac{\nabla\phi \times (\mathbf{F} \times \nabla\phi)}{(\nabla\phi)^2}$$

Hence, the result.

EXERCISES

(In the following, **r** denotes the position vector of any point $r = |\mathbf{r}|$ and **a, b,** denote constant vectors. Also x, y, z denote rectangular cartesian co-ordinates).

1. If

$$\mathbf{f} = \mathbf{i}\,(x + y + 1) + \mathbf{j} + \mathbf{k}\,(-x - y),$$

prove that

$$\mathbf{f} \,.\, \text{curl}\, \mathbf{f} = 0.$$

2. Find div **F** and curl **F** where

$$\mathbf{F} = \text{grad } (x^3 + y^3 + z^3 - 3xyz).$$

3. If $\phi = x/r^3$, show that $\nabla^2\phi = 0$.

4. Show that

(i) $\text{div } \mathbf{r} = 3$. (*Ajmer 98, Kumaon 2000*)

(ii) $\text{curl } \mathbf{r} = 0$. (*Ajmer 98, Kumaon 2000*)

(iii) $\nabla^2 r^{-1} = 0$. (*Gorakhpur 99*)

(iv) $\text{div } (r^n \mathbf{r}) = (n + 3) r^n$. (*Rohilkhand 2003, Avadh 2000*)

5. Show that

(i) $\text{div } (\mathbf{r} \times \mathbf{a}) = 0$, (ii) $\text{curl } (\mathbf{r} \times \mathbf{a}) = -2\mathbf{a}$

(iii) $\text{grad } (\mathbf{r} \cdot \mathbf{a}) = \mathbf{a}$.

6. Show that

(i) $\text{grad } [\mathbf{r}\ \mathbf{a}\ \mathbf{b}] = \mathbf{a} \times \mathbf{b}$ (ii) $\text{div } [(\mathbf{r} \times \mathbf{a}) \times \mathbf{b}] = 0$. (*Ajmer 98*)

(iii) $\text{curl } [(\mathbf{r} \times \mathbf{a}) \times \mathbf{b}] = \mathbf{b} \times \mathbf{a}$. (*U.P.P.C.S. 2000*)

(iv) $\text{grad } [(\mathbf{r} \times \mathbf{a}) \cdot (\mathbf{r} \times \mathbf{b})] = (\mathbf{b} \times \mathbf{r}) \times \mathbf{a} + (\mathbf{a} \times \mathbf{r}) \times \mathbf{b}$.

7. Evaluate

$$\nabla \cdot (\mathbf{a}\ \mathbf{b} \cdot \mathbf{r}) \text{ and } \nabla \times [\mathbf{a} \times (\mathbf{b} \times \mathbf{r})].$$

8. Expand

$$\nabla\left(\mathbf{a} \cdot \frac{\mathbf{r}}{r^n}\right),\ \nabla \cdot \left(\mathbf{a} \times \frac{\mathbf{r}}{r^n}\right),\ \nabla \times \left(\mathbf{a} \times \frac{\mathbf{r}}{r^n}\right).$$

9. Show that

$$\text{curl } \mathbf{r}\phi\ (r) = 0 \text{ and } (\mathbf{a} \cdot \mathbf{r})\ \mathbf{a} = 0.$$

Also find

$$\text{curl } \mathbf{a}\phi\ (r) \text{ and } (\nabla\phi \cdot \nabla)\phi.$$

10. If

$$\mathbf{f} = (\mathbf{a} \times \mathbf{r})\ r^n,$$

show that

$$\text{div } \mathbf{f} = 0,\ \text{curl } \mathbf{f} = (n + 2)\ r^n\ \mathbf{a} - nr^{n-2}\ (\mathbf{a} \cdot \mathbf{r})\ \mathbf{r}.$$

11. A vector function **f** is the product of a scalar function and the gradient of a scalar function, show that

$$\mathbf{f} \cdot \text{curl } \mathbf{f} = 0.$$

12. If $\mathbf{f} = \nabla\ (\mathbf{a} \cdot \nabla^{r-1})$, show that $\text{div } \mathbf{f} = 0$, and $\mathbf{f} = \text{curl } \mathbf{g}$, where

$$\mathbf{g} = -\ \mathbf{a} \times \nabla\ (r^{-1}).$$

13. Show that

$$\text{curl } (\psi\nabla\phi) = \nabla\phi \times \nabla\psi = -\ \text{curl } (\phi\nabla\psi)$$

where ϕ and ψ are two scalar functions.

14. ϕ_1 and ϕ_2 are two scalar functions such that

$$\left(\nabla^2 + k^2\right)\phi_1 = 0 \text{ and } \left(\nabla^2 + k^2\right)\phi_2 = 0$$

and $\mathbf{f} = \nabla \times [\mathbf{r}\phi_1 + \mathbf{r} \times \nabla\phi_2]$, show that $\text{div}\, \mathbf{f} = 0$, $(\nabla^2 + k^2)\,\mathbf{f} = 0$.

15. ρ and p are two scalar point functions such that ρ is a function of p; show that

$$\nabla\rho = \frac{d\rho}{dp}\nabla p.$$

16. Show that

$$\mathbf{q} \cdot \nabla\mathbf{q} = \frac{1}{2}\nabla\mathbf{q}^2 - \mathbf{q} \times \text{curl}\ \mathbf{q}.$$

17. Show that

$$\text{div. curl curl } (\mathbf{a}\phi) + \nabla^2 \text{ div } (\mathbf{a}\phi) = \mathbf{a}\ .\ \text{grad}\ \nabla^2\phi.$$

where ϕ is a scalar point function.

10.18. DIFFERENTIAL OPERATIONS IN TERMS OF CURVILINEAR CO-ORDINATES

Curvilinear co-ordinates. Consider three continuously differentiable scalar functions

$$u = \phi_1(x, y, z),\quad v = \phi_2(x, y, z),\quad w = \phi_3(x, y, z)$$

defined in any region R of space; x, y, z denoting the rectangular cartesian co-ordinates of any point P of R. These three functions associate a triad of numbers u, v, w to each point of the region and we suppose that these functions are such that to two different points of the region correspond different triads. Thus, to the aggregate of the points P of the region R, corresponds an aggregate of triads u, v, w such that to each point P of the region corresponds one and only one triad (u, v, w). We say that (u, v, w) *are the curvilinear co-ordinates of P* in as much as the point uniquely determines and is uniquely determined by the triad.

10.18.1. Co-ordinate Surfaces and Curves

The loci of the equations

$$u = u_0,\quad v = v_0,\quad w = w_0$$

[Curvilinear co-ordinates]

which are the same as those of the equations

$$\phi_1(x, y, z) = u_0,\quad \phi_2(x, y, z) = v_0,\quad \phi_3(x, y, z) = w_0,$$

[Cartesian co-ordinates]

u_0, v_0, w_0, being constants, are called co-ordinate surfaces. Through each point there passes a system of three co-ordinate surfaces.

The three curves of intersection of three co-ordinate surfaces through any point, taken in pairs, are called *co-ordinate curves* through the point. The three co-ordinate curves through any point (u_0, v_0, w_0) are

$$v = v_0,\ \ w = w_0,\ \ w = w_0,\ \ u = u_0;\ \ u = u_0,\ \ v = v_0.$$

10.18.2. Orthogonal Curvilinear Co-ordinates

A system of co-ordinates is said to be orthogonal, if the three co-ordinate surfaces through every point are mutually orthogonal, *i.e.*, if the three tangent planes, and accordingly the three normals to the three surfaces, through every point are orthogonal in pairs. The line of intersection of the tangent planes to any two co-ordinate surfaces is the normal to the third. The tangent to any co-ordinate curve through a point, being the line of intersection of the tangent planes to the two co-ordinate surfaces through the curve, lies along the normal to the third surface. We shall concerned with orthogonal co-ordinates only.

10.18.3. Condition for Orthogonality

Suppose that on solving

$$u = \phi_1(x, y, z),\ \ v = \phi_2(x, y, z),\ \ w = \phi_3(x, y, z)$$

for x, y, z we obtain

$$x = \psi_1(u, v, w),\ \ y = \psi_2(u, v, w),\ \ z = \psi_3(u, v, w).$$

We write

$$\begin{aligned}\mathbf{r} &= \mathbf{i}x + \mathbf{j}y + \mathbf{k}z\\ &= \mathbf{i}\psi_1(u, v, w) + \mathbf{j}\psi_2(u, v, w) + \mathbf{k}\psi_3(u, v, w)\\ &= \mathbf{f}(u, v, w), \text{ say.}\end{aligned}$$

The co-ordinate curve $v = v_0$, $w = w_0$ through (u_0, v_0, w_0) is the same as

$$\mathbf{r} = \mathbf{f}(u, v_0, w_0);$$

u being the parameter.

The tangent to this co-ordinate curve is parallel to the vector

$$\frac{\partial \mathbf{r}}{\partial u}.$$

The tangents to the other two co-ordinate curves are parallel to the vectors

$$\frac{\partial \mathbf{r}}{\partial v},\ \frac{\partial \mathbf{r}}{\partial w}.$$

Thus, for the orthogonality of the curvilinear co-ordinate system, we must have

$$\frac{\partial \mathbf{r}}{\partial u}\cdot\frac{\partial \mathbf{r}}{\partial v} = 0,\ \ \frac{\partial \mathbf{r}}{\partial v}\cdot\frac{\partial \mathbf{r}}{\partial w} = 0,\ \ \frac{\partial \mathbf{r}}{\partial w}\cdot\frac{\partial \mathbf{r}}{\partial u} = 0,$$

which are the required conditions.

10.19. DIFFERENTIAL OF LENGTH

We have

$$d\mathbf{r} = \frac{\partial \mathbf{r}}{\partial u} du + \frac{\partial \mathbf{r}}{\partial v} dv + \frac{\partial \mathbf{r}}{\partial w} dw. \qquad \text{(Lucknow 98)}$$

$$\therefore \quad d\mathbf{r} \cdot d\mathbf{r} = \left(\frac{\partial \mathbf{r}}{\partial u}\right)^2 (du)^2 + \left(\frac{\partial \mathbf{r}}{\partial v}\right)^2 (dv)^2 + \left(\frac{\partial \mathbf{r}}{\partial w}\right)^2 (dw)^2 +$$

$$2\frac{\partial \mathbf{r}}{\partial u} \cdot \frac{\partial \mathbf{r}}{\partial u} du\, dv + 2\frac{\partial \mathbf{r}}{\partial v} \cdot \frac{\partial \mathbf{r}}{\partial w} dv\, dw + 2\frac{\partial \mathbf{r}}{\partial w} \cdot \frac{\partial \mathbf{r}}{\partial u} dw\, du.$$

Hence,

$$(ds)^2 = \left(\frac{\partial \mathbf{r}}{\partial u}\right)^2 (du)^2 + \left(\frac{\partial \mathbf{r}}{\partial v}\right)^2 (dv)^2 + \left(\frac{\partial \mathbf{r}}{\partial w}\right)^2 (dw)^2 \qquad ...(1)$$

by virtue of the conditions for the orthogonality of the curvilinear co-ordinates.

If ds_1 denote the differential of length along the co-ordinate curve

$$v = v_0,\ w = w_0$$

so that

$$dv = 0, \quad dw = 0,$$

we have

$$(ds_1)^2 = \left(\frac{\partial \mathbf{r}}{\partial u}\right)^2 (du)^2. \qquad ...(2)$$

Similarly, for the other two coordinate curves.

10.20. FUNDAMENTAL TRIAD OF MUTUALLY ORTHOGONAL UNIT VECTORS THROUGH ANY POINT

By § 10.1.1, the vectors

$$\nabla u, \nabla v, \nabla w$$

lie along the normals to the co-ordinate surfaces which are the level surfaces of the functions u, v, w.

Because of the assumed orthogonality of the curvilinear co-ordinate system, we have

$$\nabla u \,.\, \nabla v = 0,\ \nabla v \,.\, \nabla w = 0,\ \nabla w \,.\, \nabla u = 0.$$

We suppose that the mutually orthogonal unit vectors

$$\frac{\nabla u}{|\nabla u|}, \frac{\nabla v}{|\nabla v|}, \frac{\nabla w}{|\nabla w|},$$

in this order, form a right handed system, for otherwise we could change the order of consideration of the three surfaces.

Now, by § 10.1.1, $|\nabla u|$ is the directional derivatives of u along the direction of the normal to the surface $u = u_0$, *i.e.*, along the tangent to the curve $v = v_0$, $w = w_0$. A ds_1 denotes the differential to length along this curve, we see that

$$|\nabla u| = \frac{du}{ds_1}$$

$$= \frac{1}{\sqrt{\left(\frac{\partial \mathbf{r}}{\partial u}\right)^2}}. \qquad \text{[By (2) above]}$$

We write

$$h_1 = \frac{1}{|\nabla u|} = \sqrt{\left(\frac{\partial \mathbf{r}}{\partial u}\right)^2},$$

$$h_2 = \frac{1}{|\nabla v|} = \sqrt{\left(\frac{\partial \mathbf{r}}{\partial v}\right)^2},$$

$$h_3 = \frac{1}{|\nabla w|} = \sqrt{\left(\frac{\partial \mathbf{r}}{\partial w}\right)^2},$$

and see that

$$\mathbf{a} = h_1 \nabla u, \quad \mathbf{b} = h_2 \nabla v, \quad \mathbf{c} = h_3 \nabla w$$

is a right handed system of mutually orthogonal unit vectors. These systems of vectors are different for different points.

10.21. DIFFERENTIAL OPERATORS IN TERMS OF ORTHOGONAL CURVILINEAR CO-ORDINATES

10.21.1. Gradient

Consider any scalar point function

$$\phi\,(u, v, w)$$

given in terms of orthogonal curvilinear co-ordinates u, v, w.

We think of u, v, w as functions of rectangular cartesian co-ordinates x, y, z. We have

$$\frac{\partial \phi}{\partial x} = \frac{\partial \phi}{\partial u}\frac{\partial u}{\partial x} + \frac{\partial \phi}{\partial v}\frac{\partial v}{\partial x} + \frac{\partial \phi}{\partial w}\frac{\partial w}{\partial x},$$

$$\frac{\partial \phi}{\partial y} = \frac{\partial \phi}{\partial u}\frac{\partial u}{\partial y} + \frac{\partial \phi}{\partial v}\frac{\partial v}{\partial y} + \frac{\partial \phi}{\partial w}\frac{\partial w}{\partial y},$$

$$\frac{\partial \phi}{\partial z} = \frac{\partial \phi}{\partial u}\frac{\partial u}{\partial z} + \frac{\partial \phi}{\partial v}\frac{\partial v}{\partial z} + \frac{\partial \phi}{\partial w}\frac{\partial w}{\partial z}.$$

Multiplying with the vectors **i**, **j**, **k** and adding, we get

$$\nabla\phi = \frac{\partial\phi}{\partial u}\nabla u + \frac{\partial\phi}{\partial v}\nabla v + \frac{\partial\phi}{\partial w}\nabla w$$

$$= \frac{1}{h_1}\frac{\partial\phi}{\partial u}\mathbf{a} + \frac{1}{h_2}\frac{\partial\phi}{\partial v}\mathbf{b} + \frac{1}{h_3}\frac{\partial\phi}{\partial w}\mathbf{c},$$

which is the required expression. Thus, the components of grad ϕ along the vectors **a, b, c** are

$$\frac{1}{h_1}\frac{\partial\phi}{\partial u}, \ \frac{1}{h_2}\frac{\partial\phi}{\partial v}, \ \frac{1}{h_3}\frac{\partial\phi}{\partial w}$$

respectively.

10.21.2. Divergence

Consider any vector point function

$$\mathbf{f}\ (u, v, w).$$

Let

$$\mathbf{f} = f_1\mathbf{a} + f_2\mathbf{b} + f_3\mathbf{c},$$

so that f_1, f_2, f_3 are the components of **f** along **a, b, c.**

We have

$$\mathbf{f} = f_1\mathbf{b} \times \mathbf{c} + f_2\mathbf{c} \times \mathbf{a} + f_3\mathbf{a} \times \mathbf{b}$$

$$= f_1 h_2 h_3\ \nabla v \times \nabla w + f_2 h_3 h_1\ \nabla w \times \nabla u + f_3 h_1 h_2\ \nabla u \times \nabla v.$$

$$\therefore \quad \nabla \cdot \mathbf{f} = \nabla \cdot (f_1 h_2 h_3\ \nabla v \times \nabla w) + \nabla \cdot (f_2 h_3 h_1\ \nabla w \times \nabla u) + \nabla \cdot (f_3 h_1 h_2\ \nabla u \times \nabla v) \quad (\S\ 10.14) \quad ...(1)$$

By § 10.15, III,

$$\nabla \cdot (f_1 h_2 h_3\ \nabla v \times \nabla w) = f_1 h_2 h_3 \cdot \nabla\ (\nabla v \times \nabla w) + \nabla v \times \nabla w \cdot \nabla\ (f_1 h_2 h_3) \quad ...(2)$$

By § 10.15, IV,

$$\nabla \cdot (\nabla v \times \nabla w) = \nabla w \cdot \text{curl}\ \nabla v - \nabla v \cdot \text{curl}\ \nabla w = 0, \quad ...(3)$$

for by § 10.16,

$$\text{curl}\ \nabla v = \text{curl grad}\ v = \mathbf{0},\ \text{curl}\ \nabla w = \mathbf{0}.$$

Also as in § 10.21,

$$\nabla\left(f_1 h_2 h_3\right) = \frac{\partial\left(f_1 h_2 h_3\right)}{\partial u}\nabla u + \frac{\partial\left(f_1 h_2 h_3\right)}{\partial v}\nabla v + \frac{\partial\left(f_1 h_2 h_3\right)}{\partial w}\nabla w. \quad ...(4)$$

From (2), (3) and (4)

$\nabla \,.\, (f_1 h_2 h_3 \, \nabla v \times \nabla w)$

$$= \nabla v \times \nabla w \cdot \nabla u + \frac{\partial (f_1 h_2 h_3)}{\partial u}$$

$$= \frac{\mathbf{b} \times \mathbf{c} \,.\, \mathbf{a}}{h_1 h_2 h_3} \frac{\partial (f_1 h_2 h_3)}{\partial u} = \frac{1}{h_1 h_2 h_3} \frac{\partial (f_1 h_2 h_3)}{\partial u}. \qquad ...(5)$$

Finally, from (1) and (5),

$$\nabla . \mathbf{f} = \frac{1}{h_1 h_2 h_3} \left[\frac{\partial (f_1 h_2 h_3)}{\partial u} + \frac{\partial (f_2 h_3 h_1)}{\partial v} + \frac{\partial (f_3 h_1 h_2)}{\partial w} \right],$$

which is the required expression for div **f**. *(Lucknow 97)*

10.21.3. Curl

Consider a vector point function **f** (u, v, w).

Let

$$\mathbf{f} = f_1 \mathbf{a} + f_2 \mathbf{b} + f_3 \mathbf{c} = h_1 f_1 \nabla u + h_2 f_2 \nabla v + h_3 f_3 \nabla w.$$

$$\therefore \quad \nabla \times \mathbf{f} = \nabla \times (h_1 f_1 \nabla u) + \nabla \times (h_2 f_2 \nabla v) + \nabla \times (h_3 f_3 \nabla w)$$

(§ 10.14) ...(1)

By § 10.16, V,

$$\nabla \times (h_1 f_1 \nabla u) = \nabla (h_1 f_1) \times \nabla v + h_1 f_1 \nabla \times \nabla u$$

$$= \nabla (h_1 f_1) \times \nabla u, \text{ for } \nabla \times \nabla u = 0, \qquad (\text{§ } 10.16, \text{ II})$$

$$= \left[\frac{\partial (h_1 f_1)}{\partial u} \nabla u + \frac{\partial (h_1 f_1)}{\partial v} \nabla v + \frac{\partial (h_1 f_1)}{\partial w} \nabla w \right] \times \nabla u$$

$$= \frac{\partial (h_1 f_1)}{\partial v} \nabla v \times \nabla u + \frac{\partial (h_1 f_1)}{\partial w} \nabla w \times \nabla u$$

$$= \frac{1}{h_1 h_2} \frac{\partial (h_1 f_1)}{\partial v} \mathbf{b} \times \mathbf{a} + \frac{1}{h_3 h_1} \frac{\partial (h_1 f_1)}{\partial w} \mathbf{c} \times \mathbf{a}$$

$$= -\frac{1}{h_1 h_2} \frac{\partial (h_1 f_1)}{\partial v} \mathbf{c} + \frac{1}{h_3 h_1} \frac{\partial (h_1 f_1)}{\partial w} \mathbf{b}. \qquad ...(2)$$

From (1) and (2),

$$\nabla \times \mathbf{f} = \frac{1}{h_2 h_3} \left[\frac{\partial (h_3 f_3)}{\partial v} - \frac{\partial (h_2 f_2)}{\partial w} \right] \mathbf{a} + \frac{1}{h_3 h_1} \left[\frac{\partial (h_1 f_1)}{\partial w} - \frac{\partial (h_3 f_3)}{\partial u} \right] \mathbf{b} + \frac{1}{h_1 h_2} \left[\frac{\partial (h_2 f_2)}{\partial u} - \frac{\partial (h_1 f_1)}{\partial v} \right] \mathbf{c}.$$

$$= \frac{1}{h_1 h_2 h_3} \begin{vmatrix} h_1 \mathbf{a} & h_2 \mathbf{b} & h_3 \mathbf{c} \\ \dfrac{\partial}{\partial u} & \dfrac{\partial}{\partial v} & \dfrac{\partial}{\partial w} \\ h_1 f_1 & h_2 f_2 & h_3 f_3 \end{vmatrix}$$

10.22.4. Laplacian, ∇^2

We have

$$\nabla^2\phi = \nabla \cdot \nabla\phi = \nabla \cdot \left[\frac{1}{h_1}\frac{\partial\phi}{\partial u}\mathbf{a} + \frac{1}{h_2}\frac{\partial\phi}{\partial v}\mathbf{b} + \frac{1}{h_3}\frac{\partial\phi}{\partial w}\mathbf{c}\right] \qquad (\S\ 10.22)$$

$$= \frac{1}{h_1h_2h_3}\left[\frac{\partial}{\partial u}\left(\frac{h_2h_3}{h_1}\frac{\partial\phi}{\partial u}\right) + \frac{\partial}{\partial v}\left(\frac{h_3h_1}{h_2}\frac{\partial\phi}{\partial v}\right) + \frac{\partial}{\partial w}\left(\frac{h_1h_2}{h_3}\frac{\partial\phi}{\partial w}\right)\right].$$

10.22.5. The Operator u.

∇; **u** being any vector.

We have seen that

$$\text{grad}\ \phi = \nabla\phi = \frac{1}{h_1}\frac{\partial\phi}{\partial u}\mathbf{a} + \frac{1}{h_2}\frac{\partial\phi}{\partial v}\mathbf{b} + \frac{1}{h_3}\frac{\partial\phi}{\partial w}\mathbf{c},$$

so that the directional derivatives of ϕ along the directions of **a, b, c** are

$$\frac{1}{h_1}\frac{\partial\phi}{\partial u}, \quad \frac{1}{h_2}\frac{\partial\phi}{\partial v}, \quad \frac{1}{h_3}\frac{\partial\phi}{\partial w}$$

respectively. Thus, if

$$\mathbf{u} = u_1\mathbf{a} + u_2\mathbf{b} + u_3\mathbf{c},$$

be any unit vector, the directional derivative $\mathbf{u} \cdot \nabla\phi$ of ϕ along **u** is

$$= \frac{u_1}{h_1}\frac{\partial\phi}{\partial u} + \frac{u_2}{h_2}\frac{\partial\phi}{\partial v} + \frac{u_3}{h_3}\frac{\partial\phi}{\partial w}$$

$$= \left(\mathbf{u}\cdot\frac{\mathbf{a}}{h_1}\frac{\partial}{\partial u} + \mathbf{u}\cdot\frac{\mathbf{b}}{h_2}\frac{\partial}{\partial v} + \mathbf{u}\cdot\frac{\mathbf{c}}{h_3}\frac{\partial}{\partial w}\right)\phi.$$

From this it follows that if **u** be any vector whatsoever, we have

$$\mathbf{u}\cdot\nabla = \frac{\mathbf{u}.\mathbf{a}}{h_1}\frac{\partial}{\partial u} + \frac{\mathbf{u}.\mathbf{b}}{h_2}\frac{\partial}{\partial v} + \frac{\mathbf{u}.\mathbf{c}}{h_3}\frac{\partial}{\partial w}.$$

10.23. SPECIAL CURVILINEAR SYSTEMS. SPHERICAL POLAR AND CYLINDRICAL POLAR SYSTEMS

Spherical Polar. The equations of transformation between the rectangular cartesian co-ordinates x, y, z and the spherical polar co-ordinates r, θ, ϕ are

$$x = r \sin\theta \cos\phi,\ y = r \sin\theta \cos\phi,\ z = r \cos\theta.$$

The co-ordinate surface $r = r_0$ is the sphere with its centre at the origin and radius r_0; the co-ordinate surface $\theta = \theta_0$ is the right circular semi-cone with its vertex at the origin and semi-vertical angle θ_0; finally the co-ordinate surface $\phi = \phi_0$ is the semi-plane through z-axis inclined at angle ϕ_0 to the XOZ plane.

We have

$$dx = \frac{\partial x}{\partial r}dr + \frac{\partial x}{\partial \theta}d\theta + \frac{\partial x}{\partial \phi}d\phi$$

$$= \sin\theta\cos\phi\, dr + r\cos\theta\cos\phi\, d\theta - r\sin\theta\sin\phi\, d\phi,$$

$$dy = \sin\theta\sin\phi\, dr + r\cos\theta\sin\phi\, d\theta + r\sin\theta\cos\phi\, d\phi,$$

$$dz = \cos\theta\, dr - r\sin\theta\, d\theta + 0\, d\phi.$$

$$\therefore \quad (ds)^2 = (dx)^2 + (dy)^2 + (dz)^2$$

$$= (dr)^2 + r^2 (d\theta)^2 + r^2\sin^2\theta\,(d\phi)^2.$$

Thus, for this system

$$h_1 = 1,\ h_2 = r,\ h_3 = r\sin\theta.$$

If ψ be any scalar point function, we obtain

$$\nabla^2\psi = \frac{1}{r^2\sin\theta}\left[\frac{\partial}{\partial r}\left(r^2\sin\theta\frac{\partial\psi}{\partial r}\right) + \frac{\partial}{\partial\theta}\left(\sin\theta\frac{\partial\psi}{\partial\theta}\right) + \frac{\partial}{\partial\phi}\left(\frac{1}{\sin\theta}\frac{\partial\psi}{\partial\phi}\right)\right]$$

[§ 10.22.4]

$$= \frac{\partial^2\psi}{\partial r^2} + \frac{1}{r^2}\frac{\partial^2\psi}{\partial\theta^2} + \frac{1}{\sin\theta}\frac{\partial^2\psi}{\partial\phi^2} + \frac{2}{r}\frac{\partial\psi}{\partial r} + \frac{\cot\theta}{r^2}\frac{\partial\psi}{\partial\theta}.$$

Cylindrical Polar. The equations of transformation between rectangular cartesian co-ordinates x, y, z and cylindrical polar co-ordinates r, θ, z are

$$x = r\cos\theta,\quad y = r\sin\theta,\quad z = z.$$

We may show that

$$(ds)^2 = (dr)^2 + r^2(d\theta)^2 + (dz)^2,$$

so that

$$h_1 = 1,\ h_2 = r,\ h_3 = 1.$$

Thus, in this case if ψ be any scalar point function,

$$\nabla^2\psi = \frac{1}{r}\left[\frac{\partial}{\partial r}\left(r\frac{\partial\psi}{\partial r}\right) + \frac{\partial}{\partial\theta}\left(\frac{1}{r}\frac{\partial\psi}{\partial\theta}\right) + \frac{\partial}{\partial z}\left(r\frac{\partial\psi}{\partial z}\right)\right]$$

$$= \frac{\partial^2\psi}{\partial r^2} + \frac{1}{r^2}\frac{\partial^2\psi}{\partial\theta^2} + \frac{\partial^2\psi}{\partial z^2} + \frac{1}{r}\frac{\partial\psi}{\partial r}.$$

Ex. Describe the co-ordinate surfaces in the case of cylindrical polar co-ordinates.

EXAMPLES

Example 1. *If u_1, u_2, u_3 are orthogonal curvilinear co-ordinates, show that $\partial\mathbf{r}/\partial u_1$, $\partial\mathbf{r}/\partial u_2$, $\partial\mathbf{r}/\partial u_3$ and ∇u_1, ∇u_2, ∇u_3 are reciprocal systems of vectors.*

Solution. We know that $\nabla u_1 = e_1/h_1$, $\nabla u_2 = e_2/h_2$, $\nabla u_3 = e_3/h_3$.

$\therefore\quad [\nabla u_1 + \nabla u_2 + \nabla u_3] = (\nabla u_1 \times \nabla u_2) . \nabla u_3$

$$= \left(\frac{1}{h_1}\mathbf{e}_1 \times \frac{1}{h_2}\mathbf{e}_2\right) \times \frac{1}{h_3}\mathbf{e}_3$$

$$= \frac{1}{h_1 h_2 h_3}\mathbf{e}_3 \cdot \mathbf{e}_3 = \frac{1}{h_1 h_2 h_3} \qquad \text{...(1)}$$

Also, $$\frac{\partial \mathbf{r}}{\partial u_1} = h_1 \mathbf{e}_1,\ \frac{\partial \mathbf{r}}{\partial u_2} = h_2 \mathbf{e}_2,\ \frac{\partial \mathbf{r}}{\partial u_3} = h_3 \mathbf{e}_3 \qquad \text{...(2)}$$

Now, $$\frac{\nabla u_2 \times \nabla u_3}{[\nabla u_1\ \nabla u_2\ \nabla u_3]} = \frac{(1/h_2)\,\mathbf{e}_2 \times (1/h_3)\,\mathbf{e}_3}{1/h_1 h_2 h_3}$$

$$= h_1 \mathbf{e}_1 = \frac{\partial \mathbf{r}}{\partial u_1} \qquad \text{by (2)}$$

Proceeding similarly, we get

$$\frac{\partial \mathbf{r}}{\partial u_2} = \frac{\nabla u_3 \times \nabla u_1}{[\nabla u_1\ \nabla u_2\ \nabla u_3]},\ \frac{\partial \mathbf{r}}{\partial u_3} = \frac{\nabla u_1 \times \nabla u_2}{[\nabla u_1\ \nabla u_2\ \nabla u_3]}$$

Hence, $\partial\mathbf{r}/\partial u_1$, $\partial\mathbf{r}/\partial u_2$, $\partial\mathbf{r}/\partial u_3$ *and* ∇u_1, ∇u_2, ∇u_3 form reciprocal system of vectors.

Example 2. *Express the vector $\mathbf{f} = z\mathbf{i} - 2x\mathbf{j} + y\mathbf{k}$ in (i) cylindrical polar, (ii) spherical polar co-ordinates.* *(Lucknow, 97)*

Solution. (*i*) Cartesian and cylindrical polar co-ordinates (x, y, z) and (ρ, ϕ, z) are related as follows :

$$x = \rho \cos\phi,\quad y = \rho\sin\phi,\quad z = z$$

$$\mathbf{i} = \cos\phi\,\mathbf{e}_\rho - \sin\phi\,\mathbf{e}_\phi,\quad \mathbf{j} = \sin\phi\,\mathbf{e}_\rho + \cos\phi\,\mathbf{e}_\phi,\quad k = k.$$

$\therefore\quad z\mathbf{i} - 2x\mathbf{j} + y\mathbf{k}$

$$= z(\cos\phi\,\mathbf{e}_\rho - \sin\phi\,\mathbf{e}_\phi) - 2\rho\cos\phi\,(\sin\phi\,\mathbf{e}_\rho + \cos\phi\,\mathbf{e}_\phi) + \rho\sin\phi\,\mathbf{k}$$

$$= z(\cos\phi - 2\rho\sin\phi\cos\phi)\,\mathbf{e}_\rho - (z\sin\phi + 2\rho\cos^2\phi)\,\mathbf{e}_\rho + \rho\sin\phi\,\mathbf{k}.$$

(*ii*) Cartesian and spherical polar co-ordinates (x, y, z) and (r, θ, ϕ) are related as follows :

$$x = r \sin\theta \cos\phi,\ \ y = r \sin\theta \sin\phi,\ \ z = r\cos\phi$$

and $\quad \mathbf{i} = \sin\theta\cos\phi\ \mathbf{e}_r + \cos\theta\cos\phi\ \mathbf{e}_\theta - \sin\phi\ \mathbf{e}_\phi,$

$\quad \mathbf{k} = \cos\theta\ \mathbf{e}_r - \sin\theta\ \mathbf{e}_\theta$

$\therefore \quad z\mathbf{i} - 2x\mathbf{j} + y\mathbf{k}$

$$= r\cos\theta\,(\sin\theta\cos\phi\ \mathbf{e}_r + \cos\theta\cos\phi\ \mathbf{e}_\theta - \sin\phi\ \mathbf{e}_\phi)$$
$$- 2r\sin\theta\cos\theta\,(\sin\theta\cos\phi\ \mathbf{e}_r + \cos\theta\sin\phi\ \mathbf{e}_\theta$$
$$+ \cos\phi\ \mathbf{e}_\phi) + r\sin\theta\cos\phi\,(\cos\theta\ \mathbf{e}_r - \sin\theta\sin\ \mathbf{e}_\theta)$$
$$= (r\sin\theta\cos\phi - 2r\sin^2\theta\sin\phi\cos\theta + r\sin\theta\sin\phi\cos\theta)$$
$$\mathbf{e}_r + (r\cos^2\theta\cos\phi - 2r\cos\theta\cos\phi\sin\theta\sin\phi$$
$$- r\sin^2\theta\sin\phi)\ \mathbf{e}_\phi$$
$$- (r\cos\theta\cos\phi - 2r\sin\theta\cos^2\phi)\ \mathbf{e}_\phi.$$

Example 3. *Transform the function* $\mathbf{f} = \rho\ \mathbf{e}_\rho + \rho\ \mathbf{e}_\phi$ *from cylindrical to cartesian system.*

Solution. Given $\quad \mathbf{f} = \rho\ \mathbf{e}_\rho + \rho\ \mathbf{e}_\phi$

Cylindrical and cartesian co-ordinates are related by

$$x = \rho\cos\phi,\ \ y = \rho\sin\phi,\ \ z = z$$

and $\quad \mathbf{e}_\rho = \cos\phi\ \mathbf{i} + \sin\phi\ \mathbf{j},\ \mathbf{e}_\phi = -\sin\phi\ \mathbf{i} + \cos\phi\ \mathbf{j}$

$\therefore \quad \mathbf{f} = \rho\,(\cos\phi\ \mathbf{i} + \sin\phi\ \mathbf{j}) + \rho\,(-\sin\phi\ \mathbf{i} + \cos\phi\ \mathbf{j})$

$$= x\mathbf{i} + y\mathbf{j} - y\mathbf{i} + x\mathbf{j}$$
$$= (x - y)\ \mathbf{i} + (x + y)\ \mathbf{j}.$$

EXERCISES

1. Show that the cylindrical and spherical co-ordinate systems (bases) are self-reciprocal.
2. Express wave equation $\dfrac{\partial^2 u}{\partial t^2} = c^2\nabla^2 u$ in spherical co-ordinates if u is independent of ϕ.
3. Prove that in any orthogonal curvilinear co-ordinate system div curl $\mathbf{f} = 0$ and curl grad $\mathbf{f} = 0$.
4. If $u_1 = 2x + 3,\ u_2 = y - 4,\ u_3 = z + 2$, show that u_1, u_2, u_3 are orthogonal.
5. If $x = u_1 u_2 \cos u_3,\ y = u_1 u_2 \sin u_3,\ z = \dfrac{1}{2}\left(u_1^2 - u_2^2\right)$, obtain h_1, h_2, h_3 and the unit vectors $\mathbf{e}_1, \mathbf{e}_2, \mathbf{e}_3$. Determine grad ψ, div $\mathbf{F}$, curl $\mathbf{F}$ and $\nabla^2\psi$ for the above system of co-ordinates.
6. If (r, θ, ϕ) are spherical co-ordinates, show that

   (*i*) $\nabla(1/r) = \nabla \times (\cos\theta\ \nabla\phi)$ $\qquad$ (*ii*) $\nabla\phi = \nabla \times (r \operatorname{cosec}\theta\ \nabla\theta)$

7. Express the following vectors in terms of (*i*) cylindrical polar (*ii*) spherical polar co-ordinates.

(*a*) $\mathbf{f} = 3y\mathbf{i} + x^2\mathbf{j} - z^2\mathbf{k}$

(*b*) $\mathbf{f} = xyz^2\mathbf{i} + yz\mathbf{j} + xy\mathbf{k}$

(*c*) $\mathbf{f} = xy\mathbf{i} + 2yz\,\mathbf{j} + (z^2 - yz)\,\mathbf{k}$.

8. Transform the following function from spherical to cartesian system :

$\mathbf{f} = 3ar^2 \sin\theta \cos\phi\, \mathbf{e}_r + 2a^2r \cos\theta \sin\phi\, \mathbf{e}_\theta + r^3\, \mathbf{e}_\phi$.

OBJECTIVE QUESTIONS

For each of the following questions, four alternatives are given for the answer. Only one of them is correct. Choose the correct alternative.

1. Value of grad r is

(*a*) $\mathbf{r}$ (*b*) $\hat{\mathbf{r}}$ (*c*) $-\dfrac{\mathbf{r}}{r^2}$ (*d*) $\dfrac{\mathbf{r}}{r^2}$

2. Value of $\nabla\,(1/r)$ is

(*a*) $\mathbf{r}$ (*b*) $\hat{\mathbf{r}}$ (*c*) $-\dfrac{\mathbf{r}}{r^2}$ (*d*) $\dfrac{-\mathbf{r}}{r^3}$

3. Value of $\mathbf{a}\cdot\nabla\left(\dfrac{1}{r}\right)$ is

(*a*) $-\dfrac{\mathbf{a}\,.\,\mathbf{r}}{r^2}$ (*b*) $\dfrac{\mathbf{a}\,.\,\mathbf{r}}{r^2}$ (*c*) $-\dfrac{\mathbf{a}\,.\,\mathbf{r}}{r^3}$ (*d*) $\dfrac{\mathbf{a}\,.\,\mathbf{r}}{r^3}$

4. Value of $\mathbf{f}(u)\,\nabla u$ is

(*a*) $u\,\nabla\mathbf{f}(u)$ (*b*) $\nabla u\,\mathbf{f}(u)$

(*c*) $\nabla\int\mathbf{f}(u)\,du$ (*d*) None of these

5. Value of grad ($\mathbf{r}\,.\,\mathbf{a}$) is

(*a*) $\mathbf{r}$ (*b*) $\mathbf{a}$ (*c*) 0 (*d*) $\hat{\mathbf{a}}\,.\,\hat{\mathbf{r}}$

6. A vector whose divergence is zero, is called

(*a*) Solenoidal vector (*b*) Irrotational vector

(*c*) Polar vector (*d*) Axial vector

7. If curl $\mathbf{a} = 0$ then $\mathbf{a}$ is

(*a*) Solenoidal (*b*) Irrotational

(*c*) Polar (*d*) Axial

8. Value of div $\mathbf{r}$ is (Garhwal 2001)

(*a*) 0 (*b*) 1 (*c*) 3 (*d*) 2

9. Value of curl $\mathbf{r}$ is (Garhwal 2002)

(*a*) 0 (*b*) $\hat{\mathbf{r}}$ (*c*) $\mathbf{r}$ (*d*) None of these

10. Value of curl grad **f** is

(*a*) $\nabla^2 \mathbf{f}$ (*b*) 0 (*c*) grad div **f** (*d*) None of these

11. Value of div $\hat{\mathbf{r}}$ is

(*a*) $1/r$ (*b*) $2/r$ (*c*) $-1/r$ (*d*) $1/r^2$

12. Value of div ($\mathbf{r} \times \mathbf{a}$), where **a** is a constant vector is

(*a*) 0 (*b*) $-2\mathbf{a}$ (*c*) $2\mathbf{a}$ (*d*) **a**

13. If **a** and **b** are irrotational, then $\mathbf{a} \times \mathbf{b}$ is

(*a*) irrotational

(*b*) solenoidal

(*c*) irrotational and solenoidal both

(*d*) neither irrotational and solenoidal

14. If **v** is a vector having fixed direction, then value of curl **v** is

(*a*) 0 (*b*) along **v**

(*c*) perpendicular to **v** (*d*) None of these

15. For the circular motion of a particle, we have $\mathbf{v} = \omega \times \mathbf{r}$, assuming that ω is constant, we have div $\mathbf{v}$ =

(*a*) 2ω (*b*) ω

(*c*) -2ω (*d*) 0

16. For the surface $\phi(x, y, z) = c$, c being a constant, grad ϕ represents

(*a*) position vector (*b*) vector tangent

(*c*) vector normal (*d*) None of these

17. Two surfaces $\phi(x, y, z) = c_1$ and $\phi(x, y, z) = c_2$ will cut orthogonally, if

(*a*) $\nabla\phi \, . \, \nabla\psi = 0$ (*b*) $\nabla^2\phi = 0, \nabla^2\psi = 0$

(*c*) $\nabla^2\phi = 0, \nabla^2\psi \neq 0$ (*d*) $\nabla^2\phi \neq 0, \nabla^2\psi = 0$

18. Curl of the velocity of any particle of a rigid body is equal to

(*a*) the linear velocity vector of the body

(*b*) the twice the linear velocity vector of the body

(*c*) the angular velocity vector of the body

(*d*) the twice the angular velocity vector of the body

19. Function $\mathbf{f} = \rho \mathbf{e}_\rho + \rho \mathbf{e}_\phi$, when transformed from cylindrical to cartesian system, yields

(*a*) $(x + y)\,\mathbf{i} + (x - y)\,\mathbf{j}$ (*b*) $(x - y)\,\mathbf{i} + (x + y)\,\mathbf{j}$

(*c*) $(x + y)\,\mathbf{i} - (x - y)\,\mathbf{j}$ (*d*) $(x - y)\,\mathbf{i} - (x + y)\,\mathbf{j}$

20. If $\mathbf{e}_\rho$, $\mathbf{e}_\phi$, $\mathbf{e}_z$, be the basis vectors along the co-ordinate curves in cylindrical co-ordinates, then which of the following is correct

(*a*) $\mathbf{e}_\rho \times \mathbf{e}_\phi = \mathbf{i}$ (*b*) $\mathbf{e}_\phi \times \mathbf{e}_z = \mathbf{k}$

(*c*) $\mathbf{e}_\phi \times \mathbf{e}_z = \cos\phi\,\mathbf{i} + \cos\phi\,\mathbf{j}$ (*d*) $\mathbf{e}_z \times \mathbf{e}_\rho = \sin\phi\,\mathbf{i} - \sin\phi\,\mathbf{j}$

21. If $\mathbf{e}_r$, $\mathbf{e}_\theta$, $\mathbf{e}_\phi$, denote the basis vectors for a spherical co-ordinate system, then

(*a*) curl curl $\mathbf{e}_r = \mathbf{0}$ (*b*) curl curl $\mathbf{e}_r = (\cot\theta/r^2)\,\mathbf{e}_r$

(*c*) $-(2/r^2)\,\mathbf{e}_r$ (*d*) $-(\text{cosec}^2\theta/r^2)\,\mathbf{e}_\theta$

ANSWERS

1. (*b*)	**2.** (*d*)	**3.** (*c*)	**4.** (*c*)	**5.** (*b*)
6. (*a*)	**7.** (*b*)	**8.** (*c*)	**9.** (*a*)	**10.** (*b*)
11. (*b*)	**12.** (*a*)	**13.** (*b*)	**14.** (*c*)	**15.** (*d*)
16. (*c*)	**17.** (*a*)	**18.** (*d*)	**19.** (*b*)	**20.** (*c*)
21. (*a*)				

11

Integral Transformations

Introduction. This chapter will deal with the different types of Line, Surface and Volume Integrals and with the transformation theorems of Gauss and Stokes. Conditions for vector point functions to be Irrotational and Solenoidal will also be obtained.

11.1. LINE INTEGRALS

(*Ajmer 98, 99*)

A curve $\mathbf{r} = \mathbf{f}(t)$ is called *smooth* if $\mathbf{f}(t)$ is continuously differentiable. In the following study of line integrals, it will be always understood that any curve which we consider is smooth or, at the most, it consists of a finite number of **connected** smooth arcs.

11.1.1. Tangential Line Integral

Let $\mathbf{F}(\mathbf{r})$ be a continuous vector point function and $\mathbf{r} = \mathbf{f}(t)$ a curve.

Consider any arc C of the curve enclosed between two points A, B and let a, b be the values of the parameter t for these points. The sense of description from A to B is a part of the name C of the arc and we denote the arc whose sense of description is from B to A by $-C$.

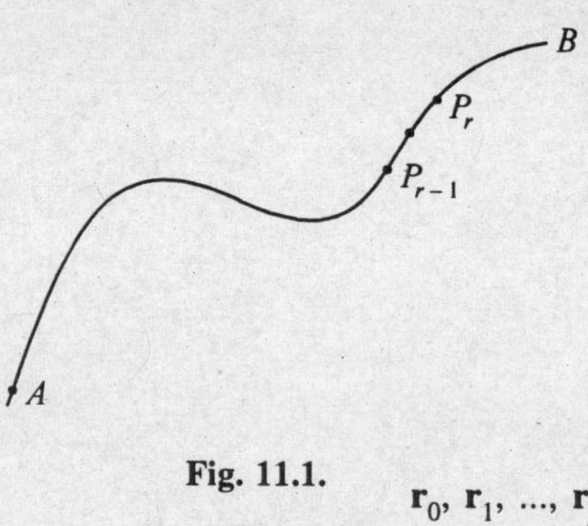

Fig. 11.1.

Take any system of points

$$P_0, P_1, ..., P_{p-1}, P_p, ..., P_n$$

on the arc C. The points $P_0, ..., P_n$ are the same as A and B respectively. Let

$$\mathbf{r}_0, \mathbf{r}_1, ..., \mathbf{r}_{p-1}, \mathbf{r}_p, ..., \mathbf{r}_n$$

be the position vectors of these points which divide the arc C in a finite number of sub-arcs.

We take arbitrary points, one on each sub-arc, with position vectors

$$\xi_1,\ \xi_2\ ...,\ \xi_{p-1}, \xi_p, ..., \xi_n.$$

Consider now the sum S of the scalar products given by

$$S = \sum_{p=1}^{n} \mathbf{F}(\xi_p) \cdot (\mathbf{r}_p - \mathbf{r}_{p-1}) = \sum_{p=1}^{n} \mathbf{F}(\xi_p) \cdot \delta \mathbf{r}_p,$$

where the summation extends to all the sub-arcs and

$$\delta \mathbf{r}_p = \mathbf{r}_p - \mathbf{r}_{p-1}.$$

Then the limit of this sum, if it exists and is independent of the arbitrary points ξ_p when n tends to infinity and each of

$$| \mathbf{r}_p - \mathbf{r}_{p-1} |$$

tends to zero, is called the *tangential line integral* of $\mathbf{F}(\mathbf{r})$ along C and is denoted by the symbol

$$\int \mathbf{F}(\mathbf{r}) \cdot d\mathbf{r} \text{ or simply } \int_C \mathbf{F} \cdot d\mathbf{r}.$$

Existence and Scalar Cartesian Expression. Introduce any system of rectangular axes.

Let

$$\mathbf{F}(\mathbf{r}) = \mathbf{F}(x, y, z) = \mathbf{i}\, F_1\,(x, y, z) + \mathbf{j}\, F_2\,(x, y, z) + \mathbf{k}\, F_3\,(x, y, z)$$

and

$$\mathbf{r}_p = \mathbf{i}x_p + \mathbf{j}y_p + \mathbf{k}z_p,$$

$$\xi_p = \mathbf{i}\alpha_p + \mathbf{j}\beta_p + \mathbf{k}\gamma_p.$$

$$\begin{aligned}
\therefore \quad S &= \Sigma \mathbf{F}\,(\xi_p)\,.\,(\mathbf{r}_p - \mathbf{r}_{p-1}) \\
&= \Sigma[\mathbf{i}F_1\,(\alpha_p, \beta_p, \gamma_p) + \mathbf{j}F_2\,(\alpha_p, \beta_p, \gamma_p) + \mathbf{k}F_3\,(\alpha_p, \beta_p, \gamma_p)]. \\
&\qquad [\mathbf{i}\,(x_p - x_{p-1}) + \mathbf{j}\,(y_p - y_{p-1}) + \mathbf{k}\,(z_p - z_{p-1})] \\
&= \Sigma F_1\,(\alpha_p, \beta_p, \gamma_p)\,(x_p - x_{p-1}) + \Sigma F_2\,(\alpha_p, \beta_p, \gamma_p)\,(y_p - y_{p-1}) + \\
&\qquad + \Sigma F_3\,(\alpha_p, \beta_p, \gamma_p)\,(z_p - z_{p-1}) \\
&= S_1 + S_2 + S_3, \text{ say.}
\end{aligned}$$

As $\mathbf{F}\,(x, y, z)$ is continuous, therefore, $F_1\,(x, y, z)$, $F_2\,(x, y, z)$, $F_3\,(x, y, z)$ are also continuous.

From Scalar Calculus, we know that under the given conditions, the sums S_1, S_2, S_3 tend to limits denoted by the symbols

$$\int_C F_1(x, y, z)\, dx,\ \int_C F_2(x, y, z)\, dy,\ \int_C F_3(x, y, z)\, dz.$$

Hence, lim S, also exists. Also we see that

$$\int_C \mathbf{F} \cdot d\mathbf{r} = \int_C (F_1 dx + F_2 dy + F_3 dz). \qquad ...(1)$$

If s denote the actual length of any point (x, y, z) on the curve from a fixed point, we have

$$\int_C (F_1 dx + F_2 dy + F_3 dz) = \int_{s_1}^{s_2} \left(F_1 \frac{dx}{ds} + F_2 \frac{dy}{ds} + F_3 \frac{dz}{ds} \right) ds$$

$$= \int_{s_1}^{s_2} \mathbf{F} \cdot \mathbf{t} \, ds. \qquad \text{...(2)}$$

Here s_1, s_2 are the values of s, for the extremities A, B of the arc C and

$$\mathbf{t} = \mathbf{i}\frac{dx}{ds} + \mathbf{j}\frac{dy}{ds} + \mathbf{k}\frac{dz}{ds},$$

is the unit tangent vector to the curve.

We also have

$$\int_C \mathbf{F} \cdot d\mathbf{r} = \int_C (F_1 dx + F_2 dy + F_3 dz)$$

$$= \int_a^b \left(F_1 \frac{dx}{dt} + F_2 \frac{dy}{dt} + F_3 \frac{dz}{dt} \right) dt$$

$$= \int_a^b \mathbf{F} \cdot \frac{d\mathbf{r}}{dt} dt. \qquad \text{...(3)}$$

The equations (2) and (3) express the line integral in terms of ordinary definite integrals of single variable.

11.1.2. Other Types of Line Integrals.

$$\int_C \mathbf{F} \times d\mathbf{r}, \ \int_C \phi d\mathbf{r},$$

where $\mathbf{F}$ is a continuous vector and ϕ a continuous scalar point function.

In terms of the notation of the preceding § 11.1.1, we consider the vector sum

$$S = \sum_{p=1}^{n} \mathbf{F}(\xi_p) \times (\mathbf{r}_p - \mathbf{r}_{p-1})$$

$$= \sum_{p=1}^{n} \mathbf{F}(\xi_p) \times \delta \mathbf{r}_p.$$

The limit of $\mathbf{S}$, when n tends to infinity and each of

$$| \mathbf{r}_p - \mathbf{r}_{p-1} |$$

tends to zero, if it exists and does not depend on the choice of ξ_p, is denoted by the symbol

$$\int_C \mathbf{F}(\mathbf{r}) \times d\mathbf{r} \text{ or simply } \int_C \mathbf{F} \times d\mathbf{r}.$$

Existence and expression in terms of components.

We have

$$\Sigma \mathbf{F}(\xi_p) \times (\mathbf{r}_p - \mathbf{r}_{p-1})$$

$$= \Sigma[\mathbf{i}F_1(\alpha_p, \beta_p, \gamma_p) + \mathbf{j}F_2(\alpha_p, \beta_p, \gamma_p) + \mathbf{k}F_3(\alpha_p, \beta_p, \gamma_p)] \times [\mathbf{i}(x_p - x_{p-1}) + \mathbf{j}(y_p - y_{p-1}) + \mathbf{k}(z_p - z_{p-1})]$$

$$= \Sigma \{[F_2(z_p - z_{p-1}) - F_3(y_p - y_{p-1})]\,\mathbf{i} + [F_3(x_p - x_{p-1}) - F_1(z_p - z_{p-1})]\,\mathbf{j} + [F_1(y_p - y_{p-1}) - F_2(x_p - x_{p-1})]\,\mathbf{k}\},$$

where, for the sake of brevity, we have written F_1, F_2, F_3 instead of $F_1(\alpha_p, \beta_p, \gamma_p)$, $F_2(\alpha_p, \beta_p, \gamma_p)$ and $F_3(\alpha_p, \beta_p, \gamma_p)$ respectively.

From scalar calculus, we know that under the given conditions

$$\Sigma [F_2(\alpha_p, \beta_p, \gamma_p)(z_p - z_{p-1}) - F_3(\alpha_p, \beta_p, \gamma_p)(y_p - y_{p-1})] \to \int_C (F_2 dz - F_3 dy),$$

$$\Sigma [F_3(\alpha_p, \beta_p, \gamma_p)(x_p - x_{p-1}) - F_1(\alpha_p, \beta_p, \gamma_p)(z_p - z_{p-1})] \to \int_C (F_3 dx - F_1 dz),$$

$$\Sigma [F_1(\alpha_p, \beta_p, \gamma_p)(y_p - y_{p-1}) - F_2(\alpha_p, \beta_p, \gamma_p)(x_p - x_{p-1})] \to \int_C (F_1 dy - F_2 dx).$$

Hence, lim **S** exists and we have

$$\int_C \mathbf{F} \times d\mathbf{r} = \mathbf{i}\int_C (F_2 dz - F_3 dy) + \mathbf{j}\int_C (F_3 dx - F_1 dz) + \mathbf{k}\int_C (F_1 dy - F_2 dx).$$

As before, it can also be seen that

$$\int_C \mathbf{F} \times d\mathbf{r} = \int_{s_1}^{s_2} \mathbf{F} \times \mathbf{t}\, ds.$$

Finally, in terms of the notation adopted above,

$$\int_C \phi d\mathbf{r} = \lim \Sigma \phi(\xi_p)(\mathbf{r}_p - \mathbf{r}_{p-1})$$

and we may show that

$$\int_C \phi d\mathbf{r} = \mathbf{i}\int_C \phi dx + \mathbf{j}\int_C \phi dy + \mathbf{k}\int \phi dz.$$

11.2. CIRCULATION, IRROTATIONAL VECTOR POINT FUNCTIONS

Def. 1. *The line integral*

$$\int_C \mathbf{F} \cdot d\mathbf{r}$$

of a continuous vector point function **F** *along a closed curve C is called the circulation of* **F** *along C.*

Def. 2. *A vector point function is said to be irrotational in a region, if its circulation along every closed curve in the region is zero.*

We shall now prove a theorem which gives a necessary and sufficient condition for a vector function to be irrotational.

11.2.1. Theorem.

The necessary and sufficient condition for a continuous vector point function to be irrotational in a simply connected region R is that it is the gradient of a scalar point function.

The condition is necessary. Let **F** be irrotational in a simply connected region R. Take any two points A and P in the region and join them by any two curves ALP and AMP in the region. Let these be denoted by C_1 and C_2 and let C denote the closed curve $ALPMA$.

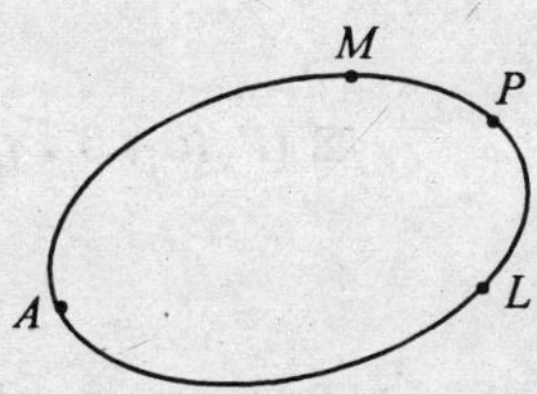

Fig. 11.2.

We have

$$0 = \int_C \mathbf{F} \cdot d\mathbf{r} = \int_{C_1} \mathbf{F} \cdot d\mathbf{r} + \int_{-C_2} \mathbf{F} \cdot d\mathbf{r}$$

$$= \int_{C_1} \mathbf{F} \cdot d\mathbf{r} - \int_{C_2} \mathbf{F} \cdot d\mathbf{r}.$$

$$\therefore \qquad \int_{C_1} \mathbf{F} \cdot d\mathbf{r} = \int_{C_2} \mathbf{F} \cdot d\mathbf{r}.$$

Thus, the line integral

$$\int \mathbf{F} \cdot d\mathbf{r},$$

along any curve joining A to P depends only on the point B if we keep A fixed. We write

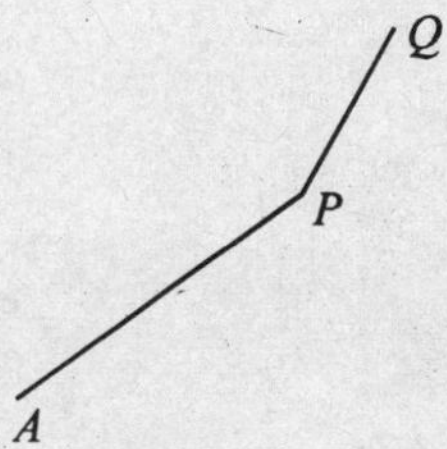

Fig. 11.3.

$$\phi(P) = \int_A^P \mathbf{F} \cdot d\mathbf{r}.$$

Take a point Q near P and join PQ by a straight line. Let **a** be the unit vector along AQ. We have

$$\phi(Q) = \int_A^Q \mathbf{F} \cdot d\mathbf{r} = \int_A^P \mathbf{F} \cdot d\mathbf{r} + \int_P^Q \mathbf{F} \cdot d\mathbf{r}.$$

$$\therefore \quad \phi(Q) - \phi(P) = \int_P^Q \mathbf{F} \cdot d\mathbf{r} = \int_0^{PQ} \mathbf{F} \cdot \frac{d\mathbf{r}}{ds} ds,$$

where s denotes the distance of any point on PQ from P.

$$\therefore \quad \phi(Q) - \phi(P) = \int_0^{PQ} \mathbf{F} \cdot \mathbf{a}\, ds$$

$= [\mathbf{F}(R) \,.\, \mathbf{a}]\, PQ$, by Mean Value Theorem

R being a point of the line PQ.

$$\therefore \quad \frac{\phi(Q) - \phi(P)}{PQ} = \mathbf{F}(R) \cdot \mathbf{a}$$

Let $Q \to P$ so that $R \to P$.

$$\therefore \quad \lim \frac{\phi(Q) - \phi(P)}{PQ} = \mathbf{F}(P) \cdot \mathbf{a}.$$

But

$$\lim \frac{\phi(Q) - \phi(P)}{PQ} = \mathbf{a} \nabla \phi,$$

being the directional derivative of ϕ along the direction of $\mathbf{a}$. Thus, for any point P,

$$\mathbf{a} \,.\, \nabla \phi = \mathbf{a} \,.\, \mathbf{F}$$

or

$$\mathbf{a} \,.\, (\nabla \phi - \mathbf{F}) = 0$$

i.e.,

$$\nabla \phi = \mathbf{F};$$

since $\mathbf{a}$ is a any arbitrary vector.

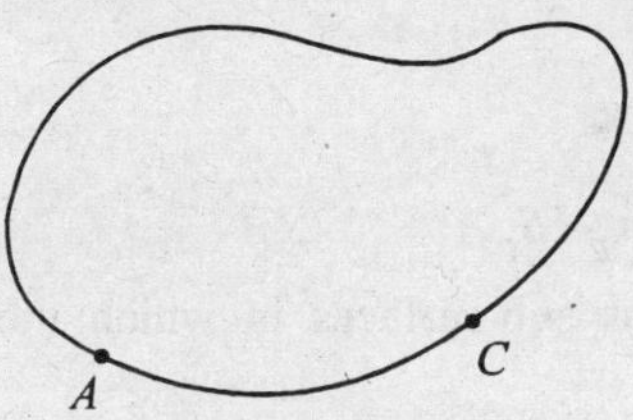

Fig. 11.4.

The condition is sufficient.

Suppose now that

$$\mathbf{F} = \nabla \phi.$$

If C be any closed curve,

$$\int_C \mathbf{F} \cdot d\mathbf{r} = \int_C \left(\frac{\partial \phi}{\partial x} \frac{dx}{ds} + \frac{\partial \phi}{\partial y} \frac{dy}{ds} + \frac{\partial \phi}{\partial z} \frac{dz}{ds} \right) ds$$

$$= \int_C \frac{d\phi}{ds} ds = \left| \phi \right|_A^A = \phi(A) - \phi(A) = 0.$$

A being any point on the curve C.

Note 1. The student should notice that the condition that line integral along every closed curve is zero is equivalent to the condition that the line integral along a curve joining any two given points is independent of the particular curve and depends on the position vectors of end points only.

Note. 2. Later on we shall obtain another form of necessary and sufficient condition for a vector point function to be irrotational.

11.3. SURFACE INTEGRALS

A surface $\mathbf{r} = \mathbf{f}(u, v)$ is called smooth if $\mathbf{f}(u, v)$ possesses continuous first order partial derivatives. In the following, it will always be understood that any surface which we consider is smooth or it consists of a finite number of connected smooth pieces.

11.3.1. Normal Surface Integral

Let $\mathbf{F}(\mathbf{r})$ be a continuous vector point function and $\mathbf{r} = \mathbf{f}(u, v)$ a surface.

Consider any portion S of the surface, closed or not. We shall distinguish between the two sides of the surface and suppose that the particular side considered is a part of the name S.

Divide S into any finite number of sub-surfaces and consider any sub-surface. Take a point ξ_p, in the sub-surface and let $\mathbf{n}_p$ denote the unit vector along the normal to the sub-surface at ξ_p drawn on the side of S.

Let δS_p denote the magnitude of the area of the sub-surface. Then

$$\delta \mathbf{a}_p = \mathbf{n}_p \, \delta S_p,$$

denotes the vector area of the sub-surface.

Consider now the sum

$$\Sigma \, \mathbf{F}(\xi_p) \cdot \delta \mathbf{a}_p = \Sigma \, \mathbf{F}(\xi_p) \cdot \mathbf{n}_p \, \delta S_p,$$

where the summation extends to the various sub-surfaces in which the surface S has been divided.

The limit of this sum when the number of sub-surfaces tends to infinity and the area of each sub-surface tends to zero is called the *normal surface integral* of $\mathbf{F}(\mathbf{r})$ over S and is denoted by

$$\int_S \mathbf{F}(\mathbf{r}) \cdot d\mathbf{a} = \int_S \mathbf{F}(\mathbf{r}) \cdot \mathbf{n} \, dS$$

or simply

$$\int_S \mathbf{F} \cdot d\mathbf{a} = \int_S \mathbf{F} \cdot \mathbf{n} \, dS.$$

The value of the integral will change in sign but not in magnitude when we carry the integral on one side than on the other, for in that case, in each term the sign of $\mathbf{n}_p$ would change.

Existence and Scalar Cartesian Expression

Let

$$\mathbf{F}(\mathbf{r}) = \mathbf{F}\,(x, y, z) = \mathbf{i}\,F_1(x, y, z) + \mathbf{j}\,F_2(x, y, z) + \mathbf{k}\,F_3(x, y, z),$$

$$\xi_p = \mathbf{i}\alpha_p + \mathbf{j}\beta_p + \mathbf{k}\gamma_p,$$

$$\delta\mathbf{a}_p = \mathbf{n}_p\delta S_p = \mathbf{i}\delta A_p + \mathbf{j}\delta B_p + \mathbf{k}\delta C_p.$$

Here δA_p, δB_p, δC_p, denote the projections of the vector area $\mathbf{n}_p\delta S_p$ on the three co-ordinate planes.

$$\therefore \quad \Sigma\mathbf{F}(\xi_p)\,.\,\delta A_p = \Sigma\,[F_1(\alpha_p, \beta_p, \gamma_p)\,\delta A_p + F_2(\alpha_p, \beta_p, \gamma_p)\,\delta B_p + F_3(\alpha_p, \beta_p, \gamma_p)\,\delta C_p] \qquad ...(1)$$

Now from Scalar Calculus, the sum on the right of (1) tends, under the conditions stated, to the limit.

$$\int_S \int \left[F_1(x, y, z)\,dydz + F_2(x, y, z)\,dzdx + F_3(x, y, z)\,dxdy\right] \qquad ...(2)$$

Thus,

$$\int_S \mathbf{F}\cdot d\mathbf{a} = \int_S \mathbf{F}\cdot\mathbf{n}\,dS.$$

exists and equals (2).

By the rule of change of variables for integrals, the integral (2) can be re-written as an ordinary double integral

$$\iint\left\{F_1\frac{\partial(y, z)}{\partial(u, v)} + F_2\frac{\partial(z, x)}{\partial(u, v)} + F_3\frac{\partial(x, y)}{\partial(u, v)}\right\} dudv$$

carried over the region in the u-v plane which corresponds to the surface S given by $\mathbf{r} = \mathbf{f}\,(u, v)$.

Here $\dfrac{\partial(y, z)}{\partial(u, v)} = \begin{vmatrix} \dfrac{\partial y}{\partial u} & \dfrac{\partial y}{\partial v} \\ \dfrac{\partial z}{\partial u} & \dfrac{\partial z}{\partial v} \end{vmatrix}$ and so on.

11.3.2. Other Forms of Surface Integrals

$$\int_S \mathbf{F}\times d\mathbf{a}, \quad \int_S \phi d\mathbf{a}$$

where **F** is a continuous vector and ϕ a continuous scalar point function.

In terms of the notation of the preceding sub-section § 11.3.1, we have

$$\int_S \mathbf{F}\times d\mathbf{a} = \lim \Sigma\mathbf{F}\left(\xi_p\right)\times\delta\mathbf{a}_p$$

and

$$\int_S \phi d\mathbf{a} = \lim \Sigma\, \phi\left(\xi_p\right) \delta \mathbf{a}_p.$$

It can be shown that under the conditions stated, these integrals exist and

$$\int_S \mathbf{F} \times d\mathbf{a} = \mathbf{i} \int_S \int (F_2 dydz - F_3 dzdx) +$$

$$\mathbf{j} \int_S \int (F_3 dydz - F_1 dxdy) + \mathbf{k} \int_S \int (F_1 dzdx - F_2 dydx)$$

$$\int_S \phi d\mathbf{a} = \mathbf{i} \int_S \int \phi dydz + \mathbf{j} \int_S \int \phi dzdx + \mathbf{k} \int_S \int \phi dxdy.$$

The proofs are left as exercises for the reader.

11.4. FLUX ACROSS A SURFACE : SOLENOIDAL VECTOR POINT FUNCTIONS

Def. 1. The normal surface integral

$$\int_S \mathbf{F} \cdot d\mathbf{a},$$

of a continuous vector point function **F** *over a closed surface S is called the Flux of* **F** *across S.*

Def. 2. *A vector point function is said to be solenoidal in a region if its flux across every closed surface in the region is zero.*

Note. We shall later on obtain a necessary and sufficient condition for a vector function to be solenoidal in a region.

11.5. VOLUME INTEGRALS

$$\int_V \mathbf{F}\, dv.$$

Consider a continuous vector point function **F**(**r**) and a region V enclosed by a surface $\mathbf{r} = \mathbf{f}\,(u, v)$.

Divide the region into a finite number of sub-regions and consider any sub-region. Let δv_p denote the volume of this sub-region and ξ_p the position vector of any point of the same.

Consider the sum

$$\Sigma\, \mathbf{F}\, (\xi_p)\, \delta v_p,$$

where the summation extends to all the sub-regions. The limit of this sum if it exists when the number of sub-regions tends to infinity and the volume of each sub-region tends to zero is denoted by the symbol

$$\int_V \mathbf{F}(\mathbf{r})\, dv \text{ or briefly } \int_V \mathbf{F}\, dv.$$

Introducing rectangular cartesian co-ordinates, we may see that the limit, in question, exists and

$$\int_V \mathbf{F}\, dv = \mathbf{i} \iiint_V F_1 dxdydz + \mathbf{j} \iiint_V F_2 dxdydz + \mathbf{k} \iiint_V F_3 dxdydz.$$

EXAMPLES

Example 1. *Evaluate*

$$\int_C \mathbf{F} \cdot d\mathbf{r},$$

where

$$\mathbf{F} = c\,[-3a \sin^2 \theta \cos \theta\, \mathbf{i} + a\,(2 \sin \theta - 3 \sin^3 \theta)\, \mathbf{j} + b \sin 2\theta\, \mathbf{k}]$$

and the curve C is given by

$$\mathbf{r} = a \cos \theta\, \mathbf{i} + a \sin \theta\, \mathbf{j} + b\theta\, \mathbf{k};$$

θ varying from $\pi/4$ to $\pi/2$.

Solution. We have

$$\frac{d\mathbf{r}}{d\theta} = -a \sin \theta \mathbf{i} + a \cos \theta \mathbf{j} + b\mathbf{k}$$

$$\therefore \quad \int_C \mathbf{F} \cdot d\mathbf{r} = \int_{\pi/4}^{\pi/2} \mathbf{F} \cdot \frac{d\mathbf{r}}{d\theta}\, d\theta \qquad \text{[Refer (3), § 11.1.1]}$$

$$= c\left(a^2 + b^2\right) \int_{\pi/4}^{\pi/2} \sin 2\theta\, d\theta = \frac{1}{2} c\left(a^2 + b^2\right).$$

Example 2. *Evaluate*

$$\int_C \mathbf{F} \cdot d\mathbf{r}$$

where $\mathbf{F} = (x^2 + y^2)\, \mathbf{i} - 2xy\mathbf{j}$ and the curve C is the rectangle in the xy-plane bounded by $y = 0$, $x = a$, $y = b$, $x = 0$.

Solution. In the xy-plane $z = 0$.

$$\therefore \quad \mathbf{r} = x\mathbf{i} + y\mathbf{j}$$

$$d\mathbf{r} = dx\mathbf{i} + dy\mathbf{j}$$

$$\therefore \quad \int_C \mathbf{F} \cdot d\mathbf{r}$$

$$= \int_C \left\{\left(x^2 + y^2\right) dx - 2xy\, dy\right\} \qquad \text{...(1)}$$

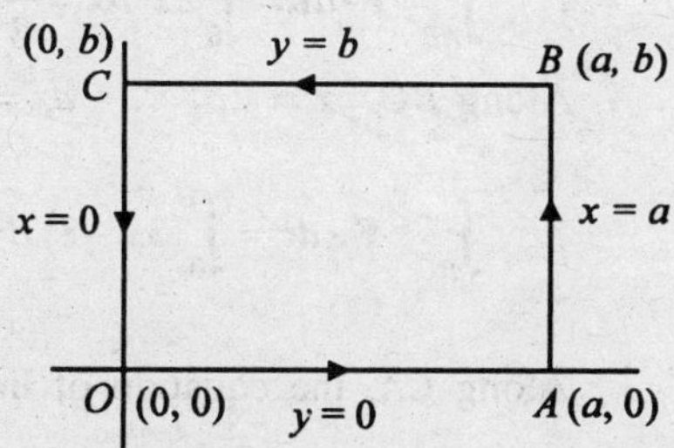

Fig. 11.5.

$$\int_C \mathbf{F}\cdot d\mathbf{r} = \int_{OA} \mathbf{F}\cdot d\mathbf{r} + \int_{AB} \mathbf{F}\cdot d\mathbf{r} + \int_{BC} \mathbf{F}\cdot d\mathbf{r} + \int_{CO} \mathbf{F}\cdot d\mathbf{r} \quad ...(2)$$

Along OA, $y = 0$ $\therefore$ $dy = 0$ and x varies from 0 to a.

Along AB, $x = a$ $\therefore$ $dx = 0$ and y varies from 0 to b.

Along BC, $y = b$ $\therefore$ $dy = 0$ and x varies from a to 0.

Along CO, $x = 0$ $\therefore$ $dx = 0$ and y varies from b to 0.

Hence, from (1) and (2),

$$\int_C \mathbf{F}\cdot d\mathbf{r} = \int_0^a x^2 dx - \int_0^b 2ay\, dy + \int_a^0 \left(x^2 + b^2\right) dx + \int_b^0 0.dy$$

$$= \frac{a^3}{3} - 2a\frac{b^2}{2} + \left[\frac{x^3}{3} + b^2 x\right]_a^0$$

$$= \frac{a^3}{3} - ab^2 - \frac{a^3}{3} - b^2 a = -2ab^2.$$

Example 3. *If* $\mathbf{F} = \left(2x^2 + y^2\right)\mathbf{i} + (3y - 4x)\mathbf{j}$, *evaluate* $\int_C \mathbf{F}\cdot d\mathbf{r}$ *around the triangle ABC whose vertices are A (0, 0); B = (2, 0); C = (2, 1).*

Solution. $\int_C \mathbf{F}\cdot d\mathbf{r}$

$$= \int_C \left(2x^2 + y^2\right) dx + (3y - 4x)\, dy \quad ...(1)$$

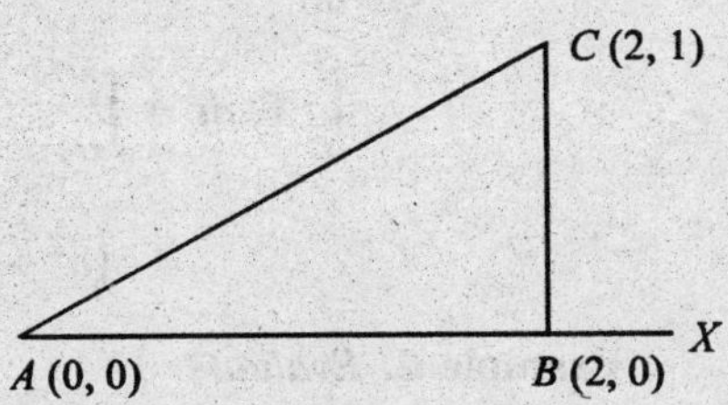

Fig. 11.6.

$r = x\mathbf{i} + y\mathbf{j}$

so, $d\mathbf{r} = dx\mathbf{i} + dy\mathbf{j}$

Along AB, $y = 0$, $\therefore$ $dy = 0$ and x varies from 0 to 2.

$$\therefore \int_{AB} \mathbf{F}\cdot d\mathbf{r} = \int_0^2 2x^2 dx = \frac{2}{3}\left[x^3\right]_0^2 = \frac{16}{3}.$$

Along BC, $x = 2$, $\therefore$ $dx = 0$ and y varies from 0 to 1.

$$\therefore \int_{BC} \mathbf{F}\cdot d\mathbf{r} = \int_0^1 (3y - 8)\, dy = \left[\frac{3y^2}{2} - 8y\right]_0^1 = -\frac{13}{2}$$

Along CA, the equation of lines is $y = \frac{1}{2}x$ or $x = 2y$.

$\therefore$ $dx = 2dy$ and y varies from 1 to 0.

$$\therefore \quad \int_C \mathbf{F}\cdot d\mathbf{r} = \int_1^0 \left[2(2y)^2 + y^2\right] 2dy + \left[3y - 4(2y)\right] dy$$

$$= \int_1^0 \left(18y^2 - 5y\right) dy = \left[6y^3 - \frac{5y^2}{2}\right]_1^0 = -\frac{7}{2}$$

$$\therefore \quad \int_C \mathbf{F}\cdot d\mathbf{r} = \frac{16}{3} - \frac{13}{2} - \frac{7}{2} = -\frac{14}{3}$$

Example 4. *Find the circulation of* **F** *round the curve C, where* $\mathbf{F} = \left(2x + y^2\right)\mathbf{i} + (3y - 4x)\mathbf{j}$ *and C is the curve* $y = x^2$ *from (0, 0) to (1, 1) and the curve* $y^2 = x$ *from (1, 1) to (0, 0).*

Solution.

$$\int_C \mathbf{F}\cdot d\mathbf{r}$$

$$= \int_C \left[\left(2x + y^2\right) dx + (3y - 4x)\, dy\right]$$

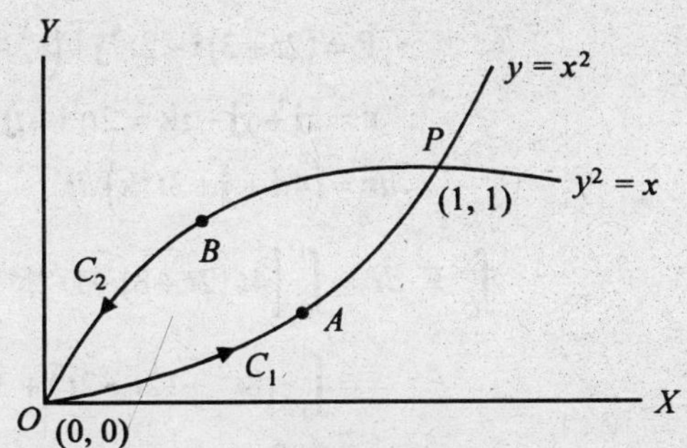

Fig. 11.7.

We have to evaluate this integral along two curves C_1 and C_2.

For C_1, $y = x^2$ $\therefore$ $dy = 2xdx$ and x varies from 0 to 1.

For C_2, $x = y^2$ $\therefore$ $dx = 2ydy$ and y varies from 1 to 0.

$$\int_{C_1} \mathbf{F}\cdot d\mathbf{r} = \int_0^1 \left(2x + x^4\right) dx + \left(3x^2 - 4x\right) 2xdx$$

$$= \int_0^1 \left(2x - 8x^2 + 6x^3 + x^4\right) dx$$

$$= 2\cdot\frac{1}{2} - 8\cdot\frac{1}{3} + 6\cdot\frac{1}{4} + \frac{1}{5} = \frac{1}{30}$$

$$\int_{C_2} \mathbf{F}\cdot d\mathbf{r} = \int_1^0 \left(2y^2 + y^2\right) 2y\, dy + \left(3y - 4y^2\right) dy$$

$$= \int_1^0 \left(3y - 4y^2 + 6y^3\right) dy$$

$$= -\left[\frac{3}{2} - \frac{4}{3} + \frac{6}{4}\right] = \frac{5}{3}$$

$$\therefore \quad \int_C \mathbf{F}\cdot d\mathbf{r} = \int_{C_1} \mathbf{F}\cdot d\mathbf{r} + \int_{C_2} \mathbf{F}\cdot d\mathbf{r}$$

$$= \frac{1}{30} - \frac{5}{3} = -\frac{49}{30}.$$

Example 5. *If* $\mathbf{F} = (2y+3)\mathbf{i} + xz\,\mathbf{j} + (yz-x)\,\mathbf{k}$, *evaluate* $\int_C \mathbf{F}\cdot d\mathbf{r}$ *along the following paths C :*

(*a*) $x = 2t^2,\ y = t,\ z = t^3$ *from* $t = 0$ *to* $t = 1$.

(*b*) *The straight lines from (0, 0, 0) to (0, 0, 1) then to (0, 1, 1) and then to (2, 1, 1).*

(*c*) *The straight line joining (0, 0, 0) to (2, 1, 1).*

Solution. (*a*) On putting the values of *x, y, z* in terms of *t*.

$$\mathbf{F} = (2t+3)\mathbf{i} + 2t^5\mathbf{j} + \left(t^4 - 2t^2\right)\mathbf{k}$$

$$\mathbf{r} = x\mathbf{i} + y\mathbf{j} + z\mathbf{k} = 2t^2\mathbf{i} + t\mathbf{j} + t^3\mathbf{k}$$

$$\therefore \quad d\mathbf{r} = \left(4t\mathbf{i} + \mathbf{j} + 3t^2\mathbf{k}\right) dt.$$

$$\int_C \mathbf{F}\cdot d\mathbf{r} = \int_0^1 \left[4t\,(2t+3) + 1\cdot 2t^5 + 3t^2\left(t^4 - 2t^2\right)\right] dt$$

$$= \int_0^1 \left(8t^2 + 12t + 2t^5 + 3t^6 - 6t^4\right) dt$$

$$= \frac{288}{35}.$$

(*b*) $$\int_C \mathbf{F}\cdot d\mathbf{r} = \int_C (F_1 dx + F_2 dy + F_3 dz)$$

$$= \int \left[(2y+3)\,dx + xz\,dy + (yz-x)\,dz\right] \qquad \text{...(1)}$$

Along C_1 the line joining (0, 0, 0) to (0, 0, 1), $x = 0,\ y = 0$.

$\therefore\ dx = 0,\ dy = 0$ and z varies from 0 to 1.

$$\therefore \quad \int_C \mathbf{F}\cdot d\mathbf{r} = \int_0^1 (yz-x)\,dz = 0, \ \because\ x = 0,\ y = 0.$$

Along C_2 the line joining (0, 0, 1) to (0, 1, 1), $x = 0,\ z = 1$.

$\therefore\ dx = 0,\ dz = 0$ and y varies from 0 to 1.

$$\therefore \quad \int_2 \mathbf{F}\cdot d\mathbf{r} = \int_0^1 xz\,dy = 0, \ \because\ x = 0.$$

Along C_3, the line joining (0, 1, 1) to (2, 1, 1), $y = 1,\ z = 1$.

$\therefore\ dy = 0,\ dz = 0$ and x varies from 0 to 2.

$$\therefore \quad \int_{C_3} \mathbf{F}\cdot d\mathbf{r} = \int_0^2 (2y+3)\,dx = \int_0^2 5dx$$

$$= 10.$$

$$\therefore \quad \int_C \mathbf{F}\cdot d\mathbf{r} = \int_{C_1} \mathbf{F}\cdot d\mathbf{r} + \int_{C_2} \mathbf{F}\cdot d\mathbf{r} + \int_{C_3} \mathbf{F}\cdot d\mathbf{r}$$

$$= 0 + 0 + 10 = 10.$$

(*c*) The parametric equations of the line joining (0, 0, 0) to (2, 1, 1) is $\frac{x}{2} = \frac{y}{1} = \frac{z}{1} = t.$

or $\quad x = 2t, \; y = t, \; z = t$ and t varies from 0 to 1.

$$\therefore \quad \int_C \mathbf{F}\cdot d\mathbf{r} = \int_C (2y+3)\, dx + xz\, dy + (yz - x)\, dz$$

$$= \int_0^1 \left[(2t+3)\,2 + 2t^2 + \left(t^2 - 2t\right)\right] dt$$

$$= \int_0^1 \left(3t^2 + 2t + 6\right) dt$$

$$= \left[t^3 + t^2 + 6t\right]_0^1 = 8.$$

Example 6. *Evaluate*

$$\int_S \frac{\mathbf{r}}{r^3} \cdot d\mathbf{a},$$

where S denotes the sphere of radius a with centre at the origin.

Solution. At any point of S,

$$\mathbf{n} = \frac{\mathbf{r}}{r},$$

for the normal lies along the line joining the centre to the point.

$$\therefore \quad \int_S \frac{\mathbf{r}}{r^3} \cdot d\mathbf{a} = \int_S \frac{\mathbf{r}}{r^3} \cdot \frac{\mathbf{r}}{r}\, dS$$

$$= \int_S \frac{\mathbf{r}^2}{r^4} dS = \frac{1}{a^2} \int_S dS = \frac{1}{a} \times 4\pi a = 4\pi.$$

Example 7. *Evaluate the surface integral*

$$\int_S (yz\mathbf{i} + zx\mathbf{j} + xy\mathbf{k}) \cdot d\mathbf{a}, \qquad \text{(Avadh 2004)}$$

where S is the surface of the sphere $x^2 + y^2 + z^2 = 1$ in the first octant.

Solution. The sphere is parametrically given by

$$x = \sin\theta \cos\phi, \;\; y = \sin\theta \sin\phi, \;\; z = \cos\theta$$

and the part in the first octant corresponds to

$$0 \le \theta \le \frac{1}{2}\pi, \;\; 0 \le \phi \le \frac{1}{2}\pi.$$

Thus, the required surface integral

$$= \iint_S (yz\, dy\, dz + zx\, dz\, dx + xy\, dx\, dy)$$

$$= \int_0^{\frac{1}{2}\pi} \int_0^{\frac{1}{2}\pi} \left[yz \frac{\partial(y, z)}{\partial(\theta, \phi)} + zx \frac{\partial(z, x)}{\partial(\theta, \phi)} + xy \frac{\partial(x, y)}{\partial(\theta, \phi)} + \right] d\theta\, d\phi$$

Now

$$\frac{\partial(y, z)}{\partial(\theta, \phi)} = \begin{vmatrix} \frac{\partial y}{\partial \theta} & \frac{\partial y}{\partial \phi} \\ \frac{\partial z}{\partial \theta} & \frac{\partial z}{\partial \phi} \end{vmatrix} = \begin{vmatrix} \cos\theta \sin\phi & \sin\theta \cos\phi \\ -\sin\theta & 0 \end{vmatrix}$$

$$= \sin\theta \cos^2 \phi.$$

$$\frac{\partial(z, x)}{\partial(\theta, \phi)} = \begin{vmatrix} \frac{\partial z}{\partial \theta} & \frac{\partial z}{\partial \phi} \\ \frac{\partial x}{\partial \theta} & \frac{\partial x}{\partial \phi} \end{vmatrix} = \begin{vmatrix} -\sin\theta & 0 \\ \cos\theta \cos\phi & -\sin\theta \sin\phi \end{vmatrix}$$

$$= \sin^2 \theta \sin\phi.$$

$$\frac{\partial(x, y)}{\partial(\theta, \phi)} = \begin{vmatrix} \frac{\partial x}{\partial \theta} & \frac{\partial x}{\partial \phi} \\ \frac{\partial y}{\partial \theta} & \frac{\partial y}{\partial \phi} \end{vmatrix} = \begin{vmatrix} \cos\theta \cos\phi & -\sin\theta \sin\phi \\ \cos\theta \sin\phi & -\sin\theta \cos\phi \end{vmatrix}$$

$$= \sin\theta \cos\theta.$$

$\therefore$ integral

$$= \int_0^{\frac{1}{2}\pi} \int_0^{\frac{1}{2}\pi} [\sin\theta \cos\theta \sin\phi \sin^2\theta \cos\phi +$$

$$\sin\theta \cos\theta \cos\phi \sin^2\theta \sin\phi + \sin^2\theta \sin\phi \cos\phi \sin\theta \cos\theta]\, d\theta\, d\phi$$

$$= \int_0^{\frac{1}{2}\pi} \int_0^{\frac{1}{2}\pi} \left[2\sin^3\theta \cos\theta \sin\phi \cos\phi + \sin^3\theta \cos\theta \sin\phi \cos\phi\right] d\theta\, d\phi$$

$$= \frac{7}{20}.$$

Example 8. *If* $\mathbf{f} = y\mathbf{i} + (x - 2xz)\,\mathbf{j} - xy\mathbf{k}$, *evaluate* $\int_S (\nabla \times \mathbf{f}) \cdot \mathbf{n}\, dS$, *where S is the surface of the sphere* $x^2 + y^2 + z^2 = a^2$ *above the xy-plane.*

(Rohilkhand 2000, 2002)

Solution. Let $\mathbf{F} = \nabla \times \mathbf{f} = \begin{vmatrix} \mathbf{i} & \mathbf{j} & \mathbf{k} \\ \dfrac{\partial}{\partial x} & \dfrac{\partial}{\partial y} & \dfrac{\partial}{\partial z} \\ y & x-2xz & -xy \end{vmatrix}$

$$= x\mathbf{i} + y\mathbf{j} - 2z\mathbf{k}$$

$\phi = c$ is given by $x^2 + y^2 + z^2 = a^2$

grad ϕ is along the normal and grad $\phi = \Sigma \mathbf{i} \dfrac{\partial \phi}{\partial x}$

$$= 2x\mathbf{i} + 2y\mathbf{j} + 2z\mathbf{k}$$

$\therefore$ $\mathbf{n}$ = a unit normal $= \dfrac{2x\mathbf{i}+2y\mathbf{j}+2z\mathbf{k}}{\sqrt{(4x^2+4y^2+4z^2)}}$

$$= \frac{x\mathbf{i}+y\mathbf{j}+z\mathbf{k}}{a}$$

$\therefore$ $$\mathbf{F}\cdot\mathbf{n} = (x\mathbf{i}+y\mathbf{j}-2z\mathbf{k})\cdot\frac{(x\mathbf{i}+y\mathbf{j}+z\mathbf{k})}{a}$$

$$= \frac{x^2+y^2+2z^2}{a}$$

Also, we know that

$$\int_S \mathbf{F}\cdot\mathbf{n}\, dS = \iint_{S_3} \mathbf{F}\cdot\mathbf{n}\frac{dx\, dy}{\mathbf{n}\cdot\mathbf{k}}$$

where S_3 is the projection of S on xy-plane.

$$\mathbf{n}\cdot\mathbf{k} = \frac{x\mathbf{i}+y\mathbf{j}+z\mathbf{k}}{a}\cdot\mathbf{k} = \frac{z}{a}$$

$$= \frac{\sqrt{(a^2-x^2-y^2)}}{a}$$

Also $$\mathbf{F}\cdot\mathbf{n} = \frac{x^2+y^2-2z^2}{a} = \frac{x^2+y^2-2(a^2-x^2-y^2)}{a}$$

$$= \frac{3(x^2+y^2)-2a^2}{a}$$

$\therefore$ $$\int_S \mathbf{F}\cdot\mathbf{n}\, dS = \iint_{S_3} \frac{3(x^2+y^2)-2a^2}{a}\cdot\frac{dx\, dy}{\sqrt{(a^2-x^2-y^2)}}a \qquad ...(1)$$

Now, S_3 is the projection of $x^2+y^2+z^2=a^2$ in the xy-plane and is given by

$$x^2 + y^2 = a^2.$$

In order to integrate (1) put $x = r\cos\theta$, $y = r\sin\theta$.

$\therefore$ $dxdy = rd\theta dr$ and limits of θ are from 0 to 2π and that of r from 0 to a.

$$\therefore \quad \int_S \mathbf{F}\cdot\mathbf{n}\, dS = \int_0^{2\pi}\int_0^a \frac{3r^2-2a^2}{\sqrt{(a^2-r^2)}}\, rd\theta dr$$

$$= 2\pi \int_0^a \frac{3r^2-2a^2}{\sqrt{(a^2-r^2)}}\, rdr.$$

Put $a^2 - r^2 = t$, $\therefore$ $-2rdr = 2tdt$

$$\therefore \quad \int_S \mathbf{F}\cdot\mathbf{n}\, dS = 2\pi \int_0^a \frac{3\left(a^2-t^2\right)-2a^2}{t}(-t)\, dt$$

$$= 2\pi \int_0^a \left(a^2 - 3t^2\right) dt$$

$$= 2\pi\left(a^3 - a^3\right) = 0.$$

Example 9. *Evaluate* $\int_V (2x+y)\, dV$, *where V is the closed region bounded by the cylinder* $z = 4 - x^2$ *and the plane* $x = 0$, $y = 0$, $y = 2$ *and* $z = 0$.

Solution. The given cylinder $z = 4 - x^2$ meets the x-axis ($y = 0, z = 0$) at $x^2 = 4$ or $x = 2$, *i.e.*, at the point (2, 0, 0). It meets z-axis at $z = 4$, *i.e.*, at (0, 0, 4). Therefore, the limits of integration are as under.

$$z = 4 \text{ to } z = 4 - x^2, \quad y = 0 \text{ to } y = 2 \text{ and } x = 0 \text{ to } x = 2.$$

Also $dV = dx\, dy\, dz$.

$$\therefore \quad \int_V (2x+y)\, dV = \iiint (2x+y)\, dx\, dy\, dz$$

$$= \iint (2x+y)\left[z\right]_0^{4-x^2} dx\, dy$$

$$= \int_{x=0}^{2}\int_{y=0}^{2} (2x+y)\left(4-x^2\right) dx\, dy$$

$$= \int_0^2 \left[2x\left(4-x^2\right)y + \left(4-x^2\right)\frac{y^2}{2}\right]_{y=0}^{2} dx$$

$$= \int_0^2 \left[4x\left(4-x^2\right)+2\left(4-x^2\right)\right] dx$$

$$= 2\int_0^2 \left(4+8x-x^2-2x^3\right) dx$$

$$= \frac{80}{3}.$$

EXERCISES

1. Evaluate

$$\int_C \mathbf{F}\cdot d\mathbf{r} \text{ when } \mathbf{F} = xy\mathbf{i} + yz\,\mathbf{j} + zx\mathbf{k}$$

where C is the curve $\mathbf{r} = \mathbf{i}t + \mathbf{j}t^2 + \mathbf{k}t^3$; t varying from -1 to $+1$.

2. Calculate

$$\int_C \left[\left(x^2+y^2\right)\mathbf{i} + \left(x^2-y^2\right)\mathbf{j}\right]\cdot d\mathbf{r}$$

where C is the curve

(i) $y^2 = x$ joining (0, 0) to (1, 1).

(ii) $x^2 = y$ joining (0, 0) to (1, 1).

(iii) consisting of two straight lines joining (0, 0) to (1, 0) and (1, 0) to (1, 1).

(iv) consisting of three straight lines joining (0, 0) to (2, – 2); (2, – 2) to (0, – 1) and (0, – 1) to (1, 1).

3. Show that

$$\int_C \left(\frac{y}{x^2+y^2}\mathbf{i} + \frac{x}{x^2+y^2}\mathbf{j}\right)\cdot d\mathbf{r} = 2\pi,$$

where C is the circle $x^2 + y^2 = 1$ in the z-plane described in the counter-clockwise sense.

4. Evaluate

$$\int_C \frac{-y^3\mathbf{i} + x^3\mathbf{j}}{\left(x^2+y^2\right)^2}\cdot d\mathbf{r},$$

where C is the

(i) boundary of the square $x = \pm a,\ y = \pm a$;

(ii) the circle $x^2 + y^2 = a^2$; both in the counter-clockwise sense.

5. Show that

$$\int_C \left[\left(y^2+z^2\right)\mathbf{i} + \left(z^2+x^2\right)\mathbf{j} + \left(x^2+y^2\right)\mathbf{k}\right]\cdot d\mathbf{r} = -2\pi ab^2$$

where the curve C is that part for which $z \geq 0$ of the intersection of the surfaces $x^2 + y^2 + z^2 = 2ax$ and $x^2 + y^2 = 2bx$, $a > b > 0$; the curve begins at the origin and runs at first in the positive octant,

$$\left[x = b(1+\cos\theta),\ y = b\sin\theta,\ z = 2\sqrt{(a-b)}\,b\cos\frac{\theta}{2}\cdot(-\pi \le \theta \le \pi)\right].$$

6. Evaluate

$$\int_S \mathbf{r}\cdot d\mathbf{a}.$$

where S is the part of the surface of the ellipsoid

$$\frac{x^2}{a^2}+\frac{y^2}{b^2}+\frac{z^2}{c^2}=1.$$

lying above the z-plane, the normal at any point being directed outward.

7. Evaluate

$$\int_S \left(x^3\mathbf{i}+y^3\mathbf{j}+z^3\mathbf{k}\right)\cdot d\mathbf{a}$$

where S is the surface of the sphere

$$x^2 + y^2 + z^2 = 1.$$

8. Evaluate

$$\int_S \mathbf{F}\cdot\mathbf{n}\,dS$$

where $\mathbf{F} = 4xz\mathbf{i} - y^2\mathbf{j} + yz\mathbf{k}$ where S is the surface of the cube bounded by $x = 0,\ x = 1,\ y = 0,\ y = 1,\ z = 0,\ z = 1.$

9. If

$$\mathbf{F} = \left(2x^2 - 3z\right)\mathbf{i} - 2xy\mathbf{j} - 4x\mathbf{k}$$

then evaluate $\int_V \nabla\cdot\mathbf{F}\,dV$ and $\iiint_V \nabla\times\mathbf{F}\,dV$, where V is closed region bounded by the plane $x = 0,\ y = 0,\ z = 0$ and $2x + 2y + z = 4.$

(*Kumaon, 98*)

10. Evaluate

$$\int_S \phi\,\mathbf{n}\,dS,$$

where $\phi = \frac{3}{8}xyz$ and S is the surface of the cylinder $x^2 + y^2 = 16$ included in the first octant between $z = 0$ and $z = 5$.

11.6. REDUCTION OF VOLUME TO SURFACE INTEGRALS

Gauss's Divergence Theorem. *If* $\mathbf{F}$ *is a continuously differentiable vector point function and S is a closed surface enclosing a region V, then*

$$\int_S \mathbf{F}\cdot\mathbf{n}\,dS = \int_V \text{div}\,\mathbf{F}\,dv. \qquad ...(1)$$

where $\mathbf{n}$ *is the unit outward drawn normal vector.*

(*U.P.P.C.S. 2000, Garhwal 2000, Nagpur 2005, Utkal 2003, Gorakhpur 99, Srivenkateshwara 2005*)

Briefly, the theorem states that the normal surface integral of a function $\mathbf{F}$ over the boundary of a closed region is equal to the volume of div $\mathbf{F}$ taken throughout the region.

To start with, we shall suppose that the region V is such that it is possible to choose a rectangular cartesian co-ordinate system such that each line parallel to any co-ordinate axis which has internal points in common with the region meets the boundary S in two points.

In terms of cartesian co-ordinates, the equation (1) is stated as follows :

$$\iint_S (F_1 dy\, dz + F_2 dz\, dx + F_3 dx\, dy)$$

$$= \iiint_V \left(\frac{\partial F_1}{\partial x} + \frac{\partial F_2}{\partial y} + \frac{\partial F_3}{\partial z}\right) dx\, dy\, dz$$

Consider now the volume integral

$$\iiint_V \frac{\partial F_3}{\partial z}\, dx\, dy\, dz.$$

Let R_3 denote the projection of the region V, on the OXY plane. Every line through a point $(x, y, 0)$ of R_3 meets the boundary S in two points. Let the z, co-ordinates of these points be

$$z = \phi\,(x, y), \quad z = \psi\,(x, y)$$

where

$$\phi\,(x, y) \geq \psi\,(x, y).$$

Now

$$\iiint_V \frac{\partial F_3}{\partial z}\, dx\, dy\, dz = \iint_{R_3}\left[\int_\psi^\phi \frac{\partial F_3}{\partial z}\, dz\right] dx\, dy$$

$$= \iint_{R_3}\left[F_3(x, y, \phi) - F_3(x, y, \psi)\right] dx\, dy$$

$$= \iint_{R_3} F_3(x, y, \phi)\, dx\, dy -$$

$$\iint_{R_3} F_3(x, y, \psi)\, dx\, dy \qquad ...(1)$$

Let us now denote the parts of the surface S corresponding to $z = \phi\,(x, y)$ and $z = \psi\,(x, y)$ by S_1 and S_2 respectively.

If $\mathbf{n}$ denotes the *outward* drawn unit normal vector at any point of S, we have

$$\iint_{R_3} F_3(x, y, \phi)\, dx\, dy = \int_{S_1} F_3\, \mathbf{n} \cdot \mathbf{k}\, dS \qquad ...(2)$$

$$\iint_{R_3} F_3(x, y, \psi)\, dx\, dy = -\int_{S_2} F_3\, \mathbf{n} \cdot \mathbf{k}\, dS \qquad ..(3)$$

for the outward drawn normal at any point of S_1 makes an acute angle with positive direction of z-axis and that at any point of S_2 makes an obtuse angle with z-axis.

From (1), (2) and (3), we have

$$\iiint_V \frac{\partial F_3}{\partial z} dx\, dy\, dz = \int_{S_1} F_3 \,\mathbf{n} \,.\, \mathbf{k}\, dS + \int_{S_2} F_3 \,\mathbf{n} \,.\, \mathbf{k}\, dS$$

$$= \int_S F_3 \,\mathbf{n} \,.\, \mathbf{k}\, dS. \quad \text{...(4)}$$

Similarly it may be proved that

$$\iiint_V \frac{\partial F^2}{\partial y} dx dy dz = \int_S F_2 \mathbf{n} \cdot \mathbf{j} ds \quad \text{...(5)}$$

$$\iiint_V \frac{\partial F_1}{\partial x} dx\, dy\, dz = \int_S F_1 \,\mathbf{n} \,.\, \mathbf{i}\, dS. \quad \text{...(6)}$$

Adding (4), (5) and (6), we get

$$\iiint_V \left(\frac{\partial F_1}{\partial x} + \frac{\partial F_2}{\partial y} + \frac{\partial F_3}{\partial z} \right) dx\, dy\, dz = \int_S (F_1 \mathbf{i} + F_2 \mathbf{j} + F_3 \mathbf{k}) \cdot \mathbf{n}\, dS$$

i.e.,

$$\int_V \text{div}\, \mathbf{F}\, dv = \int_S \mathbf{F} \cdot \mathbf{n}\, dS = \int_S \mathbf{F} \cdot d\mathbf{a}. \quad \text{...(7)}$$

We now extend the result to the case of a general region.

Suppose that the region is such that it can be split up into a finite number of sub-regions each of which is met by a line parallel to any co-ordinate axis in two points only. We shall now apply the theorem to each sub-region and add the corresponding results. The volume integrals over separate sub-regions will combine to give the volume integral over the whole region V. Also each boundary surface of two sub-regions appears twice as parts of the two regions and the two corresponding surface integrals over the same will cancel, for the two normals at any point are oppositely directed. Thus, the soul of the surface integrals also combine to give a surface integral over the entire surface S. Thus, in the general case also we have

$$\int_V \text{div}\, \mathbf{F}\, dv = \int_S \mathbf{F} \cdot \mathbf{n}\, dS = \int_S \mathbf{F} \cdot d\mathbf{a}.$$

Note. The theorem also holds good for a region V (shaded) such as that shown in the figure. It is bounded by two closed surface S_1 and S_2 one of which lies within the other. It should, however, be understood that the *outward* drawn normal at any point means the normal directed away from the region. Thus, the normals at points of S_1 and S_2 are to be directed as shown in the figure.

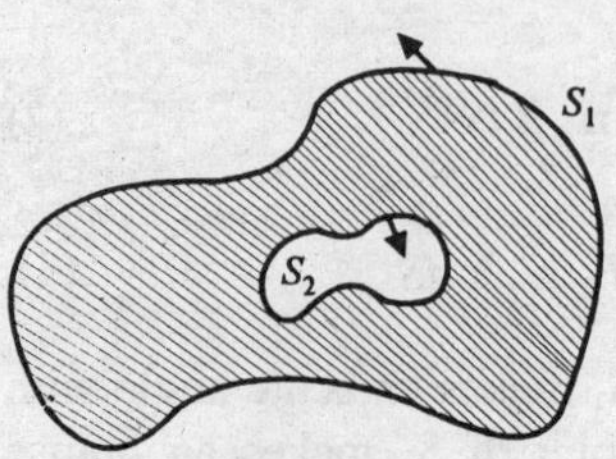

Fig. 11.8.

The theorem can be applied to regions enclosed by several surfaces.

Cor. 1. The Gauss's theorem obtained above enables us to express a normal surface integral as a volume integral. We shall now see that with the help of this theorem we are able to express the other two types of surface integrals also as volume integrals.

It will be shown that

$$(i) \quad \int_S \mathbf{F} \times \mathbf{n}\, dS = \int_S \mathbf{F} \times d\mathbf{a} = -\int_V \text{curl}\, \mathbf{F}\, dv$$

$$(ii) \quad \int_S \phi\, \mathbf{n}\, dS = \int_S \phi\, d\mathbf{a} = \int_V \nabla\phi\, dr$$

We write

$$\mathbf{f} = \mathbf{a} \times \mathbf{F},$$

where **a** is any constant vector and apply Gauss's theorem to the vector function **f**. Thus, we have

$$\int_S \mathbf{a} \times \mathbf{F} \cdot \mathbf{n}\, dS = \int_V \text{div}\,(\mathbf{a} \times \mathbf{F})\, dv. \qquad ...(1)$$

But

$$\mathbf{a} \times \mathbf{F} \,.\, \mathbf{n} = \mathbf{a} \,.\, \mathbf{F} \times \mathbf{n}, \qquad ...(2)$$

and

$$\text{div}\,(\mathbf{a} \times \mathbf{F}) = \nabla \,.\, (\mathbf{a} \times \mathbf{F}) = -\,\mathbf{a} \,.\, \nabla \times \mathbf{F}. \qquad ...(3)$$

From (1), (2) and (3),

$$\int_S \mathbf{a} \cdot \mathbf{F} \times \mathbf{n}\, dS = -\int_V \mathbf{a} \cdot \nabla \times \mathbf{F}\, dv.$$

or

$$\mathbf{a} \cdot \int_S \mathbf{F} \times \mathbf{n}\, dS = -\,\mathbf{a} \cdot \int_V \nabla \times \mathbf{F}\, dv$$

or

$$\mathbf{a} \cdot \left[\int_S \mathbf{F} \times \mathbf{n}\, dS + \int_V \nabla \times \mathbf{F}\, dv\right] = 0 \qquad ...(4)$$

As, however, **a** is any arbitrary vector, we have from (4)

$$\int_S \mathbf{F} \times \mathbf{n}\, dS + \int_V \nabla \times \mathbf{F}\, dv = 0,$$

i.e.,

$$\int_S \mathbf{F} \times \mathbf{n}\, dS = -\int_V \nabla \times \mathbf{F}\, dv = -\int_V \text{curl}\, \mathbf{F}\, dv.$$

Thus, we have proved (*i*).

To prove (*ii*) we write

$$\mathbf{f} = \mathbf{a}\phi$$

where **a** is a constant vector. Applying Gauss's theorem to the function **f**, we have

$$\int_S \mathbf{f} \cdot \mathbf{n}\, dS = \int_S \text{div}\, \mathbf{f}$$

i.e.,

$$\int_S \mathbf{a}\phi \cdot \mathbf{n}\, dS = \int_V \operatorname{div}(\mathbf{a}\phi)\, dv = \int_V \nabla \cdot (\mathbf{a}\phi)\, dv$$

or

$$\mathbf{a} \cdot \int_S \phi\mathbf{n}\, dS = \mathbf{a} \cdot \int_V \nabla\phi\, dv$$

or

$$\mathbf{a} \cdot \left[\int_S \phi\mathbf{n}\, dS - \int_V \nabla\phi\, dv\right] = 0.$$

The vector **a** being arbitrary, we obtain

$$\int_S \phi\mathbf{n}\, dS - \int_V \nabla\phi\, dv = 0,$$

or

$$\int_S \phi\mathbf{n}\, dS - \int_V \nabla\phi\, dv,$$

which is (*ii*).

Note. The three results obtained above can be re-written as follows :

$$\int_S \mathbf{n} \cdot \mathbf{F}\, dS = \int_V \nabla \cdot \mathbf{F}\, dv,$$

$$\int_S \mathbf{n} \times \mathbf{F}\, dS = \int_V \nabla \times \mathbf{F}\, dv,$$

$$\int_S \mathbf{n}\phi\, dS = \int_V \nabla\phi\, dv.$$

It will thus, be seen that each of these results can be written by just changing **n** to ∇.

The integrand in each case has been so written that **n** appears *before* the function and not after.

Cor. 2. With the help of the results obtained above, we can state the following integral expressions for grad ϕ, div **F** and curl **F**. These expressions are also often adopted as definitions for the differential operators.

$$\operatorname{grad}\phi = \lim_{v\to 0} \frac{\int_S \mathbf{n}\phi\, dS}{V},$$

$$\operatorname{div}\mathbf{F} = \lim_{v\to 0} \frac{\int_S \mathbf{n} \cdot \mathbf{F}\, dS}{V},$$

$$\operatorname{curl}\mathbf{F} = \lim_{v\to 0} \frac{\int_S \mathbf{n} \times \mathbf{F}\, dS}{V}.$$

Take any point P and enclose the same in a small region volume V enclosed by a surface S.

As proved in Cor. 1, above

$$\int_S \mathbf{n}\phi \, dS = \int_V \nabla\phi \, dv$$

$$= V\left[(\nabla\phi)_0 + \varepsilon\right],$$

where $(\nabla\phi)_0$ denotes the value of $\nabla\phi$ at P and $\varepsilon \to 0$ as $V \to 0$.

$$\therefore \qquad \lim_{v\to 0} \frac{\int_S \mathbf{n}\phi \, ds}{V} = (\nabla\phi)_0.$$

Similarly for div **F** and curl **F**.

11.7. PHYSICAL INTERPRETATION OF GAUSS'S THEOREM

Let the vector point function **q**, denote the velocity vector of an incompressible fluid of unit density and let S denote any closed surface drawn in the fluid. By Gauss's Theorem,

$$\int_S \mathbf{q}\cdot\mathbf{n} \, ds = \int_V \text{div}\, \mathbf{q} \, dv. \qquad ...(1)$$

Now $\mathbf{q} \cdot \mathbf{n}$ is the component of the velocity at any point of S in the direction of the outward drawn normal so that $\mathbf{q} \cdot \mathbf{n}\, \delta S$ denotes the amount of fluid that flows out in unit time through the element δS. Hence *the left hand side of (1) denotes the amount of fluid flowing across the surface S in unit time from the inside to the outside.* This amount may be positive, negative or zero.

Now the total amount flowing outwards must be continuously supplied so that inside the region we must have sources producing fluid. We have already seen that div **q** at any point denotes the amount of fluid per unit time per unit volume that goes through any point. Thus div **q** may be thought of as the *source-intensity* of the incompressible fluid at any point P. Hence *the right hand side of (1) denotes the amount of fluid per unit time supplied by the sources within S.* Thus, the equality (1) appears intuitively evident.

EXAMPLES

Example 1. *Show that*

$$\frac{1}{3}\int_S \mathbf{r}\cdot d\mathbf{a} = V.$$

where V is the volume enclosed by the surface S.

Solution. We have

$$\int_S \frac{1}{3}\mathbf{r}\cdot d\mathbf{a} = \int_V \text{div}\,\frac{1}{3}\mathbf{r}\,dv$$

$$= \int_V dv = V, \text{ for div } \mathbf{r} = 3.$$

Example 2. *If* $\overrightarrow{OA} = a\mathbf{i}$, $\overrightarrow{OB} = a\mathbf{j}$, $\overrightarrow{OC} = a\mathbf{k}$, *form three coterminous edges of a cube and S denotes the surface of the cube, evaluate*

$$\int_S \left\{\left(x^3 - yz\right)\mathbf{i} - 2x^2y\mathbf{j} + 2\mathbf{k}\right\}\cdot\mathbf{n}\,dS,$$

by expressing it as a volume integral. Also verify the result by direct evaluation of the surface integral. (Garhwal 2004)

Solution. Let

$$\mathbf{F} = \left(x^3 - yz\right)\mathbf{i} - 2x^2y\mathbf{j} + 2\mathbf{k}$$

$$\therefore \quad \text{div}\,\mathbf{F} = \frac{\partial\left(x^3 - yz\right)}{\partial x} + \frac{\partial\left(-2x^2y\right)}{\partial y} + \frac{\partial(2)}{\partial z}$$

$$= 3x^2 - 2x^2 = x^2.$$

$$\therefore \quad \int_S \mathbf{F}\cdot\mathbf{n}\,dS = \int_V \text{div}\,\mathbf{F}\,dv = \iiint_V x^2 dx\,dy\,dz$$

$$= \left[\int_0^a dz\right]\left[\int_0^a dy\right]\left[\int_0^a x^2 dx\right]$$

$$= a\cdot a\cdot\frac{1}{2}a^3 = \frac{1}{2}a^5.$$

We shall now evaluate the surface integral directly. The surface S consists of *six* faces.

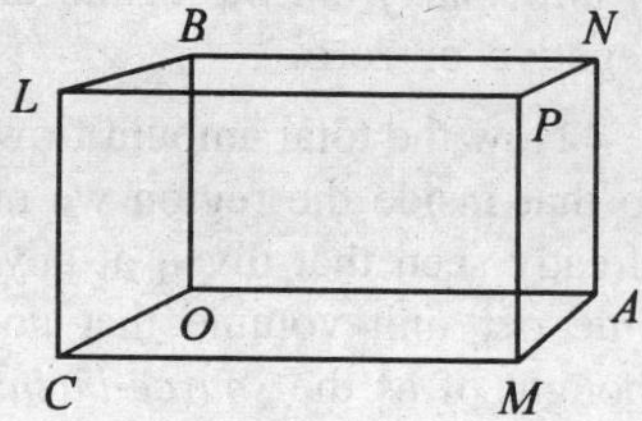

Fig. 11.9.

Over the surface $AOBN$

$$\int_S \mathbf{F}\cdot\mathbf{n}\,dS$$

$$= \int_S \left\{\left(x^3 - yz\right)\mathbf{i} - 2x^2y\mathbf{j} + 2\mathbf{k}\right\}\cdot(-\mathbf{k})\,dS$$

$$= \int_0^a \int_0^a -2dx\,dy = -2a^2.$$

Over the surface $PLCM$,

$$\int_S \mathbf{F}\cdot\mathbf{n}\,dS = \int_S \left\{\left(x^3 - yz\right)\mathbf{i} - 2x^2y\mathbf{j} + 2\mathbf{k}\right\}\cdot\mathbf{k}\,dS$$

$$= \int_0^a \int_0^a 2dx\,dy = 2a^2$$

Similarly over the faces *NPMA, BLCO, AOCM, NBLP,* the corresponding surface integrals are respectively

$$\int_S \mathbf{F}\cdot\mathbf{i}\, dS = \iint_S \left(x^3 - yz\right) dy\, dz = \int_0^a \int_0^a \left(a^3 - yz\right) dy\, dz$$

$$= a^5 - \frac{1}{4}a^4,$$

$$\int_S \mathbf{F}\cdot(-\mathbf{i})\, dS = -\iint_S \left(x^3 - yz\right) dy\, dz = \int_0^a \int_0^a yz\, dy\, dz = \frac{1}{4}a^4$$

$$\int_S \mathbf{F}\cdot(-\mathbf{j})\, dS = \int_0^a \int_0^a 2x^2 y\, dx\, dy = 0$$

$$\int_S \mathbf{F}\cdot\mathbf{j}\, dS = -\int_0^a \int_0^a 2x^2 y\, dx\, dy = -2a \int_0^a \int_0^a x^2 dx\, dz$$

$$= \frac{-2}{3}a^5.$$

Adding we see that over the whole surface

$$\int_S \mathbf{F}\cdot\mathbf{n}\, dS = -2a^2 + 2a^2 + a^5 - \frac{1}{4}a^4 + \frac{1}{4}a^4 + 0 - \frac{2}{3}a^5 = \frac{1}{3}a^5.$$

Hence the verification.

Example 3. *Evaluate*

$$\int_S -\gamma \frac{\mathbf{r}}{r^3}\cdot d\mathbf{a},$$

where S denotes any closed surface enclosing a region V.

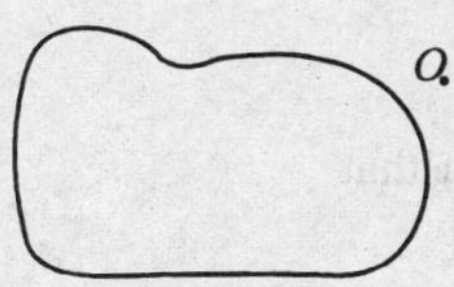

Fig. 11.10.

Solution. Two cases arise according as the origin lies within S or without S.

Suppose first that the origin lies without S. In this case

$$\mathbf{F} = -\gamma \frac{\mathbf{r}}{r^3},$$

is continuously differentiable throughout the region enclosed by S.

$$\therefore \qquad \int_S -\frac{\gamma\mathbf{r}}{r^3}\cdot d\mathbf{a} = \int_V \operatorname{div}\left(-\frac{\gamma\mathbf{r}}{r^3}\right) dv.$$

It may be easily shown that

$$\operatorname{div}\frac{-\gamma\mathbf{r}}{r^3} = 0.$$

Thus, the surface integral is 0.

Suppose now that the origin O lies within S. In this case $\mathbf{F}$ is not defined at the point O within S so that we cannot apply Gauss's theorem as such.

We now enclose O in a small sphere with its centre at O and lying within S. Let its radius be a. Let its surface be denoted by S_1. In relation to the application of Gauss's theorem, the normal at any point of the surface S_1 of the sphere has to be directed inwards towards O.

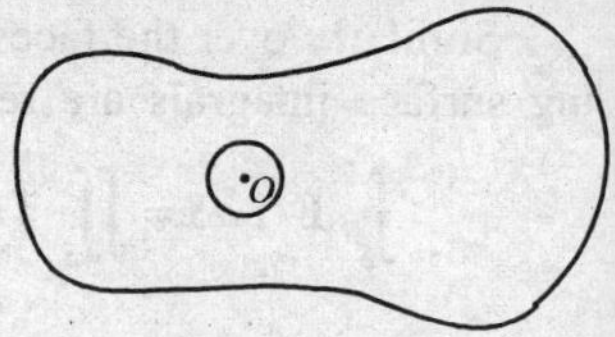

Fig. 11.11.

The function **F** is continuously differentiable at all points of the region V' enclosed between S and S_1.

$\therefore$ By Gauss's theorem

$$\int_S -\frac{\gamma \mathbf{r}}{r^3} \cdot d\mathbf{a} + \int_{S_1} -\frac{\gamma \mathbf{r}}{r^3} \cdot d\mathbf{a} = \int_{V'} \text{div}\left(-\frac{\gamma \mathbf{r}}{r^3}\right) \cdot dv = 0.$$

As in Ex. 6, on Page 334, we have

$$-\gamma \int_{S_1} \frac{\mathbf{r}}{r^3} \cdot d\mathbf{a} = -4\pi\gamma.$$

$$\therefore \qquad \int_S -\gamma \frac{\mathbf{r}}{r^3} \cdot d\mathbf{a} = 4\pi\gamma.$$

Example 4. *Convert*

$$\int_S \mathbf{F} \times d\mathbf{a},$$

taken over a closed surface into a volume integral and verify the result in the case

$$\mathbf{F} = xy\mathbf{k}$$

where S is the surface of a cube whose base is in the XY plane and which is bounded by the planes

$$x = 0,\ x = a;\ y = 0,\ y = a.$$

Solution. In Cor. 1 to § 11.5, it has been shown that

$$\int_S \mathbf{F} \times d\mathbf{a} = -\int_S \text{curl } \mathbf{F}\, dv.$$

Now when $\mathbf{F} = xy\mathbf{k}$, we have

$$\text{curl } \mathbf{F} = \mathbf{i}x - \mathbf{j}y.$$

Now $\displaystyle\int_V \text{curl } \mathbf{F}\, dv = \mathbf{i}\int_V x\, dv - \mathbf{j}\int_V y\, dv$

$$= \mathbf{i}\int_0^a \int_0^a \int_0^a x\, dx\, dy\, dz - \mathbf{j}\int_0^a \int_0^a \int_0^a y\, dx\, dy\, dz$$

$$= \mathbf{i}\frac{a^4}{2} - \mathbf{j}\frac{a^4}{2} = \frac{1}{2}a^4(\mathbf{i} - \mathbf{j}).$$

[Refer Fig. Ex. 2, Page 345]

We shall now calculate

$$\int_S \mathbf{F} \times d\mathbf{a},$$

over the six faces of the cube.

Over the face *AOCM*,

$$\int \mathbf{F} \times d\mathbf{a} = \iint (xy\mathbf{k}) \times -\mathbf{j}dx\, dz = \mathbf{i} \int_0^a \int_0^a xy\, dx\, dz = 0,$$

for y is zero at every point of this face.

Over the face *PNBL*,

$$\int \mathbf{F} \times d\mathbf{a} = \iint (xy\mathbf{k}) \times \mathbf{j}dx\, dz = -\mathbf{i} \int_0^a \int_0^a xy\, dx\, dz$$

$$= -\mathbf{i}a \int_0^a \int_0^a x\, dx\, dz = -\mathbf{i}\frac{a^4}{2}.$$

Similarly, over the face *NPMA, BLCO, AOBN, MCLP,* the corresponding surface integrals respectively are :

$$\int \mathbf{F} \times d\mathbf{a} = \iint xy\mathbf{k} \times \mathbf{i}\, dy\, dz = \mathbf{j}a \int_0^a \int_0^a y\, dy\, dz = \mathbf{j}\frac{a^4}{2}$$

$$\int \mathbf{F} \times d\mathbf{a} = \iint xy\mathbf{k} \times (-\mathbf{i}\, dy\, dz) = 0, \text{ for } \mathbf{F} = \mathbf{0} \text{ at every point.}$$

$$\int \mathbf{F} \times d\mathbf{a} = \iint xy\mathbf{k} \times -\mathbf{k}\, dx\, dy = 0,$$

$$\int \mathbf{F} \times d\mathbf{a} = \iint xy\mathbf{k} \times \mathbf{k}\, dx\, dy = 0,$$

$\therefore$ Over the six faces,

$$\int \mathbf{F} \times d\mathbf{a} = 0 - \mathbf{i}\frac{a^4}{2} + \mathbf{j}\frac{a^4}{2} + 0 + 0 + 0 = -\frac{1}{2}a^4\,(\mathbf{i} - \mathbf{j}).$$

Hence, the verification.

Example 5. *Prove that*

$$\int_S \mathbf{n} \times (\mathbf{a} \times \mathbf{r})\, dS = 2\mathbf{a}V$$

where **a** *is a constant vector.*

Solution. By Gauss's divergence theorem

$$\int_S \mathbf{n} \times (\mathbf{a} \times \mathbf{r})\, dS = \int_V \nabla \times (\mathbf{a} \times \mathbf{r})\, dV$$

Now, $$\nabla\times(\mathbf{a}\times\mathbf{r})=\Sigma\mathbf{i}\times\frac{\partial}{\partial x}(\mathbf{a}\times\mathbf{r})$$

$$=\Sigma\mathbf{i}\times\left(\frac{\partial\mathbf{a}}{\partial x}\times\mathbf{r}+\mathbf{a}\times\frac{\partial\mathbf{r}}{\partial x}\right)$$

$$=\Sigma\mathbf{i}\times(0+\mathbf{a}\times\mathbf{i})$$

$$=\Sigma\mathbf{i}\times(\mathbf{a}\times\mathbf{i})$$

$$=\Sigma\,(\mathbf{i}\cdot\mathbf{i}\,\mathbf{a}-\mathbf{i}\cdot\mathbf{a}\,\mathbf{i})$$

$$=\Sigma\,(\mathbf{a}-\mathbf{a}_1\mathbf{i})=\Sigma\mathbf{a}-\Sigma\mathbf{a}_1\mathbf{i}$$

$$=3\mathbf{a}-\mathbf{a}=2\mathbf{a}$$

$$\therefore\qquad \int_S \mathbf{n}\times(\mathbf{a}\times\mathbf{r})\,dS=\int_V 2a\,dV=2aV.$$

Example 6. *Prove that*

$$\int_V(\mathbf{g}\cdot curl\ curl\ \mathbf{f}-\mathbf{f}\cdot curl\ curl\ \mathbf{g})\,dV$$

$$=\int_S\{(\mathbf{f}\times\text{curl }\mathbf{g})-(\mathbf{g}\times\text{curl }\mathbf{f})\}\cdot d\mathbf{a}$$

Solution. $$\int_S(\mathbf{f}\times\text{curl }\mathbf{g})\cdot d\mathbf{a}=\int_S(\mathbf{f}\times\text{curl }\mathbf{g})\cdot\mathbf{n}\,dS$$

$$=\int_V\text{div}\,[\mathbf{f}\times\text{curl }\mathbf{g}]\cdot dV$$

$$=\int_V[\text{curl }\mathbf{g}\cdot\text{curl }\mathbf{f}\;-\mathbf{f}\cdot\text{curl curl }\mathbf{g}]\,dV \quad ...(1)$$

$$\int_S(\mathbf{g}\times\text{curl }\mathbf{f})\cdot d\mathbf{a}=\int_S(\mathbf{g}\times\text{curl }\mathbf{f})\cdot\mathbf{n}\cdot dS$$

$$=\int_V\text{div}\,(\mathbf{g}\times\text{curl }\mathbf{f})\,dV$$

$$=\int[\text{curl }\mathbf{f}\cdot\text{curl }\mathbf{g}-\mathbf{g}\cdot\text{curl curl }\mathbf{f}]\,dV \quad ...(2)$$

Subtracting (1) from (2), we get the required result.

EXERCISES

1. Evaluate

$$\int_S(\mathbf{i}x+\mathbf{j}y+\mathbf{k}z)\cdot d\mathbf{a}$$

where S denotes the surface of the cube bounded by the planes $x = 0$, $x = a$; $y = 0$, $y = a$; $z = 0$, $z = a$ by the application of Gauss's theorem and verify by direct manipulation.

2. Evaluate

$$\int_S \left(y^2z^2\mathbf{i}+z^2x^2\mathbf{j}+z^2y^2\mathbf{k}\right)\cdot d\mathbf{a},$$

where S is the part of the sphere $x^2 + y^2 + z^2 = 1$ above the xy-plane.

3. Show that

$$\int_S \left(x^2\mathbf{i}+y^2\mathbf{j}+z^2\mathbf{k}\right)\cdot \mathbf{n}\, dS = 0,$$

where S denotes the surface of the ellipsoid

$$\frac{x^2}{a^2}+\frac{y^2}{b^2}+\frac{z^2}{c^2}=1.$$

4. Show that

$$\int_S (ax\mathbf{i}+by\mathbf{j}+cz\mathbf{k})\cdot d\mathbf{a} = \frac{4}{3}\pi\,(a+b+c),$$

where S is the surface of the sphere $x^2 + y^2 + z^2 = 1$.

(*Ajmer 98, Rohilkhand 99*)

5. Evaluate

$$\int_S \left(x\mathbf{i}+y\mathbf{j}+z^2\mathbf{k}\right)\cdot d\mathbf{a},$$

where S denotes the closed surface bounded by the cone $x^2 + y^2 = z^2$ and the plane $z = 1$.

6. Show that

$$\int_S (x+y+z)\,(lx+my+nz)\, dS = 2a^2h^2\pi,$$

where S is the boundary of the region

$$x^2+y^2 \le a^2,\ 0 \le z \le h$$

and l, m, n are the direction cosines of the outward drawn normal.

7. Show that

$$\int_S d\mathbf{a} = 0,\ \int_S \mathbf{r}\times d\mathbf{a} = 0,$$

over any closed surface S.

8. If $\mathbf{F}$ is normal to the surface S at each point, show that

$$\int_V \text{curl}\,\mathbf{F}\, dv = 0.$$

9. Show that

$$\int_S \nabla\phi \times \mathbf{n}\, dS = 0.$$

10. Prove that

$$\int_V \mathbf{F}\cdot\nabla\phi\, dv = \int_S \phi\,\mathbf{F}\cdot d\mathbf{a} - \int_V \phi\nabla\cdot\mathbf{F}\, dv.$$

(*Ajmer, 99*)

11. Prove that

$$\int_V \nabla\phi \cdot \text{curl } \mathbf{F} dv = \int_S (\mathbf{F} \times \nabla\phi) \cdot d\mathbf{a}.$$

12. Show that

$$\int_S \nabla\phi \times \nabla\psi \cdot d\mathbf{a} = 0.$$

13. If $\mathbf{W} = \frac{1}{2}\nabla \times \mathbf{V}$ and $\mathbf{V} = \nabla \times \mathbf{U}$, show that

$$\frac{1}{2}\int \mathbf{V}^2 dv = \frac{1}{2}\int \mathbf{U} \times \mathbf{V} \cdot d\mathbf{a} + \int \mathbf{U} \cdot \mathbf{W}\, dv.$$

14. Prove that

$$\int \mathbf{f} \cdot \text{curl } \mathbf{F} \text{ dv} = \int \mathbf{F} \times \mathbf{f} \cdot d\mathbf{a} + \int \mathbf{F} \cdot \text{curl } \mathbf{f}\, dv.$$

(Gorakhpur 2000)

15. **f** and **g** are two continuously differentiable vector point functions in a region V such that

$$\text{curl } \mathbf{f} = 0, \quad \text{div } \mathbf{g} = 0$$

in V and

$$\mathbf{g} \cdot \mathbf{n} = 0$$

on the boundary S of V where **n** is a unit vector along the normal to S.

16. Show that

$$\int_S \mathbf{f} \cdot \mathbf{g}\, dv = 0.$$

[Show that there exists a scalar point function ϕ such that $\mathbf{F} = \nabla\phi$.]

REDUCTION OF SURFACE TO LINE INTEGRALS

11.8. STOKE'S THEOREM

If **F** *is any continuously differentiable vector point function and S is a surface bounded by a curve C, then*

$$\int_C \mathbf{F} \cdot d\mathbf{r} = \int_S \text{curl } \mathbf{F} \cdot \mathbf{n}\, dS,$$

where the unit normal vector **n** *at any point of S is drawn in the sense in which a right handed screw would move when rotated in the sense of description of C.* *(Ajmer 99, Garhwal 99, U.P.P.C.S. 99, 2001 Osmania 2005, Srivenkateshwara 2005)*

Stoke's Theorem for Plane. Firstly we shall prove the theorem for the case of a plane surface. Stoke's theorem in a plane is also often referred to as Green's theorem. Take a system of cartesian rectangular co-ordinate axes such that the plane of the given surface S is the OXY plane and the z-axis lies along the direction of the normal vector **n**. In the present case of plane the normal vector is constant.

Let

$$\mathbf{F} = \mathbf{i}F_1 + \mathbf{j}F_1 + \mathbf{k}F_3.$$

We have

$$\int_C \mathbf{F}\cdot d\mathbf{r} = \int_C \mathbf{F}\cdot\mathbf{t}\, ds$$

$$= \int_C (\mathbf{i}F_1 + \mathbf{j}F_2 + \mathbf{k}F_3)\cdot\left(\mathbf{i}\frac{dx}{ds} + \mathbf{j}\frac{dy}{ds} + \mathbf{k}\frac{dz}{ds}\right) ds$$

$$= \int_C \left(F_1\frac{dx}{ds} + F_2\frac{dy}{ds}\right) ds = \int_C (F_1 dx + F_2 dy).$$

for $dz / dx = 0$; the tangent at any point lying in the OXY plane.

Also $$\int_C \text{curl } \mathbf{F} \,.\, \mathbf{n}\, dS = \int_C \text{curl } \mathbf{F} \,.\, \mathbf{k}\, dS$$

$$= \iint_S \left(\frac{\partial F_2}{\partial x} - \frac{\partial F_1}{\partial y}\right) dx\, dy.$$

Thus, for the case of a plane surface the theorem is equivalent to showing

$$\int_C (F_1 dx + F_2 dy) = \iint_C \left(\frac{\partial F_2}{\partial x} - \frac{\partial F_1}{\partial y}\right) dx\, dy.$$

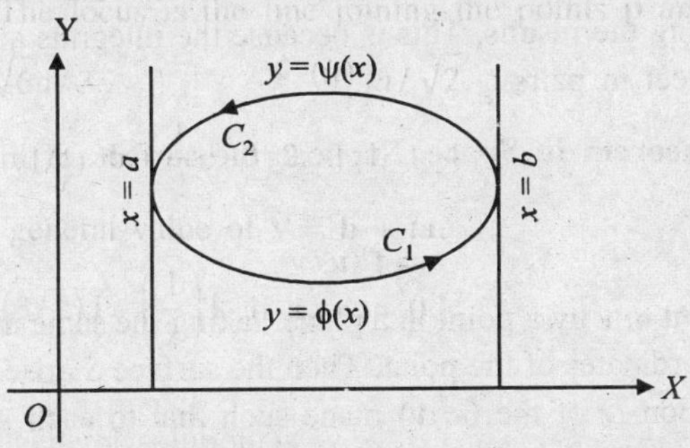

Fig. 11.12.

We shall now prove it. Firstly suppose that the region S is such that any line parallel to either axis meets the boundary C in at the most two points. Let the region be included between the lines $x = a$, $x = b$ and any line parallel to y-axis with abscissa x meet C in the points given by

$$y = \psi(x), \quad y = \phi(x)$$

where $\psi(x) \geq \phi(x)$. Thus, the boundary curve C is split up into two arcs C_1 and C_2 as shown in the figure.

We have

$$\iint_S \frac{\partial F_1}{\partial y} dx\, dy = \int_a^b \left[\int_{\phi(x)}^{\psi(x)} \frac{\partial F_1}{\partial y} dy\right] dx$$

$$= \int_a^b \left[F_1[x, \psi(x)] - F_1[x, \phi(x)]\right] dx$$

$$= -\int_{C_2} F_1(x, y)\, dx - \int_{C_1} F_1(x, y)\, dx$$

$$= -\int_C F_1(x, y)\, dx. \qquad ...(1)$$

Similarly, we may show that

$$\iint_S \frac{\partial F_2}{\partial x} dx\, dy = \int_C F_2(x, y)\, dy. \qquad ...(2)$$

From (1) and (2)ve

$$\int_C (F_2 dy + F_1 dx) = \iint_S \left(\frac{\partial F_2}{\partial x} - \frac{\partial F_1}{\partial y}\right) dx\, dy.$$

Now if a plane region be such that it can be divided into a finite number of sub-regions such that the boundary of each is met in at the most two points by any line parallel to either axis, then we can see the truth of the theorem for the region by applying the result obtained to each sub-region and adding the results. This is because the integrals along boundary curves will cancel in pairs.

Stoke's Theorem in Space. Suppose the surface of integration S is given by

$$\mathbf{r} = \mathbf{f}(u, v).$$

We represent u, v by a point in a plane, taking the same as the cartesian rectangular co-ordinates of the point. Then the surface S arises as the image of a certain region D of the (u, v) plane such that to each point (u, v) of D, there corresponds one and only one point of S. Let K be the boundary of D whose sense of description corresponds to that of the curve C bounding the surface S.

We have

$$\int_C \text{curl}\, \mathbf{F} \cdot \mathbf{n}\, dS = \iint_S \left(\frac{\partial F_3}{\partial y} - \frac{\partial F_2}{\partial z}\right) dy\, dz + \left(\frac{\partial F_1}{\partial z} - \frac{\partial F_3}{\partial x}\right) dz\, dx + \left(\frac{\partial F_2}{\partial x} - \frac{\partial F_1}{\partial y}\right) dx\, dy$$

$$= \iint_D \left[\left(\frac{\partial F_3}{\partial y} - \frac{\partial F_2}{\partial z}\right)\frac{\partial (y, z)}{\partial (u, v)} + \left(\frac{\partial F_1}{\partial z} - \frac{\partial F_3}{\partial x}\right)\frac{\partial (z, x)}{\partial (u, v)} + \left(\frac{\partial F_2}{\partial x} - \frac{\partial F_1}{\partial y}\right)\frac{\partial (x, y)}{\partial (u, v)}\right] du\, dv.$$

In the integrand corresponding to the double integral over the domain D, in the u, v plane, the terms involving F_1 are

$$\frac{\partial F_1}{\partial z}\frac{\partial(z, x)}{\partial(u, v)} - \frac{\partial F_1}{\partial y}\frac{\partial(x, y)}{\partial(u, v)}$$

$$= -\frac{\partial F_1}{\partial z}\frac{\partial(x, z)}{\partial(u, v)} - \frac{\partial F_1}{\partial y}\frac{\partial(x, y)}{\partial(u, v)} - \frac{\partial F_1}{\partial x}\frac{\partial(x, x)}{\partial(u, v)},$$

the last term introduced being zero. Again these terms are

$$= -\frac{\partial F_1}{\partial z}\left[\frac{\partial x}{\partial u}\frac{\partial z}{\partial v} - \frac{\partial x}{\partial v}\frac{\partial z}{\partial u}\right] - \frac{\partial F_1}{\partial y}\left[\frac{\partial x}{\partial u}\frac{\partial y}{\partial v} - \frac{\partial x}{\partial v}\frac{\partial y}{\partial u}\right] - \frac{\partial F_1}{\partial x}\left[\frac{\partial x}{\partial u}\frac{\partial x}{\partial v} - \frac{\partial x}{\partial v}\frac{\partial x}{\partial u}\right]$$

$$= -\frac{\partial x}{\partial u}\left[\frac{\partial F_1}{\partial z}\frac{\partial z}{\partial v} + \frac{\partial F_1}{\partial y}\frac{\partial y}{\partial v} + \frac{\partial F_1}{\partial x}\frac{\partial x}{\partial v}\right] + \frac{\partial x}{\partial v}\left[\frac{\partial F_1}{\partial z}\frac{\partial z}{\partial u} + \frac{\partial F_1}{\partial y}\frac{\partial y}{\partial u} + \frac{\partial F_1}{\partial x}\frac{\partial x}{\partial u}\right]$$

$$= -\frac{\partial x}{\partial u}\frac{\partial F_1}{\partial v} + \frac{\partial x}{\partial v}\frac{\partial F_1}{\partial u} = \frac{\partial F_1}{\partial u}\frac{\partial x}{\partial v} - \frac{\partial F_1}{\partial v}\frac{\partial x}{\partial u}.$$

Similarly the terms involving F_2, F_3 are

$$\frac{\partial F_2}{\partial u}\frac{\partial y}{\partial v} - \frac{\partial F_2}{\partial v}\frac{\partial y}{\partial u}, \quad \frac{\partial F_3}{\partial u}\frac{\partial z}{\partial v} - \frac{\partial F_3}{\partial v}\frac{\partial z}{\partial u}$$

Thus, the double integral over the domain in the (u, v) plane is

$$= \iint_D \left[\left(\frac{\partial F_1}{\partial u}\frac{\partial x}{\partial v} - \frac{\partial F_1}{\partial v}\frac{\partial x}{\partial u}\right) + \left(\frac{\partial F_2}{\partial u}\frac{\partial y}{\partial v} - \frac{\partial F_2}{\partial v}\frac{\partial y}{\partial u}\right) + \left(\frac{\partial F_3}{\partial u}\frac{\partial z}{\partial v} - \frac{\partial F_3}{\partial v}\frac{\partial z}{\partial u}\right)\right] du\, dv.$$

By Stoke's theorem in plane,

$$\iint_D \left(\frac{\partial F_1}{\partial u}\frac{\partial x}{\partial v} - \frac{\partial F_1}{\partial v}\frac{\partial x}{\partial u}\right) du\, dv$$

$$= \int_C \left(F_1 \frac{\partial x}{\partial u} du + F_1 \frac{\partial x}{\partial v} dv\right)$$

$$= \int_C F_1 dx,$$

and similarly for the other two parts of the double integral over D. Thus,

$$\int_S \text{curl } \mathbf{F} \cdot \mathbf{n}\, dS = \int_C (F_1 dx + F_2 dy + F_3 dz)$$

$$= \int_C \mathbf{F} \cdot d\mathbf{r}.$$

Note. The theorem also holds good for a surface S (shaded) such as that shown in the figure. It is enclosed by two curves C_1 and C_2. We have

$$\int_S \text{curl } \mathbf{F} \cdot \mathbf{n}\, dS = \int_{C_1} \mathbf{F} \cdot d\mathbf{r} + \int_{C_2} \mathbf{F} \cdot d\mathbf{r},$$

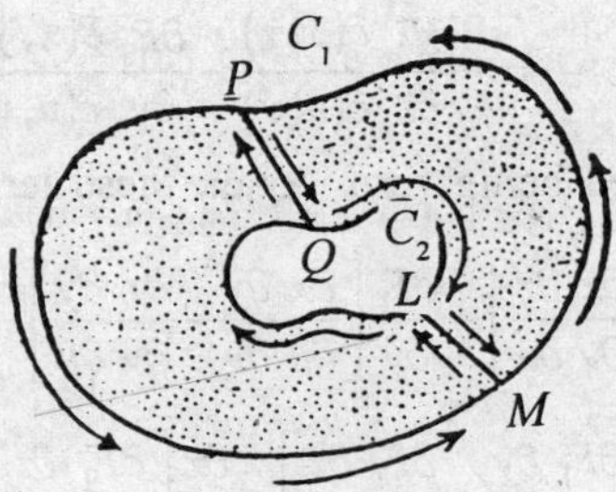

Fig. 11.13.

where the line integrals on the right are taken along the directions of C_1 and C_2 as shown in the figure. The result will follow, if we join the curves by any lines PQ and LM as shown and apply the theorem to the two surfaces thus obtained. The line integrals along PQ and LM will be taken in opposite directions and will consequently vanish on addition.

Cor. (*i*) We write

$$F = \mathbf{a}\phi,$$

where $\mathbf{a}$ is a constant vector and ϕ, a continuously differentiable scalar point function. Applying Stoke's theorem, we have

$$\int_C \mathbf{a}\phi \cdot d\mathbf{r} = \int_S \text{curl } (\mathbf{a}\phi) \cdot \mathbf{n}\, dS. \quad ...(1)$$

Also

$$\text{curl } (\mathbf{a}\phi) = \nabla \times (\mathbf{a}\phi)$$

$$= \nabla\phi \times \mathbf{a}. \quad [\S\ 10.15,\ \text{p. } 295] \quad ...(2)$$

From (1) and (2),

$$\mathbf{a} \cdot \int_C \phi\, d\mathbf{r} = \int_S \nabla\phi \times \mathbf{a} \cdot \mathbf{n}\, dS$$

$$= -\,\mathbf{a} \cdot \int_S \nabla\phi \times \mathbf{n}\, dS$$

or

$$\mathbf{a} \cdot \left[\int_C \phi \, d\mathbf{r} + \int_S \nabla\phi \times \mathbf{n} \, dS \right] = 0.$$

$$\therefore \qquad \int_C \phi \, d\mathbf{r} + \int_S \nabla\phi \times \mathbf{n} \, dS = 0;$$

a being any arbitrary vector.

$$\therefore \qquad \int_C \phi \, d\mathbf{r} = \int_S \mathbf{n} \times \nabla\phi \, dS.$$

(*ii*) Again, we write

$$\mathbf{f} = \mathbf{a} \times \mathbf{F},$$

where **F** is a constant vector and apply Stoke's theorem to the function **f**.

$$\therefore \qquad \int_C \mathbf{a} \times \mathbf{F} \cdot d\mathbf{r} = \int_S \text{curl} \, (\mathbf{a} \times \mathbf{F}) \cdot \mathbf{n} \, dS.$$

Also

$$\text{curl} \, (\mathbf{a} \times \mathbf{F}) = \nabla \times (\mathbf{a} \times \mathbf{F}) = \mathbf{a} \, (\nabla \cdot \mathbf{F}) - (\mathbf{a} \cdot \nabla) \, \mathbf{F}$$

and

$$[(\mathbf{a} \cdot \nabla) \, \mathbf{F}] \cdot \mathbf{n} = \mathbf{a} \cdot \nabla \, (\mathbf{F} \cdot \mathbf{n}),$$

where it is understood that ∇ does not operate on **n** and for the purpose of ∇, **n** is thought of as constant. Thus,

$$\int_C \mathbf{a} \cdot \mathbf{F} \times d\mathbf{r} = \int_S \left[\mathbf{a} \cdot (\nabla \cdot \mathbf{F}) \mathbf{n} - \mathbf{a} \cdot \nabla (\mathbf{F} \cdot \mathbf{n}) \right] dS$$

or

$$\mathbf{a} \cdot \int_C \mathbf{F} \times d\mathbf{r} = \mathbf{a} \cdot \int_S \left[(\nabla \cdot \mathbf{F}) \, \mathbf{n} - \nabla \, (\mathbf{F} \cdot \mathbf{n}) \right] dS$$

so that, as before, **a** being a constant vector,

$$\int_C \mathbf{F} \times d\mathbf{r} = \int_S \left[(\nabla \cdot \mathbf{F}) \, \mathbf{n} - \nabla \, (\mathbf{F} \cdot \mathbf{n}) \right] dS.$$

Looking upon ∇ as a vector, we may write

$$(\nabla \cdot \mathbf{F}) \, \mathbf{n} - \nabla \, (\mathbf{F} \cdot \mathbf{n}) = -\, (\mathbf{n} \times \nabla) \times \mathbf{F}.$$

Hence

$$\int_C d\mathbf{r} \times \mathbf{F} = \int_S (\mathbf{n} \times \nabla) \times \mathbf{F} \, dS.$$

The three results connecting line and surface integrals which we have arrived at may now be restated as follows :

$$\int_C d\mathbf{r} \cdot \mathbf{F} = \int_S (\mathbf{n} \times \nabla) \cdot \mathbf{F} \, dS,$$

$$\int_C d\mathbf{r} \times \mathbf{F} = \int_S (\mathbf{n} \times \nabla) \times \mathbf{F}\, dS,$$

$$\int_C d\mathbf{r}\phi = \int_S (\mathbf{n} \times \nabla)\, \phi dS,$$

so that we have replaced $d\mathbf{r}$ by $\mathbf{n} \times \nabla$. It should also be understood that ∇ on the right does not operate on $\mathbf{n}$.

Note. The truth of Stoke's theorem in three dimensions can also be shown to be a simple *intuitively* perceived consequence of that in plane, as we shall now see.

Let C be any curve and S any surface through the same. Divide S into any finite number of sub-surfaces. The boundary curve of each sub-surface is supposed to be described in the sense of rotation of a right handed screw when it is moved along the direction of the normal.

Let δS_p denote the magnitude of the area of any sub-surface and $\mathbf{n}_p$ the unit normal vector drawn at any point of the sub-surface. Then, by def.

$$\int_S \text{curl}\, \mathbf{F} \cdot \mathbf{n}\, dS = \lim \Sigma \text{ curl}\, \mathbf{F} \cdot \mathbf{n}_p\, \delta S_p, \qquad \text{...(1)}$$

where curl **F** denotes the value at any point of the sub-surface.

Also by the theorem for the plane surface,

$$\int_{\delta S_p} \text{curl}\, \mathbf{F} \cdot \mathbf{n}\, dS = \int_{C_p} \mathbf{F} \cdot d\mathbf{r}_p \qquad \text{...(2)}$$

for we can regard each sub-surface as approximately a plane region. As δS_p is small, we can re-write (2) as

$$\text{curl}\, \mathbf{F} \cdot \mathbf{n}_p\, \delta S_p = \int_{C_p} \mathbf{F} \cdot d\mathbf{r}. \qquad \text{...(3)}$$

On addition, we now get, from (1) and (3)

$$\int_S \text{curl}\, \mathbf{F} \cdot \mathbf{n}\, dS = \int_C \mathbf{F} \cdot d\mathbf{r},$$

for the line integrals along bounding arcs of sub-surfaces cancel in pairs.

EXAMPLES

Example 1. *Verify Stoke's theorem for the function*

$$\mathbf{F} = x\,(\mathbf{i}x + \mathbf{j}y),$$

integrated round the square in the plane $z = 0$ whose sides are along the lines

$$x = 0, \quad y = 0, \quad x = a, \quad y = a.$$

Solution. We have

curl $x\,(\mathbf{i}x + \mathbf{j}y) = \mathbf{k}y$.

$$\therefore \qquad \int_S \text{curl}\, x\,(\mathbf{i}x + \mathbf{j}y) \cdot d\mathbf{a}$$

$$= \int_0^a \int_0^a \mathbf{k} y.\mathbf{k}\, dx\, dy$$

$$= \int_0^a \int_0^a y\, dx\, dy = \frac{a^3}{2}$$

Again

$$+ \int_{BC} \mathbf{F} \cdot d\mathbf{r} + \int_{CO} \mathbf{F} \cdot d\mathbf{r}.$$

$$+ \int_{BC} \mathbf{F} \cdot d\mathbf{r} + \int_{CO} \mathbf{F} \cdot d\mathbf{r}.$$

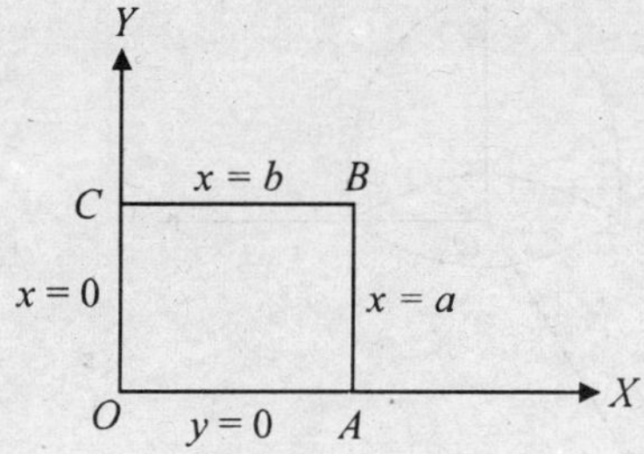

Fig. 11.14.

Now

$$\int_{OA} \mathbf{F} \cdot d\mathbf{r} = \int_0^a x\left(\mathbf{i}x + \mathbf{j}y\right) \cdot \mathbf{i}dx = \int_0^a x^2 dx = \frac{1}{3}a^3,$$

$$\int_{AB} \mathbf{F} \cdot d\mathbf{r} = \int_0^a x\left(\mathbf{i}x + \mathbf{j}y\right) \cdot \mathbf{j}dy = \int_0^a ay\, dy = \frac{1}{2}a^3,$$

$$\int_{BC} \mathbf{F} \cdot d\mathbf{r} = \int_a^0 x\left(\mathbf{i}x + \mathbf{j}y\right) \cdot \mathbf{i}dx = -\int_0^a x^2 dx = -\frac{1}{3}a^3,$$

$$\int_{CO} \mathbf{F} \cdot d\mathbf{r} = \int_a^0 x\left(\mathbf{i}x + \mathbf{j}y\right) \cdot \mathbf{j}dy = 0.$$

$$\therefore \quad \int_C \mathbf{F} \cdot d\mathbf{r} = \frac{1}{3}a^3 + \frac{1}{2}a^3 - \frac{1}{3}a^3 + 0 = \frac{1}{2}a^3.$$

Hence the verification.

Example 2. *Find the value of*

$$\int \text{curl } \mathbf{F} \cdot d\mathbf{a},$$

taken over the portion of the surface

$$x^2 + y^2 - 2ax + az = 0,$$

for which $z \geq 0$, *when*

$$\mathbf{F} = \left(y^2 + z^2 - x^2\right)\mathbf{i} + \left(z^2 + x^2 - y^2\right)\mathbf{j} + \left(x^2 + y^2 - z^2\right)\mathbf{k}.$$

Solution. Rewriting the equation

$$x^2 + y^2 - 2ax + az = 0$$

as

$$(x - a)^2 + y^2 = -a\,(z - a),$$

we see that the surface is a paraboloid with its vertex at (a, o, a) and axis parallel to z-axis and turned towards the negative direction of the same. It meets the plane $z = 0$ in the circle C, given by

$$x^2 + y^2 - 2ax = 0, \quad z = 0.$$

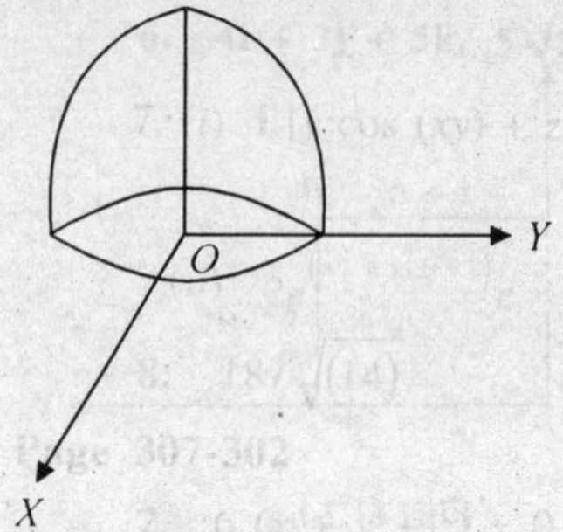

Fig. 11.15.

(In the Fig., O is the point (a, o, o), and ox, oy, oz, are the lines parallel to the co-ordinate axes).

By Stoke's theorem, the given surface integral is equal to the line integral

$$\int_C \mathbf{F} \cdot d\mathbf{r}.$$

The circle C is given by

$$x = a\,(1 + \cos\theta), \quad y = a \sin\theta, \quad z = 0.$$

Along C,

$$\mathbf{F} = \left[a^2 \sin^2\theta - a^2(1+\cos\theta)^2\right]\mathbf{i} + \left[a^2(1+\cos\theta)^2 - a^2\sin^2\theta\right]\mathbf{j} + \left[a^2(1+\cos\theta)^2 + a^2\sin^2\theta\right]\mathbf{k}$$

$$\therefore \quad \mathbf{F} \cdot \left(\frac{dx}{d\theta}\mathbf{i} + \frac{dy}{d\theta}\mathbf{j} + \frac{dz}{d\theta}\mathbf{k}\right)$$

$$\therefore \quad = [a^2 \sin^2\theta - a^2(1+\cos\theta)^2]\,[-a\sin\theta - a\cos\theta].$$

$$\therefore \quad \int_C \mathbf{F} \cdot d\mathbf{r} = \int_0^{2\pi} \mathbf{F} \cdot \frac{d\mathbf{r}}{d\theta}\, d\theta = 2a^3\pi.$$

Another method. By a further application of Stoke's theorem, we see that the given integral

$$= \int_{S_1} \text{curl } \mathbf{F} \cdot \mathbf{n}\, dS,$$

where S_1 is the plane region bounded by the circle C.

Here

$$\mathbf{n} = \mathbf{k}$$

Thus, $\quad \text{curl } \mathbf{F} \,.\, \mathbf{n} = 2(x - y).$

$\therefore$ The integral $= 2\iint (x - y)\, dx\, dy,$

taken over S_1. Changing to polar co-ordinates, so that

$$x = a + r\cos\theta, \quad y = r\sin\theta,$$

we see that the integral

$$= 2\int_0^a \int_0^{2\pi} (a + r\cos\theta - r\sin\theta)\, r d\theta\, dr$$

$$= 2a^3\pi.$$

Example 3. *Applying Stoke's theorem to prove that*

$$\int_C (ydx + zdy + xdz) = -2\sqrt{2\pi a^2},$$

where C is the curve given by

$$x^2 + y^2 + z^2 - 2ax - 2ay = 0, \ \ x + y = 2a$$

and begins at the point (2*a*, 0, 0) *and goes at first below the z-plane.*

Solution. It may be easily seen that the curve C is the circle drawn on AB as diameter and lying in the plane $x + y = 2a$ which passes through A, B and is perpendicular to the z-plane.

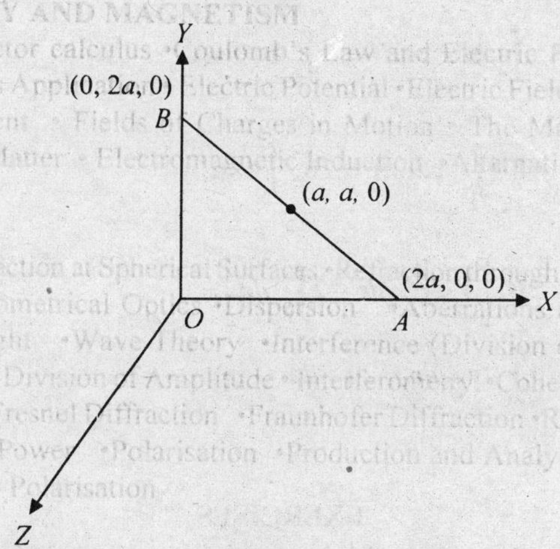

Fig. 11.16.

The given line integral

$$= \int_C (y\mathbf{i} + z\mathbf{j} + x\mathbf{k}) \cdot d\mathbf{r} = \int_S \text{curl}\,(y\mathbf{i} + z\mathbf{j} + x\mathbf{k}) \cdot \mathbf{n}\, dS$$

where S is the surface of the circle referred to above.

We have

$$\text{curl}\,(y\mathbf{i} + z\mathbf{j} + x\mathbf{k}) = -(\mathbf{i} + \mathbf{j} + \mathbf{k}),$$

$$\mathbf{n} = \frac{1}{\sqrt{2}}\mathbf{i} + \frac{1}{\sqrt{2}}\mathbf{j}.$$

$\therefore$ The given line integral

$$= \int_S -(\mathbf{i} + \mathbf{j} + \mathbf{k}) \cdot \left(\frac{1}{\sqrt{2}}\mathbf{i} + \frac{1}{\sqrt{2}}\mathbf{j}\right) dS$$

$$= -\frac{2}{\sqrt{2}} \int_S dS = -\sqrt{2\left(2a^2\pi\right)} = -2\sqrt{2\pi a^2}.$$

Example 4. *Prove that*

$$\int_C \mathbf{r} \times d\mathbf{r} = 2\int_S d\mathbf{a},$$

where S is a diaphragm enclosing a circuit C. *(Lucknow, 98)*

Solution. We apply Stoke's theorem to the function $\mathbf{b} \times \mathbf{r}$ where $\mathbf{b}$ is any constant vector.

Thus,

$$\int_C \mathbf{b} \times \mathbf{r} \cdot d\mathbf{r} = \int_S \text{curl}\left(\mathbf{b} \times \mathbf{r}\right) \cdot d\mathbf{a}.$$

Now $$\nabla \times \left(\mathbf{b} \times \mathbf{r}\right) = \mathbf{b}\nabla \cdot \mathbf{r} - \mathbf{b} \cdot \nabla\mathbf{r} = 3\mathbf{b} - \mathbf{b} = 2\mathbf{b},$$

$$\therefore \quad \int_C \mathbf{b} \times \mathbf{r} \cdot d\mathbf{r} = 2\int_S \mathbf{b} \cdot d\mathbf{a},$$

or

$$\mathbf{b} \cdot \int_C \mathbf{r} \times d\mathbf{r} = 2\mathbf{b} \cdot \int_S d\mathbf{a},$$

or

$$\mathbf{b} \cdot \left[\int_C \mathbf{r} \times d\mathbf{r} - 2\int_S d\mathbf{a}\right] = 0,$$

$$\therefore \quad \int_C \mathbf{r} \times d\mathbf{r} = 2\int_S d\mathbf{a},$$

for $\mathbf{b}$ is an arbitrary vector.

EXERCISES

1. Evaluate

$$\int \left(xy\,dx + xy^2 dy\right)$$

taken round the positively oriented square with vertices (± 1, 0), (0, ± 1) and verify by using Stoke's theorem.

2. Show that

$$\int_S \text{curl}\,\mathbf{f} \cdot \mathbf{n}\,ds = 0,$$

S being a closed surface.

[This question may be done both by Gauss's as well as Stoke's theorem.]

3. If the normal surface integral of a vector point function **G** over every open surface is equal to the tangential line integral of another **F** round its boundary, prove that

$$\mathbf{G} = \text{curl}\ \mathbf{F}.$$

4. Show that

(*i*) $\int_C \mathbf{r} \cdot d\mathbf{r} = 0.$

(ii) $\int_C \phi\nabla\phi \cdot d\mathbf{r} = 0.$

(iii) $\int_C \phi\nabla\psi \cdot d\mathbf{r} = -\int_C \psi\nabla\phi \cdot d\mathbf{r}.$

5. Show that

$$\phi(p) = \int_A^P \mathbf{F} \cdot d\mathbf{r}$$

6. $\mathbf{f} = \nabla\phi$, $\mathbf{g} = \nabla\psi$ are two functions such that $\nabla^2\phi = 0$, $\nabla^2\psi = 0$, show that

$$\int_S \phi \text{ curl } \mathbf{f} \cdot d\mathbf{a} = \int_C \phi \mathbf{f} \, d\mathbf{r} - \int_S \text{grad } \phi \times \mathbf{f} \cdot d\mathbf{a}.$$

11.9. CONDITION FOR IRROTATIONAL VECTOR FUNCTIONS

We have already obtained one form of condition for a vector point function to be irrotational and we shall now obtain another with the help of Stoke's theorem.

Theorem. *The necessary and sufficient condition for a vector point function* **F** *to be irrotational in a simply connected region is that curl* **F** *= 0 at every point of the region.*

The condition is necessary. Let **F** be irrotational so that, as shown in § 11.2.1, Page 325, there exists a scalar ϕ such that

$$\mathbf{F} = \nabla\phi.$$

$$\therefore \qquad \text{curl } \mathbf{F} = \text{curl } \nabla\phi = \nabla \times \nabla\phi = 0.$$

The condition is sufficient. Let curl **F** = 0. We have, by Stoke's theorem

$$\int_C \mathbf{F} \cdot d\mathbf{r} = \int_S \text{curl } \mathbf{F} \cdot d\mathbf{a} \equiv 0.$$

Hence, **F** is irrotational.

Note. The gradient of a scalar point function is a vector point function. It has now been seen that a given vector function **F** is the gradient of a scalar function if and only if

$$\text{curl } \mathbf{F} = 0$$

11.9.1. *The determination of a scalar function ϕ such that*

$$\mathbf{F} = \nabla\phi \; ; \qquad ...(1)$$

F *being a given irrotational vector function.*

Firstly we notice that ϕ is not unique. In fact, if ϕ be any solution of (1), then ϕ + **c** is the *general* solution; *c* being any arbitrary constant.

[Refer § 10.14, Page 295].

We have seen that [Refer § 11.2.1, Page 325]

$$\phi(p) = \int_A^P \mathbf{F} \cdot d\mathbf{r} \qquad ...(2)$$

A being any fixed point.

In terms of rectangular cartesian axes, let A be (x_0, y_0, z_0) and P be any point (x, y, z). The integral being independent of the curve joining A to P, we choose the same as consisting of 3 portions of straight lines AB, BC, CP respectively parallel to the three co-ordinate axes. Thus, we have from (2),

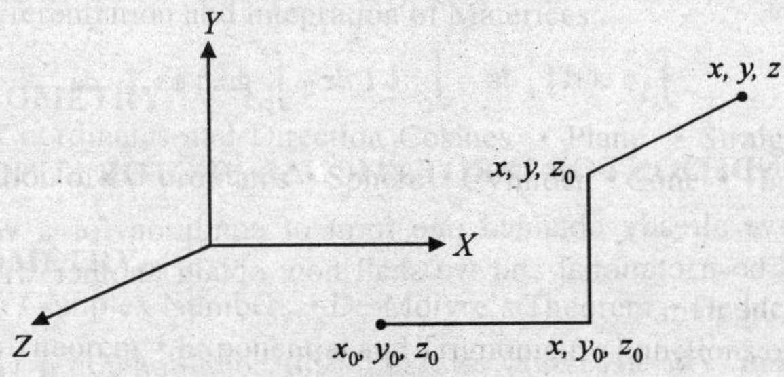

Fig. 11.17.

$$\phi(x, y, z) = \int_{(x_0, y_0, z_0)}^{(x, y, z)} (F_1 dx + F_2 dy + F_3 dz)$$

$$= \int_{(x_0, y_0, z_0)}^{(x, y_0, z_0)} F_1 dx + \int_{(x, y_0, z_0)}^{(x, y, z_0)} F_2 dy + \int_{(x, y, z_0)}^{(x, y, z)} F_3 dz,$$

for $\int F_2 dy$ and $\int F_3 dz$ are zero for lines parallel to x-axis and so on.

For different choices of the point (x_0, y_0, z_0), the values of $\phi(x, y, z)$ will differ from each other by constants.

Illustration. Let

$$\mathbf{F} = (y \sin z - \sin x)\,\mathbf{i} + (x \sin z + 2yz)\,\mathbf{j} + (xy \cos z + y^2)\,\mathbf{k}.$$

We may verify that curl $\mathbf{F} = 0$. We may take $(0, 0, 0)$ for (x_0, y_0, z_0).

$$\therefore \quad \phi = \int_{(0,0,0)}^{(x,0,0)} (y \sin z - \sin x)\, dx + \int_{(x,0,0)}^{(x,y,0)} (x \sin z + 2yz)\, dy + \int_{(x,y,0)}^{(x,y,z)} \left(xy \cos z + y^2\right) dz$$

$$= \Big| xy \sin z + \cos x \Big|_{(0,0,0)}^{(x,0,0)} + \Big| xy \sin z + y^2 z \Big|_{(x,0,0)}^{(x,y,0)} + \Big| xy \sin z + y^2 z \Big|_{(x,y,0)}^{(x,y,z)}$$

$= -1 + \cos x + xy \sin z + y^2 z.$

$\therefore \quad \mathbf{F} = \nabla (\cos x + xy \sin z + y^2 z)$, omitting the constants – 1.

The result may be easily verified.

Ex. 1. Show that the following vector functions **F** are irrotational and find the corresponding scalars ϕ such that $\mathbf{F} = \nabla\phi$

(*i*) $yz\mathbf{i} + zx\mathbf{j} + xy\mathbf{k}$.

(*ii*) $(\sin y + z \cos x)\, \mathbf{i} + (x \cos y + \sin z)\, \mathbf{j} + (y \cos z + \sin x)\, \mathbf{k}$.

(*iii*) $(x^2 - yz)\, \mathbf{i} + (y^2 - zx)\, \mathbf{j} + (z^2 - xy)\, \mathbf{k}$.

11.10. CONDITION FOR SOLENOIDAL VECTOR POINT FUNCTIONS

Theorem. *The necessary and sufficient condition for a vector point function* **F** *to be solenoidal in a simply connected region is that*

$$\text{div } \mathbf{F} = 0,$$

at every point of the region.

The condition is necessary. We have see in Cor. 2, § 11.6, Page 339, that at any point

$$\text{div } \mathbf{F} = \lim_{v \to 0} \frac{\int_S \mathbf{F} . d\mathbf{a}}{V} = 0,$$

if **F** is solenoidal [§ 11.4, Page 329].

The condition is sufficient. The condition is sufficient, for if div **F** = 0, we have by Gauss's theorem

$$\int_S \mathbf{F} \cdot d\mathbf{a} = \int_V \text{div } \mathbf{F} dv = 0,$$

so that the flux

$$\int_S \mathbf{F} \cdot d\mathbf{a}$$

across every closed surface is zero.

Hence the theorem.

11.10.1. We know that

$$\text{div curl } \mathbf{f} = 0,$$

so that the divergence of every curl is zero, *i.e.*, the curl of every function is solenoidal. It will now be shown that every solenoidal vector point function is the curl of some vector function.

Theorem. *If* **F** *is a continuously differentiable vector point function such that div* **F** *= 0, then there exists a vector point function* **f** *such that*

$$\mathbf{F} = \text{curl } \mathbf{f}.$$

Firstly, we shall see *that if* **f** *is any function whose curl is* **F**, *then the general function whose curl is* **F** *is*

$$\mathbf{f} + \nabla\phi,$$

where ϕ is any continuously differentiable scalar point function.

Firstly, if

$$\text{curl } \mathbf{f} = \mathbf{F} = \text{curl } \mathbf{g},$$

so that

$$\text{curl } (\mathbf{g} - \mathbf{f}) = 0,$$

we see that $\mathbf{g} - \mathbf{f}$ is the gradient of some scalar ϕ, [§ 11.9, Page 362].

i.e., $\mathbf{g} - \mathbf{f} = \nabla\phi$ or $\mathbf{g} = \mathbf{f} + \nabla\phi$.

Also if ϕ be *any* scalar point function, then

$$\text{curl } (\mathbf{f} + \nabla\phi) = \text{curl } \mathbf{f} + \text{curl } \nabla\phi$$
$$= \mathbf{F} + \nabla \times \nabla\phi = \mathbf{F}.$$

Hence the result.

Now we attend to the theorem as stated.

Let

$$\mathbf{F} = F_1\mathbf{i} + F_2\mathbf{j} + F_3\mathbf{k}, \quad \mathbf{f} = f_1\mathbf{i} + f_2\mathbf{j} + f_3\mathbf{k}.$$

$$\therefore \quad \text{curl } \mathbf{f} = \left(\frac{\partial f_3}{\partial y} - \frac{\partial f_2}{\partial z}\right)\mathbf{i} + \left(\frac{\partial f_1}{\partial z} - \frac{\partial f_3}{\partial x}\right)\mathbf{j} + \left(\frac{\partial f_2}{\partial x} - \frac{\partial f_1}{\partial y}\right)\mathbf{k}.$$

Then

$$\mathbf{F} = \text{curl } \mathbf{f}$$

gives

$$F_1 = \frac{\partial f_3}{\partial y} - \frac{\partial f_2}{\partial z}, \quad F_2 = \frac{\partial f_1}{\partial z} - \frac{\partial f_3}{\partial x}, \quad F_3 = \frac{\partial f_2}{\partial x} - \frac{\partial f_1}{\partial y}. \qquad \text{...(1)}$$

Also we have

$$\frac{\partial F_1}{\partial x} + \frac{\partial F_2}{\partial y} + \frac{\partial F_3}{\partial z} = 0. \qquad \text{...(2)}$$

We have to establish the existence of some **f** such that

$$\text{curl } \mathbf{f} = \mathbf{F}.$$

We suppose that $f_1 = 0$.

Then we have, from (1)

$$F_2 = -\frac{\partial f_3}{\partial x} \text{ which gives } f_3 = -\int_{x_0}^{x} F_3 dx + \phi(y, z), \qquad \text{...(3)}$$

$$F_3 = -\frac{\partial f_2}{\partial x} \text{ which gives } f_2 = -\int_{x_0}^{x} F_3 dx + \psi(y, z). \qquad ...(4)$$

In these integrations y and z have to be thought of as parameters. From (3) and (4),

$$\frac{\partial f_3}{\partial x} = -\int_{x_0}^{x} \frac{\partial F_2}{\partial y} dx + \frac{\partial \phi}{\partial y} \qquad ...(5)$$

$$\frac{\partial f_2}{\partial z} = \int_{x_0}^{x} \frac{\partial F_3}{\partial z} dx + \frac{\partial \psi}{\partial z} \qquad ...(6)$$

$$\therefore \quad F_1 = \frac{\partial f_3}{\partial y} - \frac{\partial f_2}{\partial z} = -\int_{x_0}^{x} \left(\frac{\partial F_2}{\partial y} + \frac{\partial F_3}{\partial z}\right) dx + \frac{\partial \phi}{\partial y} - \frac{\partial \psi}{\partial z}$$

$$= \int_{x_0}^{x} \frac{\partial F_1}{\partial x} dx + \frac{\partial \phi}{\partial y} - \frac{\partial \psi}{\partial z} \qquad \text{[By (2)]}$$

$$= F_1(x, y, z) - F_1(x_0, y, z) + \frac{\partial \phi}{\partial y} - \frac{\partial \psi}{\partial z},$$

or
$$F_1(x_0, y, z) = \frac{\partial \phi}{\partial y} - \frac{\partial \psi}{\partial z}.$$

We now *suppose* that $\psi = 0$.

$$\therefore \quad \frac{\partial \phi}{\partial y} = F_1(x_0, y, z) \text{ or } \phi = \int_{y_0}^{y} F_1(x_0, y, z)\, dy.$$

Thus, we have

$$f_1 = 0, \; f_2 = \int_{x_0}^{x} F_3 dx, \; f_3 = -\int_{x_0}^{x} F_2 dx + \phi(y, z),$$

where

$$\phi(y, z) = \int_{y_0}^{y} F_1(x_0, y, z)\, dy.$$

It may now be easily shown that f_1, f_2, f_3 as determined above, are the components of a vector **f** whose curl is the given function **F**.

Illustration. Let

$$\mathbf{F} = (y - z)\,\mathbf{i} + (z - x)\,\mathbf{j} + (x - y)\,\mathbf{k}.$$

We may verify that div **F** = 0. We take $x_0 = 0, \; y_0 = 0$.

Now

$$f_2 = \int_0^x (x - y)\, dx = \frac{1}{2}x^2 - xy,$$

$$f_3 = -\int_0^y (z-x)\,dx + \phi(y,z) = \frac{1}{2}x^2 - xz + \phi(y,z),$$

$$\phi(y,z) = \int_0^y (y-z)\,dy = \frac{1}{2}y^2 - yz.$$

$$\therefore \quad \mathbf{f} = f_1\mathbf{i} + f_2\mathbf{j} + f_3\mathbf{k}$$

$$= \left(\frac{1}{2}x^2 - xy\right)\mathbf{j} + \left[\frac{1}{2}\left(x^2+y^2\right) - (x+y)\,z\right]\mathbf{k}.$$

The *general* function **f** whose curl is **F** is

$$\left(\frac{1}{2}x^2 - xy\right)\mathbf{j} + \left[\frac{1}{2}\left(x^2+y^2\right) - (x+y)\,z\right]\mathbf{k} + \nabla g,$$

where g is any scalar point function.

Note. The curl of a vector point function is a vector point function but every vector point function is not necessarily the curl of a vector function. It follows from above that a vector point function **F** is the curl of a function if and only if

$$\text{div}\,\mathbf{F} = 0,$$

i.e., every solenoidal vector function is the curl of a function.

Ex. Show that the following vector functions **F** are solenoidal and find the functions **f** such that **F** = curl **f**.

(*i*) $(y+z)\,\mathbf{i} + (z+x)\,\mathbf{j} + (x+y)\,\mathbf{k}$.

(*ii*) $x\,(y-z)\,\mathbf{i} + y\,(z-x)\,\mathbf{j} + z\,(x-y)\,\mathbf{k}$.

(*iii*) $yz\mathbf{i} + zx\mathbf{j} + xy\mathbf{k}$.

11.11. GREEN'S THEOREM

If ϕ and ψ are two continuously differentiable scalar point functions such that $\nabla\phi$ and $\nabla\psi$ are also continuously differentiable, then

$$\int_V \left(\phi\nabla^2\psi - \psi\nabla^2\phi\right) dv = \int_S (\phi\nabla\psi - \psi\nabla\phi)\cdot\mathbf{n}\,dS,$$

where V is the region enclosed by a surface S.

(*Utkal 2003 ;Gorakhpur 98, 2000; Lucknow 98*)

Applying Gauss's Theorem to the function $\phi\nabla\psi$, we obtain

$$\int_S (\phi\nabla\psi\cdot\mathbf{n}dS) = \int_V \text{div}\,(\phi\nabla\psi)\,dv.$$

Also

$$\text{div}\,(\phi\nabla\psi) = \nabla\cdot(\phi\nabla\psi)$$
$$= \nabla\phi\cdot\nabla\psi + \phi\nabla^2\psi,$$

$$\therefore \qquad \int_S \phi\nabla\psi\cdot\mathbf{n}\,dS = \int_V \nabla\phi\cdot\nabla\psi dv + \int_V \phi\nabla^2\psi dv, \qquad ...(1)$$

or
$$\int_V \nabla\phi\cdot\nabla\psi dv = \int_S \phi\nabla\psi.\mathbf{n}dS - \int_V \phi\nabla^2\psi dv. \qquad ...(2)$$

The result (2) is known as Green's first identity.

Interchanging ϕ and ψ in (1), we get

$$\int_S \psi\nabla\phi\cdot\mathbf{n}dS = \int_V \nabla\psi.\nabla\phi dv + \int_V \psi\nabla^2\phi dv. \qquad ...(3)$$

Subtracting (3) from (1), we get

$$\int_S (\phi\nabla\psi - \psi\nabla\phi)\cdot\mathbf{n}dS = \int_V \left(\phi\nabla^2\psi - \psi\nabla^2\phi\right) dv.$$

Hence the theorem.

The result of the theorem can also be stated as

$$\int_V \left(\phi\nabla^2\psi - \psi\nabla^2\phi\right) dv = \int_S \left(\phi\frac{\partial\psi}{\partial n} - \psi\frac{\partial\phi}{\partial n}\right) dS,$$

where $\partial\phi/\partial n$ and $\partial\psi/\partial n$ denote the directional derivatives of ϕ and ψ respectively along the outward drawn normal at any point of S.

11.11.1. Green's Formula.

Suppose that ϕ is a scalar point function which is twice continuously differentiable in a region V enclosed by a surface S.

Take any fixed point O, *within* the region and let $\mathbf{r}$ denote the position vector of any point P relative to O and let

$$OP = |\,\mathbf{r}\,| = r.$$

We write

$$\psi = \frac{1}{r}$$

so that, except at O, ψ is a twice continuously differentiable scalar function.

We surround O by a small sphere of radius $\mathbf{r}$. Let S_1 denote the surface of this sphere and V_1 the region bounded by S and S_1.

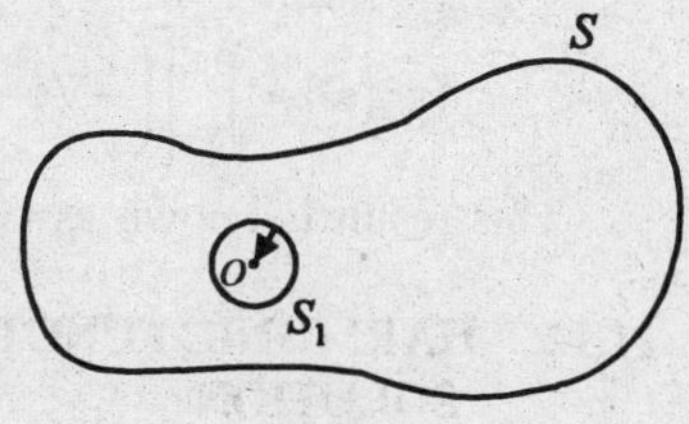

Fig. 11.18.

We have

$$\nabla\psi = -\frac{\mathbf{r}}{r^2},\ \nabla^2\psi = 0.$$

We apply Green's theorem to the region V_1 enclosed by the surfaces S, S_1. Thus

$$\int_V -\frac{1}{r}\nabla^2\phi dv = \int_S \left[\phi\frac{\partial}{\partial n}\left(\frac{1}{r}\right) - \frac{1}{r}\frac{\partial\phi}{\partial n}\right] dS + \int_{S_1} \left[\phi\frac{\partial}{\partial n}\left(\frac{1}{r}\right) - \frac{1}{r}\frac{\partial\phi}{\partial n}\right] dS, \qquad ...(1)$$

where the unit normal at any point of the sphere S_1 is directed toward O.

At any point of S_1,

$$\frac{\partial}{\partial n}\left(\frac{1}{r}\right) = \nabla\left(\frac{1}{r}\right)\cdot\mathbf{n}$$

$$= -\frac{\mathbf{r}}{r^3}\cdot -\frac{\mathbf{r}}{r} = \frac{\mathbf{r}^2}{r^4} = \frac{1}{r^2} = \frac{1}{r^2}.$$

Now

$$\int_{S_1} \phi\frac{\partial}{\partial n}\left(\frac{1}{r}\right) dS = \frac{1}{\varepsilon^2}\int_{S_1} \phi dS,$$

$$\lim_{\varepsilon\to 0}\int_{S_1} \phi\frac{\partial}{\partial n}\left(\frac{1}{r}\right) dS = 4\pi\phi(O)$$

$$\lim_{\varepsilon\to 0}\int_{S_1} \frac{1}{r}\frac{\partial\phi}{\partial n} dS = 0$$

Thus, from (1), taking the limits when $\varepsilon \to 0$, we obtain

$$\int_V -\frac{1}{r}\nabla^2\phi dv = \int_S \left[\phi\frac{\partial}{\partial n}\left(\frac{1}{r}\right) - \frac{1}{r}\frac{\partial\phi}{\partial n}\right] dS + 4\pi\phi(O)$$

or $$4\pi\phi(O) = \int_S \left[\frac{1}{r}\frac{\partial\phi}{\partial n} - \phi\frac{\partial}{\partial n}\left(\frac{1}{r}\right)\right] dS - \int_V \frac{1}{r}\nabla^2\phi dv$$

which may be re-written as

$$4\pi\phi(O) = \int_S \left[\left(\frac{1}{r}\nabla\phi - \phi\nabla\frac{1}{r}\right)\right]\cdot d\mathbf{a} - \int_V \frac{1}{r}\nabla^2\phi dv.$$

This result is known as Green's formula.

11.12. HARMONIC FUNCTIONS, LAPLACE'S EQUATION AND ITS SOLUTION

Def. *A function ϕ is said to be harmonic in a region if it is twice continuously differentiable in the region and*

$$\nabla^2\phi = 0$$

at every point of the region.

Also the equation $\nabla^2\phi = 0$ is known as Laplace's equation.

Thus, if ϕ be a harmonic function in a region V enclosed by a surface S then by Green's formula, we see that for any point O within the region

$$4\pi\phi(O) = \int_S \left(\frac{1}{r}\nabla\phi - \phi\nabla\frac{1}{r}\right)\cdot d\mathbf{a}, \qquad ...(1)$$

so that it is proved *that the value of a harmonic function ϕ at any point within a region is known in terms of the values of ϕ and $\partial\phi/\partial n$ at any point of the surface enclosing the region.*

11.13. POISSON'S EQUATION AND SOLUTION

The equation

$$\nabla^2\phi = -4\pi\rho,$$

where ϕ is a scalar point function vanishing outside a finite region is known as *Poisson's equation.*

Consider a region V bounded by a surface S. Applying Green's formula to the region V; ϕ being a scalar point function satisfying (1), we have

$$4\pi\phi(O) = \int_S \left(\frac{1}{r}\nabla\phi - \phi\nabla\frac{1}{r}\right)\cdot d\mathbf{a} + 4\pi\int_V \frac{\rho}{r}dv.$$

We now let the region V tend to infinity in all directions so that the surface S also recedes to infinity.

We now assume that for sufficiently large values of r, ϕ is of the form k/r where k remains bounded. Then for sufficiently large values of r, $|\nabla\phi|$ is of the form k/r^2.

$$\therefore \qquad \int_S \left(\frac{1}{r}\nabla\phi - \phi\nabla\frac{1}{r}\right)\cdot d\mathbf{a} \to 0,$$

as S recedes to infinity in all directions.

Hence, we see that

$$\phi(O) = \int_V \frac{\rho}{r}dv,$$

where the volume integral is carried over the whole space. The volume integral carried over the whole space is the same as the volume integral carried over the region outside which ρ is zero.

Cor. Consider

$$\nabla^2\mathbf{F} = -4\pi\mathbf{f}. \qquad ...(1)$$

Let

$$\mathbf{f} = f_1\mathbf{i} + f_2\mathbf{j} + f_3\mathbf{k}.$$

We write

$$\mathbf{F} = F_1\mathbf{i} + F_2\mathbf{j} + F_3\mathbf{k}.$$

The equation (1) is equivalent to

$$\nabla^2 F_1 = -4\pi f_1, \quad \nabla^2 F_2 = -4\pi f_2, \quad \nabla^2 F_3 = -4\pi f_3.$$

Under suitable conditions imposed on f_1, f_2, f_3, we see that for any point O,

$$F_1(O) = \int \frac{f_1}{r} dv, \quad F_2(O) = \int \frac{f_2}{r} dv, \quad F_3(O) = \int \frac{f_3}{r} dv.$$

These give

$$\mathbf{F}(O) = \int \frac{\mathbf{f}}{r} dv.$$

EXAMPLES

Example 1. *Show that two functions which are harmonic in a region enclosed by a surface and take on the same values at any point of the surface are identical.*

Solution. Let ϕ and ψ be two functions harmonic in a region bounded by a surface S and let $\phi = \psi$ at every point of S.

We have

$$\int_V \theta \nabla^2 \theta dv + \int_V \nabla\theta \cdot \nabla\theta dv = \int_S \theta \nabla\theta \cdot d\mathbf{a}.$$

Take $\theta = \phi - \psi$ so that $\nabla^2\theta = \nabla^2\phi - \nabla^2\psi = 0$ at every point of V and $\theta = 0$ at every point of S.

$$\therefore \qquad \int_V (\nabla\theta)^2 dv = 0.$$

$$\therefore \qquad \nabla\theta = 0,$$

i.e., θ is a constant.

Also θ is zero at every point of S, we see that θ is everywhere zero.

Hence the result.

Example 2. *Evaluate by Green's theorem in the plane*

$$\oint_C \left[\left(x^2 - \cosh y\right) dx + \left(y + \sin x\right) dy\right],$$

where C is the rectangle with vertices (0, 0) (π, 0), (π, 1), (0, 1).

Solution. The path of integration C is the rectangle OA, BC, as shown in figure. Let R be the plane region bounded by C.

Using Green's theorem in plane, we have

$$\oint_C \left[\left(x^2 - \cosh y\right) dx + \left(y + \sin x\right) dy\right]$$

$$= \iint_R \left[\frac{\partial}{\partial x}(y + \sin x) - \frac{\partial}{\partial y}\left(x^2 - \cosh y\right)\right] dx\, dy$$

$$= \int_{x=0}^{\pi} \int_{y=0}^{1} (\cos x + \sinh y)\, dx\, dy$$

$$= \int_0^{\pi} \left[y \cos x + \cosh y\right]_0^1 dx$$

$$= \int_0^{\pi} \left(\cos x + \cosh 1 - 1\right) dx$$

$$= \pi \cosh 1 - \pi = \pi \left(\cosh 1 - 1\right).$$

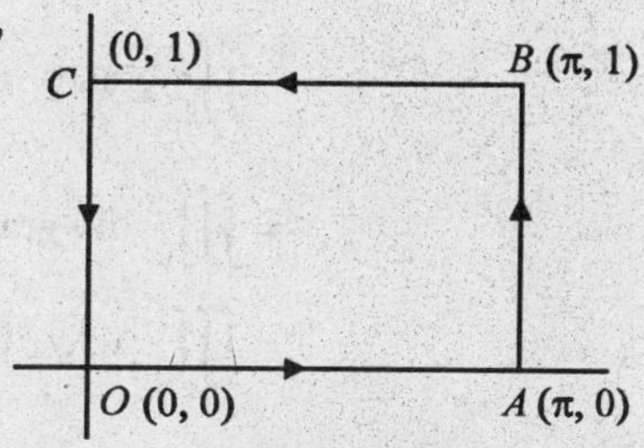

Fig. 11.19.

Example 3. *Show that the area bounded by a simple closed curve C is given by* $\frac{1}{2}\oint_C (x\, dy - y\, dx)$. *Hence, find the area of the ellipse* $x = a \cos\theta$, $y = b \sin\theta$. *(Garhwal, 92)*

Solution. If R is a closed region in the *xy*-plane bounded by a simple closed curve C, then by Green's theorem

$$\oint_C (\psi dx + \phi dy) = \iint_R \left(\frac{\partial \phi}{\partial x} - \frac{\partial \psi}{\partial y}\right) dx\, dy \qquad ...(1)$$

Putting $\psi = -y$ and $\phi = x$, we get

$$\oint_C (x\, dy - y\, dx) = \iint_R \left[\frac{\partial x}{\partial x} - \frac{\partial(-y)}{\partial y}\right] dx\, dy$$

$$\Rightarrow \qquad \frac{1}{2}\oint_C (x\, dy - y\, dx) = \iint_R dx\, dy = A.$$

where A is the area of the region R bounded by C.

Parametric equations of ellipse are

$$x = a \cos\theta, \quad y = b \sin\theta,$$

$$\Rightarrow \qquad dx = -a \sin\theta\, d\theta, \quad dy = b \cos\theta\, d\theta.$$

$\therefore$ Required area of ellipse

$$= \frac{1}{2}\int_0^{2\pi} \left(ab \cos^2\theta + ab \sin^2\theta\right) d\theta$$

$$= \frac{1}{2} ab \left[\theta\right]_0^{2\pi} = \pi\, ab.$$

Example 4. *If ϕ is harmonic in V, then show that $\iint_S (\partial\phi/\partial n)\, dS = 0$, where S is the surface enclosing V.*

Solution.
$$\iint_S \frac{\partial\phi}{\partial n} dS = \iint_S \left(\frac{\partial\phi}{\partial n}\mathbf{n}\right)\cdot \mathbf{n}\, dS$$

$$= \iint_S \text{grad}\ \phi \cdot \mathbf{n}\, dS \qquad \left[\because\ \text{grad}\ \phi = \frac{\partial\phi}{\partial n}\mathbf{n}\right]$$

$$= \iiint_V \text{div grad}\ \phi\, dV, \text{ by Gauss's divergence theorem.}$$

$$= \iiint_V \nabla^2\phi\, dV$$

$$= 0 \qquad [\because\ \phi \text{ is harmonic in } V \Rightarrow \nabla^2\phi = 0]$$

Example 5. *Show that*

$$\mathbf{F} = (\sin y + z)\,\mathbf{i} + (x \cos y - z)\,\mathbf{j} + (x - y)\,\mathbf{k}$$

is irrotational and find a function ϕ such that $\mathbf{F} = \nabla\phi$. *(Rohilkhand 2005*

Solution.
$$\text{curl}\ \mathbf{F} = \begin{vmatrix} \mathbf{i} & \mathbf{j} & \mathbf{k} \\ \partial/\partial x & \partial/\partial y & \partial/\partial z \\ \sin y + z & x\cos y - z & x - y \end{vmatrix}$$

$$= \mathbf{i}\,(-1+1) - \mathbf{j}\,(1-1) + \mathbf{k}\,(\cos y - \cos y) = 0.$$

$\therefore$ The given vector is irrotational and so $\mathbf{F} = \nabla\phi$. [§ 11.9]

Hence, $(\sin y + z)\,\mathbf{i} + (x \cos y - z)\,\mathbf{j} + (x - y)\,\mathbf{k}$

$$= \frac{\partial\phi}{\partial x}\mathbf{i} + \frac{\partial\phi}{\partial y}\mathbf{j} + \frac{\partial\phi}{\partial z}\mathbf{k}$$

$$\therefore \quad \frac{\partial\phi}{\partial x} = \sin y + z, \text{ hence, } \phi = x \sin y + xz + f_1(y, z) \qquad \text{...(1)}$$

$$\frac{\partial\phi}{\partial y} = x\cos y - z, \text{ hence, } \phi = x \sin y - yz + f_2(x, z) \qquad \text{...(2)}$$

$$\frac{\partial\phi}{\partial z} = x - y, \text{ hence, } \phi = xz - yz + f_3(x, y) \qquad \text{...(3)}$$

(1), (2) and (3) each represents ϕ. These agree if we choose $f_1(y, z) = -yz$, $f_2(x, z) = xz$ and $f_3(x, y) = x \sin y$. Hence, the required ϕ is given by

$$\phi = x \sin y + xz - yz + c,$$

c being a constant.

EXERCISES

1. Verify Green's theorem in the plane for

$$\oint_C \left[\left(x^2+y^2\right)dx - 2xy\,dy\right],$$

where C is the rectangle bounded by $y = 0, x = 0, y = b, x = a$.

2. Verify Green's theorem in the plane for

$$\oint_C \left[\left(xy+y^2\right)dx + x^2 dy\right],$$

where C is the closed curve of the region bounded by $y = x$ and $y = x^2$.

3. Evaluate by Green's theorem

$$\oint_C \left[(\cos x \sin y - xy)\,dx + \sin x \cos y\,dy\right],$$

where C is the circle $x^2 + y^2 = 1$. (*Sri Venkateshwara 2005*)

4. If ϕ is harmonic in V, then prove that

$$\iint_S \phi \frac{\partial \phi}{\partial n} dS = \iiint_V |\nabla \phi|^2\, dV.$$

5. If ϕ and ψ are harmonic in V and $\partial\phi / \partial n = \partial\psi / \partial n$ on S, then show that $\phi = \psi + c$, where c is a constant.

6. If ϕ is harmonic in V and $\partial\phi / \partial n = 0$ on S, then prove that ϕ is a constant in V.

7. Show that the vector

$$\mathbf{F} = (x^2 - yz)\,\mathbf{i} + (y^2 - zx)\,\mathbf{j} + (z^2 - xy)\,\mathbf{k}$$

is irrotational. Find a scalar ϕ such that $\mathbf{F} = \nabla\phi$.

OBJECTIVE QUESTIONS

For each of the following questions, four alternatives are given for the answer. Only one of them is correct. Choose the correct alternative.

1. A force $\mathbf{F}$ acts upon a particle which is displaced along a given curve c in space from the point P whose position vector is $\mathbf{r}$, the workdone by the force is

(*a*) $\mathbf{F} \cdot d\mathbf{r}$ (*b*) $\mathbf{F} \cdot \dfrac{d\mathbf{r}}{ds}$

(*c*) $\int_O^P \mathbf{F} \cdot d\mathbf{r}$ (*d*) $\int_C \mathbf{F} \cdot d\mathbf{r}$

2. If $\mathbf{F} = 3xy\mathbf{i} - y^2\mathbf{j}$, then value of $\int_C \mathbf{F} \cdot d\mathbf{r}$, where c is the curve $y = 2x^2$ in the xy-plane, from (0, 0) to (1, 2) is

(*a*) – 6/7 (*b*) 7/6 (*c*) – 7/6 (*d*) 6/7

3. If $\mathbf{t}$ is the unit tangent vector and c is the unit circle in xy-plane with centre at origin, then value of $\int_C \mathbf{t} \cdot d\mathbf{r}$

(a) 2π (b) π (c) $\pi/2$ (d) 0

4. Value of $\int (x\,dy - y\,dx)$ around the circle $x^2 + y^2 = 1$ is

(a) 0 (b) $\pi/2$ (c) π (d) 2π

5. If c be a closed curve, then $\oint_C \mathbf{r} \cdot d\mathbf{r} =$

(a) r (b) r^2 (c) $1/r$ (d) 0

6. Area bounded by a simple closed curve c is

(a) $\oint_C (x\,dy - y\,dx)$

(b) $\frac{1}{2}\oint_C (x\,dy - y\,dx)$

(c) $\oint_C (x\,dy + y\,dx)$

(d) $\frac{1}{2}\oint_C (x\,dy + y\,dx)$

7. Value of $\oint_C \mathbf{r} \times d\mathbf{r} =$

(a) $\iint_S \mathbf{n}\,ds$ (b) 0

(c) $2\iint_S \mathbf{n}\,ds$ (d) None of these

8. Value of $\iint_S \mathbf{n} \times \mathbf{F}\,ds$ is

(a) $\iiint_V \text{div}\,\mathbf{F}\,dV$ (b) $\iiint_V \text{curl}\,\mathbf{F}\,dV$

(c) $\iiint_V \mathbf{F}\,dV$ (d) zero

9. Value of $\iint_S \nabla r^2 \cdot ds$ is

(a) V (b) $3\,V$ (c) $6\,V$ (d) $12\,V$

10. For any closed surface S, prove that $\iint_S \text{curl}\,\mathbf{F}\,.\,\mathbf{n}\,dS$ is

(a) 0 (b) V (c) $2\,V$ (d) $3\,V$

11. For any closed surface S, which of the following is true :

(*a*) $\iint_S \mathbf{r} \times \mathbf{n}\, dS = 0$ (*b*) $\iiint_V \dfrac{dV}{r^2} = 0$

(*c*) $\iint_S \mathbf{r} \cdot \mathbf{n}\, dS = 0$ (*d*) $\iint_S \nabla r^2 \cdot dS = 0$

12. The necessary and sufficient condition that $\int_C \mathbf{F} \cdot d\mathbf{r}$ should be independent of the path **c** joining A and B is

(*a*) $\mathbf{F} \times d\mathbf{F} = 0$ (*b*) div $\mathbf{F} = 0$

(*c*) curl $\mathbf{F} = 0$ (*d*) Nones of these

ANSWERS

1. (*d*)	**2.** (*c*)	**3.** (*a*)	**4.** (*d*)	**5.** (*d*)
6. (*b*)	**7.** (*c*)	**8.** (*b*)	**9.** (*c*)	**10.** (*a*)
11. (*a*)	**12.** (*c*)			

Answers

Page 14-15

1. (*i*) $\vec{BC} = \vec{AC} - \vec{AB}$. $\vec{AD} = \frac{1}{2}\vec{AB} + \frac{1}{2}\vec{AC}$.

$\vec{BE} = \frac{1}{2}\vec{AC} - \vec{AB}$. $\vec{CF} = \frac{1}{2}\vec{AB} - \vec{AC}$.

(*ii*) $\vec{AB} = -\frac{4}{3}\vec{BE} + \frac{4}{3}\vec{CF}$. $\vec{BC} = \frac{2}{3}\vec{BE} - \frac{2}{3}\vec{CF}$.

$\vec{CA} = \frac{2}{3}\vec{BE} + \frac{4}{3}\vec{CF}$ $\vec{AD} = -\vec{BE} - \vec{CF}$.

(*iii*) $\vec{AC} = 2\vec{AB} + 2\vec{BE}$. $\vec{BC} = \vec{AB} + 2\vec{BE}$.

$\vec{AD} = \frac{3}{2}\vec{AB} + \vec{BE}$. $\vec{CF} = -\frac{3}{2}\vec{AB} - 2\vec{BE}$.

2. (*i*) $\vec{AC} = \vec{AB} + \vec{AD}$. $\vec{BD} = -\vec{AB} + \vec{AD}$.

(*ii*) $\vec{AB} = \frac{1}{2}\vec{AC} - \frac{1}{2}\vec{BD}$. $\vec{AD} = \frac{1}{2}\vec{AC} + \frac{1}{2}\vec{BD}$.

(*iii*) $\vec{AB} = \vec{AD} - \vec{BD}$. $\vec{AC} = 2\vec{AD} - \vec{BD}$.

3. (*i*) $\vec{BC} = \vec{OC} - \vec{OB}$, $\vec{CA} = \vec{OA} - \vec{OC}$.

$\vec{AB} = \vec{OB} - \vec{OA}$.

(*ii*) $\vec{OA} = \vec{OC} - \vec{AB} - \vec{AC}$. $\vec{OB} = \vec{OC} - \vec{BC}$.

$\vec{CA} = -\vec{AB} - \vec{BC}$

4. $\mathbf{a} = 5\mathbf{c} + 3\mathbf{d}$, $\mathbf{b} = 3\mathbf{c} + d$

5. (*i*) $\mathbf{a} = \frac{2\mathbf{c} + 3\mathbf{d}}{4}$, $\mathbf{b} = \frac{10\mathbf{c} + 3\mathbf{d}}{8}$. (*ii*) $\mathbf{a} = \mathbf{d}$, $\mathbf{b} = \mathbf{d}$.

7. (*c*) is the only correct statement.

8. (*b*), (*d*) are correct statements.

Page 22-23

1. (*i*) (4, – 2, 2), (*ii*) (10, 0, 7), (*iii*) (3, – 4, 8).

3. (*i*) $-\mathbf{b} + 2\mathbf{c} = 2(\mathbf{a} - 2\mathbf{b} + 3\mathbf{c}) + (-2\mathbf{a} + 3\mathbf{b} - 4\mathbf{c})$.

(*ii*) $3\mathbf{a} + 20\mathbf{b} + 5\mathbf{c} = 2(5\mathbf{a} + 6\mathbf{b} + 7\mathbf{c}) - (7\mathbf{a} - 8\mathbf{b} + 9\mathbf{c})$.

5. Linearly dependent.

6. (0, – 7/5, 1/5).

7. $x' = \frac{1}{2}x - y + \frac{1}{2}z,\ y' = -\frac{1}{2}x + \frac{1}{2}z,\ z' = y.$

8. $\frac{1}{2}(-x + y + z),\ \frac{1}{2}(x - y + z),\ \frac{1}{2}(x + y - z).$

9. (*i*) $(7/2)(\mathbf{a} - \mathbf{b} + \mathbf{c}) + (\mathbf{b} + \mathbf{c} - \mathbf{a}) - \frac{1}{2}(\mathbf{c} + \mathbf{a} + \mathbf{b}) = 2\mathbf{a} - 3\mathbf{b} + 4\mathbf{c}.$

(*ii*) $-8(\mathbf{a} - 2\mathbf{b} + \mathbf{c}) + 6(-\mathbf{b} + 2\mathbf{c} + \mathbf{a}) - 3(\mathbf{c} - \mathbf{a} + 3\mathbf{b})$
$= \mathbf{a} + \mathbf{b} + \mathbf{c}.$

10. (*i*) $h = -4/3,\ k = -5/3.$

(*ii*) $h = -\frac{1}{2},\ k = 1.$

11. (*i*) $h = -13/5,\ k = -2/5,\ l = -4.$

(*ii*) $h = 4/3,\ k = -4/3,\ l = 1.$

Page 43-44

1. (*i*) $2\mathbf{b} - 2\mathbf{a}$. (*ii*) Origin. (*iii*) $5\mathbf{a} + 3\mathbf{b}$.
(*iv*) Parallel. (*v*) $\mathbf{a} + \mathbf{b}$.

2. (*i*) Collinear, (*ii*) Non-collinear,
(*iii*) Collinear, (*iv*) Non-collinear.

3. $\mathbf{a} + \mathbf{b} + \mathbf{c}$.

4. $OC/CE = 4/1,\ BC/CF = 3/2.$

5. $5\mathbf{a} - \mathbf{b},\ \mathbf{r} = \mathbf{b} - \mathbf{a} + \lambda(\mathbf{a} + \mathbf{b}),$
$\mathbf{r} = 2\mathbf{a} - 4\mathbf{b} + \mu(\mathbf{b} - 3\mathbf{a}),\ -4\mathbf{a} - 2\mathbf{b}.$

6. 2 : 1, 2 : 1, 2 : 1.

7. $3\mathbf{a} - \mathbf{b},\ 3\mathbf{b} - \mathbf{a},\ \frac{3\mathbf{a} - \mathbf{b}}{4}.$

8. – 1 : 6.

13. 1 : 2.

Page 51-52

1. $\mathbf{r} = t\mathbf{a} + p\mathbf{b},\ \mathbf{r} = t\mathbf{b} + p\mathbf{c}.$
$\mathbf{r} = t\mathbf{a} + p\mathbf{c},\ \mathbf{r} = \mathbf{a} + t(\mathbf{b} - \mathbf{a}) + p(\mathbf{c} - \mathbf{a}).$

2. $\mathbf{r} = t\mathbf{a} + p\mathbf{b}$, $\mathbf{r} = \mathbf{c} + t\mathbf{a} + p\mathbf{b}$.
$\mathbf{r} = t\mathbf{b} + p\mathbf{c}$, $\mathbf{r} = \mathbf{a} + t\mathbf{b} + p\mathbf{c}$.
$\mathbf{r} = t\mathbf{a} + p\mathbf{c}$, $\mathbf{r} = \mathbf{b} + t\mathbf{a} + p\mathbf{c}$.

3. (*i*) $\mathbf{r} = (2 + t + 5p)\,\mathbf{a} + (2 + 2t - 2p)\,\mathbf{b} + (-2 + 4t + 8p)\mathbf{c}$.
(*ii*) $\mathbf{r} = (1 + 4p)\,\mathbf{a} + (1 + t + 2p)\,\mathbf{b} + (1 + 3p)\,\mathbf{c}$.

4. (*i*) Coplanar (*ii*) Coplanar (*iii*) Non-Coplanar.

7. Line lies in each of the planes.

8. $2\mathbf{a} + 3\mathbf{b} - 4\mathbf{c}$.

9. (*i*) The lines intersect at the point $\mathbf{a} + 2\mathbf{b} - \mathbf{c}$.
(*ii*) The lines intersect at the point $3\mathbf{a} + 5\mathbf{b} - 2\mathbf{c}$.
(*iii*) The lines intersect at the point $\mathbf{b} - 2\mathbf{c}$.

10. (*i*) $\mathbf{r} = t\,(2\mathbf{a} + 3\mathbf{b} - \mathbf{c}) + (\mathbf{b} - \mathbf{c})$.
(*ii*) Planes are parallel.
(*iii*) Planes are parallel.
(*iv*) $\mathbf{r} = -2\mathbf{a} + t\,(\mathbf{b} + \mathbf{c})$.

11. $5\mathbf{a} - 7\mathbf{b} + 6\mathbf{c}$.

13. $\frac{1}{2}(\mathbf{a} - \mathbf{b} - 3\mathbf{c})$ is the point of intersection.

Miscellaneous Exercises 1

Page 66-73

1. (*d*) is correct.

4. Parallelogram

5. $\overrightarrow{AB} = \mathbf{b} - \mathbf{a},\ \overrightarrow{PA} = \frac{1}{2}\mathbf{a},\ \overrightarrow{PQ} = \frac{1}{2}\mathbf{b},\ \overrightarrow{AR}\frac{1}{2}\mathbf{b} - \mathbf{a},$

where $\overrightarrow{OA} = \mathbf{a}$ and $\overrightarrow{OB} = \mathbf{b}$.

6. $\overrightarrow{PQ} = \overrightarrow{OQ} - \overrightarrow{OP},\ \overrightarrow{SR} = \frac{1}{2}\left(\overrightarrow{OQ} = \overrightarrow{OP}\right),$

$\overrightarrow{SQ} = \overrightarrow{OQ} - \frac{1}{2}\overrightarrow{OP},\ \overrightarrow{RP} = \overrightarrow{OP} - \frac{1}{2}\overrightarrow{OQ}.$

7. $\overrightarrow{BE} = (1/3)(-2\mathbf{a} + \mathbf{b}),\ \overrightarrow{DC} = 2\mathbf{a}.$

9. (*i*) $(1/4)\,\mathbf{a} + (1/3)\,\mathbf{b}$, (*ii*) $(1/3)\,\mathbf{b} - (3/4)\,\mathbf{a}$.
$\lambda = 4/3,\ \mu = 4/9,\ AP/PQ = 3/1,\ OQ/OB = 4/5.$

11. B is $\mathbf{a} + \mathbf{p}$, C is $\mathbf{a} + \mathbf{p} + \mathbf{q}$, D is $\mathbf{a} + \mathbf{q}$, E is $\frac{1}{2}(2\mathbf{a} + \mathbf{p} + \mathbf{q})$.

12. R is $(1/11)(\mathbf{a} + 6\mathbf{b} + 4\mathbf{c})$, $6 : 1$.

13. $\overrightarrow{AQ} = 3\mathbf{b} - \mathbf{a}$, $\overrightarrow{BP} = 2\mathbf{a} - \mathbf{b}$.

S divides AB in the ratio $3 : 4$.

14. $AF : FB = 3 : 4$, $AL \div LC = -2$.

15. $\overrightarrow{OS} = \frac{1}{2}(\mathbf{a} + h\mathbf{b})$, $\overrightarrow{AM} = h\mathbf{b} - \mathbf{a}$.

(*i*) $\overrightarrow{OS} = (3/4)\left[\mathbf{b}k + (1-k)\,\mathbf{a}\right]$, $\overrightarrow{OL} = \mathbf{b}k + (1-k)\,\mathbf{a}$

$h = \frac{1}{2}$, $k = \frac{1}{3}$,

(*ii*) $AL / LB = \frac{1}{2}$, $OM / OB = 1$, $ON / NA = 2$.

16. (*i*) $\overrightarrow{BC} = 2(\mathbf{b} - \mathbf{a})$, (*ii*) $\overrightarrow{PQ} = \mathbf{b} - \mathbf{a}$, (*iii*) $\overrightarrow{PC} = 2\mathbf{b} - \mathbf{a}$.

(*iv*) $\overrightarrow{BQ} = \mathbf{b} - 2\mathbf{a}$, $BC \parallel PQ$ and $BC - 2PQ$.

17. $\overrightarrow{OY} = \mathbf{a} + 2\mathbf{b}$, $OX / XY = \frac{1}{2}$.

18. $\overrightarrow{PQ} = (1/3)(2\mathbf{c} - \mathbf{b} - \mathbf{a})$, $\overrightarrow{SR} = (1/3)(\mathbf{d} + \mathbf{c} - 2\mathbf{a})$

19. $\overrightarrow{AD} = \mathbf{d} - \mathbf{a}$, $\overrightarrow{OC} = \frac{1}{2}\mathbf{a} + 3\mathbf{d}$, $\overrightarrow{DC} = \frac{1}{2}\mathbf{a} + 2\mathbf{d}$.

$OX/XC = 2/5$, $AX/XD = 6/1$.

20. P is the mid-point of AB, $XP/PY = 3/1$.

21. $2LM = NL$.

23. (*i*) $\overrightarrow{OP} = \frac{1}{2}\mathbf{a}$, $\overrightarrow{OQ} = \mathbf{b}$, $\overrightarrow{OR} = \frac{1}{2}(\mathbf{b} + \mathbf{c})$, $\overrightarrow{OS} = \frac{1}{2}(\mathbf{a} + \mathbf{c})$.

(*iv*) $\overrightarrow{OG} = \frac{1}{4}(\mathbf{a} + \mathbf{b} + \mathbf{c}) = \overrightarrow{OH}$.

PR and QS bisect each other.

24. $\overrightarrow{AE} = \frac{1}{2}(\mathbf{a} + \mathbf{b})$, $\overrightarrow{BD} = \mathbf{b} - \mathbf{a}$, $\overrightarrow{MB} = \frac{1}{2}(\mathbf{a} - 2\mathbf{b})$.

26. (*ii*) $\overrightarrow{PR} = \mathbf{a} + \mathbf{b},\ \overrightarrow{SM} = \frac{1}{2}(4\mathbf{a} - \mathbf{b}),\ \overrightarrow{SX} = (2/5)(4\mathbf{a} - \mathbf{b}),$

$\overrightarrow{PX} = (3/5)(\mathbf{a} + \mathbf{b}).$

(*i*) $PX/PR = 3/5$. (*iii*) P, X, R are collinear.

28. $\overrightarrow{AB} = \mathbf{b} - \mathbf{a},\ \overrightarrow{AC} = (k - 1)\,\mathbf{a} + l\mathbf{b},$

$\overrightarrow{BC} = k\mathbf{a} + (1 - 1)\,\mathbf{b},$

A, B, C are collinear, $AC/CB = l/k$,

$\overrightarrow{XY} = (2/3)(3\mathbf{b} - \mathbf{a}).$

29. $\overrightarrow{AB} = \mathbf{b} - \mathbf{a},\ \overrightarrow{BC} = -(3\mathbf{a} + \mathbf{b}),$

$\overrightarrow{AD} = -(3\mathbf{b} + \mathbf{a}),\ \overrightarrow{DC} = 3(\mathbf{b} - \mathbf{a}),\ \overrightarrow{AB} \parallel \overrightarrow{DC}.$

30. $(1/4)\ \mathbf{a} + (1/3)\ \mathbf{b},\ (1/3)\ \mathbf{b} - (3/4)\ \mathbf{a}$

$h = 4/3,\ k = 4/9,$

$\frac{AP}{PQ} = 3/1,\ \frac{OQ}{OB} = 4/5.$

31. $\overrightarrow{OD} = \frac{1}{2}(\mathbf{a} + \mathbf{b}),\ \overrightarrow{OE} = \frac{1}{3}(\mathbf{b} + 2\mathbf{c}),\ \overrightarrow{OF} = 4(\mathbf{a} + \mathbf{b} + 2\mathbf{c})$

$AF : FE = 3 : 1.$

32. $OY/YA = 2/7,\ XC/CY = 9/2.$

34. (*i*) $\mathbf{m} = \frac{1}{2}(\mathbf{p} + \mathbf{q}).$

(*ii*) $\mathbf{t} = \frac{1}{3}(\mathbf{p} + \mathbf{q} + \mathbf{s}),$

(*iii*) $\mathbf{r} = \mathbf{q} - \mathbf{p} - \mathbf{s}.$

Page 91–93

1. $\sqrt{14}, \sqrt{5}, \sqrt{45}.$

2. (*i*) $\frac{\pi}{2}$, (*ii*) $\cos^{-1}\frac{8}{5\sqrt{29}}$, (*iii*) $\cos^{-1}\frac{1}{\sqrt{39}}$.

3. (*i*) $a_1\mathbf{i} + a_2\mathbf{j} + a_3\mathbf{k}$, where $a_1 = -1/\sqrt{14};\ a_2 = -2/\sqrt{14};$
$a_3 = -3/\sqrt{14}.$

(*ii*) $a_1\mathbf{i} + a_2\mathbf{j} + a_3\mathbf{k}$, where $d_1 = -1/\sqrt{6};\ a_2 = -1/\sqrt{6},$

$a_3 = 2/\sqrt{6}.$

(*iii*) $\frac{1}{\sqrt{6}}(\mathbf{i} - \mathbf{j} - 2\mathbf{k})$, (*iv*) $\frac{1}{\sqrt{35}}(3\mathbf{i} - \mathbf{j} - 5\mathbf{k})$.

4. $(-\mathbf{j} + \mathbf{k})/\sqrt{2}$ and $\frac{\sqrt{2}\,(4\mathbf{i} - \mathbf{j} + \mathbf{k})}{6}$.

6. $\frac{1}{\sqrt{2}}(\mathbf{i} + \mathbf{j})$; $\mathbf{k}$; Yes.

7. (*i*) $c = \frac{1}{\sqrt{6}}(\mathbf{i} + \mathbf{j} + 2\mathbf{k})$. $\mathbf{d} = \frac{1}{\sqrt{2}}(\mathbf{i} + \mathbf{j} - \mathbf{k})$.

(*ii*) $\mathbf{c} = l(\mathbf{i} + \mathbf{j} + \mathbf{k}) + m(\mathbf{i} + \sqrt{2}\mathbf{j} - \sqrt{6}\mathbf{k})$.

where l and m satisfy the relations

$$3l = \left(\sqrt{6} - \sqrt{2} - 1\right)m \text{ and } 9m^2 - 3l^2 = 1,$$

$$\mathbf{d} = \frac{1}{k}\left[\left(-\sqrt{6} - \sqrt{2}\right)\mathbf{i} + \left(1 + \sqrt{6}\right)\mathbf{j} + \left(\sqrt{2} - 1\right)\mathbf{k}\right]$$

where $k^2 = 18 + 2\left(\sqrt{12} + \sqrt{6} - \sqrt{2}\right)$.

8. (*i*) 1. (*ii*) – 2k (*iii*) 7.

9. (*i*) $\frac{19 - 9\sqrt{2} + 8\sqrt{6}}{54}$, $\frac{12 - 9\sqrt{2} + 8\sqrt{6}}{51}$.

(*ii*) $\frac{\sqrt{3} + 1}{4}$, $\frac{3 + \sqrt{3}}{3}$.

Page 98-100

1. (*i*) $\sqrt{65}$, (*ii*) $3\sqrt{14}$.

2. (*i*) $\sqrt{26}$, (*ii*) $3\sqrt{2}$.

3. (*i*) $\cos^{-1}\left(\frac{91}{3\sqrt{6}}\right)$, (*ii*) $\cos^{-1}\left(\sqrt{\frac{7}{12}}\right)$.

4. (*i*) 1/3, 2/3, – 2/3, (*ii*) 3/7, – 6/7, 2/7.

6. The lines are parallel.

12. All statements are false.

13. $\cos^{-1}\left(\frac{1}{3}\right)$.

15. (*ii*) (1, 3, 5).

Page 110

2. $(\mathbf{r}\,.\,\mathbf{n}_1 - 1)\,(\mathbf{n}_2\,.\,\mathbf{n}_3) = (\mathbf{r}\,.\,\mathbf{n}_2 - 1)\,(\mathbf{n}_1\,.\,\mathbf{n}_3)$.

3. $\mathbf{a} + 2\,\frac{q - \mathbf{a}\,.\,\mathbf{n}}{\mathbf{n}^2}\,\mathbf{n}$.

$$\mathbf{r} = \mathbf{a} + 2\,\frac{q - \mathbf{a}\,.\,\mathbf{n}}{\mathbf{n}^2}\,\mathbf{n} + t\left[\mathbf{b} - 2\,\frac{\mathbf{b}\,.\,\mathbf{n}}{\mathbf{n}^2}\,\mathbf{n}\right].$$

4. $2\mathbf{b} - \mathbf{a} + 2\,\frac{(\mathbf{a} - \mathbf{b})\,.\,\mathbf{c}}{\mathbf{c}^2}\,\mathbf{c}$.

7. $\frac{|q - \mathbf{a}\,.\,\mathbf{n}|}{|\mathbf{c}\,.\,\mathbf{n}|}\,|\mathbf{c}|$.

Page 138-139

1. (*a*) $-\mathbf{i} + \mathbf{j} + 3\mathbf{k}, \sqrt{11}$. (*b*) $3\mathbf{j} + 2\mathbf{k}, \sqrt{13}$.
   (*c*) $4\mathbf{i} - 3\mathbf{j} - 7\mathbf{k}, \sqrt{74}$.
2. (*a*) 12 (*b*) – 14.
3. (*a*) $-15\mathbf{i} - 25\mathbf{j} + 15\mathbf{k}$. (*b*) $\mathbf{i} + 16\mathbf{j} - 9\mathbf{k}$. **6.** 7

Page 144-145

7. (*a*) $\frac{\sqrt{150}}{2}$, (*b*) $\frac{\sqrt{3}}{2}$,
   (*c*) $\frac{\sqrt{89}}{2}$, (*d*) $\frac{\sqrt{(a^2b^2 + b^2c^2 + c^2a^2)}}{2}$.
8. (*a*) 2/3, (*b*) 6.
10. [**b c d**] + [**c a d**] + [**a b d**] = [**a b c**]

Page 158-160

1. $\mathbf{r} = \mathbf{a} + t\mathbf{n}_1 \times \mathbf{n}_2$.
2. $\mathbf{r} = \mathbf{a} + t\mathbf{n}_1 \times \mathbf{n}_2$.
3. $\mathbf{r} = \mathbf{c} + t\mathbf{b} \times \mathbf{n}$.
4. $\mathbf{r} = \mathbf{a} + t\left[(\mathbf{a} - \mathbf{b}) - \frac{(\mathbf{a} - \mathbf{b})\cdot\mathbf{n}}{\mathbf{c}\cdot\mathbf{n}}\,\mathbf{c}\right]$
5. $[\mathbf{r}\;\mathbf{n}_1\;\mathbf{n}_2] = [\mathbf{a}\;\mathbf{n}_1\;\mathbf{n}_2]$.

6. $[\mathbf{r\ n\ c}] = [\mathbf{a\ n\ c}]$.

7. $\mathbf{r} \cdot [\mathbf{b} - \mathbf{a}] \times \mathbf{n} = [\mathbf{a\ b\ n}]$

8. $[\mathbf{r\ b\ n}] = [\mathbf{a\ b\ n}]$.

9. $[\mathbf{n}_2\ \mathbf{n}_3\ \mathbf{n}_4]\ (\mathbf{r} \cdot \mathbf{n}_1 - \mathbf{q}_1) = [\mathbf{n}_1\ \mathbf{n}_3\ \mathbf{n}_4]\ (\mathbf{r} \cdot \mathbf{n}_2 - \mathbf{q}_2)$.

10. $\mathbf{r} \cdot (\mathbf{b} - \mathbf{a}) \times \mathbf{c} = [\mathbf{a\ b\ c}]$.

11. $a.c = b.c;\ \dfrac{1}{[a\,b\,c]}\{(b.a)(b \times c) + (a.b)(c \times b) + (a.c)\,(a \times b)\}$.

12. $\mathbf{a} + \mathbf{b}$.

13. $\mathbf{r} = \mathbf{a} + t\,\{(\mathbf{b} - \mathbf{a}) \times \mathbf{c}\} \times \{(\mathbf{d} - \mathbf{a}) \times \mathbf{c}\}$.

14. $[\mathbf{r\ b\ d}] = [\mathbf{a\ b\ d}]$. The required shortest distance is the distance of any point, say **c**, on the line $\mathbf{r} = \mathbf{c} + p\mathbf{d}$ from this plane so that it is

$$\frac{\left|[\mathbf{c\ b\ d}] - [\mathbf{a\ b\ d}]\right|}{|\mathbf{b} \times \mathbf{d}|}.$$

16. (*i*) The line *AB* is normal to the plane *OCD*.

(*ii*) The line *AB* is parallel to the plane *OCD*.

23. $\cos^{-1}\dfrac{x^2}{\sqrt{(x^2+z^2)}\sqrt{(x^2+y^2)}};\ \cos^{-1}\dfrac{xz}{\sqrt{(x^2y^2+y^2z^2+z^2x^2)}}$.

27. The given ratio should be 1 : 1.

Page 174-176

1. (*i*) (15, – 13, – 2); (*ii*) (– 9, 9, 12); (*iii*) (7, – 10, – 3).

2. (*i*) (6, 4, 1); (12, 8, 2); (*ii*) (1, 1, 3), (2, 2, 6);

(*iii*) (2, 3, 4) (4, 6, 8).

3. (9, 5, – 2).

5. (*a*) Co-planar, (*b*) Co-planar, (*c*) Non-coplanar.

6. (*i*) $3x - 2y + 3z = 4$, (*ii*) $x - y - 7z + 23 = 0$.

9. (*i*) $\sqrt{90}$, (*ii*) $(6/7)\sqrt{14}$, (*iii*) 6, (*iv*) $3/5\sqrt{10}$, (*v*) $\sqrt{6}$.

10. $2p^3/3lmn$.

11. $x - y - z = 2$.

13. $\left(\frac{1}{2}, -\frac{1}{2}, -3/2\right); x - 2y + z = 0$.

14. $(a + a', b + b', c + c');\ \Sigma x\,(bc' - b'c) = 0$.

15. (5, – 7, 6); $11x = 6y + 5z + 67$.

16. (*i*) $x = y = z$; $4\sqrt{3}$,

(*ii*) $(x - 4) = (y - 2) / 3 = - z\ (z + 3)/5$; $\sqrt{35}$.

17. (11, 11, 31) and (3, 5, 7).

18. 0.

Miscellaneous Exercises II

Page 177-183

7. 27 : 1.

12. $\frac{1}{2}\mathbf{a}, \frac{1}{2}\mathbf{b}, \frac{1}{2}\mathbf{c}, \frac{1}{2}(\mathbf{b}+\mathbf{c}), \frac{1}{2}(\mathbf{c}+\mathbf{a}), \frac{1}{2}(\mathbf{a}+\mathbf{b})$;

$$r = \frac{1}{2}\mathbf{a} + t\left(\mathbf{b}+\mathbf{c}-\mathbf{a}\right), \quad \mathbf{r} = \frac{1}{2}\mathbf{b} + t\left(\mathbf{c}+\mathbf{a}-\mathbf{b}\right),$$

$$\left(\frac{1}{4}\right)(\mathbf{a}+\mathbf{b}+\mathbf{c}).$$

14. (3, 7, 11), (0, 5, 4).

15. $p = 1$.

16. P is $\mathbf{i} + 2\mathbf{j} + \mathbf{k}$.

17. $p = \sqrt{3/2},\ q = \sqrt{3}/\sqrt{2}, r = 0$.

18. $\pm(1/13)(-12\mathbf{i}+3\mathbf{j}+4\mathbf{k}),\ \cos^{-1}\left(1/\sqrt{3}\right),\ \pi - \cos^{-1}\left(1/\sqrt{3}\right)$.

20. $\mathbf{a} = \left(3/\sqrt{2}\right)(\mathbf{i}+\mathbf{j}),\ \mathbf{b} = \left(1/\sqrt{2}\right)(\mathbf{i}-\mathbf{j}),\ \mathbf{c} = l\mathbf{i} - l\mathbf{j} - \sqrt{6}k$

where $l = 5\sqrt{\frac{3}{2}}$.

21. $\mathbf{x} = \mathbf{j} + t\,(\mathbf{i} + 2\mathbf{j} + \mathbf{k})$ where t is any scalar.

23. $\cos^{-1}\sqrt{\frac{1}{30}},\ \cos^{-1}\left(\frac{1}{3\sqrt{30}}\right)$.

25. B is (4, 1), C is (4, 6), C, E, F are collinear and E is the mid-point of CF.

27. $\mathbf{r} \cdot (\mathbf{i} + \mathbf{j}) = 2$, $(\mathbf{r} - \mathbf{j})\ (\mathbf{i} + 2\mathbf{j} - \mathbf{k}) = 0$.

29. $l_1 : x = 1 - y = z,\ l_2 : x - 1 = y + 1 = z$.

$\pi_1 : x + 2y + z = 2,\ \pi_2 : x - 2y + z = 1.\ AB : 1/\sqrt{2}$.

The equation of AB is

$$\mathbf{r} = (1/4)\,(5\mathbf{i} - \mathbf{j} + 5\mathbf{k}) + t\,(\mathbf{i} - \mathbf{k}).$$

30. $\mathbf{r} \,.\, (3\mathbf{i} + \mathbf{j}) = 6,\ \mathbf{r} \,.\, (\mathbf{i} + 3\mathbf{j}) = 6.$

31. $\pi_1 : \mathbf{r} \cdot (\mathbf{i} - \mathbf{j} + \mathbf{k}) = 4,\quad \pi_2 : \mathbf{r} \cdot\ (\mathbf{i} + 2\mathbf{j} + \mathbf{k}) = 6,$

$\pi_2 : \mathbf{r} \cdot (\mathbf{i} - 4\mathbf{j} + \mathbf{k}) = 2,\ \sqrt{2}/3.$

32. $\cos^{-1}\left(-\frac{1}{4}\right),\ \cos^{-1}\left(\frac{1}{4}\right).$

35. The vectors **p, q** are parallel.

36. $x = \mathbf{q} - \frac{1}{2\mathbf{p}^2}(\mathbf{p} \cdot \mathbf{q})\,\mathbf{p}$

39. $\lambda = d/|\mathbf{a}|^2,\ \mu = 0,\ \gamma = -1/|\mathbf{a}|.$

40. (*i*) $\overrightarrow{OB} = \mathbf{c} - (2/3)\,\mathbf{a},\ \overrightarrow{AB} = \mathbf{c} - \frac{1}{3}\mathbf{a},\ \overrightarrow{AM} = \frac{1}{2}\mathbf{c} - \mathbf{a},$

(*iv*) 4.6 cm.

41. $\left(\frac{1}{3}\right)\mathbf{b} - \mathbf{a},\ (3/4)\,\mathbf{a} - \mathbf{b},\ 0.922.$

44. (*i*) The locus is the circle on the line joining the points **p** and **q** as diameter.

(*ii*) The locus is the line joining the points **p** and **q.**

45. (*i*) $\sqrt{6}a/3$ (*ii*) $a/\sqrt{2}.$

46. N is (1/4) $\mathbf{b} + \frac{1}{2}\mathbf{c},\ 3:2,\ 3:1,\ 1:4$

52. The general value of $\mathbf{v} = \mathbf{b} + t\mathbf{a},$

$\mathbf{v} = (3/2)\,\mathbf{i} + \left(\frac{1}{2}\right)\mathbf{k}$ if $\mathbf{a} \,.\, \mathbf{v} = 0.$

Page 195

1. The general solution is

$$\mathbf{r} = \frac{\mathbf{n}_2{}^2 - \mathbf{n}_2 \,.\, \mathbf{n}_1}{(\mathbf{n}_2 \times \mathbf{n}_1)^2}\mathbf{n}_1 + \frac{\mathbf{n}_1{}^2 - \mathbf{n}_1 \,.\, \mathbf{n}_2}{(\mathbf{n}_1 \times \mathbf{n}_2)^2} + t\mathbf{n}_1 \times \mathbf{n}_2;$$

being any scalar.

(Start with assuming $\mathbf{r} = x\mathbf{n}_1 + y\mathbf{n}_2 + t\mathbf{n}_1 \times \mathbf{n}_2$.

2. $\mathbf{b} \,.\, \mathbf{c} + \mathbf{a} \,.\, \mathbf{d} = 0,\ \frac{1}{[\mathbf{a\,b\,c}]}\left[-(\mathbf{d} \cdot \mathbf{b})\,\mathbf{a} + (\mathbf{d} \cdot \mathbf{a})\,\mathbf{b} + (\mathbf{b} \cdot \mathbf{b})\,\mathbf{c}\right],$

If $[\mathbf{a\,b\,c}] \neq 0$. [Assume $\mathbf{r} = x\mathbf{a} + y\mathbf{b} + z\mathbf{c}$].

Page 197-198

2. $\sqrt{(3/8)}k, \ \frac{\sqrt{6}}{12}k$; k being length of each edge.

6. $\frac{1}{2\sqrt{2}}\sqrt{\left(\mathbf{a}^2+\mathbf{b}^2+\mathbf{c}^2\right)},$

$$\frac{\sqrt{\left[\left(\mathbf{b}^2+\mathbf{c}^2-\mathbf{a}^2\right)\left(\mathbf{c}^2+\mathbf{a}^2-\mathbf{b}^2\right)\left(\mathbf{a}^2+\mathbf{b}^2-\mathbf{c}^2\right)\right]}}{2\sqrt{2}\sqrt{\left(2\mathbf{a}^2\mathbf{b}^2+2\mathbf{b}^2\mathbf{c}^2+2\mathbf{c}^2\mathbf{a}^2-\mathbf{a}^4-\mathbf{b}^4-\mathbf{c}^4\right)}}.$$

7. Take O, as the origin of reference. Then the axis is given by

$$\mathbf{r}=t\,\frac{|\mathbf{b}\times\mathbf{c}|\,\mathbf{a}+|\mathbf{c}\times\mathbf{a}|\,\mathbf{b}+|\mathbf{a}\times\mathbf{b}|\,\mathbf{c}}{|\mathbf{b}\times\mathbf{c}|+|\mathbf{c}\times\mathbf{a}|+|\mathbf{a}\times\mathbf{b}|}.$$

where $\overrightarrow{\mathbf{OA}}=\mathbf{a},\ \overrightarrow{\mathbf{OB}}=\mathbf{b},\ \overrightarrow{\mathbf{OC}}=\mathbf{c}.$

Page 235-237

26. $\frac{5}{3}(-2\mathbf{i}+\mathbf{j}).$ 27. $-2\,(2\mathbf{i}+\mathbf{j}).$

28. $\frac{1}{\sqrt{3}}, \frac{-1}{\sqrt{3}}, \frac{1}{\sqrt{3}}, \frac{-1}{\sqrt{3}}, \frac{1}{\sqrt{3}}, \frac{-1}{\sqrt{3}}.$

Page 259-261

2. $a^2 \sec\alpha;\ a^3\tan\alpha.$ 3. 0, 0.

4. (*i*) $\left[\mathbf{r}\,\frac{d\mathbf{r}}{dt}\,\frac{d^3\mathbf{r}}{dt^3}\right],$ (*ii*) $\frac{d\mathbf{r}}{dt}\times\left(\frac{d\mathbf{r}}{dt}\times\frac{d^2\mathbf{r}}{dt^2}\right)+\mathbf{r}\times\left(\frac{d\mathbf{r}}{dt}\times\frac{d^3\mathbf{r}}{dt^3}\right),$

(*iii*) $\frac{d^2\mathbf{r}}{dt^2}\times\left(\frac{d\mathbf{r}}{dt}\times\frac{d^2\mathbf{r}}{dt^2}\right)+2\frac{d\mathbf{r}}{dt}\times\left(\frac{d\mathbf{r}}{dt}\times\frac{d^3\mathbf{r}}{dt^3}\right)$

$$+\mathbf{r}\times\left(\frac{d^2\mathbf{r}}{dt^2}\times\frac{d^3\mathbf{r}}{dt^3}\right)+\mathbf{r}\times\left(\frac{d\mathbf{r}}{dt}\times\frac{d^4\mathbf{r}}{dt^4}\right)$$

5. (*i*) $\frac{\mathbf{r}\,d\mathbf{r}-\mathbf{r}\,dr}{r^2};$ (*ii*) $nr^{n-1}\,dr\,\mathbf{r}+r^n\,d\mathbf{r}$

(*iii*) $\frac{(\mathbf{a}\cdot\mathbf{r})\,d\mathbf{r}-(\mathbf{a}\cdot d\mathbf{r})\,\mathbf{r}}{(\mathbf{a}.\mathbf{r})^2}-\frac{r\,(d\mathbf{r}\times\mathbf{a})-(\mathbf{r}\times\mathbf{a})\,dr}{r^2}$

(*iv*) $\frac{2\left(r^4-1\right)}{r^3}\,dr.$

6. $\sqrt{(11)}, \dfrac{8}{\sqrt{11}}$; 7. $\left(2+\dfrac{3\pi}{2}\right) \mathbf{i}+5\mathbf{j}+6\mathbf{k}$.

12. $e^t\ \mathbf{i}+\dfrac{1}{2}\left(e^{2t}+1\right)\mathbf{j}+t\mathbf{k}$.

13. $\mathbf{r}=\dfrac{1}{a^2}\left[m\,\mathbf{a}-\dfrac{1}{2}(\mathbf{a}\times\mathbf{b})\,t^2-(\mathbf{a}\times\mathbf{c})\,t\right]-(\mathbf{a}\times\mathbf{d})$,

where $\mathbf{a}\ .\ \mathbf{r} = m$ and $|\ \mathbf{a}\ | = a$.

14. 0; 16. $\mathbf{r}=\dfrac{1}{2}\mathbf{a}t^2$; 17. $\mathbf{r}=\dfrac{1}{6}\mathbf{a}t^3+\dfrac{1}{2}\mathbf{b}t^2+\mathbf{u}t$

19. $x=\dfrac{CE}{H}t-\dfrac{mc^2E}{cH^2}\sin\dfrac{eHt}{mc}$, $y=\dfrac{mc^2E}{eH^2}\left(1-\cos\dfrac{eHt}{mc}\right)$, $z=0$.

Page 268

1. (*i*) $am\sin ve^{mv}\mathbf{i}-am\cos ve^{mk}\ \mathbf{j}+u\mathbf{k}$

(*ii*) $\dfrac{\mathbf{i}\,b(u+v)-\mathbf{j}\,a(u-v)-2ab\mathbf{k}}{\sqrt{\left\{b^2(u+v)^2+a^2(u-v)^2+4a^2b^2\right\}}}$.

(*iii*) $\dfrac{1}{u}\left[-c(\cos v\ \mathbf{i}+\sin a\mathbf{j})+\sqrt{\left(u^2-c^2\right)}\ \mathbf{k}\right]$

2. (*i*) $0,\ -ame^{mv},\ aum^2\ e^{mv}$

(*ii*) $0,\ -\dfrac{1}{4}ab, 0$

(*iii*) $\dfrac{-cu^2}{\left(u^2-c^2\right)^{3/2}},\ 0,\ \dfrac{cu^2}{\sqrt{\left(u^2-c^2\right)}}$.

Page 289-290

1. (*i*) $\dfrac{-\mathbf{r}}{r^3}$, (*ii*) $2\mathbf{r}$, (*iii*) $-12\ \mathbf{i}-9\mathbf{j}-16\ \mathbf{k}$,

(*iv*) $-4\mathbf{i}+12\ \mathbf{j}+9\ \mathbf{k}$.

2. (*i*) $\dfrac{21}{\sqrt{6}}$, (*ii*) $\dfrac{10}{3}$, (*iii*) $\dfrac{8}{\sqrt{6}}$, max. **d . d.** $= 2\sqrt{(21)}$.

3. (*i*) $\dfrac{1}{3}(2i+2j+k)$; (*ii*) $-\dfrac{1}{\sqrt{(21)}}(2\mathbf{i}+4\mathbf{j}+\mathbf{k})$

4. $\cos^{-1}\dfrac{16}{6\sqrt{(21)}}$; 5. $a = 0,\ b = 2,\ c = 0$.

6. $4\mathbf{i} + 3\mathbf{j} + 5\mathbf{k},\ 5\sqrt{2}$, no.

7. (*i*) $\mathbf{i}\,[y \cos (xy) + z \cos (zx)] + \mathbf{j}\,[x \cos (xy) + z \cos (yz)] + \mathbf{k}\,[y \cos (yz) + x \cos (zx)]$

(*ii*) $2e^{\left(x^2+y^2+z^2\right)}\mathbf{r}$.

8. $18/\sqrt{(14)}$.

Page 307-302

2. $6\,(x + y + z) = 0$.

8. $\dfrac{\mathbf{a}}{r^n} - (\mathbf{a}\cdot\mathbf{r})\,\mathbf{r}, 0, (2-n)\,r^{-n}\,\mathbf{a} + nr^{-n-2}\,(\mathbf{a}\cdot\mathbf{r})\,\mathbf{r}$.

Page 318-319

2. $\dfrac{\partial^2 u}{\partial t^2} = c^2\left[\dfrac{1}{r^2}\dfrac{\partial}{\partial r}\left(r^2\dfrac{\partial u}{\partial r}\right) + \dfrac{1}{r^2 \sin\theta}\dfrac{\partial}{\partial\theta}\left(\sin\theta\,\dfrac{\partial u}{\partial\theta}\right)\right]$

5. $h_1 = h_2 = \sqrt{\left(u_1{}^2 + u_2{}^2\right)};\ h_3 = u_1 u_2,$

$\mathbf{e}_1 = \left(u_2 \cos u_3\,\mathbf{i} + u_2 \sin u_3\,\mathbf{j} + u_1\,\mathbf{k}\right)/\sqrt{\left(u_1{}^2 + u_2{}^2\right)},$

$\mathbf{e}_2 = \left(u_2 \cos u_3\,\mathbf{i} + u_1 \sin u_3\,\mathbf{j} - u_2\,\mathbf{k}\right)/\sqrt{\left(u_1{}^2 + u_2{}^2\right)},$

$\mathbf{e}_3 = -\sin u_3\,\mathbf{i} + \cos u_3\,\mathbf{j}.$

7. (*i*) (*a*) $\mathbf{f} = \mathrm{r} \sin\phi \cos\phi\,(3 + R\cos\phi)\,\mathbf{e}_\rho + R\left(R\cos^3\phi - 3\sin^2\phi\right)\mathbf{e}_\phi - z^2\,\mathbf{k}.$

(*b*) $\mathbf{f} = \rho z \sin\phi\left(Rz\cos^2\phi + \sin\phi\right)\mathbf{e}_\rho + \rho z \sin\phi\cos\phi\,(1-\rho z)\sin\phi)\,\mathbf{e}_\phi + \rho^2 \sin\phi\cos\phi\,\mathbf{k}.$

(*c*) $\mathbf{f} = \rho\sin\phi\left(\rho\cos^2\phi + 2z\sin\phi\right)\mathbf{e}_\rho + \rho\sin\phi\cos\phi\,(2z - \rho\sin\phi)\,\mathbf{e}_\phi + z\,(z - \rho\sin\phi)\,\mathbf{k}.$

(*ii*) (*a*) $\mathbf{f} = \left(3r\sin^2\theta\cos\phi\sin\phi + r^2\sin^3\theta\sin\phi\cos^2\phi - r^3\cos^3\theta\right)$
$\mathbf{e}_r + (3r\sin\theta\sin\phi\cos\theta\cos\phi + r^2\cos\theta\sin^2\theta\sin\phi$
$\cos^2\phi + r^2\sin\theta\cos^2\phi)\,e_\theta + (r^2\sin^2\theta\cos^2\phi -$
$3r\sin\theta\sin\phi\,\mathbf{e}_\phi).$

(*b*) $\mathbf{f} = (r^4 \cos^2\theta \sin^2\theta \sin\phi \cos^2\phi + r^2 \cos\theta \sin^2\theta \sin^2\phi + r^2 \sin^2\theta \cos\theta \cos\phi \sin\phi)\, \mathbf{e}_r + (r^4 \cos^3\theta \cos^2\phi \sin^2\theta \sin\phi + r^2 \cos^2\theta \cos\phi \sin^2\theta - r^2 \sin^3\theta \cos\phi \sin\phi)\, \mathbf{e}_\theta + (r^2 \cos\theta \cos\phi \sin\theta \sin\phi - r^4 \cos^2\theta \sin^2\phi \sin\theta)\, \mathbf{e}_\phi.$

(*c*) $\mathbf{f} = [r^2 \sin^2\theta \cos^2\phi \sin\phi + 2r^2 \cos\theta \sin^2\theta \sin^2\phi + r^2 \cos^2\theta\,(\cos\theta - \sin\theta \sin\phi)]\, \mathbf{e}_r + [r^2 \sin^2\theta \cos\theta \cos^2\phi + 2r^2 \cos^2\theta \sin\theta \sin^2\phi - r^2 \cos\theta \sin\theta\,(\cos\theta - \sin\theta \sin\phi)]\, \mathbf{e}_\theta + (2r^2 \cos\theta \cos\phi \sin\theta \sin\phi - r^2 \sin^2\theta \cos\phi \sin^2\phi)\, \mathbf{e}_\phi.$

8. $$\mathbf{f} = \left[3ax^2 + \left(2a^2z^2\, xy\right)/\left(rR^2\right) - \left(yr^3\right)/R\right]\mathbf{i} + \left[3axy + \left(2a^2y^2z^2\right)/\left(rR^2\right) + \left(xr^3\right)/R\right]\mathbf{j} + \left[3axy - \left(2a^2zy\right)/r\right]\mathbf{k},$$

where $r = \left(x^2 + y^2 + z^2\right)^{1/2}, \quad R = \left(x^2 + y^2\right)^{1/2}.$

Page 339-340

1. $\frac{10}{7}$; **2.** (*i*) 19/35, (*ii*) 19/35, (*iii*) 1, (*iv*) – 22/3.

4. (*i*) 0, (*ii*) 0. **6.** $4\pi abc$. **7.** $\frac{12}{5}\pi$.

Page 350-352

1. $3a^3$. **2.** $\pi/12$. **5.** 320π.

Page 362-363

1. 4/3

Page 375

3. 0. **7.** $\phi = \frac{1}{3}\left(x^3 + y^3 + z^3\right) + c.$

INTEGRAL CALCULUS

Contents: Integration • Methods of Integration • Reduction Formulae • Definite Integrals • Rectification • Area under a Given Curve • Volume and Surface of Revolution • Differential Equations • First Order Differential Equations • Trajectories of a Family of Curves • Linear Equations of the nth Order • Miscellaneous Problems

MATRICES

Contents: Addition of Matrices • Multiplication of Matrices • Special Matrices • Determinants • Inverse of Matrix • Rank or a Matrix • Linear Equations • Eigen Values and eigen Vectors • Linear Dependence or Vectors • Quadratic Forms • Differentiation and Integration of Materices .

SOLID GEOMETRY

Contents: Coordinates and Direction Cosines • Plane • Straight Line • Transformation of Co-ordinates • Sphere • Cylinder • Cone • The Conicoid

TRIGONOMETRY

Contents: • Complex Numbers • De-Moivre's Theorem • Deductions from Demoivre's Theorem • Exponential and Trignometric Functions of Complex Quanities • Logrithms of Complex Quantities • Hyperbolic Functions • Inverse Circular and Inverse Hyperbolic Functions • Gregory's Series • Summation of Series • Expansions of Trigonometrical Functions.

VECTOR ALGEBRA

Contents : Vector Addition • Centroid • Vector equations of Straight Line and Plane • Scalar Product of Two Vectors • Vectors Product of two Vectors • Multiple Products • Vector Equations of Straight Line and Plane • Solution of Vector Euqations • Volume of Tetratedron • Vector Equation of Sphere • Applicatioon of Vectors tp Mechanics.

COORDINATE GEOMETRY

Contents : Coordinates • The Straight Line • Transformation of Co-ordinates • Pair of Straight Lines • Circles • System of Circles • Parabola • Ellopse • Hyperbola • The Polar Equation of a Conic • General Equation of the Second Degree • General Conic • System or Conic • Conlocal Conics • Circle of Curvature.

DIFFERENTIAL CALCULUS

Contents: Real Numbers • Limit and Continuity • Differentiation or Derivatives • Successive Differentiation • Indetrminate Forms • Tangents amd Normals • Some General Theorems and Exapansions of Fucntions • Maxima and Minima • Curvature • Asymptotes • Partial Differentiation • Points of Inflexion and Singular Points • Curve Tracing

PHYSICS

by

C.L. ARORA

The books in this section are unique in the respect that these have adopted a new style in the form of " questions and answers". Questions set in B.Sc examination of various prominent Indian universities during the last few years have been selected and arranged chapterwise to cover almost the entire syllabus.

HEAT AND THERMODYNAMICS

Contents : Kinetic Theory of Gases • Behaviour of Real Gases •Basic Ideas of Statistical Physics • Maxwell-Boltzmann Statistics •Quantum Statistics •First Law of Thermodynamics • Second Law of Thermodynamics •Entropy •Maxwell's Thermodynamical Relations • Liquefaction of Gases •Conductivity

ELECTRICITY AND MAGNETISM

Contents : Vector calculus •Coulomb's Law and Electric Field • Gauss's Theorem and Its Application • Electric Potential •Electric Fields in Dielectrics • Electric Current • Fields of Charges in Motion • The Magnetic Field • Magnetism in Matter • Electromagnetic Induction •Alternating Currents

OPTICS

Contents : Refraction at Spherical Surfaces •Refraction through Lenses •Matrix Methods in Geometrical Optics •Dispersion •Aberrations and Eye-Pieces •Velocity of Light •Wave Theory •Interference (Division of Wave Front) •Interference by Division of Amplitude • Interferometry •Coherence and Laser •Holography •Fresnel Diffraction •Fraunhofer Diffraction •Resolving Power and Dispersive Power •Polarisation •Production and Analysis of Polarised Light •Rotatory Polarisation

ATOMIC AND MOLECULAR PHYSICS

Contents: The Electron • uantum Theory–Particle Properties of Waves • Quantum Theory–Wave Properties of Particles • Schrodinger's Theory of Quantum Mechanics • Applications of Quantum Mechanics • Schrodinger's Theory of hydrogen Atom • Atoms with One electron and Zeeman Effect • Atoms with Many Electrons • X-rays •Molecules and Raman Effect • Laser

NUCLEAR PHYSICS

Contents: Atomic Nucleus •Nuclear Models •Radioactivity •Nuclear Reactions and Fission •Interaction of Nuclear Radiation with Matter • Particle Accelerators •Radiation Detectors • Cosmic Rays and Elementary Particles

SOLID STATE PHYSICS AND ELECTRONICS

Contents: Crystal Structure • Crystal Diffraction and Reciprocal Lattice •Crystal Bonding •Lattice Vibrations •Free Electron Theory of Metals • Band Theory of Solids • Semiconductors • Super Conductivity • Magnetic and Dielectric Properties • Semi-conductor Diodes • Rectifiers and Filters • Bipolar

Junction Transistor • Transistor • Biasing • Feedback Amplifiers • Oscillators • Modulation and Detection • Radio Communication • Television • Valve Electronics

MECHANICS AND PROPERTIES OF MATTER
Contents: Vectors • Co-ordinate System • Conservation of Momentum and Energy • Moment of Inertia • Frames of Reference • Gravitation • Motion Under a Central force • Rockets and Satellites • Elastic and Inelastic Collisions • Relativity • Elasticity • Surface Tension • Viscosity

WAVES, VIBRATIONS AND SOUND
Contents: Simple Harmonic Motion • Damped Simple Harmonic Motion • Forced Oscillator and Resonance • Coupled Oscillators • Transverse Waves • Longitudinal Waves • Interference, Beats and Stationary Waves • Electromagnetic Waves • Fourier Analysis

CHEMISTRY

by

R.L. Madan

The books in this section have been written in "question-answer "style to match the style and nature of questions that are asked in the examination. Solved numericals alongwith well-labelled structures and diagrams have been given fer clear understanding of the subject. Reference of previous years questions further helps the readers to locate the most important questions those are repeatedly asked.

PHYSICAL CHEMISTRY
Contents: Gaseous State •Liquid State • Solutions and Dilute Solutions • Chemical Kinetics • Surface Chemistry • Thermodynamics • Polymers • Electro Chemistry • Electrochemical Cell • Photochemistry • Quantum Mechanics • Statistical Mechanics • Distribution Lae • Phase Rate Crystalline State • Physical Properties and Molecular Structure • chemical Equilibrium • Colloidal State.

INORGANIC CHEMISTRY
Contents: Atomic structure • Ionic Bond • Covalent Bond • Periodic Classification of Elements • Acids and Bases • Metallurgies • General Characteristics or *S*–and P–Block Elements and their Compounds • Nuclear Chemistry • Compounds of Boron and Silicon • Oxides and Oxy-acids of Nitrogen, Phosphorus • Sulphur, Chlorine • Organometallic Compounds • The *d*-block Elements. The f-block Elements • Oxidation and Reduction • Chemical Industries • Noble Gases • Coordination Compounds • Environmental Chemistry • Non-aqueous Solvents • Qualitative Inorganic Analysis • Quantitative Analysis • Some Important Compounds • Spectral Properties of Compounds • Some more Organometallic Compounds and π–acid complexes.

ORGANIC CHEMISTRY

Contents : Occurrence and Characteristics of Organic Chemistry • Quantitative Analysis • Systematic Nomenclature • General Principles of Organic Reaction Mechanism • Chromatography • Stereochemistry • Alkynes • Diens • Arenes • Aliphatic Halogen Compounds • Aryl Halides • Organometallic Compounds • Alcohols • Phenols • Ethers • Aldehydes and Ketones • Carboxylic Acids • Derivatives of Carboxylic Acids • Substituted Acid and their Derivatives • Polycarboxylic Acids and their Derivatives, Fats, Oils, Soaps and Waxes • Sulphonic Acids • Nitro Compounds • Amines • Dia Zonium Salts • Spectroscopy and Structure • Cycloakanes • Conformations • Carbohydrates • Polynuclear Hydrocarbons • Heterocytics • Amino Acids and Proteiry • Synthetic Drugs • Insecticides and Pesticides • Photochemistry • Reactive Intermediates • Rearrangements
• Oxidation and Reduction • Natural Products • Biosynthesis of Natural Products • Dyes-Vitamins

BOTANY

by B.P. Pandey

The books on this section serve as a standard guide for the B.Sc. (three-year degree) course of Indian Universities. The text is written in simple, lucid and grapable language with self explanatory well labelled diagrams.

BOTANY - I

Contents : Algae • Fungi • Plant Pathology • Microbiology • Bryophyte • Pteriodphyta • Gymnosperms • Palaeobotany.

BOTANY - II

Contents: Taxonomy of Angiosperms • Anatomy of Angiosperms • Embrology of Angiosperms • Economic Botany

ZOOLOGY

By V.K. Agarwal

The books under this category will prove to be immensely helpful to the students of B.Sc (Pass/ Honours) and those who are preparing for various competitive examination such as NET, IAS, PCS, IFS etc. Each book contains well-illustrated Illustrations numerous tables, informations boxes etc. Examination question or various shades of amlost all Indian Universities have been dealt with.

INVERTEBRATE ZOOLOGY

Contents: Introduction • Principles of classification and Nomenclature • Phylum Protozoa; Amoeba, Euglena, Trypanosoma and sleeping sickness, Paramecium; Miscellaneous Questions • Phylum Porifera : Seypha, Miscellaneous Questions • Phylum Platyhelminthes : Dynesia, Fasciola hepatica, Taenia solium, miscellaneous questions • Phylum Nematoda : Ascaris, miscellaneous Questions • Minor Phyla • Phylum Mollusca : Pila globosa, Unio, Miscellaneous questions • Phylum Anndida : Nereis, Pheretima, hirudinaria, Miscellaenous Questions • Phylum Onychophora and Peripat • Phylum Arthropida : Palemon, scorpion, Insects and Miscellaneous Questions • Phylum Echinodermata : asterias • Phylum Hemichordata : Balanoglossus

CELL BIOLOGY

Contents : Introduction and Microscopy • Cytological Techniques • Prokaryotic Cell (viruses, bacteria and Blue Green Algae) • Eukaryotic Cell (animal Cell and Plant Cell) • Protoplasam • Plasma Membrane and Cell Wall • Endoplasmic Reticulum (ER) • Golgi Apparatus • Lysosomes • Ribosomes • Mitochondria • Chloroplast • Nucelus • Chromosomes • Cell Division • Gametogenesis • Fertilization • Parthenogenesis.

GENETICS

Contents : Mendel's Laws of Inheritance • Hybridization • Linkage • Crossing Over • chromosome Maps (Gene Mapping) • Interactions of Genes (Factor Hypothesis) • Quantitative Genetics (Inheritance of Multiple genes or Polygenic Inheritance) • Multiple Alleles (Allelic Variation) • The Gene (fine Structure of Gene) • Sex-linked Inheritance • Sex Determination and Sex Differentiation • Chromosmal Mutation • Cytoplasmic Inheritance • Human Traits and syndromes • Eugenics, Euthemics and Euphenics

MOLECULAR BIOLOGY

Contents: Identification of the Genetic Materials (*i.e.,* DNA and RNA) • Replication of DNA • Transcription and RNA • Genetic Code • Protein Synthesis • Bacterial Gene Regulation • Point Multation • Genetic Engineering (Isolation, Sequencing and Synthesis of Gene DNA Fingerprinting) • Genetic Recombination and Gene Transfer (Bacterial Conjugation, Transformation, transduction, Episomes and Plasmids) • Transposable Genetic Elements (Jumping Genes or Mobile Genes) • Immunology.